THIRD EDITION

FIRE APPARATUS OPERATOR: PUMPER

THIRD EDITION

FIRE APPARATUS OPERATOR: PUMPER

Howard Sykes   Thomas B. Sturtevant

DELMAR
CENGAGE Learning

Australia • Brazil • Japan • Korea • Mexico • Singapore • Spain • United Kingdom • United States

Fire Apparatus Operator: Pumper
Third Edition
Howard Sykes and Thomas B. Sturtevant

Vice President, Career and Professional Editorial: Dave Garza

Director of Learning Solutions: Sandy Clark

Senior Acquisitions Editor: Janet Maker

Managing Editor: Larry Main

Senior Product Manager: Jennifer A. Starr

Editorial Assistant: Amy Wetsel

Vice President, Career and Professional Marketing: Jennifer Baker

Marketing Director: Deborah S. Yarnell

Senior Marketing Manager: Erin Coffin

Associate Marketing Manager: Shanna Gibbs

Production Director: Wendy Troeger

Production Manager: Mark Bernard

Senior Content Project Manager: Jennifer Hanley

Senior Art Director: Casey Kirchmayer

For product information and technology assistance, contact us at
Cengage Learning Customer & Sales Support, 1-800-354-9706

For permission to use material from this text or product, submit all requests online at **www.cengage.com/permissions.**
Further permissions questions can be e-mailed to
permissionrequest@cengage.com

Library of Congress Control Number: 2010935202

ISBN-13: 978-1-4354-3862-0
ISBN-10: 1-4354-3862-0

Delmar
5 Maxwell Drive
Clifton Park, NY 12065-2919
USA

Cengage Learning is a leading provider of customized learning solutions with office locations around the globe, including Singapore, the United Kingdom, Australia, Mexico, Brazil, and Japan. Locate your local office at: **international.cengage.com/region**

Cengage Learning products are represented in Canada by Nelson Education, Ltd.

To learn more about Delmar, visit **www.cengage.com/delmar**

Purchase any of our products at your local college store or at our preferred online store **www.CengageBrain.com**

Notice to the Reader

Publisher does not warrant or guarantee any of the products described herein or perform any independent analysis in connection with any of the product information contained herein. Publisher does not assume, and expressly disclaims, any obligation to obtain and include information other than that provided to it by the manufacturer. The reader is expressly warned to consider and adopt all safety precautions that might be indicated by the activities described herein and to avoid all potential hazards. By following the instructions contained herein, the reader willingly assumes all risks in connection with such instructions. The publisher makes no representations or warranties of any kind, including but not limited to, the warranties of fitness for particular purpose or merchantability, nor are any such representations implied with respect to the material set forth herein, and the publisher takes no responsibility with respect to such material. The publisher shall not be liable for any special, consequential, or exemplary damages resulting, in whole or part, from the readers' use of, or reliance upon, this material.

Printed in the United States of America
1 2 3 4 5 6 7 15 14 13 12 11

DEDICATION

This book is dedicated to Joyce R. Sykes, my wonderful wife, without whom many of my accomplishments would not have been possible; all the members of the Lebanon Volunteer Fire Department, full-time career, part-time, and volunteer, who endured getting the implementation right of the many skills covered in this book; Robert Glover, Chief of the Pelham Fire Department and FDNY ret., for developing my enthusiasm in the fire service; the Setauket, Port Jefferson, and Rocky Point Fire Departments, for planting the seeds of interest in the fire service as I grew up on Long Island; and to the Broward County, Florida, fire service instructors, for providing me with a model for an ideal fire service instructor.

Most important, this book is dedicated to firefighters everywhere, with the hope it will help and inspire you.

—Chief Howard Sykes

This book is dedicated to Karen, my loving wife, for her careful review and honest evaluation of the manuscript and, more important, for her love, encouragement, and commitment to me and our three wonderful children Rachel, Hannah, and James.

—Dr. Thomas B. Sturtevant

Contents

3 Driving Emergency Vehicles 62

SECTION II Pump Construction and Peripherals

4 Pump Operating Principles and Construction 98

5 Pump and Apparatus Peripherals 120

6

Hose, Appliances, and Nozzles 162

SECTION III Pump Procedures

7

Water Supplies 190

8

Pump Operations 230

11 Fireground Flow and Friction Loss Considerations 334

12 Pump Discharge Pressure Calculations 354

Preface

INTENDED AUDIENCE

Fire Apparatus Operator: Pumper, third edition, is a straightforward, reader-friendly text designed for firefighters who are aspiring to meet the professional qualifications for vehicle driver/pump operator. It is intended to be used for training in fire departments, academies, and college fire programs. Although the basic flow of the text is the same as in previous editions, the material in the third edition has been updated and expanded. In addition, a new full-color design and several new features serve to enhance the learning experience for the student. The most important aspect of this edition is that it addresses all of the requirements identified in NFPA 1002, Chapters 4 and 5, 2009 edition, for driver/pump operator. In addition, the text addresses the objectives listed in the Fire Protection Hydraulics and Water Supply model curriculum course established at the National Fire Academy's Fire and Emergency Services Higher Education (FESHE) conferences.

A Correlation Guide to the 2009 edition of NFPA Standard 1002 and FESHE course outcomes follows this Preface.

DEVELOPMENT OF THIS BOOK

This book was written for several reasons. First, and perhaps most important, it was written because I enjoy the subject matter. Of all the positions I have held in the fire service, the most memorable and enjoyable were those as a pump operator. Second, it was written because the majority of books on the subject are outdated. I have been teaching from the same textbooks that I learned from when I took classes at a community college some 15 years ago. I'm not implying that these older texts are of poor quality, rather that they are simply outdated. In fact, their extended existence pays tribute to those who wrote them. Granted, some concepts of pump operations have remained relatively unchanged over the years, but pump operations, related standards, and terminology have not been stagnant. Finally, it was written because I wanted a single resource for relevant information on pump operations. When teaching pump operations, I found the "perfect" textbook to be several existing texts combined with information I picked up during my career in the fire service. So, this book attempts to place all of the information needed to operate a pump efficiently and effectively within the same cover.

—*Dr. Thomas B. Sturtevant*

OUR REVIEW AND VALIDATION COMMITTEES

Through the dedication of our authors, content and technical reviewers, as well as our Fire Advisory Board members and validation committee, the third edition of *Fire Apparatus Operator: Pumper* continues to remain up to date with the changing landscape of the fire service world. As part of the development process, each chapter is carefully reviewed by a select number of practicing individuals in the fire service who offer their expertise and insight as we revise the content. Additionally, technical reviewers thoroughly check the manuscript in detail for clarity and accuracy.

We are excited to announce that for the third edition of this book, we have created a validation committee to ensure that the content meets 100% of the NFPA Standard 1002, Chapters 4 and 5. Learning objectives, which have been validated by this committee of subject-matter experts, are included at the beginning of each chapter. The learning objectives tie all components of the learning solution (e.g., text, curriculum, test bank, supplements) to the NFPA standard, providing instructors and students a pathway to meet the intent of the job performance requirements outlined in the NFPA standard.

The Validation Process

The National Fire Protection Association (NFPA) professional qualifications standards identify the minimum job performance requirements (JPRs) for fire service positions. A JPR states the behaviors required to perform a specific skill(s) on the job. The JPR statements must be converted into instructional objectives with behaviors, conditions, and standards that can be measured within the teaching/learning environment.

Our process includes the development of the learning objectives from the JPR by a subject-matter expert. The learning objectives are then reviewed and validated by a committee. The committee reviews the learning objectives to ensure that the JPR was correctly interpreted and also makes recommendations for additional learning objectives within our materials. Our authors are provided with the validated learning objectives to develop the materials. This ensures that 100% of the standard is met within our materials.

The learning objectives are used throughout the entire development process. This includes the development of the book, curriculum, and the certification test question banks. This process ensures consistency among all materials and offers the best possible materials available on the market.

For a complete list of our review and validation committee members, as well as other contributors to this book, please refer to the Acknowledgments section of this Preface.

HOW TO USE THIS BOOK

The basic goal of this text is to provide one location for the knowledge required to carry out the duties of a pump operator efficiently and effectively. Therefore, the textbook includes all of the requirements of NFPA 1002, *Fire Apparatus Driver/Operator Professional Qualifications*, 2009 edition, for drivers/pump operators. Other relevant NFPA standards are also introduced and discussed. In addition, the textbook covers the objectives contained in the FESHE's Fire Protection Hydraulics and Water Supply model core curriculum course. Because of this, the textbook can be used for state or local pump operator training and certification as well as in college-level courses.

The position of pump operator is vital to the mission of the fire service. It is a position that requires knowledge and skills gained from classroom lecture, practical hands-on training, and experience. To that end, the location of the training/education, whether it be a fire department, state training agency, college, or vocational school, doesn't matter as much as the quality of information provided. Apparatus equipped with pumps are expensive pieces of equipment that require extensive knowledge and skills to operate them safely.

The text is divided into four sections:

- *Section I* serves as an introduction to the duties and responsibilities of the pump operator and includes preventive maintenance and driving of emergency vehicles.
- *Section II* focuses on the operating principles, theories, and construction of pumps as well as the systems and components typically used in conjunction with fire pumps.
- *Section III* presents the three interrelated fire pump operation tasks or activities of securing a water supply, operating the pump, and maintaining discharge pressures. This section builds on the foundation laid in the first two sections of the text and provides basic, step-by-step procedures for operating the pump and related components.
- *Section IV* focuses on water flow calculations, including hydraulic theory, friction loss principles, and fireground pump discharge pressure calculations. This section brings to light the importance of understanding the relationship between the amount of water needed and the amount that is actually flowing.

The purpose for dividing the text into the four sections is to present like subjects together. For example, the first section focuses on the driver and emergency apparatus. The theory, construction, and operating principles of pumps and related components, for the most part, are all contained within the second section. The operation of pumps and related components is primarily covered within the third section of the text. Finally, most calculations are contained within the fourth section. By structuring the book in this way, the text can easily be rearranged depending on the focus and preference of the student and instructor.

NEW TO THIS EDITION

It is our goal to continually strive to meet and exceed expectations for firefighter training, as well as to remain current with standards, practices, and initiatives in the fire service. For this reason, this book was carefully reviewed and updated for the third edition to include the following:

- Compliant with NPFA Standard 1002, 2009 Edition. Revised to meet the intent of the 2009 edition of NFPA Standard 1002, including new learning objectives validated by a committee of experts to ensure 100% compliance with the standard.
- Additional Emphasis on Safety. Safety is emphasized throughout the book, including a special focus on driving safely during emergency response in Chapter 3 and separate *Safety* boxes highlighted throughout the chapters to instill a constant attention to safety.
- Full-color Design and New Features. With a new 8-1/2 x 11 design, this book features photographs and line art in full color as well as new features to enhance learning:
  - Street Stories at the beginning of each chapter relate actual experiences from drivers/operators across the nation and highlight important lessons learned.
  - Skills have been added with photographic step-by-step instructions for performing common driver/operator tasks.
  - StreetSmart tip boxes have been added to provide practical advice and applications for completing tasks in the field.
  - Metric equivalents accompany English/customary units in mathematical examples and calculations as well as in reference charts and tables.
  - Key Terms are listed at the end of each chapter and have been expanded to include many new terms.
- Comprehensive and Current Information on Essential Topics. With this new edition, every effort was made to ensure that students are thoroughly introduced to the most current information on pump operations, including:
  - Expanded information on rural water supplies and operations, including nurse feeding.
  - Expanded information on annual performance tests, including information on new tests now required by NFPA 1911, such as the interlock tests and intake relief valve tests.
  - Expanded operator maintenance information, including information on acceptable gauge readings and recovery procedures for warning conditions.
  - Information on operating with newer vehicle systems, such as auxiliary braking, anti-lock braking, automatic traction control, and roll-stability and enhanced roll-stability systems.
  - Additional driving information, such as rollover prevention and maneuvering through driver skills courses.
  - Information on newly approved equipment types, such as air primers, along with tips on the differences in using oil-less primers.
  - Expanded information on priming, with emphasis on how to resolve priming problems quickly and effectively.
  - Expanded information on each of the primary pump subsystems: intake manifold, discharge manifold, pump casting, and pump transmission.
  - Expanded information on foam systems, including CAFS.
  - Expanded information on the NFPA 291 hydrant coloring system.
  - Expanded information on determining hydrant capacity, including details of the first-digit method, percentage method, and the squaring-the-lines method of determining hydrant capacity.
  - Expanded information on apparatus positioning, including rationales, so that operators can make sound decisions in unique situations.

FEATURES OF THIS BOOK

The following features serve to enhance learning and help you gain competence and confidence in mastering fire pump operations.

NFPA 1002 and FESHE Correlation Guides

These grids provide a correlation between *Fire Apparatus Operator: Pumper* and the requirements for NFPA Standard 1002, 2009 edition, Chapters 4 and 5. A second grid provides the correlation to the FESHE course curriculum for Fire Protection Hydraulics and Water Supply.

Street Stories

Each chapter opens with personal experiences contributed by drivers/operators from across the nation. These personal accounts bring to life the chapter contents that follow, and highlight important lessons learned.

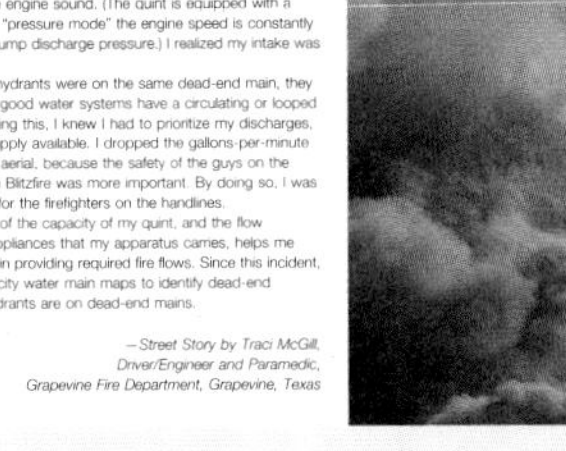

STREET STORY

We had a residential structure fire involving a 3,000-square-foot (900-square-metre) two-story house, and I was a newly promoted driver assigned to driving a quint. My apparatus was the first on the scene. We immediately deployed our 200-foot, 2-inch (60-metre, 50 mm) handline to the front of the house. When the battalion chief arrived on-scene, he requested the aerial ladder be set for a potential defensive attack. Once the ladder was set, I was able to operate it from the ground with a remote control, instead of on the turnstile, and be at the pump panel getting the lines flowing. Knowing this quint has a 2,000-gallon-per-minute (8,000-litre-per-minute) capacity, I was able to pump to near maximum capacity with approximately 1,250 gallons per minute (5,000 litres per minute) from the aerial, 225 gallons per minute (900 litres per minute) from the 200 feet (60 metre) of 2-inch (50 mm) hose, 150 gallons (550 litres) per minute from 150 feet (45 metre) of 1¾-inch (45 mm) hose, and 300 gallons (1,200 litres) per minute from 200 feet (60 metre) of 3-inch (77 mm) hose on the Blitzfire (portable master stream).

When the second-in apparatus arrived, they hooked me in to the hydrant at the end of the cul-de-sac. I knew that I could maintain this output as long as my intake from the hydrant wasn't compromised. I just needed to watch my intake to make sure that I was maintaining a 20-psi (140 kPa) residual pressure from the hydrant. Within a few minutes, our CAFS engine arrived on-scene and hooked up to a different hydrant at the beginning of the street. Strangely, once they started flowing from their pump, my pressures started fluctuating. I could see this on the gauges and could hear the changes in the engine sound. (The quint is equipped with a pressure governor; when this is in "pressure mode" the engine speed is constantly adjusted to maintain the desired pump discharge pressure.) I realized my intake was affected.

It turns out that because both hydrants were on the same dead-end main, they shared the same water line. Most good water systems have a circulating or looped system, but this one didn't. Realizing this, I knew I had to prioritize my discharges, since there was no other water supply available. I dropped the gallons-per-minute (litres-per-minute) output from the aerial, because the safety of the guys on the ground with the handlines and the Blitzfire was more important. By doing so, I was able to maintain safe fire streams for the firefighters on the handlines.

Knowledge and understanding of the capacity of my quint, and the flow capabilities of all the hoses and appliances that my apparatus carries, helps me identify my abilities and limitations in providing required fire flows. Since this incident, our fire crew has also utilized the city water main maps to identify dead-end hydrants, as not all cul-de-sac hydrants are on dead-end mains.

—Street Story by Traci McGill,
Driver/Engineer and Paramedic,
Grapevine Fire Department, Grapevine, Texas

Validated Learning Objectives

Each chapter includes a list of objectives. Satisfy yourself that you have learned each objective so that you are assured of gaining the appropriate competency for certification. Information that goes beyond the validation is identified in blue typeface.

LEARNING OBJECTIVES

After completing this chapter, the reader should be able to:

9-1. Produce an effective hand stream from the internal tank of a pumping apparatus, so that all safety considerations are addressed, the pressure control device is set, the rated nozzle flow is achieved and maintained, and the apparatus is monitored for potential problems.

Key Terms

Throughout the text you will find key terms and notes of particular importance. These important items will provide you with the knowledge and terminology for effectively communicating with your superiors, subordinates, and peers.

Note

This feature highlights and outlines important points for you to learn and understand. Based on key concepts, this content is an excellent source for review.

It is important to make sure that the units of volume are the same as those given in the constant used for density.

Safety

These note boxes offer advice on how to react and, more important, be proactive in protecting the safety of yourself and your crew.

Tanker operations should be improved by decreasing fill and dump timings, *not* by questionable driving techniques. Both tanker weight and the fact that the weight (water) moves make safe tanker driving difficult. About a quarter of firefighter deaths occur while in transit, with tankers being the apparatus type most often involved with fatalities.

NFPA has mandated changes that help reduce such incidents, including better brakes than commercial vehicles, lower top speeds, additional tank baffles, and seatbelt warning systems. Nothing, however, is more important than how the driver operates the apparatus.

Caution

These note boxes indicate a hazardous situation that, if not avoided, could result in injury or alert against unsafe practices that can result in property damage.

While service testing ensures that the engine is still able to perform at the appropriate level, it is also the most stressful operation most engines go through all year. Extreme vigilance in monitoring apparatus gauges is required during the test to make sure that the apparatus is not run outside of acceptable operating ranges, which would result in excessive wear and/or damage requiring repairs.

StreetSmart

These tips offer practical advice for completing tasks and succeeding in the duties of a driver/operator.

All manufacturers must meet the same NFPA specifications; however, the designs they have implemented to meet those specifications vary significantly. For example, Hale ball valves contain a grease fitting and need to be greased routinely to operate smoothly. Akron Brass and Elkhart Brass ball valves do not contain grease fittings and, furthermore, if greased, will not last as long as they should. Therefore, pump operators should obtain information from the manufacturer of their equipment and follow those instructions. Differences do exist in how different vendors' equipment should be maintained and operated.

Skills

Step-by-step photo sequences illustrating important procedures are located at the end of the chapters. These are intended to be used as a guide in mastering the job performance skills and to serve as an important review.

SKILL 8-3

Using a Four-way Hydrant Valve

A The first-in engine should stop just past the hydrant on the same side of the street as the hydrant and allow a firefighter to disembark with the engine's hydrant tool bag, four-way hydrant bag, and hose. Once the firefighter has wrapped the hose around the hydrant, the firefighter should signal the driver to proceed to the fire while deploying a forward lay from the hose bed on the same side of the street as the hydrant.

B Once the engine has laid at least 100 ft (30 m) of hose or stopped at the fire scene, the firefighter should flush the hydrant, connect the four-way hydrant valve to the hydrant, connect the supply hose for the first engine to the four-way hydrant valve discharge, and confirm that the four-way hydrant valve handle is in the correct position. The hydrant should be opened when the first engine indicates that it is ready for water.

C When the second engine arrives at the hydrant, a hose should be connected between its intake and the port on the four-way hydrant valve designed for its intake. Another line should be connected between the second pump's discharge and the four-way hydrant valve's pump discharge connection. When the second engine's pump operator is ready, the valve handle on the four-way hydrant valve should be turned to the position that will charge the line to the second engine without interrupting the flow to the first engine.

D Once the operator of the second engine has bled air from the intake line and charged the discharge line with a pressure above the hydrant's pressure, the check valve in the four-way hydrant valve will automatically switch to flow to the first engine from the hydrant to the second engine's discharge.

Review Questions and Practice Problems

In addition to Review Questions, there are Practice Problems integrated throughout the hydraulics sections, carefully testing your knowledge by building on each concept presented in the book. Additional practice problems can be found at the end of each chapter, as well as a list of formulas at the end of each water flow calculation chapter.

Additional Resources

Each chapter includes a list of additional resources. Each recommended listing offers additional information on the topic covered in that particular chapter. The goal is to give you as much information as possible to help you in your duties as a driver/operator.

This information will help you better perform your duties, not just to meet the requirements of the standard, as important as that is. Whether you are reading this book to improve your skills or taking a structured program leading to certification, we hope that you will find this book informative, interesting, and useful.

CURRICULUM PACKAGE

This book was created not only as a stand-alone manual for firefighters, drivers/operators, and fire officers, but as a special package of materials for the full instructional experience. The supplement package provides a variety of tools for students and instructors to enhance the learning experience.

Instructor's Curriculum CD-ROM

The Instructor's Curriculum CD-ROM is designed to allow instructors to run programs according to the standards set by the authority having jurisdiction where the course is conducted. It contains the information necessary to conduct driver/operator courses. It is divided into sections to facilitate its use for training:

- Administration. Provides the instructor with an overview of the various courses, student and instructor materials, and practical advice on how to set up courses and run skill sessions.
- Equipment Checklist. Offers a quick guide for ensuring the necessary equipment is available for hands-on training.
- Lesson Plans. Ideal for instructors, whether they are teaching at fire departments, academies, or longer-format courses, each Lesson Plan correlates to the corresponding PowerPoint® Presentation.
- Answers to Review Questions. Include answers to questions and problems in the book in order to evaluate student learning.
- PowerPoint® Presentations. Outline key concepts from each chapter, and contain graphics and photos from the book, to bring the content to life.

- Computerized Test Banks. Contain hundreds of questions in ExamView 6.0 to help instructors prepare candidates to take the written portion of the certification exam for driver/operator.
- Skill Sheets. Outline important steps of each skill that candidates must master to meet requirements for certification.
- Progress Logs. Provide a system to track the progress of individual candidates as they complete the required skills.
- Quick Reference Guides. Contain valuable information for instructors. Included are the following grids:
  - *2009 Edition of NFPA Standard 1002 Correlation Grid* used to cross-reference *Fire Apparatus Operator: Pumper* with the standard.
  - *New Edition Correlation Guide* used to cross-reference the revisions between the second and third editions of this book.
  - *Fire and Emergency Services Higher Education (FESHE) Correlation Guide* correlates the model curriculum course Fire Protection Hydraulics and Water Supply requirements to the textbook chapters.
- Additional Resources. Offer supplemental resources for important information on various topics presented in the book.
- Image Gallery. Contains hundreds of graphics and photos from the book and offers an additional resource for instructors to enhance classroom presentations.

Order #: 978-1-4354-3863-7

ABOUT THE AUTHORS

Howard Sykes

Howard Sykes edited the third edition. He is the paid full-time chief of the Lebanon Volunteer Fire Department, a combination department with full-time, part-time, and volunteer members. The department serves a suburban community in northern Durham County, North Carolina. Chief Sykes started his fire service career as a volunteer member of the Pelham Fire Department, a combination department just north of New York City, and was a member of the Pompano Highlands Fire Department in Florida before joining Lebanon. Howard teaches various Firefighter I and II classes, emergency vehicle driver, and driver/operator pumps classes through the North Carolina Community College system.

Howard Sykes has a BS in Computer Science with an unofficial minor in electrical engineering from the State University of New York at Stony Brook and a Masters in Project Management from George Washington University. While a volunteer firefighter, he worked a full career at IBM, was a member of the industrial fire brigade, and is listed as an inventor on various patents.

Howard is certified as an ASE (Automotive Service Excellence) technician, EVT (Emergency Vehicle Technician)/Fire Mechanic, and enjoys apparatus maintenance and repair.

Howard has lectured worldwide, worked in an executive briefing center, and has been responsible for resolving critical customer problems.

Dr. Thomas B. Sturtevant

Dr. Thomas B. Sturtevant wrote the first and edited the second edition. He also reviewed and provided valuable feedback during the development of the third edition. Dr. Sturtevant is a program manager for the Emergency Services Training Institute (ESTI) within the Texas Engineering Extension Service, itself a member of the Texas A&M University System. He currently manages the Emergency Management Administration online bachelor degree program with West Texas A&M University and the Department of Defense Emergency Services Training and Education program. He manages ESTI's curriculum development and accreditation/certification with the National Professional Qualification System.

He was a tenured assistant professor at Chattanooga State Technical Community College, Tennessee, where he held the positions of dean of Distance Education and coordinator of Fire Science Technology. Dr. Sturtevant was a fire protection specialist with the Tennessee Valley Authority and held various firefighting positions with the San Onofre Nuclear Generating Station, California, and the United States Air Force. He is a Certified Fire Protection Specialist and has an education doctorate in Leadership for Teaching and Learning and a Masters in Public Administration from the University of Tennessee. His research and consulting efforts focus on program evaluation and emergency service professional development.

ACKNOWLEDGMENTS

The author and publisher would like to extend our gratitude to the following individuals who participated in the development of the third edition of this book.

Our Reviewers

Technical Advisor/Author: Dr. Thomas B. Sturtevant, Program Director, Emergency Services Training Institute (ESTI), Texas A&M–TEEX, College Station, TX

Tom Bentley, Fire Science/Homeland Security Coordinator, John Wood Community College, Quincy, IL

Chris Best, Supervisor, Research and Program Development, Office of State Fire Marshal, Raleigh, NC

Tie Burtlow, Lieutenant, Fairfax County Fire and Rescue Department, Fairfax, VA

Mike Feaster, Technician, Fairfax County Fire and Rescue, Stephens City, VA

Chris Gilbert, Captain/Training Officer, Alachua County Fire Rescue, Gainesville, FL

Tom Hand, Training Coordinator, Mesa Fire Department, Mesa, AZ

Carnie Earl Hedgepath, Fire Rescue Training Specialist, Office of State Fire Marshal, Raleigh, NC

Steve Malley, Department Chair, Public Safety Professions, Weatherford College, Weatherford, TX

Our Advisory Board Members

To those experts who work behind the scenes and provide us with continuing guidance on this book as well as other learning materials on our emergency services list, we extend our gratitude:

Douglas Carter, Raleigh-Durham International Airport Emergency Services, NC

Steve Chikerotis, Battalion Chief, Chicago Fire Department, Chicago, IL

Doug Fry, Fire Chief, City of San Carlos Fire Department, San Carlos, CA

Tom Labelle, Executive Director, New York State Association of Fire Chiefs, East Schodack, NY

Jerry Laughlin, Deputy Director of Education Services, Alabama Fire College, Tuscaloosa, AL

Pat McCauliff, Director of Fire Science and EMS, Colin County Community College, McKinney, TX

Michael Petroff, Western Director for Fire Department Safety Officers Association, St. Louis, MO

Jeff Pindelski, Battalion Chief, Downers Grove Fire Department, Downers Grove, IL

Adam Piskura, Director of Training, Connecticut Fire Academy, Windsor Locks, CT

Peter Sells, District Chief of Operations Training, Toronto Fire Services, Toronto, Canada

Billy Shelton, Executive Director, Virginia Department of Fire Programs, Glen Allen, VA

Theresa Staples, Program Manager, Firefighter and HazMat Certifications, Centennial, CO

Ray Vernon, President, NTI Fire, Walnut Cove, NC

Lewis Womack, Bethesda Fire Department, Durham, NC

Our Validation Committee Members

Russ Kerns, Battalion Chief of Special Operations, Metropolitan Washington Airports Authority Fire and Rescue Department, Washington DC

Chris Larson, Fire Operations Battalion Chief, Metropolitan Washington Airports Authority Fire and Rescue Department, Washington DC

Steve Malley, Program Director, Public Safety, Weatherford College, Weatherford, TX

Jarett Metheny, Major, Oklahoma Fire Department, Midwest City, OK

Jerry Schroeder, Program Manager, Emergency Services Training, Boise, ID

Dr. Thomas B. Sturtevant, Program Manager, Emergency Services Training Institute (ESTI), Texas A&M, College Station, TX

Photography

Brad Allison, Chief, Caldwell Fire Department, Rougemont, NC

Robert Andrews, Chief, Bethesda Fire Department, Durham, NC

Christopher G. Blackburn, Firefighter, City of King Fire Department, King, NC

Arthur Boone, Firefighter, Lebanon Volunteer Fire Department, Durham, NC

Chad Burrow, Firefighter, City of King Fire Department, King, NC

Dennis English, Captain, Parkwood Volunteer Fire Department, Parkwood, NC

Jim Feely, Chief; Larry Strayer, Assistant Chief; and B-shift; Durham Highway Volunteer Fire Department, NC

Preston George, Chief, Francisco Volunteer Fire Department, Westfield, NC

Glen Gillette, Firefighter and IAFF Local 4209 President, City of King Fire Department, King, NC

John Hamlett, City of Roxboro Fire Department, Roxboro, NC

Wesley D. Hutchins, Dean, Emergency Services Programs, Forsyth Technical Community College, Winston-Salem, NC

Cecil Johnson, Firefighter, Timberlake Volunteer Fire Department, Timberlake, NC

Andrew J. Mathys III, Firefighter, City of King Fire Department, King, NC

Leeanna R. Mims, Chief; Paula J. Ritchey, Lieutenant and PIO; Gregory Kirby, Fire and Life Safety Education Coordinator; Corey Steff, Former Firefighter, Seminole County Fire Department, Sanford, FL

Len Needham, Chief, Bahama Volunteer Fire Department, Bahama, NC

Joseph R. Ramsey, Captain, City of Winston-Salem Fire Department, Winston-Salem, NC

Stan A. Roberson, Director, Sauratown Fire Rescue, Sauratown, NC

Steven Roberson, Chief, City of King Fire Department, King, NC

Kevin P. Terry, Chief, Fuller Road Fire Department (Colonie, NY), and Investigator, Town of Colonie Police Department, NY

Street Story Contributors

Chapter 1: Thomas B. Sturtevant, Author and Curriculum Director, Emergency Services, Training Institute (ESTI) of Texas A&M University, College Station, TX

Chapter 2: Howard Sykes, Author and Chief, Lebanon Volunteer Fire Department, Durham, NC

Chapter 3: J. Christopher Milne, Apparatus Division Captain, Salt Lake City Fire Department, Salt Lake City, UT

Chapter 4: Shannon Askew, Firefighter (Apparatus Operator), Calgary Fire Department, Engine Company #14, Calgary, Alberta, Canada

Chapter 5: Steve Saksa, Firefighter/Paramedic, Engine Co. 14, 2 Unit Columbus Division of Fire Columbus, OH

Chapter 6: Tracy Burrus, Apparatus Engineer, Engine Company #3, City of Madison Fire Department, Madison, WI

Chapter 7: Traci McGill, Driver/Engineer and Paramedic, Grapevine Fire Department, Grapevine, TX

Chapter 8: Mike Feaster, Apparatus Technician, Fairfax County Fire & Rescue, Fairfax, VA

Chapter 9: Scott Shawaluk, Engineer, Chicago Fire Department, 4th District Relief, Chicago, IL

Chapter 10: Chris George, Apparatus Engineer, Quincy Fire Department, Engine Company 3, Quincy, IL

Chapter 11: Tie Burtlow, Driver, Fairfax County Fire and Rescue, Engine 404 Fairfax, VA

Chapter 12: Louis Ramos, Firefighter/Fire Inspector, Hudson Oaks Fire Department, Parker County ESD #3, Hudson Oaks, TX

Thanks to Margaret Magnarelli, who interviewed these fine firefighters so that they could share their stories.

Special Thanks

The author would like to add a special thank you to Ken Kroeker, Emergency Services Officer, Office of the Fire Commissioner, Manitoba Emergency Services College, Brandon, MB, Canada. In addition to reviewing the book and providing numerous helpful recommendations, he provided all of the metric conversions.

The third edition was built upon a foundation of the previous editions. For this reason, we would like to extend our appreciation to those who previously contributed to this book:

Steve Aranbasich, Henderson Fire Department, Henderson, NV

Richard Arwood, Memphis Fire Department, Memphis, TN

Dennis Childress, Orange County Fire Authority, Rancho Santiago College, Orange, CA

Gary Courtney, New Hampshire Technical College, Laconia, NH

Victor Curtis, Mesa Fire Department, Mesa, AZ

Jim Duffy, Henderson Fire Department, Henderson, NV

Doug Hall, Red Rocks Community College, Westminster, CO

Keith Heckler, Rothfuss Engineering Company, Jessup, MD

Attila Hertelendy, University of Nevada, Fire Science Academy, Carlin, NV

Robert Kinniburgh, Charlotte Fire Department, Charlotte, NC

Ric Koonce, J Sargeant Reynolds Community College, Richmond, VA

Chief Dave Leonardo, Loudonville Fire Department, Loudonville, NY

Chief Dave Leonardo, Verdoy Fire Department, Latham, NY

Captain Bob Sanborn, Bowling Green Fire Department, Bowling Green, KY

Clarence E. White, Maryland Fire Rescue Institute, University of Maryland, College Park, MD

Douglas Whittaker, Onondaga Community College, Syracuse, NY

We would also like to thank the many manufacturers who willingly provided information, artwork, and photographs.

And Dr. Charles Waggoner for his support and encouragement of this project, as well as for his review of several manuscript sections.

DELMAR CENGAGE LEARNING EMERGENCY SERVICES TEAM

This team is the group of employees that develops, produces, and markets this book and performs countless behind-the-scene activities. These group members not only set an example for getting the job done but also have the creativity and fortitude to go above and beyond. As author, it is not possible for me to overstate how much their wisdom has helped set directions to ensure a high-quality, high-appeal finished publication. It is a pleasure to recognize Janet Maker, Rich Hall, Jennifer Starr, Jennifer Hanley, and Amy Wetsel.

SUGGESTIONS ENCOURAGED

Those who reviewed and commented on this text during its preparation provided valuable insight and suggestions, which in turn produced a better text. Those who read and use the text will also, undoubtedly, have valuable insight and suggestions. I strongly encourage any and all comments and feel that to have someone comment on this text is truly an honor, which will, in turn, produce a better text. E-mail comments can be sent to http://www.delmar.fire@cengage.com.

NFPA 1002, 2009 Edition Correlation Guide

General Requirements (Chapter 4)

4.2.1 Task Statement	Learning Objective(s)	Chapter(s)	Page(s)
Perform routine tests, inspections, and servicing functions on the systems and components specified in the following list, given a fire department vehicle and its manufacturer's specifications, so that the operational status of the vehicle is verified:	2-10	2	27-46, 48
Battery(ies)	2-28	2	27-46, 48
Braking system	2-26	2	46
Coolant system			
Electrical system			
Fuel	2-4	2	21-27
Hydraulic fluids			
Oil			
Tires	2-14	2	27-46, 48
Steering system			
Belts			
Tools, appliances, and equipment	2-17	2	48, 53-56

4.2.1 (A) Requisite Knowledge	Learning Objective(s)	Chapter(s)	Page(s)
Manufacturer specifications and requirements	2-1	2	21
	2-9	2	23, 27-30
	2-27	2	46-47
	2-13	2	23-24
	2-16	2	27-46, 48
Policies and procedures of the jurisdiction	2-2	2	21-23
	2-29	2	46-47
	2-30	2	47
	2-5	2	24-25

4.2.1 (B) Requisite Skills	Learning Objective(s)	Chapter(s)	Page(s)
The ability to use hand tools	2-18	2	51-52
Recognize system problems	2-15	2	28-46, 48
Correct any deficiencies noted according to policies and procedures	2-10	2	27-46, 48

4.2.2 Task Statement	Learning Objective(s)	Chapter(s)	Page(s)
Document the routine tests, inspections, and servicing functions, given maintenance and inspection forms, so that all items are checked for operation and	2-6	2	26-29
deficiencies are reported.	2-11	2	28-29

4.2.2 (A) Requisite Knowledge	Learning Objective(s)	Chapter(s)	Page(s)
Departmental requirements for documenting maintenance performed	2-6	2	26-29
and the importance of keeping accurate records.	2-7	2	26-27

4.2.2 (B) Requisite Skills	Learning Objective(s)	Chapter(s)	Page(s)
The ability to use tools and equipment and complete all related departmental forms.	2-11	2	28-29

4.3.1 Task Statement	Learning Objective(s)	Chapter(s)	Page(s)
Operate a fire department vehicle, given a vehicle and a predetermined route on a public way that incorporates the maneuvers and features, specified in the following list, that the driver/operator is expected to encounter during normal operations, so that the vehicle is operated in compliance with all applicable state and local laws, departmental rules and regulations, and the requirements of NFPA 1500, Section 4.2.	1-3	1	6-7
Four left turns and four right turns			
A straight section of urban business street or two-lane rural road at least 1.6km (1 mile) in length	3-4	3	69
One through-intersection and two intersections where a stop has to be made			
One railroad crossing			
One curve, either left or right			
A section of limited-access highway that includes a conventional ramp entrance and exit and a section of road long enough to allow two lane changes	3-2	3	66-68
A downgrade steep enough and long enough to require down-shifting and braking			
An upgrade steep enough and long enough to require gear changing to maintain speed	3-18	3	77-82, 85-91
One underpass or a low clearance or bridge			

4.3.1 (A) Requisite Knowledge	Learning Objective(s)	Chapter(s)	Page(s)
The effects on vehicle control of speed, brake reaction time, braking, and load factors	3-7	3	70-75
Effects of high center of gravity and liquid surge on roll-over potential	3-6	3	69-70, 74
Effects of general steering reactions, speed, and centrifugal force	3-11	3	73
Applicable laws and regulations	3-1	3	65-68
Principles of skid avoidance	3-12	3	73
Night driving	3-16	3	75-76
Shifting and gear patterns	3-9	3	71
Negotiating intersections, railroad crossings, and bridges	3-18	3	71-73
Weight and height limitations for both roads and bridges	3-18	3	68-70, 74
Identification and operation of automotive gauges, and operational limits	2-19	2	36-39

4.3.1 (B) Requisite Skills	Learning Objective(s)	Chapter(s)	Page(s)
The ability to operate passenger restraint devices	3-3	3	68
Maintain safe following distances			
Maintain control of the vehicle while accelerating, decelerating, and turning, given road, weather, and traffic conditions	3-8	3	70-77
Operate under adverse environmental or driving surface conditions			
Use automotive gauges and controls	2-20	2	36-39

4.3.2 Task Statement	Learning Objective(s)	Chapter(s)	Page(s)
Back a vehicle from a roadway into a restricted space on both the right and left sides of the vehicle, given a fire department vehicle, a spotter, and restricted spaces 3.7m (12 ft) in width, requiring 90-degree right-hand and left-hand turns from the roadway, so that the vehicle is parked within the restricted areas without having to stop and pull forward and without striking obstructions.	3-19	3	76, 79-80, 85-86

4.3.2 (A) Requisite Knowledge	Learning Objective(s)	Chapter(s)	Page(s)
Vehicle dimensions and turning characteristics	3-5	3	69-70, 73
Spotter signaling	3-14	3	74-75
Principles of safe vehicle operations	3-13	3	70-77

4.3.2 (B) Requisite Skills	Learning Objective(s)	Chapter(s)	Page(s)
The ability to use mirrors and judge vehicle clearance	3-15	3	74-75

4.3.3 Task Statement	Learning Objective(s)	Chapter(s)	Page(s)
Maneuver a vehicle around obstructions on a roadway while moving forward and in reverse, given a fire department vehicle, a spotter for backing, and a roadway with obstructions, so that the vehicle is maneuvering through the obstructions without stopping to change the direction of travel and without striking the obstructions.	3-20	3	80-81, 87-88

4.3.3 (A) Requisite Knowledge	Learning Objective(s)	Chapter(s)	Page(s)
Vehicle dimensions and turning characteristics	3-5	3	69-70
The effects of liquid surge	3-6	3	69-70, 74
Spotter signaling	3-14	3	74-76, 78
Principles of safe vehicle operations	3-13	3	70-77

4.3.3 (B) Requisite Skills	Learning Objective(s)	Chapter(s)	Page(s)
The ability to use mirrors and judge vehicle clearance	3-15	3	74-75

4.3.4 Task Statement	Learning Objective(s)	Chapter(s)	Page(s)
Turn a fire department vehicle 180 degrees within a confined space, given a fire department vehicle, a spotter for backing up, and an area in which the vehicle cannot perform a U-turn without stopping and backing up, so that the vehicle is turned 180 degrees without striking obstructions within the given space.	3-21	3	81, 89-90

4.3.4 (A) Requisite Knowledge	Learning Objective(s)	Chapter(s)	Page(s)
Vehicle dimensions and turning characteristics	3-5	3	69-70
The effects of liquid surge	3-6	3	69-70, 74
Spotter signaling	3-14	3	74-76, 78
Principles of safe vehicle operations	3-13	3	70-77

4.3.4 (B) Requisite Skills	Learning Objective(s)	Chapter(s)	Page(s)
The ability to use mirrors and judge vehicle clearance	3-15	3	74-75

4.3.5 Task Statement	Learning Objective(s)	Chapter(s)	Page(s)
Maneuver a fire department vehicle in areas with restricted horizontal and vertical clearances, given a fire department vehicle and a course that requires the operator to move through areas of restricted horizontal and vertical clearances, so that the operator accurately judges the ability of the vehicle to pass through the openings and so that no obstructions are struck.	3-17	3	76

4.3.5 (A) Requisite Knowledge	Learning Objective(s)	Chapter(s)	Page(s)
Vehicle dimensions and turning characteristics	3-5	3	69-70
The effects of liquid surge	3-6	3	69-70, 74
Spotter signaling	3-14	3	74-76, 78
Principles of safe vehicle operations	3-13	3	70-77

4.3.5 (B) Requisite Skills	Learning Objective(s)	Chapter(s)	Page(s)
The ability to use mirrors and judge vehicle clearance	3-15	3	74-75

4.3.6 Task Statement	Learning Objective(s)	Chapter(s)	Page(s)
Operate a vehicle using defensive driving techniques under emergency conditions, given a fire department vehicle and emergency conditions, so that control of the vehicle is maintained.	3-22	3	82-84
	3-23	3	82-84

4.3.6 (A) Requisite Knowledge	Learning Objective(s)	Chapter(s)	Page(s)
The effects on vehicle control of liquid surge, braking reaction time, and load factors	3-7	3	70-75
Effects of high center of gravity on roll-over potential	3-6	3	69-70, 74
General steering reactions, speed, and centrifugal force	3-11	3	73
Applicable laws and regulations	3-1	3	65-68
Principles of skid avoidance	3-12	3	73
Night driving	3-16	3	75-76

4.3.6 (A) Requisite Knowledge	Learning Objective(s)	Chapter(s)	Page(s)
Shifting and gear patterns	3-9	3	71
Automatic braking systems in wet and dry conditions	3-10	3	65-66
Negotiation of intersections, railroad crossings, and bridges	3-2	3	65-66
Weight and height limitations for both roads and bridges	3-2	3	36-39
Identification and operation of automotive gauges, and operational limits	2-19	2	70-75

4.3.6 (B) Requisite Skills	Learning Objective(s)	Chapter(s)	Page(s)
The ability to operate passenger restraint devices	3-3	3	68
Maintain safe following distances			
Maintain control of the vehicle while accelerating, decelerating, and turning, given road, weather, and traffic conditions	3-8	3	70-77
Operate under adverse environmental or driving surface conditions			
Use automotive gauges and controls	2-20	2	36-39

4.3.7 Task Statement	Learning Objective(s)	Chapter(s)	Page(s)
Operate all fixed systems and equipment on the vehicle not specifically addressed elsewhere in this standard, given systems and equipment, manufacturer's specifications and instructions, and departmental policies and procedures for the systems and equipment, so that each system or piece of equipment is operated in accordance with the applicable instructions and policies.	5-19	5	147-151, 156-157

4.3.7 (A) Requisite Knowledge	Learning Objective(s)	Chapter(s)	Page(s)
Manufacturer's specifications and operating procedures, and policies and procedures of the jurisdiction.	5-20	5	147-151

4.3.7 (B) Requisite Skills	Learning Objective(s)	Chapter(s)	Page(s)
The ability to deploy, energize, and monitor the system or equipment and	5-21	5	147-151
To recognize and correct system problems.	5-22 5-23	5 5	147-151 147-151

5.1.1 Task Statement	Learning Objective(s)	Chapter(s)	Page(s)
Perform the routine tests, inspections, and servicing functions specified in the following list in addition to those in 4.2.1, given a fire department pumper, its manufacturer's specifications, and procedures of the jurisdiction, so that the operational status of the pumper is verified:	2-8	2	27-28, 30
(1) Water tank and other extinguishing agent levels (if applicable)	2-21	2	43-45
	2-22	2	43-45
(2) Pumping systems	2-25	2	44-46
	2-24	2	44
	8-13	8	252, 255-256, 258-259
	2-23	2	44
(3) Foam systems	5-7	5	143-147
	5-8	5	144-145
	5-9	5	147
	5-12	5	143-147

5.1.1 (A) Requisite Knowledge	Learning Objective(s)	Chapter(s)	Page(s)
Manufacturer's specifications and requirements.	2-1	2	21
	2-9	2	23, 27-30
	2-27	2	46-47
	2-13	2	23-24
	2-16	2	27-46, 48
Policies and procedures of the jurisdiction	2-2	2	21-23
	2-29	2	46-47
	2-30	2	47
	2-5	2	24-25

5.1.1 (B) Requisite Skills	Learning Objective(s)	Chapter(s)	Page(s)
The ability to use hand tools	2-12	2	28
Recognize system problems	2-15	2	28-46, 48
	2-3	2	21-23
Correct any deficiency noted according to policies and procedures.	2-10	2	27-46, 48

5.2.1 Task Statement	Learning Objective(s)	Chapter(s)	Page(s)
Produce effective hand or master streams, given the sources specified in the following list so that the pump is engaged, all pressure control and vehicle safety devices are set, the rated flow of the nozzle is achieved and maintained, and the apparatus is continuously monitored for potential problems:	8-1	8	233-239
	8-2	8	233-234
	8-4	8	236-242
	8-7	8	246-249
	5-1	5	123-127
	5-3	5	130-133
	9-6	9	287-288
(1) Internal tank	9-7	9	288-290
(2) * Pressurized source	7-8	7	206-209
	9-5	9	287
(3) Static source	11-2	11	336-339
(4) Transfer from internal tank to external source	11-4	11	342-348
	11-5	11	348-350
	9-1	9	279-284, 298
	9-2	9	299
	8-10	8	247-248
	9-3	9	284-300

5.2.1 (A) Requisite Knowledge	Learning Objective(s)	Chapter(s)	Page(s)
Hydraulic calculations for friction loss and flow using both written formulas and estimation methods.	10-1	10	311-313
	10-3	10	312-324
	10-4	10	313-316
	10-6	10	316-320
	10-7	10	320-324
	10-8	10	325-326
	10-9	10	326-328
	11-1	11	336-339
	10-2	10	311-312
	12-1	12	356-383
Safe operation of the pump.	4-1	4	100
	4-2	4	100-101
	4-3	4	101-116
	5-2	5	128-130
	5-4	5	130-132
	5-5	5	132
	8-5	8	236-239
	8-8	8	239-242
	9-4	9	285-286
	9-8	9	290-291
	7-1	7	192-193
	7-3	7	194
Problems related to small-diameter or dead-end mains.	7-2	7	193-194
	7-4	7	194-205
Low-pressure and private water supply systems.	7-5	7	197-201
	7-7	7	204

5.2.1 (A) Requisite Knowledge	Learning Objective(s)	Chapter(s)	Page(s)
Hydrant coding systems.	7-6	7	201-204
Reliability of static sources	7-9 10-5 11-3	7 10 11	210-214 312-316 340-341

5.2.1 (B) Requisite Skills	Learning Objective(s)	Chapter(s)	Page(s)
The ability to position a fire department pumper to operate at a fire hydrant and at a static water source.	8-2 8-12	8 8	233-236, 260-264 235-236, 265-268
Power transfer from vehicle engine to pump.	8-6	8	236-239, 242, 245-248, 269-272
Draft. Operate pumper pressure control systems.	8-9	8	247-248
Operate the volume/pressure transfer valve (multistage pumps only)			
Operate auxiliary cooling systems.	8-11	8	249
Make the transition between internal and external water sources.			
Assemble hose lines, nozzles, valves and appliances	6-10 6-9	6 6	180-184 185

5.2.2 Task Statement	Learning Objective(s)	Chapter(s)	Page(s)
Pump a supply line of 65mm (21/2in.) or larger, given a relay pumping evolution, the length and size of the line and the desired flow and intake pressure, so that the correct pressure and flow are provided to the next pumper in the relay.	7-15 7-16 7-17	7 7 7	217-220 217-220 217-220

5.2.2 (A) Requisite Knowledge	Learning Objective(s)	Chapter(s)	Page(s)
Hydraulic calculations for friction loss and low using both written formulas and estimation methods.	10-1 10-3 10-4 10-6 10-7 10-8 10-9	10 10 10 10 10 10 10	311-313 312-324 313-316 316-320 320-324 325-326 326-328

5.2.2 (A) Requisite Knowledge	Learning Objective(s)	Chapter(s)	Page(s)
	11-1 10-2 10-5 11-3 12-1	11 10 10 11 12	336-339 311-312 312-316 340-341 356-383
Safe operation of the pump.	7-14	7	215-220
Problems related to small-diameter or dead-end mains.	7-10	7	215
Low-pressure and private water supply systems.	7-11	7	215
Hydrant coding systems.	7-12	7	215
Reliability of static sources.	7-13	7	216-7

5.2.2 (B) Requisite Skills	Learning Objective(s)	Chapter(s)	Page(s)
The ability to position a fire department pumper to operate at a fire hydrant and at a static water source.	8-2 8-12	8 8	233-236, 260-264 235-236, 265-268
Power transfer from vehicle engine to pump.	8-6	8	236-239, 242, 245-248, 269-272
Draft.			
Operate pumper pressure control systems.	8-9	8	247-248
Operate the volume/pressure transfer valve (multistage pumps only)			
Operate auxiliary cooling systems.	8-11	8	249
Make the transition between internal and external water sources.			
Assemble hose lines, nozzles, valves and appliances	6-10 6-9	6 6	180-184 185

5.2.3 Task Statement	Learning Objective(s)	Chapter(s)	Page(s)
Produce a foam fire stream, given foam-producing equipment, so that properly proportioned foam is provided.	5-14	5	143-144

5.2.3 (A) Requisite Knowledge	Learning Objective(s)	Chapter(s)	Page(s)
Proportioning rates and concentrations.	5-15	5	143-144
Equipment assembly procedures.	5-10	5	144-145, 152-154
	6-6	6	173-175
	5-13	5	145-147, 155
	5-18	5	152-155
Foam system limitations.	5-11	5	144-147
Manufacture's specifications.	5-6	5	144-147

5.2.3 (B) Requisite Skills	Learning Objective(s)	Chapter(s)	Page(s)
The ability to operate foam proportioning equipment.	5-16	5	143
Connect foam stream equipment.	5-17	5	152-155

5.2.4 Task Statement	Learning Objective(s)	Chapter(s)	Page(s)
Supply water to fire sprinkler and standpipe systems given specific system information and a fire department pumper, so that water is supplied to the system at the correct volume and pressure.	9-15	9	295-296
	9-20	9	296-297

5.2.4 (A) Requisite Knowledge	Learning Objective(s)	Chapter(s)	Page(s)
Calculation of pump discharge pressure.	9-12	9	291-296, 301
	12-2	12	383-384
Hose layout	9-13	9	295-296, 301
Location of fire department connection; alternative supply procedures if fire department connection is not usable.	9-16	9	295-296
Operating principles of sprinkler systems as defined in NFPA 13, NFPA 13D, and NFPA 13R; fire department operations in sprinklered properties as defined in NFPA 13E; and operation principles of standpipe systems as defined in NFPA 14.	9-9	9	291-293
	9-10	9	294-295
	9-11	9	294-295
	9-14	9	295-296
	9-17	9	296-297
	9-18	9	296-297
	9-19	9	296-297

5.2.4 (B) Requisite Skills	Learning Objective(s)	Chapter(s)	Page(s)
The ability to position a fire department pumper to operate at a fire hydrant and at a static water source.	8-2	8	233-236, 260-264
	8-12	8	235-236, 265-268
Power transfer from vehicle engine to pump.	8-6	8	236-239, 242, 245-248, 269-272
Draft.			
Operate pumper pressure control systems.	8-9	8	247-248
Operate the volume/pressure transfer valve (multistage pumps only)			
Operate auxiliary cooling systems.	8-11	8	249
Make the transition between internal and external water sources.			
Assemble hose lines, nozzles, valves and appliances	6-10	6	180-184
	6-9	6	185

FESHE Correlation Guide

In June 2001, The U.S. Fire Administration hosted the third annual Fire and Emergency Services Higher Education Conference, at the National Fire Academy campus, in Emmitsburg, Maryland. Attendees from state and local fire service training agencies, as well as colleges and universities with fire-related degree programs attended the conference and participated in work groups. Among the significant outcomes of the working groups was the development of standard titles, outcomes, and descriptions for six core associate-level courses for the model fire science curriculum that had been developed by the group the previous year. The six core courses are Fundamentals of Fire Protection, Fire Protection Systems, Fire Behavior and Combustion, Fire Protection Hydraulics and Water Supply, Building Construction for Fire Protection, and Fire Prevention.

FIRE AND EMERGENCY SERVICES HIGHER EDUCATION (FESHE) CORRELATION GUIDE

The following table correlates the Model Curriculum Course Fire Protection Hydraulics and Water Supply requirements to this textbook's chapters.

Name: Fire Protection Hydraulics and Water Supply

Course Description: This course provides a foundation of theoretical knowledge in order to understand the principles of the use of water in fire protection and to apply hydraulic principles to analyze and to solve water supply problems.

Course Requirements	Text Chapter
I. Water as an extinguishing agent	
A. Physical properties	4, 10
B. Terms and definitions	4, 10, 11
II. Math review	
A. Fractions	Practical Problems in Mathematics for Emergency Services ISBN 0-7668-0420-8
B. Ratios, proportions, and percentage	
C. Powers and roots	

Course Requirements	Text Chapter
III. Water at rest	
A. Basic principles of hydrostatics	
1. Pressure and force	4, 10
2. Six principles of fluid pressure	4, 10
3. Pressure as a function of height and density	4, 10
4. Atmospheric pressure	4, 5
B. Measuring devices for static pressure	5, 11
IV. Water in motion	
A. Basic principles of hydrokinetics	10, 11
B. Measuring devices for measuring flow	5, 11
C. Relationship of discharge velocity, orifice size, and flow	4
V. Water distribution systems	
A. Water sources	7
B. Public water distribution systems	7
C. Private water distribution systems	7
D. Friction loss in piping systems	11
E. Fire hydrants and flow testing	7
VI. Fire pumps	Section II
A. Pump theory	4
B. Pump classifications	4
C. Priming systems	4, 5
D. Pump capacity	4
E. Pump gauges and control devices	5
F. Testing fire pumps	8
VII. Fire streams	
A. Calculating fire flow requirements	11, 12
B. Effective horizontal and vertical reach	6
C. Appliances for nozzles	6
D. Performance of smooth-bore and combination nozzles	6
E. Hand-held lines	6
F. Master streams	6
G. Nozzle pressures and reaction	6, 10
H. Water hammer and cavitations	9
VIII. Friction loss	
A. Factors affecting friction loss	10
B. Maximum efficient flow in fire hose	11
C. Calculating friction loss in fire hose	11

Course Requirements	Text Chapter
D. Friction loss in appliances	11
E. Reducing friction loss	11
IX. Engine pressures	
A. Factors affecting engine pressure	11, 12
X. Standpipe and sprinkler systems	
A. Standpipe systems	
1. Classifications	9
2. Components	9
3. Supplying standpipe systems	9
B. Sprinkler systems	
1. Classifications	9
2. Components	9
3. Supplying sprinkler systems	9

Pump Operator and Emergency Vehicles

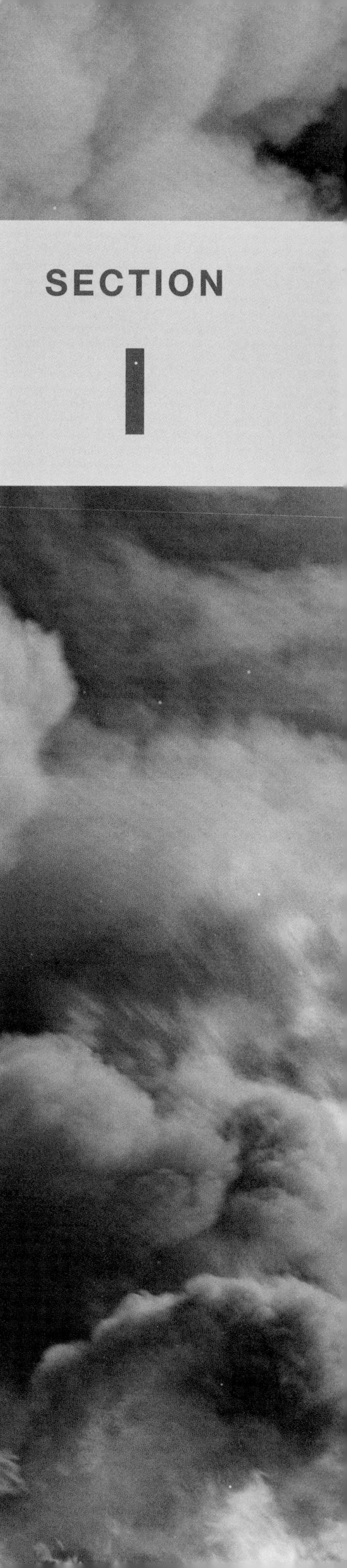

This section of the book covers the following sections of the National Fire Protection Association (NFPA) 1002 *Standard for Fire Apparatus Driver/Operator Professional Qualifications*, 2009 edition (herein referred to as NFPA 1002):

- 1.4 General Requirements
- 4.2 Preventive Maintenance
- 4.3 Driving/Operating
- 5.1 General

Chapter 1 presents the qualifications, duties, and responsibilities of the driver/operator; preventive maintenance of emergency vehicles is discussed in **Chapter 2**; and driving emergency vehicles is presented in **Chapter 3**.

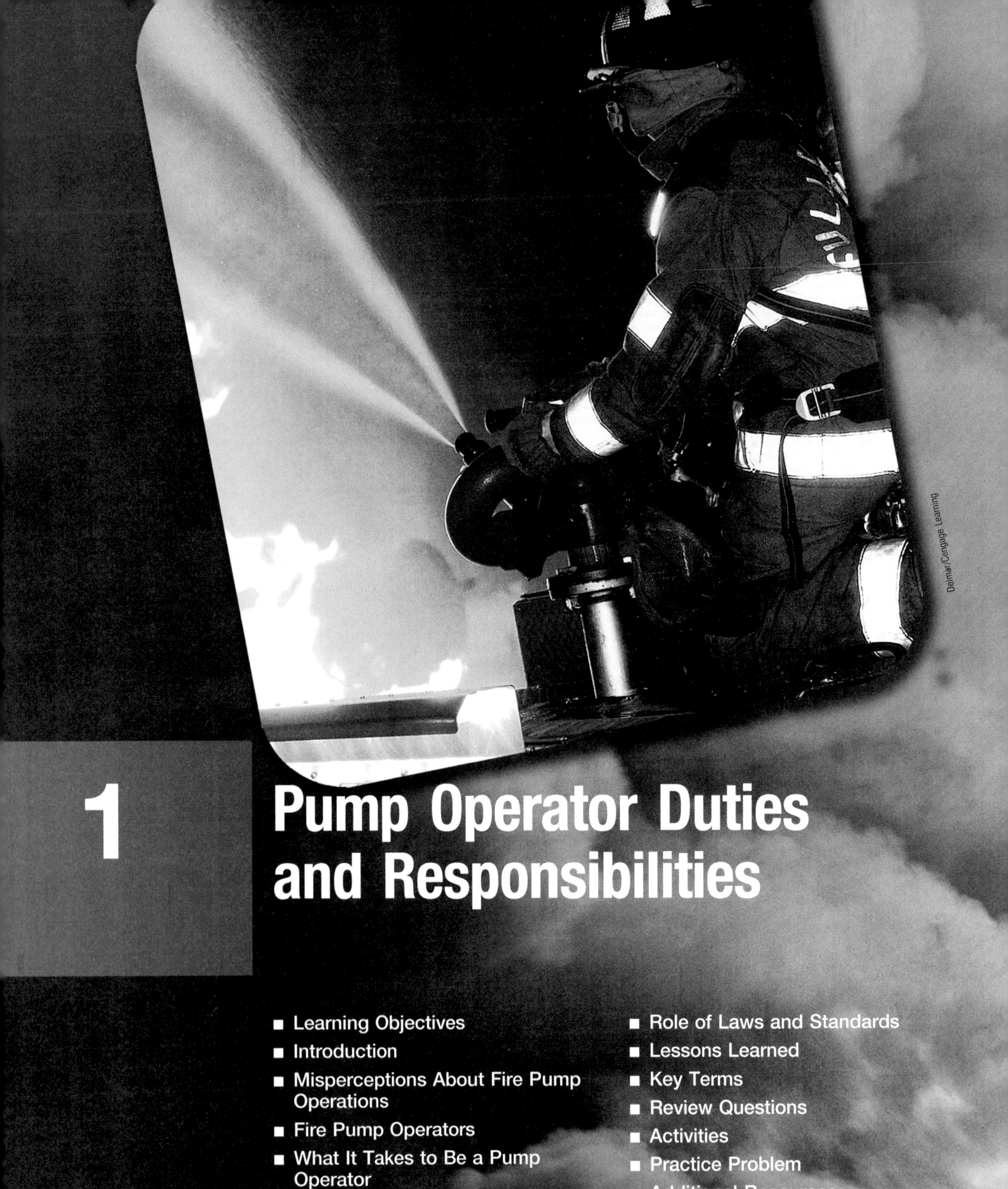

1 Pump Operator Duties and Responsibilities

My first real experience as a driver/pump operator came one day when both assigned drivers on my shift called in sick. My rank at the time was firefighter/relief driver. Still wet behind the ears, I eagerly accepted the challenge of driving Engine 3, an old 750 gallons per minute (2,900 litres per minute) pumper. After conducting the daily inspection, I thought it wise to quickly review pump procedures. I pulled out the old black notebook and began looking through the steps for basic operations. Before I finished, the house bell sounded for a reported smell of smoke at Mack's restaurant. Jumping into the driver's seat, I experienced the excitement and thrill of my first chance to prove myself as a driver/pump operator. Although excited, I handled myself in a calm manner. My captain, on the other hand, had a stressed look on his face, knowing he had to deal with a rookie driver. En route, I began to think about what I would do when we arrived on-scene.

That's when I began to lose my confidence. I couldn't remember the steps I had been taught. The more I tried to think the steps through, the more confused I became. Then I began thinking about "what if" questions. What if I had to draft, what if I had to relay pump, or, worse yet, what if I had to do hydraulic calculations? I kept thinking, "When we get on-scene, please let me get water to the nozzle." By that time I didn't care about pressures or flows as long as I was able to get water out of the nozzle. Fortunately for me, the call ended up being a false alarm and we returned without laying a line or having to engage the pump.

Back in the station, my captain praised me for my professionalism and my ability to remain calm while driving. I knew, however, how ill prepared I was for the duty of driver/pump operator. It was then that I decided to learn more about pump operations. At first I thought I could simply read a couple of books or take some classes on pump operations; however, as I began studying, I quickly became amazed at how much I didn't know and how much there was to learn. It seemed to me back then that I would never learn it all. Through the years, as both a pump operator and an educator, I have continued to learn about pump operations and can safely say that I still don't know it all. I have come to the conclusion that becoming the very best pump operator requires a constant and healthy vigilance for opportunities to learn and practice.

—Street Story by Thomas B. Sturtevant,
Author and Curriculum Director, Emergency Services,
Training Institute (ESTI) of Texas A&M University, College Station, Texas

LEARNING OBJECTIVES

After completing this chapter, the reader should be able to:

1-1. List four names used to identify the individual responsible for pump operations.

1-2. List and discuss the duties of a fire pump operator.

1-3. Describe the principles of safe vehicle operation.

1-4. Define fire pump operations and discuss its three interdependent activities.

1-5. Discuss the basic requirements every pump operator should possess, including requirements contained in NFPA 1002.

1-6. Explain the importance of studying fire pump operations.

1-7. Explain the role of laws and standards.

1-8. Explain the scope and purpose of NFPA 1002.

1-9. List and explain the additional NFPA standards that pertain to pump operations.

*The driver/operator requirements, as defined by the NFPA 1002 Standard, are identified in black; additional information is identified in blue.

INTRODUCTION

Firefighters preparing for a driver/operator position spend a considerable amount of time studying the duties and responsibilities of the pump operator. In the beginning, many feel overwhelmed by the gauges, problems, and formulas thrown at them. For some, the mystery quickly fades away as theory becomes practical. Whether in the academy, on the job, in a training class, or studying for a promotional exam, one thing seems to remain constant: the fear of hydraulics and the pump operator's duties and responsibilities. This book is written to help remove the mystery that surrounds this important topic in the fire service. This chapter focuses on the individual assigned the duties of operating the fire pump and serves as a general overview of fire pump operations. In addition, the importance of studying fire pump operations and the impact of standards and laws are discussed.

Becoming a good pump operator requires constant learning and training.

MISPERCEPTIONS ABOUT FIRE PUMP OPERATIONS

Several misperceptions about **fire pump operations** appear to be pervasive in the fire service. For example, some believe that most of the activities related to fire pump operations occur in the initial arrival on-scene. This perception may exist because during this time, visual work by the **pump operator** such as connecting supply lines, opening discharges, and setting pump pressures can be seen. It is true that initial pump operation activities are numerous and challenging in the first minutes of an operation; however, there is no autopilot on fire pumps that allows hands-free operation after the initial setup. Fire pump instrumentation must be continually observed and adjustments must be made to maintain appropriate pressures and flows. Additionally, because conditions can change rapidly on the fireground, water flow requirements can change rapidly as well. Fire pump operators must be constantly ready to adjust pumping operations to meet changing fireground needs.

Because fireground conditions can change rapidly, water flow requirements can change rapidly as well.

Another misperception is the belief that fire pump operations are more complex and unpredictable than other operations or activities on the fireground. Perhaps this perception stems from the fear and misunderstanding of scientific hydraulic theory and principles. In reality, scientific theory and principles prevail in both fire suppression and pump operations. Fire pump operations can actually be more predictable and controllable than a structural fire.

Finally, fire pump operation is often thought of as a non-glamorous and unimportant job or as a necessity for future advancement within the fire service. Images of the fireground operations rarely depict the lonely fire pump operator hard at work. Rather, the focus is on dramatic rescues and on firefighters directing elevated streams of water at the fire. Keep in mind that without pump operations, suppression efforts would be literally nonexistent, and dramatic rescues would either turn into risky rescues or would not be possible at all.

FIRE PUMP OPERATORS

The role of the fire pump operator is a noble position in the fire service. The titles given to the position range from traditional names to those reflecting the changes in the position. New technology and automation coupled with the ever-increasing responsibility of pump operators require new levels of maturity, education, and skill. Because of this, proper selection of future pump operators is vital to the overall mission of the fire service: to save lives and property.

What's in a Name?

The first fire pumps used in America were manual piston pumps. These pumps were simple devices that used manpower to operate them (**Figure 1-1**). Because of their simplicity, operating these pumps required minimal training, making unnecessary the need for a designated operator.

The steam engine was introduced in the late 1800s as a power source for fire pumps and it changed the requirements for fire pump operators (**Figure 1-2**). Steam as a power source was complicated and dangerous, and the need for a designated person to operate these sophisticated pumps was realized. Engineers familiar with the concepts of steam and pressure were assigned to each apparatus. One of the most important duties of an engineer was to develop enough pressure to power the pump when it arrived at the fire scene. The engineer's duties began to change as steam and then the internal combustion engine powered both the pump and the apparatus. This position became an integral part of the fire service, and the term *engineer* evolved into one of the more popular titles assigned to the person who operated pumps.

Today, the term *engineer* is still used to identify pump operators; however, other terms have also come into use, including wagon driver or tender, chauffeur, apparatus operator, driver/pump operator, and apparatus engineer/driver. For the purpose of this text, the term *pump operator* is used to describe the individual responsible for operating the fire pump, driving the apparatus, and conducting preventive maintenance.

Courtesy of Captain Tom Bentley, Quincy Fire Department

FIGURE 1-1 A typical early American piston fire pump. Extensive manpower was needed to operate these pumps. This model used 10 people on each side, and frequent crew replacements were needed due to exhaustion.

Courtesy National Museum of American History, Kenneth E. Behring Center, Smithsonian Institution

FIGURE 1-2 The introduction of steam to power fire pumps required trained engineers to operate them.

What Do They Do?

Pump operators have a tremendous amount of responsibility. The consequences of improper action or lack of action can be devastating. For example:

- Improper inspection and testing may result in mechanical damage and failure or reduced life span of the apparatus, pump, and equipment. In a worst-case scenario, you could find out at the time of an emergency call that the apparatus is unable to respond, or it could fail while supplying water to firefighters actively attacking a fire deep inside a commercial structure.
- Inadequate water supplies may increase the duration of suppression activities, resulting in increased risk to personnel and property.

- Careless driving techniques increase the risk of crashes, resulting in expensive damage to equipment as well as injury and loss of life to both fire service personnel and civilians.
- Inadequate training or lack of adherence to laws can find both the pump operator and the department facing criminal and/or civil repercussions.
- Improper pump operation can even damage or contaminate a municipal water supply, which at a minimum would result in expensive repairs and bad publicity.

Because the consequences are so great, one rule that a pump operator must adhere to is the constant attention to safety. With safety in mind, the duties of pump operators can be grouped in three areas: preventive maintenance, driving the apparatus, and operating the pump.

SAFETY

Because the consequences are so great, one rule that applies to all duties of a pump operator is that of constant attention to safety.

Preventive Maintenance

An often overlooked and underemphasized duty is maintaining the apparatus in a ready state (i.e., in a safe and efficient working condition). To do this, pump operators must inspect their apparatus and equipment regularly (**Figure 1-3**). In addition, pump operators must conduct periodic tests to ensure the equipment is in peak working condition. Preventive maintenance is changing due to technology changes found on newer apparatus. Items such as computers, cameras, compressed-air foam systems (CAFSs), and load managers are items requiring specialized training to understand, operate, and maintain. Many of today's batteries are maintenance free, eliminating some of the past maintenance activities. As important as understanding preventive maintenance for assigned vehicles, pump operators must understand that other apparatus will likely have different requirements due to differences in age and vendor equipment. Furthermore, the operator needs to understand the usage, operation, and maintenance of all of the equipment on his or her assigned apparatus. Only through diligent inspection and testing can the pump operator be confident that the equipment will perform in an effective and safe manner when most needed.

SAFETY

The most important duty of the pump operator is ensuring the safe transportation of personnel and equipment to the emergency scene.

Driving the Apparatus

Being the "Best Pump Operator in the World" will be of little value if the pump never makes it to the scene. Therefore, the most important duty of the pump operator is ensuring the safe transportation of personnel and equipment to the emergency scene (**Figure 1-4**). Although this sounds easy, statistics indicate that driving emergency vehicles is a dangerous activity. According to the U.S. Fire

FIGURE 1-3 The duty of preventive maintenance helps ensure the apparatus is in ready state at all times.

Delmar/Cengage Learning

FIGURE 1-4 Ensuring the safe arrival of personnel and equipment is another important duty of the pump operator.

Administration 2010 information, there is an average of 100 line-of-duty firefighter deaths per year and tens of thousands of injuries. Of the fatalities, 20–25% involve motor vehicle crashes (**Figure 1-5**), with tankers being the most frequently involved apparatus type. In fact, tankers are involved in more crashes than engines and ladders combined. Astonishingly, a mere 21% of the victims performed the basic safety function of fastening their seatbelt. These statistics do not include all crashes involving apparatus, nor do they include civilian death and injury information. Rather, they include only those reported fire department crashes (**Figure 1-6**), injuries, and fatalities that occurred while responding to or returning from an incident. In addition, they do not take into account minor occurrences or near misses.

SAFETY

Fastening your seatbelt significantly increases your chances of surviving a serious crash and may affect death benefits. Apparatus should never be moved unless everyone is seated and belted in.

The quicker emergency response crews arrive on scene, the better the chance of saving lives and property, so it makes sense that fire departments attempt to keep response times to a minimum. A few examples of attempting to keep response times to a minimum include:

- Using 9-1-1 to report and mobilize emergency forces quickly
- Strategically locating stations for shorter responses
- Continuously training to deploy suppression and rescue activities quickly
- Equipping emergency apparatus with warning devices to move through traffic
- Providing certain privileges, in terms of exemptions from various traffic regulations, for responding apparatus

The need for a quick response is an integral part of almost every emergency response activity. Unfortunately, the urge to get to the scene as quickly as possible often overshadows the safe operation of emergency vehicles. When this occurs, the chance for a crash increases. In fact, most elements of a quick response do not involve driving. An example is getting out of the station faster. Another example of improving response time is saving time "on the road" by knowing your district so you do not waste time as a result of an incorrect route or turn.

When thinking in terms of getting to the emergency scene, apparatus operators should be thinking of a quick yet safe response. Another way of looking at it is to think of a response in terms of effectiveness and efficiency. Driving the apparatus so fast that a crash occurs is not an effective response. Conversely, driving the apparatus too slowly is not an efficient response. The term *effective and efficient response* connotes the meaning of both a safe and timely response.

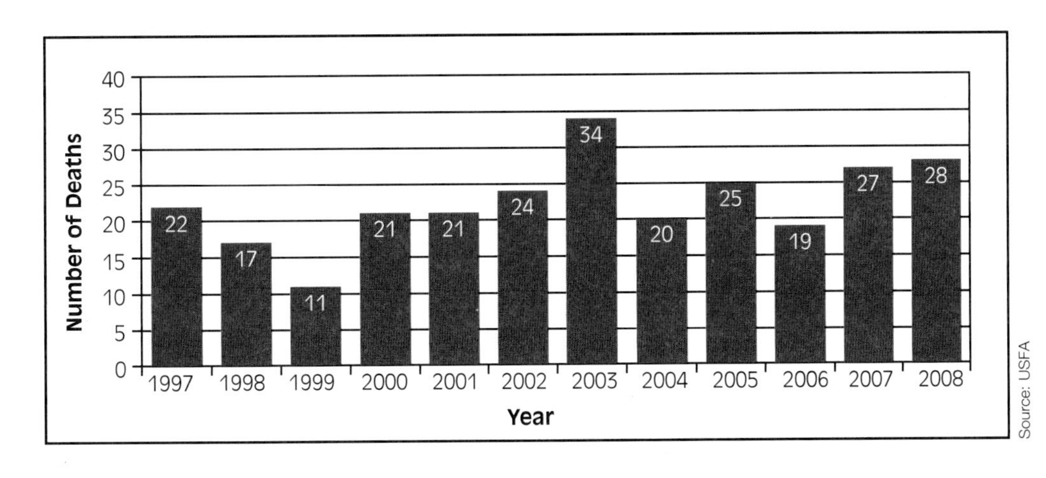

Source: USFA

FIGURE 1-5 Annual number of emergency vehicle accident fatalities from 1977 to 2008.

Photo Courtesy of the Fort Worth Fire Department

FIGURE 1-6 Crashes can cause firefighter and civilian injuries or fatalities. They create extended apparatus outages and may result in lawsuits.

Crashes involving responding apparatus can produce several significant outcomes. First, assistance to those who summoned help will be delayed because additional units must be dispatched. Not only must additional units be dispatched for the original incident, additional units must be dispatched for the second incident that involves the responding apparatus. Second, fire department personnel and civilians could be seriously or fatally injured or permanently disabled. In addition, department and civilian vehicles and property could sustain extensive damage. Time to repair or replace a damaged apparatus may take months, affecting the department's response to many calls, not just the one where the accident occurred. Third, the community, the department, and the driver could face civil and/or criminal proceedings. Finally, the image presented by the accident will last a long time in the minds of the public.

Fire department apparatus crashes occur for a variety of reasons. However, several common factors appear in a vast majority of the accidents. One such factor is associated with apparatus operators not following laws and standards related to emergency response. Emergency response laws and standards are established, in part, as a result of analyzing factors related to crashes involving emergency vehicles. The basic goal or intent is to help protect emergency responders as well as the general public. When emergency response laws and standards are not followed, the potential for a crash increases and emergency responders and the general public incur unnecessary risk. A second factor relates to operators not being fully aware of both driver and apparatus limitations. Drivers/operators must fully understand their own driving limitations as related to reaction time, vision, and attention as well as their vehicle's characteristics, such as weight, dimensions, and load.

A third factor is the lack of appreciation for driving considerations such as weather and traffic. The fourth factor is that many operators may fail to appreciate, understand, and/or compensate for emergency response driving demands and conditions. Knowledge and use of skillful safe driving techniques and practices can dramatically reduce the chance of an accident and may also help reduce wear and tear on emergency vehicles.

Pump Operations

The most obvious duty of a pump operator is operating the pump (**Figure 1-7**). *Fire pump operations* can be defined as the systematic movement of water from a supply source through a pump to a discharge point, with a resultant increase in pressure. This definition identifies the three interdependent activities of pump operations: water supply, pump procedures, and discharge maintenance (**Figure 1-8**). Each of these activities affects the outcome of the operation, as follows:

1. *Water supply*. The first activity of operating the pump encompasses securing a water supply. The focus is the movement of water from a source to the intake side of the pump. Water supplies can be secured from various sources. Water can be secured from the apparatus itself because pumping apparatus must have a minimal water supply on-board. According to NFPA 1901, *Automotive Fire Apparatus*, an on-board supply of at least 300 gallons (1,135 litres) must be provided. This water source is typically the quickest and easiest to secure, yet it is often the most limited. Water can also be secured from pressurized sources, including hydrants, relay operations, and elevated water supplies. Water can also be transported from static sources such as ponds, lakes, rivers, and tanker shuttle supplies. Each water supply source has unique characteristics and challenges for securing and maintaining adequate flows.

According to NFPA 1901, a pumper's on-board water supply must be at least 300 gallons (1,135 litres).

2. *Pump procedures*. The second activity relates to the procedures used to operate the pump. This activity moves water from the intake side of the pump and delivers it to the pump's discharge gates. Although various pump sizes and configurations exist, the general operation of most

FIGURE 1-7 One duty of the pump operator is that of operating the pump.

FIGURE 1-8 The systematic movement of water from a supply source through a pump to a discharge point defines fire pump operations.

pumps is similar. Activities include opening and closing valves and increasing and decreasing discharge pressures. Included in the operation of the pump is operating pump peripherals such as pressure relief devices and priming devices.

3. *Discharge maintenance*. The third activity is the maintenance of discharge lines. This activity maintains water flow from the discharge gate to the nozzle. Hose size, nozzles, and appliances, as well as hydraulic calculations, are used to determine appropriate flow rates and pressures. To maintain flow rates and pressures, discharge gates are opened and closed and engine speed is increased or decreased.

These three activities—water supply, pump procedures, and discharge maintenance—are interdependent in that each activity is dependent on the others as well as affected by the others. They are dependent because if one activity is missing, pump operations cannot exist. Obviously, without a water supply pump operations cannot be carried out. Likewise, without discharge lines the movement of water to a discharge point cannot occur. These activities are affected by each other because the ability to accomplish each activity influences the other. For example, with a limited water supply the pump may not be able to provide appropriate discharge flows and pressures. Proper pump procedures cannot overcome excessively long discharge lines. Pump operations can therefore be viewed as the balancing of these three activities.

Equipment Operation

The operator must be able to operate all apparatus equipment, including fixed systems, powered tools, and hand tools. Examples of fixed systems that could exist on a given apparatus include generators, light towers, hydraulic pumps, breathing air compressors, and winches. Powered tools include saws, positive pressure ventilation (PPV) fans, and extrication tools; hand tools include many type of nozzles, hose, appliances, and ladders.

In short, the operator needs to know the purpose of each switch, knob, handle, valve, and gauge or indicator on the apparatus. In addition, for gauges and indicators, besides knowing the desired reading, the operator needs to know what actions to take when an undesired reading occurs.

Firefighting

Often, on-scene apparatus operation is not required either because it is a non-fire incident or because the unit is a later-arriving apparatus needed for manpower. Operators need to maintain all of their firefighting skills for these incidents.

WHAT IT TAKES TO BE A PUMP OPERATOR

Basic Qualifications

Prior to becoming a pump operator, every individual should possess certain basic qualifications, including being qualified as either:

- A Fire Fighter Level I as defined by NFPA 1001;
- An Advanced Exterior Industrial Fire Brigade Member as defined by NFPA 1081; or
- An Interior Structural Fire Brigade Member as defined by NFPA 1081.

According to NFPA 1002, *Standard for Fire Apparatus Driver/Operator Professional Qualifications,* a medical evaluation should be conducted to ensure that the individual is medically fit to perform the duties of a pump operator. The medical evaluation should include a comprehensive physical as well a physical performance evaluation. In addition, NFPA 1002 requires pump operators to possess a valid driver's license and be licensed for all vehicles they are expected to drive. Prior to starting pump operation, individuals should possess at least a basic level of education. Math skills and communications skills such as writing and reading are essential to both learning pump operations as well as to conducting pump operator duties. Many jurisdictions have additional requirements, commonly including a minimum time in grade, Firefighter II certification, minimum age, and district knowledge. Finally, maturity and responsibility should be exhibited by pump operator candidates because the risk to lives and the potential damage to expensive equipment demand that pump operators be mature and responsible.

Knowledge

As stated earlier, pump operator duties include conducting preventive maintenance inspections, driving emergency vehicles, and operating the pump and its peripherals. Obviously, pump operators should possess the knowledge required to perform these duties. Simply memorizing the steps to perform the activities does not provide the type of knowledge pump operators should possess. A pump operator's knowledge should include the understanding of the process (the how, the why, and the variables), not just the steps. With this type of knowledge, pump operators can quickly troubleshoot problems, anticipate and correct potential problems, and utilize equipment in a safe manner.

Skills

The old saying that you cannot put a fire out with books has some validity when discussing pump operations. Although learning pump operations in the classroom is a vital step, you cannot effectively and safely operate a pump from just reading a book. Operating a pump requires skills achieved through hands-on training. Acquiring these skills is not just a one-time event. Pump operators must continually hone their skills through practice, practice, and more practice.

Learning Process

The old method of taking the pumper out and learning how to pump by following specific steps until they are memorized produces an inefficient pump operator. The need to adapt quickly to changing fireground flow demands, to troubleshoot pump problems, and to operate the pump safely requires pump operators to be intimately familiar with pump operating principles, pump construction, pump peripherals, nozzle theory, hose, and so on. Acquiring the knowledge and skills of a pump operator requires both classroom lecture and hands-on training.

Pump operators must be intimately familiar with pump operating principles, pump construction, pump peripherals, nozzle theory, hose, and so on to ensure safe operation.

The first step, however, is to ensure that the basic qualifications are met. The second step is to provide the knowledge required to perform the duties of a pump operator safely and efficiently. The third step is to provide hands-on training with a pump to develop and sharpen manipulative skills. In reality, presenting knowledge and skills can be a mixed process but should be progressive and logical in presentation.

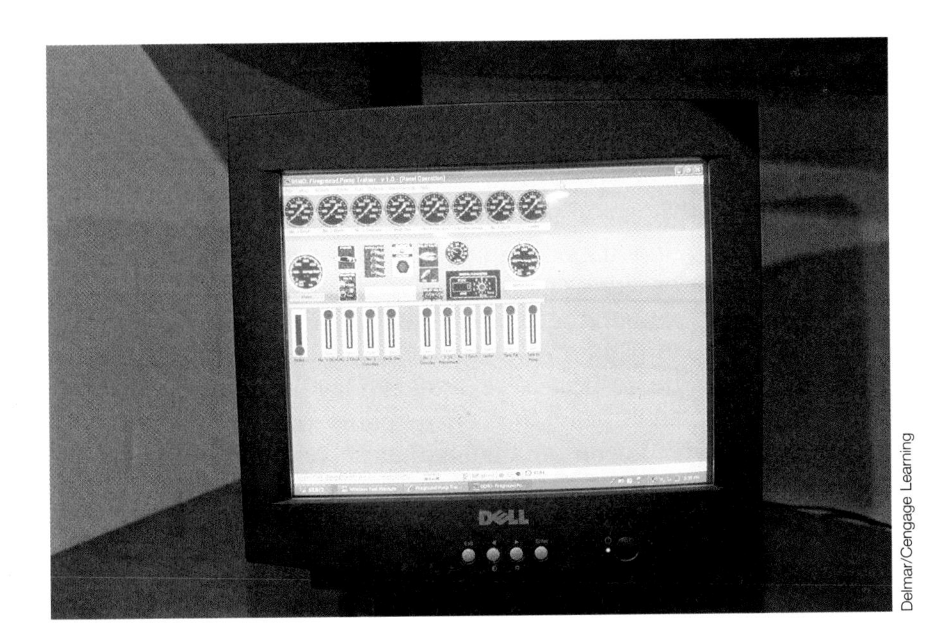

Delmar/Cengage Learning

FIGURE 1-9 Computer-controlled pump simulators provide realistic, hands-on training.

Computer-controlled pump simulators (**Figure 1-9**) and pump software provide an intermediate step between lecture and hands-on training. Pump operator skills can be practiced safely and efficiently without the threat of injury to personnel or of damaging expensive equipment. New simulators allow actual pump operation in a controlled environment. These simulators can be set up in a parking lot, attached to an apparatus, and a pump operator can be put through a variety of challenging scenarios without losing a drop of water. This provides hands-on real life training that, in the past, was not available. After demonstrating competency on the simulators, training is moved outside with actual pumping apparatus. The concept is the same as a flight simulator for training and recertifying pilots.

Selection Process

The position of pump operator is generally considered a promotion in most departments; therefore, a selection process is typically used to determine the most qualified candidate for the position. Although the process varies from one department to another, the selection process ought to include the evaluation of the candidate's basic qualifications. In addition, the process should include, as a minimum, evaluation of the requirements found in NFPA 1002.

The first step in the selection process should be to ensure that candidates have the basic qualifications. Next, a written or oral evaluation should be administered to determine knowledge of pump operations. Finally, a practical evaluation should be administered to determine the candidate's skill level. The length of service of a candidate provides little information on the ability to function in the capacity of pump operator.

WHY STUDY FIRE PUMP OPERATIONS?

Studying fire pump operations also improves the efficiency of the apparatus. Fire apparatus and their pumps are major expenditures for any fire organization. The lack of preventive inspections and maintenance as well as improper use can quickly cause unneeded expensive repairs and replacements. In reality, the apparatus and equipment belong to a community or business. Citizens of the community and shareholders of the business expect and deserve the most efficient use of their money and resources.

One of the more important reasons for studying fire pump operations is that of the safety of firefighters. Improper operation of fire pumps can cause serious injury to personnel. A simple interior attack can turn into disaster if suppression streams are inadvertently stopped, leaving attack teams with no protection from the heat and progression of the fire. Improperly managed pressures and water-pressure surges can cause hoselines to overpower firefighters, tossing them around with ease, or cause the hoselines to break loose from firefighters' hands and turn into unpredictable killers. To ensure the safety of the firefighters, pump operators must be knowledgeable about proper inspection, testing, and operating procedures for their equipment.

Another reason for studying pump operations is the desire to be promoted. As previously stated, the position of pump operator is typically considered a promotion. In most cases, this means a pay raise. In addition, it often means more responsibility with greater chances for future advancement within the department.

Finally, professionalism is another important reason for studying fire pump operations. Simply stated, it is doing the best job you can. Taking pride in your work and doing the best job you can means, in part, operating the pump in an efficient and effective manner to assist in the organization's overall goal of saving lives and property. The more efficiently and effectively pump operations are carried out, the greater the chances of realizing that goal.

To ensure the safety of the firefighter crew, which is the pump operator's greatest responsibility, pump operators must be knowledgeable about proper inspection, testing, and operating procedures for their equipment.

ROLE OF LAWS AND STANDARDS

The most significant role of laws and **standards** for fire pump operations is that of safety. Although adherence to laws and standards will not guarantee safety, it will help reduce the risk and seriousness of accidents and injuries. Lack of adherence increases the risk of accidents and injuries as well as increasing the chance that individuals, departments, and communities will be held liable. Increasingly, the fire service and its members are being challenged to defend themselves in the legal system for negligence, omissions, or wrongful acts. Knowledge of and adherence to laws and standards will assist pump operators in performing their duties in a safe and legal manner.

Laws

Two types of laws should be of concern to the pump operators: criminal law and civil law. These laws are enacted on a federal, state, and local level. The two are very different and both should be understood so the pump operator grasps the responsibility of the position.

Criminal laws are written by the U.S. Congress and state legislators and make certain behavior illegal and punishable by fines and/or imprisonment. Pump operators found criminally negligent or criminally responsible can face imprisonment. Civil law involves noncriminal cases in which one private individual or business sues another to protect, enforce, or redress private or civil rights. If the pump operator is found civilly responsible, he or she can be forced to pay a fine and face a financial burden that his or her respective department may or may not be responsible for. More often, pump operators involved in accidents are finding themselves being held accountable as defendants in criminal and civil trials. A tragic example of this occurred in Michigan in 2005. A firefighter was charged with two counts of negligent homicide for a collision he caused while driving a fire truck that killed a woman and her 11-month-old son. He later pled guilty to lesser charges, lost his job, and faced a civil lawsuit as well as possible jail time. Of particular interest to pump operators are laws enacted at the state and local level. These laws can affect pump operators on a daily basis. A major focus of these laws tends to be on emergency vehicle driving regulations. For example, nearly all state and local laws require that emergency vehicles, regardless of whether they are en route to an emergency, come to a complete stop when crossing an intersection against a red light. State and local laws also tend to identify the requirements of emergency vehicle warning devices, which include location and visibility requirements, as well as specifying when warning devices must be used. Finally, state and local laws typically identify exemptions, limitations, and immunity, if any, extended to the operation of emergency vehicles.

Standards

Standards are guidelines that are not legally binding or enforceable by law unless they are adopted by a governing body. While these standards are not always legally binding, the pump operators' actions (or inactions) will be compared to these standards. Failure to meet these standards by a pump operator can be used to show that the pump operator may be responsible in a legal action. The standards of most concern to fire pump operators are those published by the National Fire Protection Association (NFPA, see **Appendix A**). The NFPA publishes consensus standards related to equipment specifications and requirements, procedural guides, and professional qualifications. The following is an overview of NFPA standards related to pump operations.

NFPA 1002, *Fire Apparatus Driver/Operator Professional Qualifications*

NFPA 1002 identifies the minimum Job Performance Requirements (JPRs) for individuals responsible for driving and operating fire department vehicles, as well as specific JPRs for pump operators, aerial operators, tiller operators, mobile water supply operators, wildland apparatus operators, and aircraft rescue and firefighting apparatus (ARFF) operators. General requirements include those requirements common to operators of all types of fire apparatus. For example, driver/operators, like firefighters, are subject to an annual medical evaluation as required by NFPA 1500, *Fire Department Occupational Safety and Health Program*. JPRs are written as outcome objectives that describe what competencies an individual is expected to have. The prerequisite knowledge and skills required to complete the JPRs are also included when appropriate.

Chapter 5 of NFPA 1002 addresses those responsible for driving and operating apparatus equipped with fire pumps. One important requirement identified in this chapter is that, prior to being certified as a fire apparatus driver/pump operator, individuals must meet the basic firefighting requirements of either NFPA 1001 or 1081 as described later in this chapter. The JPRs listed in this chapter relate to the duties of a pump operator as discussed earlier in this text: preventive maintenance, driving, and pump operations.

NFPA 1002 is the *Standard for Fire Apparatus Driver/Operator Professional Qualifications.*

NFPA 1500, *Fire Department Occupational Safety and Health Program*

NFPA 1500 identifies the requirements for an occupational safety and health program for the fire service. This standard specifies several areas related to pump operations that must be included in an occupational safety and health plan for the fire service.

Chapter 3 of this standard requires the establishment of a training and education program. For example, this section requires that adequate initial training and retraining be provided by qualified instructors. In addition, this section requires that individuals who drive or operate emergency vehicles meet the requirements of NFPA 1002.

Chapter 4 identifies specific requirements for the safe use of emergency vehicles. One such requirement is that all new apparatus purchased by a fire department be in compliance with NFPA 1901 (discussed shortly). In addition, the standard requires that safety issues be addressed through the development of standard operating procedures (SOPs). These SOPs should include requirements for safely driving emergency vehicles, such as obeying traffic laws during emergency and nonemergency response, vehicle speed, crossing intersections, and the seating of personnel on the apparatus (see **Appendix C**).

A medical evaluation program is also required to be included in an occupational safety and health plan. This program must include a medical evaluation to ensure members meet the requirements of NFPA 1582, *Comprehensive Occupational Medical Program for Fire Departments*. In addition, members are required to demonstrate physical performance as well as participate in a physical fitness program.

Finally, a hearing protection program must be included in a department's safety and health program. Pump operators can be exposed to excessive levels of noise for extended periods of time. This standard requires hearing protection when noise levels exceed 90 decibels.

NFPA 1500 is the *Standard for a Fire Department Occupational Safety and Health Program.*

NFPA 1500 requires that safety issues be addressed through the development of standard operating procedures (SOPs).

NFPA 1582, *Comprehensive Occupational Medical Program for Fire Departments*

NFPA 1582 addresses the minimum medical requirements for firefighting personnel, including full- or part-time employees and paid or unpaid volunteers. Medical exams specified by this standard are preplacement, periodic, and return to duty. Examples of items covered are vision, hearing, heart, extremities, and specific medical conditions such as asthma and diabetes.

NFPA 1582 is the *Standard for a Comprehensive Occupational Medical Program for Fire Departments.*

NFPA 1001, *Fire Fighter Professional Qualifications*

NFPA 1001 identifies the performance requirements for individuals who perform interior structural firefighting operations. Two levels of certification are identified in this standard, Fire Fighter I and Fire Fighter II. Keep in mind that according to NFPA 1002, prior to being certified as a pump operator, individuals must be certified to the NFPA 1001 Fire Fighter I level unless they are certified according to NFPA 1081 as described below.

NFPA 1001 is the *Standard for Fire Fighter Professional Qualifications.*

NFPA 1801, *Standard for Industrial Fire Brigade Member Professional Qualifications*

NFPA 1801 identifies the minimum job performance requirements for members of an organized industrial fire brigade. Multiple levels of certification are identified in this standard, with certification at either the Advanced Exterior Industrial Fire Brigade Member or Interior Structural Fire Brigade Member required according to NFPA 1002, prior to certification as a pump operator, unless members are certified according to NFPA 1001 as described above.

NFPA 1801 is the *Standard for Industrial Fire Brigade Member Professional Qualifications*.

NFPA 1901, *Automotive Fire Apparatus*

NFPA 1901 identifies requirements for new fire apparatus. Specific requirements are included for pumper fire apparatus, initial attack fire apparatus, mobile water supply fire apparatus, aerial fire apparatus, quint fire apparatus, special service fire apparatus, and mobile foam fire apparatus. Other related fire apparatus standards include:

- NFPA 1906, *Wildland Fire Apparatus*
- NFPA 1911, *Standard for the Inspection, Maintenance, Testing, and Retirement of In-Service Automotive Fire Apparatus*
- NFPA 1912, *Fire Apparatus Refurbishing*
- NFPA 1925, *Marine Fire-Fighting Vessels*

Miscellaneous NFPA Standards

The related NFPA standards of interest to pump operators are as follows:

- NFPA 13E, *Recommended Practice for Fire Department Operations in Properties Protected by Sprinkler and Standpipe Systems*
- NFPA 291, *Recommended Practice for Fire Flow Testing and Marking of Hydrants*
- NFPA 1142, *Standard on Water Supplies for Suburban and Rural Fire Fighting*
- NFPA 1145, *Guide for the Use of Class A Foams in Manual Structural Fire Fighting*
- NFPA 1410, *Standard on Training for Initial Emergency Scene Operations*
- NFPA 1451, *Standard for a Fire Service Vehicle Operations Training Program*
- NFPA 1961, *Standard on Fire Hose*
- NFPA 1962, *Standard for the Inspection, Care, and Use of Fire Hose, Couplings, and Nozzles and the Service Testing of Fire Hose*
- NFPA 1964, *Standard for Spray Nozzles*

LESSONS LEARNED

A variety of titles are used to identify those responsible for pump operations. For the purpose of this text, *pump operator* will be used. The three major duties of a pump operator include preventive maintenance, driving, and pump operations. Pump operation is defined as the movement of water from a source through a pump to a discharge point. To ensure the most qualified individuals perform pump operator duties, the selection should be based on knowledge and skills rather than seniority. Safety is the most important reason for studying pump operations. Safety is also the primary focus of laws and standards. Knowledge of and adherence to laws and standards will assist pump operators in the safe and legal performance of their duties.

KEY TERMS

Fire pump operations The systematic movement of water from a supply source through a pump to a discharge point with a resultant increase in pressure.

Job Performance Requirements (JPRs) Objectives that describe what competencies an individual is expected to have; the prerequisite knowledge and skills required to complete the JPRs are also included when appropriate.

Laws Rules that are legally binding and enforceable.

Pump operator The individual responsible for operating the fire pump, driving the apparatus, and conducting preventive maintenance.

Standard operating procedures (SOPs) Specific information and instructions on how a task or assignment is to be accomplished.

Standards Guidelines that are not legally binding or enforceable by law unless they are adopted as such by a governing body.

REVIEW QUESTIONS

Multiple Choice

Select the most appropriate answer.

1. Which of the following is the first activity of fire pump operations?
   - **a.** driving the apparatus
   - **b.** securing a water supply
   - **c.** gaining relevant knowledge and skills
   - **d.** preventive maintenance
2. Which of the following is the correct title for NFPA 1901?
   - **a.** *Standard for Automotive Fire Apparatus*
   - **b.** *Fire Department Vehicle Driver/Operator Professional Qualifications*
   - **c.** *Initial Attack Fire Apparatus*
   - **d.** *Service Tests of Pumps on Fire Department Apparatus*
3. Which of the following is the correct title for NFPA 1002?
   - **a.** *Pumper Fire Apparatus*
   - **b.** *Standard for Fire Apparatus Driver/Operator Professional Qualifications*
   - **c.** *Initial Attack Fire Apparatus*
   - **d.** *Service Tests of Pumps on Fire Department Apparatus*

Short Answer

On a separate sheet of paper, answer/explain the following questions.

1. List at least four names used to identify the individual responsible for pump operations.
2. List the three major duties of pump operators and discuss several activities for each.
3. Define *fire pump operations.*
4. List and define the three activities of pump operations. Explain why they are interdependent.
5. Which of the three duties of pump operators do you consider to be the most important? Why?
6. Explain the role of laws and standards.
7. List and discuss four NFPA standards related to pump operations.
8. List the basic qualifications, knowledge, and skills every pump operator should possess. Identify the specific requirements contained in the 2009 edition of NFPA 1002.
9. Explain why it is important to study fire pump operations.

ACTIVITIES

1. Identify pumping apparatus in your organization. Include as much detail as you can, such as manufacturers, types and sizes of pumps, on-board water capacity, and so on.
2. Research the history of the pump operator position in your organization. Explain how that position has evolved into what it is today.
3. List and explain the requirements and selection process for fire pump operators in your organization.
4. Research specific state and local laws related to the operation of emergency vehicles that your organization must comply with. Be sure to identify standards referenced or adopted as law.
5. Identify water supply sources available for your response district or area.

PRACTICE PROBLEM

1. Your department is beginning a mandatory pump operator training course for all personnel in the position of firefighter through lieutenant. Your supervisor has asked you to make a presentation to the class on the importance as well as the responsibilities of the position. What would you present to convince your coworkers that this is an important training class for them? Provide your response.

ADDITIONAL RESOURCES

America Burning, http://www.usfa.dhs.gov/downloads/pdf/publications/fa-264.pdf

Everyone Goes Home—Firefighter Life Safety Initiatives Program, http://www.everyonegoeshome.com/

Fire Museum Network, http://www.firemuseumnetwork.org/

Memphis Museums Fire Museum of Memphis, http://www.firemuseum.com/

National Fire Fighter Near-Miss Reporting System, http://www.firefighternearmiss.com/

NFPA 1001, *Standard for Fire Fighter Professional Qualifications*. National Fire Protection Association, Quincy, MA, 2008.

NFPA 1002, *Fire Apparatus Driver/Operator Professional Qualifications*. National Fire Protection Association, Quincy, MA, 2009.

NFPA 1500, *Standard on Fire Department Occupational Safety and Health Program*. National Fire Protection Association, Quincy, MA, 2007.

NFPA 1801, *Standard for Industrial Fire Brigade Member Professional Qualifications*. National Fire Protection Association, Quincy, MA, 2007.

NFPA 1901, *Standard for Automotive Fire Apparatus*. National Fire Protection Association, Quincy, MA, 2009.

US Firefighter Fatalities, http://www.usfa.dhs.gov/fireservice/fatalities/index.shtm

Varone, Curtis. *Fire Officer's Legal Handbook*. Cengage Learning, 2007.

2 Emergency Vehicle Preventive Maintenance

- Learning Objectives
- Introduction
- Preventive Maintenance Programs
- Inspections
- Cleaning and Servicing
- Safety
- Lessons Learned
- Key Terms
- Review Questions
- Activities
- Practice Problem
- Additional Resources

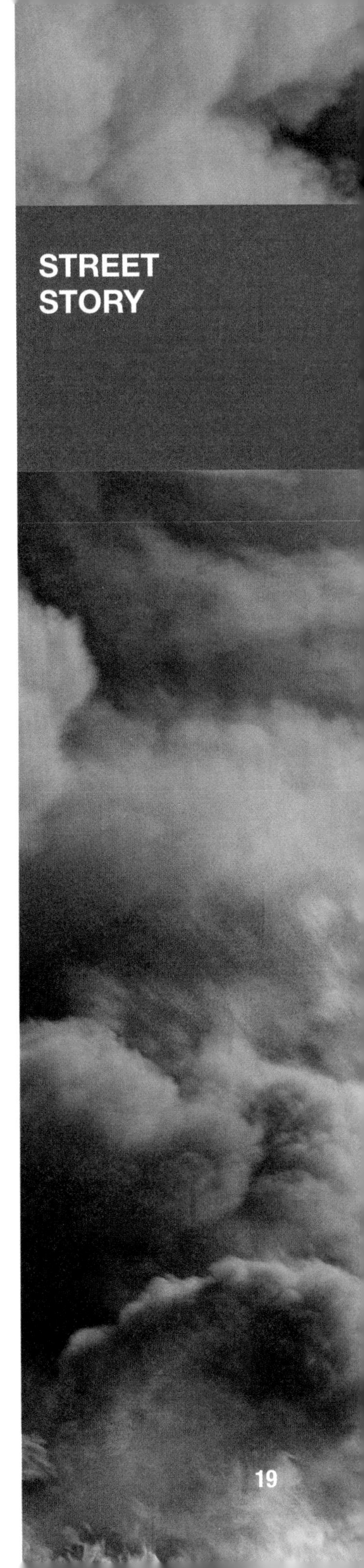

The department's pumper had a leak. Everybody knew about it. It was just through the packing gland where water is supposed to "leak" to cool and lubricate the packing. Of course everyone knew it shouldn't be such a big leak—they always had to mop the water up around the truck so no one would fall. At a fire, when the pump was under pressure, the leak was more like a stream than a drip. No one saw any real reason to repair it. A floor drain existed under the engine that caught most of the water. The district was fully hydranted, so lack of water was never a problem, and the engine developed more pressure and flow than they ever needed.

Suddenly things changed. At a house fire, with several attack lines in use, the engine suddenly froze without warning. The interior hoselines went limp, requiring the firefighters to make an emergency evacuation. By the time a backup pump was in place at the well-ventilated structure, it suffered much more fire damage than should have occurred. Fortunately, no one was injured.

When the mechanics analyzed the problem, they determined that the pressurized water stream coming through the packing gland was hitting the seal of the pump transmission. Over time, water made it into the pump transmission, displacing the transmission oil. Then the bearing started rusting. With rust and inadequate lubrication, excess friction caused components in the transmission to get extremely hot. When the transmission seized, the immediate stopping of the engine while it was under load caused its crankshaft to bend. The total repair bill, including various other items that broke, was $80,000, and the department's board replaced some high-ranking officers as they realized how much worse things could have been. The board learned that missed maintenance could cost the department in terms of expensive repairs, increased property damage to the citizens it protects, and could even affect the health of the citizens or firefighters or both. The department put in place a model maintenance program with written documentation of inspections and tracking of problems found through to their resolution. The department also instituted scheduled repairs of various frequencies based on the component involved and the manufacturer's recommendations. A computer-based program that automatically sent reminders prior to due dates so nothing would be missed was also implemented.

Even in much less extreme cases, skipping recommended maintenance may not save money. For example, replacing pump packing only half as frequently as recommended may take more than twice as long, as the mechanic removes the packing one thread at a time instead of quickly removing an entire piece.

—Street Story by Howard Sykes,
Author and Chief,
Lebanon Volunteer Fire Department,
Durham, North Carolina

LEARNING OBJECTIVES

After completing this chapter, the reader should be able to:

2-1. Explain the importance of preventive maintenance.

2-2. List and discuss the levels of responsibility with a preventive maintenance program.

2-3. Use the manufacturer's specifications and department policies and procedures to determine the operational status of a fire department pumping apparatus.

2-4. Identify the information resources used for preventive maintenance.

2-5. List the NFPA standards that contain apparatus-related preventive maintenance requirements.

2-6. Describe the procedures for accurately documenting fire apparatus maintenance and the procedure for reporting any deficiencies found to the appropriate level of maintenance or supervision.

2-7. Explain why documentation in a preventive maintenance program is so important.

2-8. Identify and describe the routine tests, inspections, and servicing functions required to determine the operational status of a fire department pumping apparatus.

2-9. List the basic steps for conducting a preventive maintenance vehicle inspection.

2-10. Conduct a hands-on, systematic inspection of the fire apparatus and its systems, correcting deficiencies, to ensure safe operating conditions.

2-11. Select the appropriate form and document the results so that all required areas of the form are properly completed, any necessary follow-up is taken, and the form is filed appropriately.

2-12. Demonstrate the safe use of hand tools required for routine tests, inspections, and vehicle servicing.

2-13. List the systems and components of a fire apparatus that are addressed during preventive maintenance.

2-14. Identify the items to inspect and explain what to look for.

2-15. Demonstrate identifying problems and potential problems when conducting preventive maintenance.

2-16. Explain how to inspect, test, and service each of the systems and components addressed in the preventive maintenance process.

2-17. Identify the types of tests to perform and describe how to perform them (daily check).

2-18. Demonstrate the safe use of tools used to conduct preventive maintenance (tire gauge, battery charger).

2-19. Identify and describe the operation of automotive gauges, and list the operational limits of fire apparatus.

2-20. Using automotive gauges and controls, maintain operation of a fire apparatus within normal operating limits.

2-21. Verify that the water level in the apparatus water supply tank is full.

2-22. Verify that the foam concentrate level in the apparatus foam concentrate tank(s) is full.

2-23. Operate all valves and pump controls to determine the operational status of a fire department pumping apparatus.

2-24. Inspect the following components: intake strainers, primer oil reservoir, discharge gauges, pressure relief valve strainer (if applicable).

2-25. List the various types of pump tests and inspections, and explain what each is assessing.

2-26. Inspect all tools and equipment stored on the fire apparatus to ensure safe operating conditions.

2-27. Explain the importance of cleaning and servicing vehicles.

2-28. Perform servicing requirements of the fire apparatus to ensure safe operating conditions.

2-29. Explain the need for diligence to safety while conducting preventive maintenance activities.

2-30. List safety concerns for preventive maintenance.

***The driver/operator requirements, as defined by the NFPA 1002 Standard, are identified in black; additional information is identified in blue.**

INTRODUCTION

As stated in **Chapter 1**, the three duties of a pump operator include conducting **preventive maintenance**, driving the apparatus, and operating the pump. Driving the apparatus is certainly a critical duty of the pump operator. The pump operator and the apparatus will be of little value if they never make it to the scene. Operating the pump is also an important duty of the pump operator. Suppression activities will be limited, and perhaps even more dangerous, if the pump is not operated properly. Each of these duties relies, in part, on the proper operation of both the apparatus and pump. The duty of preventive maintenance, then, is to ensure that the apparatus, pump, and related components are in a ready state and in peak operating efficiency. Critical to carrying out the duty of preventive maintenance is a preventive maintenance program.

A good preventive maintenance program is important because:

- Mechanical failures can jeopardize the success of an operation, cause injury to both emergency personnel and civilians, and increase property losses.
- Improper maintenance has been cited as a key factor in a number of emergency response accidents and pump failures.
- The time and money spent on preventive maintenance is significantly less than that of the potential damage likely to occur when preventive maintenance is not conducted. Poor preventive maintenance can increase the frequency and cost of repairs and reduce vehicle reliability.
- Criminal and civil liability may occur when emergency apparatus is not properly maintained.
- Manufacturers and insurance companies as well as current standards require proper preventative maintenance.

Safety cannot be ensured without proper maintenance.

There are three main levels of responsibility within a preventive maintenance program. The first level rests with the fire department. In general, the fire department is ultimately responsible for the establishment, implementation, and monitoring of a preventive maintenance program. The department should have SOPs in place that describe how preventive maintenance and repair results are to be documented, reported, filed, and retained. The next two levels of responsibility are typically divided between those activities conducted by **pump operators** and those conducted by **qualified mechanics**. The specific activities of each depend on the level of training and the type of preventive maintenance activity being conducted. In general, qualified mechanics conduct those activities that require the apparatus to be taken out of service, activities that require several hours to conduct, and activities that focus on detailed service or **repair**. Pump operators typically conduct inspections and noninvasive activities that do not require the apparatus to be taken out of service. For example, checking and adding oil to the engine may be accomplished by the pump operator, whereas changing the oil and oil filter would most likely be conducted by a mechanic.

FIGURE 2-1 Pump operators are responsible for ensuring apparatus is in a ready state at all times.

This chapter focuses on those preventive maintenance activities typically conducted by pump operators (**Figure 2-1**). First, general aspects of preventive maintenance programs are discussed. Next, specifics about conducting the vehicle inspection are presented, followed by a discussion of cleaning and servicing the apparatus. A brief discussion of pump-related inspections and testing is presented. A more detailed explanation of pump testing is provided in **Section III** of this textbook. Finally, the importance of safety while conducting preventive maintenance activities is stressed.

PREVENTIVE MAINTENANCE PROGRAMS

Preventive maintenance (PM) can be defined as proactive activities taken to ensure that apparatus, pump, and related components remain in a ready state and in peak operating condition. Typically, these activities can be grouped into **inspecting**, **servicing**, and **testing**. Inspections are conducted to verify the status of components and are the most frequently performed activity. For example, during a daily inspection, pump operators verify oil, water, and fuel levels. Servicing activities such as cleaning, lubricating, and topping off fluids help maintain vehicles in peak operating condition. Tests are conducted to determine the performance of components, typically the fire pump and related equipment. For example, during annual pump performance tests (previously called service tests), pump operators test the ability of the pump to flow its rated capacity. Each of these major activities is discussed in more detail later in this chapter.

So, what should pump operators inspect, service, and test? Basic requirements are established in

TABLE 2-1 Hale Pump Maintenance Check List

HALE **Midship Pump Maintenance Check List**

Truck Manufacturer________________

Pump Model & Serial Number____________

Year__________ Unit#______________

Recommended Weekly Procedures:

- ☐ Test relief valve system or governor at 150, 200, 250. If pump is equipped with TPM, you will need to have positive pressure.
- ☐ Operate transver valve and check clapper valves for proper operation on 2-stage pumps
- ☐ Test the priming system and check lubrication level in priming tank (if applicable)
- ☐ Operate all valves, discharge, suction, hose, drain, and multi-drain.
- ☐ Check pump shift indicator lights.

RECOMMENDED MONTHLY PROCEDURES	Jan	Feb	Mar	Apr	May	June	July	Aug	Sept	Oct	Nov	Dec
Complete weekly checks												
Lubricate threads on PM relief valve panel control and check light. DO NOT USE GREASE												
Lubricate remote valve controls and all valves												
Check controlled drip rate and adjust if necessary (8 - 10 drops per minute @ 100-150 PSI)												
Perform dry vacuum test*												
Check drive flange bolts to ensure tightness. Lubricate U Joint												
Lubricate suction tube threads. DO NOT USE EXCESSIVE GREASE												
Clean and inspect inlet strainers (Examine for loss of zinc)												
Inspect cap gskets. Replace if cracked or damaged												
Check oil level in pump gear box; add oil as necessary or replace oil with SAE EP 90 oil if contamination is found												

**Per NFPA-1911, para. 3-32, 22 inches minimum vacuum; loss not to exceed 10 inches vacuum in 5 minutes*

Recommended Annual Procedures

- ☐ Complete all previous checks
- ☐ Check gauge calibration
- ☐ Check oil level in AutoLube assembly (SEA-EP 90 oil). Pump must be drained of water prior to checking oil. See operation and maintenance manual for details.
- ☐ Lubricate power transfer cylinder, VPS shift cylinder, and shift control valve with air tool oil
- ☐ Drain pump gear box oil and refill (use SAE-EP 90 oil). Examine magnetic plug.
- ☐ Check individual drain lines from pump to multi-drain to ensure proper drainage and protection from freezing.
- ☐ Lubricate transfer valve mechanism on two stage pump. Dry moly spray is preferred.
- ☐ Perform yearly standard pump test (per NFPA-1911) to test pump performance levels.
- ☐ Repacking of pump is recommended every two or three years.
- ☐ Service ESP primer as per bulletin
- ☐ Remove and clean relief valve strainers

NOTE: The above general recommendations are provided for normal use and conditions. Extreme conditions or variables may indicate a need for increased maintenance. Good preventative maintenance lengthens pump life and ensures greater dependability. Consult service or diagnostic chart in operator's manual for detailed information

Hale Products Inc., A unit of IDEX Corporation, 700 Spring Mill Avenue, Conshohocken, PA 19428 Tel: 610/825-6300
Fax: 610/825-6440 Bulletin 889, Rev 1 11/05,

national standards and specific requirements are listed in manufacturers' recommendations such as those shown in **Table 2-1** and **Table 2-2**. In addition, both standards and manufacturer recommendations provide guidance on when preventive maintenance activities should occur and the type of documentation that should be maintained.

Table 2-3 provides a list of common apparatus systems and components. Being able to identify these and understanding their basic functions makes inspections easier and will allow more effective repair requests to be communicated to mechanics.

The Internet has become an excellent resource for preventive maintenance materials. Many manufacturers have published their operating manuals and recommended maintenance schedules on the Internet; many fire departments have published their SOPs online; and the Federal Emergency Management Agency (FEMA) has published an online guide to developing SOPs (see Additional Resources at

TABLE 2-2 Typical Chassis Inspection Items. A Mechanic Would Normally Do Most of the Chassis Maintenance Items in This Schedule

Typical Apparatus Inspection Items

Daily

All items a CDL holder is required to check in the pretrip inspection, including:

- Obvious frame or body damage or apparatus leaning abnormally
- Fluid levels—oil, coolant, transmission, power steering
- Tire conditions and pressure
- Wheel—cracks or missing lug nuts
- Running and emergency lights
- Horn
- Windshield wipers
- Equipment secured
- All gauges working properly and within normal ranges
- Adjust seats and mirrors
- Brake conditions (pad depth, no air leaks, and proper pressure)
- Operate pump and all valves, bleeders, and drains
- Booster tank level

Weekly

- Check acid level of non–maintenance-free batteries
- Check primer pump

Periodically (often every 6 months and performed by a mechanic)

- Differential fluid level
- Condition of sealed maintenance-free batteries
- Replace filters
- Change oil
- Grease chassis
- Check conditions of all belts
- Check conditions of battery cables and clamps
- Check springs and shocks
- Check air brake dryer, check valves and alarm sensors

end of chapter). Maintenance forums also exist and are an excellent source for ideas. Since many forums allow anyone to enter data, any ideas found in a forum should be confirmed with the manufacturer's representatives or other authorities as being applicable prior to implementation.

It should be noted that preventive maintenance does not occur solely during inspecting, servicing, and testing of apparatus. Rather, it is an ongoing process by the pump operator to ensure that the apparatus, pump, and related equipment are operating properly. This constant monitoring should occur during emergencies, while driving the apparatus, and while training. The individual most in tune with the apparatus is the pump operator. There is no better person to detect minor changes in the performance of the apparatus. Often, minor changes are simply warning signs for major problems looming ahead. The immediate detection of small changes can increase safety as well as the operating life of the apparatus.

TABLE 2-3 Typical Inspection Items

- Air Induction
  - Air filter
  - Intercooler
  - Turbocharger
- Braking
  - Air compressor, dryer, tanks, lines, and valves
  - Disks and rotors
  - Drums
  - Parking brake
  - Shoes and pads
  - Slack adjuster
- Cooling
  - Belts
  - Coolant hose
  - Coolant overflow reservoir
  - Coolant pumps
  - Freeze (core) plugs
  - Heat exchangers
  - Radiators
  - Thermostats
- Drive Train
  - Differential
  - Drive shaft
  - Pump transmission
  - Transmission
  - Support bearings
  - Universal joints
- Electrical
  - Alternator
  - Battery
  - Battery disconnect switch
  - Electrical motors
  - Fuses and circuit breakers
  - Gauges
  - Lighting
  - Load manager
  - Multiplexer
  - Switches
  - Voltage regulator
  - Wiring
- Exhaust
  - Catalytic converters
  - Exhaust manifold
  - Exhaust pipe (tail pipe)
  - Muffler
- Fuel
  - Fuel filters (in-tank, primary and secondary)
  - Fuel lines
  - Fuel pumps (electrical and mechanical)
  - Fuel tank(s)
  - Injectors
- Lubrication
  - Engine oil
  - Oil filters
  - Oil pumps
- Steering
  - Master steering gear
  - Power steering fluid and pump
  - Secondary island gear
  - Steering wheel

Minor changes in the performance of an apparatus can be warning signs of major problems looming ahead.

Standards

The establishment of responsibility as well as specific requirements of a preventive maintenance program can be found in several NFPA standards.

NFPA 1002, *Fire Apparatus Driver/Operator Professional Qualifications*

NFPA 1002 identifies the minimum job performance requirements for driver/operators. This standard requires that pump operators be able to conduct and document routine tests, inspections, and servicing functions to ensure that the apparatus is in a ready state. Specific driver/operator certifications provided for in this standard include (1) **pumper**, (2) **aerial**, (3) **tiller**, (4) **wildland**, and (5) **aircraft rescue and firefighting (ARFF)** apparatus.

NFPA 1071, *Emergency Vehicle Technician Professional Qualifications*

NFPA 1071 identifies specific minimum requirements that apply to all mechanics who repair and test fire apparatus.

NFPA 1500, *Fire Department Occupational Safety and Health Program*

NFPA 1500 presents requirements for inspections, maintenance, and repair of apparatus. One of the more important requirements of this section relates to the specific requirement that a fire department must establish a preventive maintenance program. As part of this program, fire department vehicles must be routinely inspected. Specifically, vehicles must be inspected at least weekly, within 24 hours after any use, when repairs or major modifications have taken place, and prior to being placed in service. If the vehicles are used on a daily basis, then the inspection should also be on a daily basis. NFPA 1500 also requires that preventive maintenance be conducted by qualified personnel. Therefore, personnel must be trained to the level of preventive maintenance they are expected to complete. Several other important requirements of this section that relate to preventive maintenance include the following:

- The use of the manufacturer's instructions as minimum criteria for inspecting, testing, and repairing of apparatus.
- The establishment of a list of major defects that automatically place a vehicle out of service.
- That fire pumps must be tested in accordance with NFPA 1911, *Inspection, Maintenance, Testing, and Retirement of In-Service Automotive Fire Apparatus.*

NFPA 1901, *Automotive Fire Apparatus*

NFPA 1901 identifies specific minimum requirements that apply to all new automotive fire apparatus. The standard first presents general requirements required by all new apparatus for items such as the chassis, electrical system, cab, and body. Next the standard provides specific requirements by apparatus type. One important requirement of this standard is the responsibility of the manufacturer to conduct and document certification inspections and tests of vital components such as the pump, water tank capacity, apparatus weight, and load analysis. The requirement ensures that the apparatus will provide the appropriate level of performance when delivered to the fire department.

NFPA 1911, *Inspection, Maintenance, Testing, and Retirement of In-Service Automotive Fire Apparatus*

NFPA 1911 requires that pumps with a rated capacity of 250 gpm (1,000 Lpm) or greater be performance tested at least annually and after any major repair or modification. This requirement ensures that the pump continues to provide the proper level of performance. The standard provides detailed information on the conditions, testing equipment, and procedures for conducting the performance test. In some departments, it is the responsibility of the pump operator to conduct this annual performance test; in others it is done by mechanics. Advantages of pump operator testing include instilling confidence in what the pump's capabilities are, and honing skills since the proper setup is required to operate the pump at its capacity successfully. Advantages of the mechanic running the test include seeing first hand any problems that may exist and repairing them. Furthermore, after any repair, tests should be performed to confirm the repair was successful.

The standard also helps ensure that fire apparatus are maintained in a ready state and in safe operating condition. The standard requires that preventive maintenance inspections be conducted as required by the manufacturer and when any defects or deficiencies are reported or suspected. This standard also requires that written criteria be established that document when an apparatus must be taken out of service. Finally, specific requirements for the inspection and maintenance of apparatus components and systems are provided within the standard.

Preventive Maintenance Schedules

STREETSMART

All manufacturers must meet the same NFPA specifications; however, the designs they have implemented to meet those specifications vary significantly. For example, Hale ball valves contain a grease fitting and need to be greased routinely to operate smoothly. Akron Brass and Elkhart Brass ball valves do not contain grease fittings and, furthermore, if greased, will not last as long as they should. Therefore, pump operators should obtain information from the manufacturer of their equipment and follow those instructions. Differences do exist in how different vendors' equipment should be maintained and operated.

The scheduling of preventive maintenance activities depends on several factors. Perhaps most important, the schedule should be based on manufacturers' recommendations. Manufacturers specify how often certain components should be inspected and tested. In addition, equipment that is suspected of having problems should be taken out of service, tested, and repaired if needed before being returned to service. The schedule depends on how often the apparatus and pump are used and may be based on time, miles (kilometres), or hours of use. This is similar to the oil change stickers placed in many cars that remind the driver to change the oil again in 3 months or 3,000 miles (5,000 kilometres), whichever occurs first. Most fire apparatus have their service done on a calendar-based time interval; however, if the apparatus is used at a week-long railcar fire, its service may be done based on the hours of operation. Finally, the schedule should be based on NFPA standards. Several examples include the following:

- NFPA 1500 suggests daily inspections for vehicles used on a daily basis and weekly inspections when vehicles are not used for extended periods.
- NFPA 1911 requires an annual pump performance test and when the pump has undergone major repair or modification. It specifies the inspections and maintenance that should be done for various apparatus systems. NFPA 1911 describes out-of-service criteria and states that all work should be done by qualified individuals.
- NFPA 1901 focuses on requirements for predelivery and acceptance inspections and test.

Often, preventive maintenance schedules include inspection and testing at the following frequencies: daily, weekly, monthly, and annually.

Daily inspections typically include the safety and fluid checks that the Department of Transportation (DOT) requires a commercial driver's license (CDL) driver to perform as part of the pretrip inspection. The **Insurance Services Office (ISO)** requires that all equipment on an apparatus be inspected at least weekly. These inspections should include: starting all portable engines such as PPV fans, generators, and saws; confirming that hoses are stowed and finished correctly; and following the manufacturer's recommended weekly inspections.

Documentation

Documenting preventive maintenance activities is important for several reasons. First, documentation and record keeping assist with keeping track of needed maintenance and repairs, which might otherwise easily be forgotten or endlessly postponed. Second, documentation provides the ability to determine maintenance trends. For example, documentation may show that, over time, engine oil is increasingly being added to a vehicle. Third, **Chapter 4** of NFPA 1500 requires that inspections, maintenance, repair, and service records be maintained for all vehicles. Fourth, preventive maintenance documentation may be required for a warranty claim. Fifth, documentation can be used to establish proper preventive maintenance in a legal dispute. Finally, when it comes time to sell the vehicle, the new purchaser will request the vehicle's maintenance history. Vehicles with incomplete maintenance records will sell for significantly less money than those with complete records. When an accident occurs, it is a good bet that preventive maintenance documentation will be closely scrutinized and the vehicle will be inspected to confirm that it has been properly maintained.

The pump operator may be required to complete several preventive maintenance documents, including:

- Daily, weekly, and periodic inspections forms
- Weekly, monthly, and annual pump test result forms
- Fuel, oil, and mileage forms
- Maintenance and repair request forms
- Equipment inventory forms

Today, most departments use computers to assist with the documentation of preventive maintenance activities (**Figure 2-2**). Preventive maintenance software allows information to be transferred to the

FIGURE 2-2 A computer can be used to maintain, track, schedule, sort, and print preventive maintenance information. With tablets, touchscreens, and wireless communications, information may be captured during the inspections, transferred to the department's server, and then printed for signatures and filing in a records retention system.

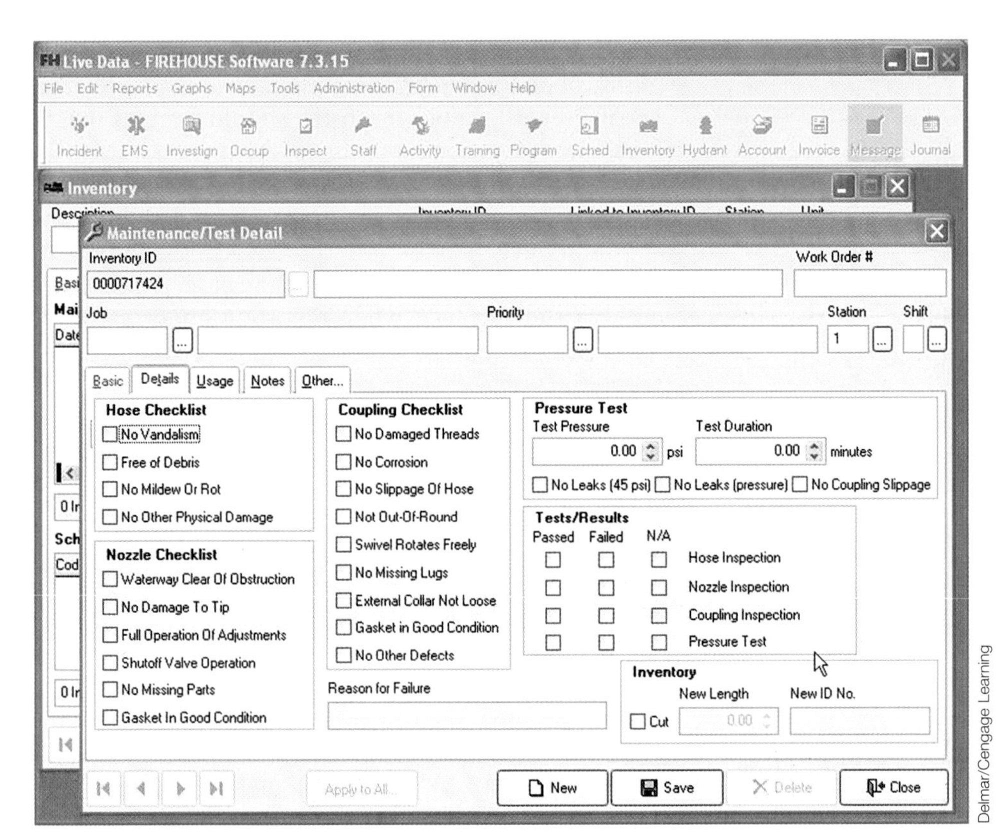

FIGURE 2-3 Preventive maintenance software can be used to maintain test information on a computer.

Delmar/Cengage Learning

computer from hard-copy forms (**Figure 2-3**). Preventive maintenance information can also be transferred to the computer through the use of handheld scanners/computers. When this technology is used, pump operators can enter data directly into the handheld device while conducting inspections or pump tests. When the inspection or pump test is completed, the data are then uploaded to a computer. After preventive maintenance information is transferred to a computer, it can be processed and printed out in a variety of formats, sorts, and lists. Preventive maintenance software can be purchased commercially or can be specifically designed for a department. In addition, public domain and shareware software programs are available and can easily be obtained through a number of online systems.

INSPECTIONS

Inspections are the most frequently conducted preventive maintenance activity. In general, inspections include checking components on a daily, weekly, or monthly basis. For example, a daily inspection might include a visual check of a tire's general condition, while a weekly inspection might include actually taking the tire's air pressure. The goal of an inspection is to ensure that the apparatus, pump, and related components are in a safe operating condition. Inspections typically include checking components for:

- Operability, position, or status
- Fluid level, leaks
- Condition, damage, wear, and corrosion

The NFPA further classifies leaks by class based on their severity. Class 1 leaks can be seen but are not big enough to cause drops. Class 2 leaks cause drops but they do not fall. Class 3 leaks cause drops that are big enough to fall.

Inspection Process

The inspection process includes three basic steps. Step one is preinspection. Prior to conducting the inspection, the pump operator should review the previous inspection report. The best way to stay current on apparatus issues is to receive a debriefing from the previous shift's pump operator. Special attention should be given to recent repairs, modifications, or changes to the apparatus. In addition, the pump operator should review the manufacturer's preventive maintenance documentation when inspecting new or unfamiliar vehicles. In doing so, the pump operator will gain insight into the current status of the apparatus.

Special attention should be given to recent repairs, modifications, or changes to the apparatus.

Step two is the actual inspection itself. The preventive maintenance information contained in national standards and provided by manufacturers does not specify the process to be used for inspecting a vehicle. The inspection, however, should be conducted in a systematic, routine process to help ensure that all components are inspected. This inspection should be conducted the same way each time. One of the best ways to accomplish this is to have a copy of the department's inspection form with you during the check-off. The inspection form should be organized in the same order used for inspections. During the inspection, the status of components is recorded on the vehicle inspection form. You should also correct any problems found that you are authorized to fix. For example, if you find a piece of equipment that was accidentally stowed in the wrong compartment, return it to its proper location. Any equipment that was stowed dirty should be cleaned. Any unresolved problems need to be documented and forwarded for resolution per the department's SOPs. A common sequence used to conduct the actual inspection includes the following:

1. Observe the apparatus on approach, looking for items such as body damage, flat tires, leaning vehicles, fluid leaks, and so forth.
2. Inspect the engine compartment.
3. Inspect the inside of the cab.
4. Inspect the outside of the vehicle (all sides, top, bottom, and compartments).
5. Inspect the pump and related components.

Additional information about what to inspect within each area is provided later in this chapter. Step three is the postinspection activity of documenting and reporting the inspection results. In some cases, this means signing of the inspection form by the pump operator and the shift officer. In other cases, it means transferring the results to an electronic preventive maintenance system as discussed previously in this chapter. In any case, all abnormal findings should be reported immediately.

(For step-by-step photos of Hands-on Inspection, please refer to page 48.)

Inspection Forms

Inspection forms are important for more than simply documenting the results of an inspection. If properly designed, the form acts as a guide to prompt the pump operator on what components to inspect. To be of greatest value, the form should be laid out in a manner that corresponds to the general steps in which the inspection will routinely be conducted (**Figure 2-4**). Finally, the form should include instructions for conducting the inspections, how to document satisfactory and unsatisfactory conditions, and the steps to take when reporting abnormal findings. See **Appendix F** for examples of vehicle inspection forms.

Inspection Tools

Common tools used for the routine inspection of an apparatus include a clipboard, flashlight, tire pressure gauge, and disposable towels or shop rags. Safety is paramount while using these tools. The tire pressure gauge should be of the proper pressure rating for the type of tire pressures that will be read. The flashlight should be used with care around flammable areas such as the engine compartment and fueling areas. If disposable towels are used, they should be disposed of according to departmental recommendations. If shop rags are used, they should be stored in appropriate containers until they can be cleaned.

What to Inspect

There are two criteria for determining what components to include in a preventive maintenance inspection. The first criterion concerns safety-related components. Safety-related components are those items that affect the safe operation of the apparatus and pump. When a problem is found, a recommendation should be made that the apparatus be taken out of service or remain in service with specified restrictions, or remain in service without restrictions based on the particular problem. If an apparatus is taken out of service, that condition should be noted by a sign on the driver's door, a special bag over the steering wheel, a large sign in the driver's window, or another highly visible indicator in the driver's area. Departments may also use additional indicators; however, one of the preceding indicators should be used to make sure that no one will attempt to operate an out-of-service apparatus because they are unaware of its status. Examples of items justifying an out-of-service condition include:

- Broken windshield
- Broken windshield wipers
- Broken door latches
- Broken foot throttle

Vehicle Driver's Safety Check

Date Odometer Reading Unit No.

☐ Pre-Trip Inspection ☐ Post-Trip Inspection

Only Items Checked Require Attention

☐ Gauges- Ammeter, Oil Pressure, Fuel, Water Temperatures, Air Pressure or Vacuum
☐ Windshield Wipers
☐ Windshield & Windows
☐ Heater & Defroster
☐ Mirrors
☐ Brakes (Foot & Parking)
☐ Engine Noises
☐ Horn & Sirens
☐ Steering
☐ Vehicle Body
☐ Wheels, Tires, Lugs
☐ Fuel Tank and Cap
☐ Leaks—Water, Fuel, Oil

☐ Head Lights
☐ Tail Lights
☐ Stop Lights
☐ Turn Signals and 4-Way Flasher
☐ Reflectors
☐ Emergency Equipment
Other- If Applicable
☐ Clearance Lights
☐ Emergency Warning Lights
☐ Side Marker Lights
☐ Brake Hoses
☐ Compartment Door Locks
☐ Drain Air Tanks of Moisture
☐ Air Systems
☐ Mounted Equipment

Remarks (explain unsatis factory items noted above)

Signature of Driver______________________

To be Completed by Repair Shop

Mechanic's Report (If defects are noted)

Signature of Repair Shop
Foreman or Mechanic______________________ Date ______________

(Use back of form for additional remarks.)

Courtesy of VFIS

FIGURE 2-4 Inspection forms can be used to guide the pump operator through the inspection.

- Tire problems
- Engine problems
- Transmission problems
- Braking problems
- Pump problems

More specific details on these problems and procedures to be followed are contained in NFPA 1911.

In essence, all safety-related components should be inspected. The following should be inspected daily:

- Tires
- Brakes
- Warning systems
- Windshield wipers
- Headlights and clearance lights
- Mirrors

The second criterion for determining what to inspect concerns the **manufacturer's inspection recommendations**. For example, manufacturers typically recommend that the following be inspected daily:

- Engine oil
- Coolant level
- Transmission oil
- Brake system
- Belts

Only the type of fluids recommended by the manufacturer should be used. To make determining the type easier, NFPA 1901 now requires the manufacturer to attach a permanent label in the cab specifying fluid types and quantities.

The Inspection Process

As each inspection is performed, the maintenance log should be updated to indicate either that the inspection was successful or to note the problem found, including any fluids added.

Safety-related components and manufacturers' recommendations make up the majority of items that should be inspected.

Engine Compartment

Gaining access to the engine compartment usually means raising the hood or the cab. In the case of a tilting cab, make sure that all cab contents that could move are secured and any exterior items that may need to be moved are moved before tilting the cab. See **Figure 2-5**. In either case, make sure to secure the hood or the cab properly prior to placing any part of your body into the engine compartment. Also, be sure to latch the hood and cab properly after the inspection is finished. To reduce injuries and damage that can occur when accessing the engine compartment, new apparatus are designed so that routine maintenance checks of lubricants and fluids can be made without tilting the cab or using hand tools.

NFPA 1901 has been recently changed to require apparatus manufacturers to construct vehicles that do not require tilting of the cab for daily checks. The change should eliminate accidents that occur while tilting the cab and climbing around the engine.

Most, if not all, of the engine compartment inspection items can and should be completed when the engine is not running. No specific sequence is suggested other than to attempt to establish a routine to ensure all items are inspected, such as following your department's inspection form. Examples of common components to inspect within the engine compartment are discussed next.

FIGURE 2-5 Before tilting a cab, make sure that no loose items exist inside the cab or obstructions outside the cab that could result in damage as a result of tilting the cab. Here, the hard suction will damage the lights if the hard suction is not moved before tilting the cab further.

Belts and Wires

Visually inspect all belts and wires for obvious damage. They should be free from dirt and debris. Belts should be checked for any nicks, cuts, or excessive wear. **Figure 2-6** shows an example of a belt ready for replacement due to cracks developing from normal wear over time. Also check to ensure no belts are missing, that they are aligned with their pulley, and that the belt is properly tensioned. Most newer belts use self-adjusting tensioners, thus requiring fewer adjustments than prior systems. Wires should be inspected for loose connections, worn or frayed insulation, and for exposed metal.

Engine Oil

Engine oil works as a lubricant for the engine and as a means of cooling and cleaning internal engine parts. Check the engine oil level with the dipstick.

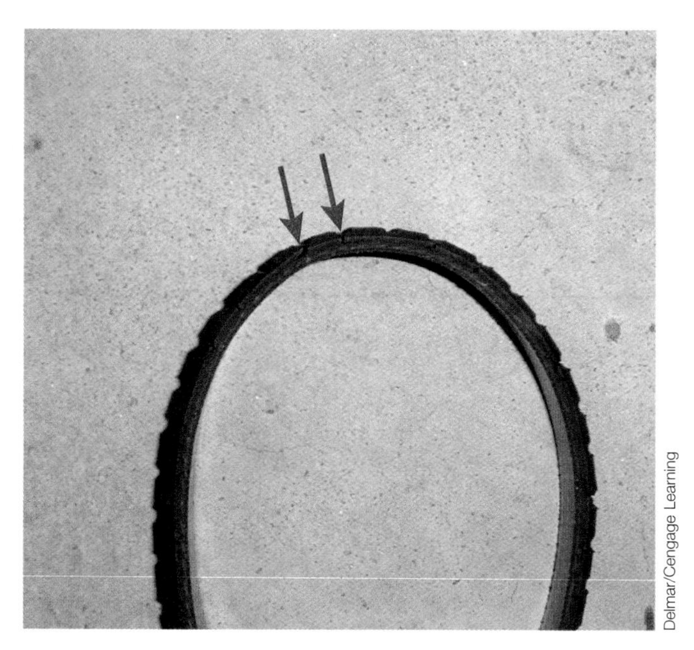

FIGURE 2-6 Inspecting belts to make sure they are replaced before they fail, creating an outage and often damaging other components. The belt shown was replaced because it started developing cracks, which indicate it is worn out and getting ready to fail.

Simple Steps to Check the Oil Level

1. Ensure the vehicle is not running and is on level ground. The engine oil can be checked prior to or after running the engine. The benefit of checking the oil prior to running the engine is that cool oil stays on the dipstick better, making it easier to measure the oil level. In addition, the cooler the engine, the less risk of incurring an accidental burn. If the oil is checked after the operation of the engine, wait several minutes to allow the oil to drain from the engine into the pan to ensure an accurate reading.
2. Remove the engine oil dipstick and wipe off the end with a clean rag (**Figure 2-7**).
3. Slowly insert the dipstick fully and then remove it again.
4. Read the level of the oil. The oil level should be within the range marked on the dipstick, usually indicated with "Full" and "Add" markings (**Figures 2-8A** and **2-8B**).
5. If the level is within the "Add" marking area, oil should be added. Slowly add oil through the fill opening as needed, being careful not to overfill. Wait a few minutes to let the oil settle and then recheck the level. Only trained individuals should add engine oil.
6. Note when and how much oil was added on the inspection form.

FIGURE 2-7 Oil is typically checked with the engine off by removing a dipstick. The dipstick is cleaned with a rag, fully reinserted using the rag as a guide, and removed again to get a reading.

In some locations, driver operators are trained to add engine oil as needed. In other locations, only qualified mechanics can add engine oil. The correct engine oil type should be used. Mixing different oil types or even different brands of oil should be avoided. Brands should not be mixed due to incompatibilities in additives used by different manufacturers. Consult the manufacturer's manual for the recommended oil type. Engine oils are classified using a rating system developed by the American Petroleum Institute (API) and the Society of Automotive Engineers (SAE).

The API classification uses a two-letter system to identify the type of engine and the service class. The first letter identifies the type of engine and is either an S (Service), indicating the oil is for use with gasoline engines, or C (Commercial), indicating use with diesel engines. The second letter indicates the service class. The current heavy-duty diesel oil classification is CJ-4. It includes performance properties of each earlier category. The SAE classification uses a numbering system to grade or rate engine oil viscosity. Engine oil viscosity ratings can range from SAE 5 (low) to SAE 50 (high). Sometimes a rating includes a W (for Winter) after the number, as in SAE low and SAE 20W, which means the oil is rated for flow at 0 degrees Fahrenheit (–18 degrees Celsius). Obviously this oil is rated for use in colder climates. Without the W, as in SAE 30 and SAE 50, the ratings are measured at 210 degrees Fahrenheit (99 degrees Celsius). Multiviscosity oils, as in SAE 20W50, include performance characteristics of a 20W oil at 0 degrees Fahrenheit (–18 degrees Celsius) and a 50 weight oil at 210 degrees Fahrenheit (99 degrees Celsius). Most apparatus use 15W-40,

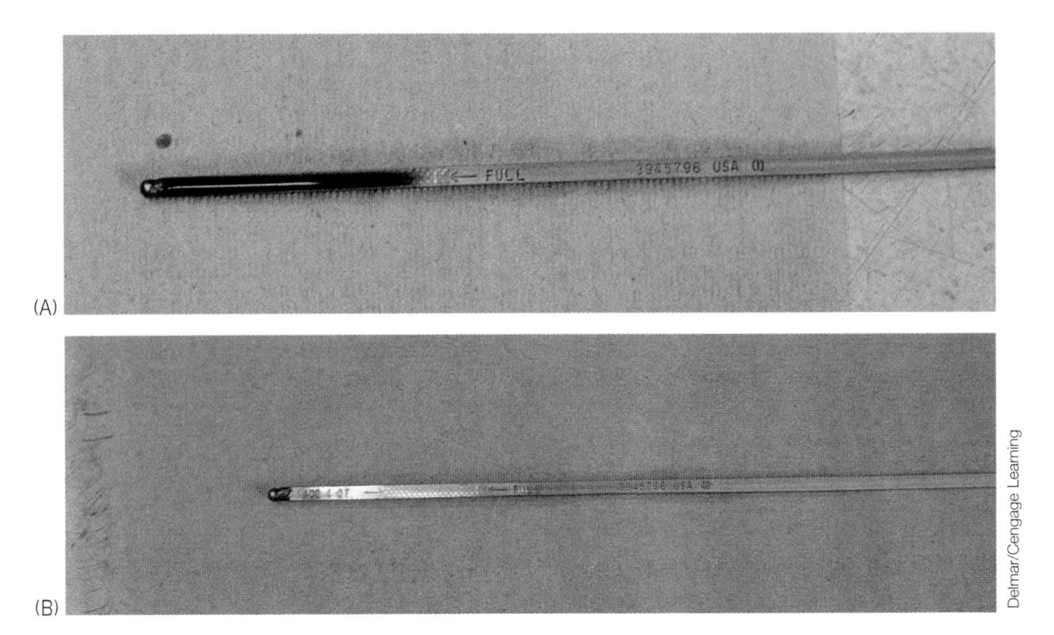

FIGURE 2-8 Oil should be within the hashed area. On lighter vehicles such as pick-ups, vans, and utility vehicles, the range of the hashed area is usually 1 quart (1 litre). On heavy-duty trucks, the range of the hashed area is usually 1 gallon (4 litres). Since keeping open containers of oils is discouraged due to contamination concerns, oil should be purchased in containers that will be used at one time unless other methods are in place to ensure contamination does not occur.

although other viscosities are used in areas with extreme temperatures.

Vehicle Batteries

Vehicle batteries can be located in the engine compartment or in a separate compartment. Ensure the battery is secure and check for obvious damage and signs of corrosion, such as a dirty battery top, corroded or swollen cables, corroded terminal clamps, loose hold-down clamps, loose cable terminals, or a leaking or damaged battery case. For older batteries, carefully remove the caps to check the electrolyte (acid) level and refill with distilled water as needed. Most newer batteries are sealed and do not require internal inspection. Battery cables should be checked for corrosion and to ensure they are securely attached to the battery post. Corrosion (**Figure 2-9**) can be removed with a rag or wire brush and cleaned with a mild base solution such as baking soda and water. Appropriate personal protective equipment, including eye and hand protection, should be worn when inspecting and cleaning batteries. Care should be exercised when recharging batteries. While charging, batteries produce hydrogen gas. Adequate ventilation and control of ignition sources should be provided while recharging batteries. The manufacturer's recommendations should be consulted before a battery is recharged with a charger or jumped from another vehicle.

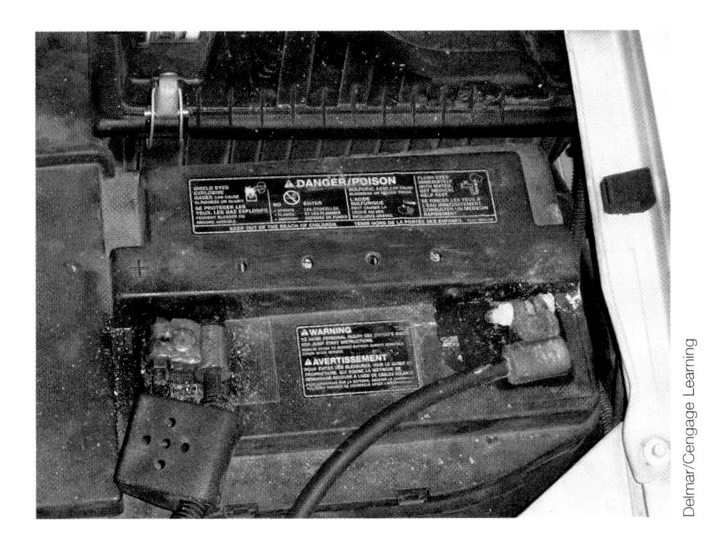

FIGURE 2-9 Corrosion, the white powder shown here on the negative (black) battery terminal, causes electrical system problems. It may be removed with a wire brush. A mechanic can apply a coating that will prevent the corrosion from reoccurring. This positive (red) battery terminal, which is normally covered, does not have a corrosion problem.

(For step-by-step photos of Jump Starting an Apparatus, please refer to pages 49–50.)

(For step-by-step photos of Connecting a Battery Charger, please refer to pages 51–52.)

CAUTION

Batteries are filled with a liquid called electrolyte, which is a mixture of sulfuric acid and water. Getting this acid on your skin can result in burns, which usually starts as itching skin. Getting it on clothing will create holes in the clothing. Protective gear to avoid acid damage includes goggles, rubber gloves, and an apron. First aid includes flushing with copious amounts of water and using baking soda to neutralize the acid. While sealed and maintenance-free batteries greatly reduce the likelihood of acid exposure, they still contain acid and should be treated accordingly. Cracked battery cases and/or signs of corrosion are indicators of increased likelihood of acid exposure requiring increased vigilance and safety precautions.

Cooling System

The cooling system, which includes the radiator, hoses, thermostats, shutters, water (coolant) pumps, and freeze plugs, should be checked for obvious signs of damage, such as dents, leaks, and debris. The coolant level should also be checked. Many systems allow the coolant level to be checked without opening the radiator fill cap. Smaller vehicles, including pickup trucks, vans, and cars, typically have a translucent plastic coolant reservoir container that is checked by looking at the fluid level in comparison to the cold or hot indicator lines. Larger vehicles typically have a sight glass on the radiator. Coolant should be visible in the sight glass as shown in **Figure 2-10**.

FIGURE 2-10 Radiators with sight glasses can be inspected hot or cold. Coolant level is OK if coolant is visible in the sight glass. Be sure that coolant is visible, not just a stained sight glass.

CAUTION

Removing a radiator cap from a hot radiator releases steam and scalding hot coolant, which can cause severe burns and other problems.

Use extreme care if the radiator fill cap must be removed to check the coolant level. The apparatus should be allowed a minimum of 30 minutes to cool down before checking the level in this manner. The radiator cap should be cool or warm (not hot) to the touch before you attempt to remove it. When using this method the coolant should be visible at the neck of the radiator. Coolant for diesel engine apparatus should contain the words "Heavy Duty Diesel" and is different from car coolant. Coolants with different colors are made with different chemicals and generally should not be mixed. Prior to adding coolant, the package should be checked to see if it is labeled "prediluted," which means it should be added directly to the system. Concentrated coolants, which are more common, should be diluted with demineralized water as specified on the container (generally 50%). Use of coolant that is either too concentrated or too diluted adversely affects operations and should be avoided. If just tap water has been added in an emergency, the cooling systems should be looked at by a mechanic to get the system back in balance. Tap or plain water does not contain the additives that are used in modern coolants and that protect the cooling system.

Windshield Washer Fluid

Check the level of the windshield washer reservoir and fill with an approved fluid as needed. Most departments require windshield washer fluid to be at least half filled.

Cooling Fan

Check the cooling fan for obvious damage such as cracks and missing blades. Ensure the fan can operate free from obstructions. Caution should be used due to some vehicles having cooling fans that operate automatically—such fans can even activate with the vehicle turned off.

Air Intake

Some systems have an automatic indicator for when to change the air filter, as shown in **Figure 2-11**. Other vehicles have an air filter light at the driver's position indicating when an air filter replacement is needed. A visual inspection should also be made to confirm that no damage has occurred that would invalidate these indicators. Such damage includes any

Delmar/Cengage Learning

FIGURE 2-11 Inspecting the air filter indicator lets the operator know if the engine will underperform due to a dirty filter.

punctures to the air intake system or problems where any components connect. Keep in mind that the purpose of the air filter is to prevent dust from entering the engine. Movement of the components without extreme care can allow dirt in the engine compartment to bypass the filter totally, invalidating its purpose and causing damage.

Power Steering

The power steering system should be checked for damage and leaks. The fluid level should be inspected at normal operating temperature and topped off when needed. Some power steering reservoirs contain a sight glass, as shown in **Figure 2-12**. In this case, the preferred maintenance check is to read the sight glass, reducing the likelihood of contaminating the fluid with dirt from the engine compartment.

Sight glasses should be replaced if staining is preventing their intended use.

Brakes

Smaller vehicles, including pickup trucks, vans, and cars, typically use hydraulic brakes with brake fluid. Larger vehicles typically use air brakes. Information for both types of systems follows. The braking systems of all apparatus must meet two NFPA requirements. Larger apparatus must be able to come to a complete stop from 20 MPH within 35 ft (10 m) on level, dry payment, and the parking brake must hold the apparatus on a 20% grade. A 20% grade is one that rises 20 feet over a 100-foot (or 20 metres over

Delmar/Cengage Learning

FIGURE 2-12 Power steering reservoirs with sight glasses should be inspected using the sight glass instead of the dipstick to prevent accidental power steering fluid contamination with dirt. Be sure that fluid is visible, not just a stained sight glass.

a 100-metre) distance and is not the same as a 20-degree angle. The NFPA also requires an auxiliary braking system on any apparatus with a gross vehicle weight in excess of 36,000 pounds (16,330 kilograms).

- *Brake fluid.* Inspect the brake lines and master cylinder for damage and leaks. The fluid level in plastic translucent reservoirs should be checked without removing the cover. Removing the cover leads to brake problems, both because dirt can enter the reservoir and because brake fluid quickly absorbs water vapor out of the air—either of these contaminants cause problems with the brake system. Dirt may be removed from the exterior of the brake reservoir, and a light may be shined through it to make reading the level easier. If additional brake fluid is required, it should be added from a new can of brake fluid and any remaining contents should *not* be saved for future use.
- *Air brakes.* Air brake systems have several items needing routine maintenance.

1. Draining water from the wet tank—Air brake systems have at least three air storage chambers (tanks), with the first tank after the compressor being called the "wet tank." Its purpose is to allow water to condense out of the air and be purged before it would cause problems with other components of the braking system. Some vehicles may contain an auto-purge valve to drain this condensate. Newer apparatus have a dryer between the compressor and wet tank that should automatically remove any moisture.

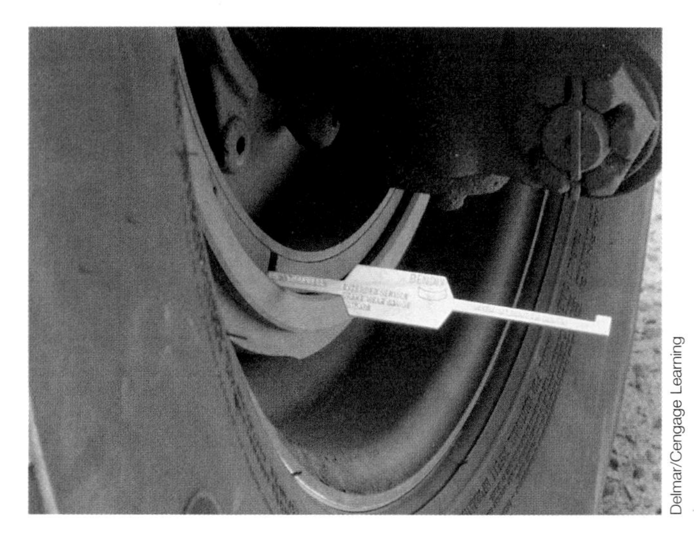

FIGURE 2-13 Drivers should confirm that they have at least ¼ in. (.6 cm) of brake lining thickness. With less brake lining, they are subject to a personal traffic ticket, may not legally drive the apparatus, and, most important, may not be able to stop safely. If this simple tool can be inserted between the brake shoe and drum, the lining is of legal thickness.

On these vehicles, if water is found in the wet tank, a qualified mechanic should be engaged to determine why. On older vehicles without a dryer, water should be drained from the wet tank on a daily basis.

2. The driver should confirm that sufficient brake lining exists to allow the apparatus to be legally and safely operated. At least ¼-in. (6-mm) thickness is required on heavy truck linings. **Figure 2-13** shows measuring with a simple tool. Many brake linings are also beveled, such that one can visually tell if they are legal without a tool.
3. The driver should confirm the stroke of the push rod. The stroke of most push rods does not exceed 2.5 in. (6 cm). This also confirms that the slack adjuster is correctly adjusted. All current apparatus have automatic slack adjusters that should adjust themselves. Some older apparatus may still have manually adjusted slack adjusters. With the brake pedal depressed, the slack-adjuster-to-push-rod angle should not exceed 90 degrees. It should not exceed the air chamber's maximum stroke (see owner's manual for maximum stroke information), nor should a red band on the diaphragm side of the push rod be visible.

Automatic Transmission Fluid

Like many fluids, if transmission fluid is either too high or too low, problems will result. If it is too low, internal parts do not receive an adequate supply and the transmission overheats. If it is too high, it aerates, causing shifting problems, overheating, and fluid may be expelled through openings in the transmission. Transmission fluid should be checked with the engine running at an idle and the transmission in neutral (or park for vehicles containing a park). The transmission temperature should be within its normal operating range (160–200ºF or 71–93°C), which may take up to an hour's operation to achieve.

NOTE

Some transmission dipsticks contain a "cold" fill mark, misleading people to believe that the transmission may be checked cold. The purpose of the cold mark is to allow a check of the transmission oil to be made after filling to see if it is safe to run at an idle until a hot check can be made.

The automatic transmission fluid may be checked and filled in the same manner as the engine oil. Some transmissions may be equipped with an optional oil sensor allowing the transmission level to be checked in the cab. (With the recent NFPA requirement that all fluid levels must be able to be checked without tilting the cab, this option is now standard on all heavy-duty transmissions.) On Allison transmissions, this check may be made by simultaneously and quickly pressing the "↑" and "↓" shift buttons. Once this has been done, the transmission will start to display a sequence of characters in the transmission digit display. (Note: Attempting this operation on an Allison transmission that does not contain the sensor will not cause problems, although it will not display the sequences shown below.) Common display sequences are:

- "o," "L," "o," "K" for Oil Level OK
- "o," "L," "H," "I," "1" for Oil Level High 1 quart (litre)
- "o," "L," "L," "o" for Oil Level Low 1 quart (litre)
- "o," "L," "-," "7," "0" for Oil Level too cold to check. Retry at normal operating temperature.
- "8," "8" ... "8"... "7," "7" ... "7" ... "1," "1," "1" may precede the display if the transmission has just been used and the oil has not yet settled.

Additional information on performing these operations is often attached to the rear of the driver's sun visor and is also contained in the transmission's owner's manual.

Inside Cab

After the outside of the vehicle and engine compartment have been inspected, it is time to inspect the inside of the cab. First, take a moment to adjust the seats and all mirrors, and inspect all glass, such as the windshield and side windows. Mirrors and glass should be clean and free of damage such as large cracks. The seatbelt restraining system should be inspected. Check for proper and unobstructed operation and any obvious sign of damage, such as worn or torn webbing. Seatbelt buckles should open, close, and latch properly. Check for operation and positioning of adjustable steering wheels. Next, make sure the parking brake is on and that the transmission is in park (automatic transmission) or neutral (manual transmission). Also, all electrical switches should be in the off position so the battery is not excessively loaded. Finally, start the vehicle, being sure to listen for any unusual noises or vibrations. The manufacturer's instructions for starting the vehicle should be followed. If the vehicle is inside, the exhaust should be vented to reduce the effects of carbon monoxide and other toxic fumes (**Figures 2-14A** and **2-14B**). After starting the vehicle, the items described next should be checked.

Gauges

Gauges should be checked to ensure that the devices or components they are measuring are operating within designed limits as specified by the manufacturer. All gauges should be checked before leaving the cab whenever the apparatus will be left running. Final determination of acceptable gauge readings and reactions should be based on the manufacturer's operating manuals and departmental SOPs. Most gauges provide visual references for normal operation.

STREETSMART

This section lists a number of dangerous conditions requiring immediate reaction. These conditions are rare and may not even occur during your career as a pump operator; however, if any of these do occur, your proficiency and prompt actions may save the lives of your crew. They are among the key reasons why a pump operator remains at the pump and monitors apparatus gauges. Gauge readings outside of acceptable ranges indicate that damage is occurring to the apparatus and can even result in the sudden failure of the apparatus. Operators should know the acceptable range and normal readings for each gauge on all of the apparatus they operate and what actions they should take as readings become too high or too low.

(A)

(B)

FIGURE 2-14 Vehicles should not be left running in buildings, allowing toxic gases to enter the air we breathe.

Examples of gauges to check include the following:

- *Air pressure gauge.* Proper air pressure is essential for safe air-brake operation. While high levels of safety and redundancy are part of the system design, part of that safety is dependent on the driver's vigilance in monitoring the gauges and taking appropriate action when the gauges indicate a developing problem. Typically, the air-brake compressor will stop when pressure reaches between 120–130 psi. The exact cut-out pressure depends on model and is within limits set by the Federal Motor Vehicle Safety Standard of 85–150 psi. The operator should be aware of both the cut-out and cut-in pressures (about 25 psi below the cut-out pressure) for their particular apparatus. They should confirm that pressure is normally within cut-out to cut-in pressure range and that the compressor is not running constantly or cycling too frequently, since either indicates a serious problem and will result in additional failures, including failure of the compressor itself.

Air pressure is required to release the parking brakes. If air pressure is less than 35 psi, the yellow diamond-shaped parking brake button (**Figure 2-15**) will not stay released. Below 60 psi it is still considered so critical that both low-air visual and audible alarms are used. Under these conditions, the vehicle should be stopped and removed from travel lanes as soon as safely possible to do so. To avoid drag, assume that at least 80 psi is needed unless you know the specific pressure that avoids drag on a specific apparatus. To help prevent drag, air accessories such as the air-operated horns must cease to function if air pressure drops below 80 psi. Even with this protection, operators need to use devices such as air horns that use air faster than the compressor can provide it judiciously. Their maximum use combined with braking may result in brake drag, brake fade, and cracked drums, and may even increase the likelihood of an accident.

FIGURE 2-15 Parking brake control valve.

STREETSMART

Drag is a condition that occurs when the brake pads touch the brake drums. Driving when the brakes are dragging creates heat from the rubbing between the brake pads and drums. The heat causes a second, worse condition known as brake fade. This occurs when brake components have become so hot that the surface between the pads and drums actually melts. As everyone knows from stepping on a wet floor, the friction between two solids disappears when separated by a liquid. With brake fade, it means stepping on the brakes will not stop the vehicle and should be avoided at all costs.

NOTE

NFPA 1901 requires new apparatus to have quick buildup times that can reach operating pressure within 60 seconds, which is much quicker than the DOT requirement of 3 minutes.

- *Oil pressure gauge.* Without adequate oil pressure, an engine will quickly destroy itself. Because of their extreme importance, oil pressure gauges are located both in the driver's gauges and on the pump panel. Oil pressure should increase with increased engine RPM. At an idle it should read 10 psi and increase to about 75 psi at full throttle. If oil pressure fails to maintain at least 5 psi, the apparatus should be taken out of service immediately, provided doing so would not create an immediate life-threatening situation such as would occur when firefighters are using an interior attack line being supplied by that engine. In such cases, an immediate evacuation call should be requested following your department's SOPs. As part of normal inspections, the operator should confirm that both gauges display the same readings.
- *Fuel level gauge.* Ensure adequate fuel is available and refill the tanks as directed by department policy or, if no policy exists, refill when the gauge reaches the three-quarter mark. Refueling should occur as soon after the vehicle is inspected as possible. According to NFPA 1901, the fuel tank must be sufficient in size to drive the pump for at least 2½ hours at its rated capacity when pumping at draft. The tank fill opening should also have a label indicating the appropriate fuel type.

Many departments have arrangements for fuel deliveries to scenes involving extended operations. Pump operators should be aware of the lead time for such deliveries and request such a delivery through their chain of command in time to get fuel before running out if it appears that operations may continue long enough that a fuel problem could exist. This requires the operator to both monitor the fuel level and understand the rate of consumption.

- *Transmission oil temperature gauge.* Proper transmission oil temperature is essential for proper transmission operation. Even the checking of the transmission oil level is dependent on its temperature. Therefore, to allow the operator to take appropriate actions based on the transmission oil's temperature, a transmission oil temperature gauge is included with the instruments on the driver's dash. Many departments also place a second gauge at the pump panel.

  At very low temperatures (transmission oil below 20ºF or –7°C), the transmission will even restrict its use to lower gears. While operations at these temperatures do not normally occur for fire departments since fire trucks are stored indoors, they are frequently driven under severe conditions before transmissions have reached their normal operating temperature range of 160–200ºF (71–93°C). On some vehicles, drivers may notice that shifting is not as smooth prior to the transmission reaching is normal operating range, which can take 30 to 60 minutes of operation.

  Of more concern to drivers/operators are overheated transmissions. Transmissions are considered to be overheated if their temperature reaches 250ºF (120°C) and should not be allowed to exceed 262ºF (128°C) except in a true emergency involving life safety; however, all operations over 220ºF (105°C) shorten the life of the transmission oil.

  Diligently observing the transmission temperature will enable the driver/operator to make proactive changes to prevent the transmission from overheating. While driving, use less aggressive driving techniques, including lower speeds and less acceleration and transmission braking. While pumping, this can include requiring the transmission to do less work. One way to accomplish this is to request additional pumpers to handle some of the workload.

  Should a transmission become overheated, a qualified mechanic should be consulted as soon as possible. At a minimum the mechanic will want to change the transmission oil since once overheated, it loses some of the properties it needs to work properly and prevent additional transmission wear and damage. In addition, the mechanic can make changes, such as increasing the size of the transmission cooler, if a reoccurrence seems likely.

- *Engine coolant temperature gauge.* Along with oil, engine coolant is the lifeblood of the engine and is required to prevent the engine from destroying itself. Due to this importance, it is one of the few gauges that must be placed both with the driver's gauges and at the pump panel. When the engine is too cold, it undergoes excessive wear. To limit excessive wear, the engine should be given several minutes to warm up before operating above an idle whenever an emergency response is not required. Even when an emergency response is required, starting the engine prior to completing other necessary actions will greatly help the engine. Normal engine operating temperatures should be within 180–195ºF (82–90°C). Operation above 212ºF (100°C) will cause engine damage. If engine temperature exceeds 212ºF (100°C), the apparatus should be taken out of service immediately, as long as doing so would not create an immediate life-threatening situation such as would occur when firefighters are using an interior attack line being supplied by that engine. In such cases, an immediate evacuation call should be requested following your department's SOPs. The operator can generally avoid such emergencies through the use of timely corrective actions. When pumping, if the temperature appears to be rising above the normal operating range, the engine cooler valve should be opened and the workload can be split between multiple engines. The operator should also confirm that no debris is preventing air from getting to the radiator. As part of normal inspections, the operator should confirm that both gauges display the same readings.

- *Voltmeter.* The voltage of a fully charged 12-V automotive battery not connected to a charger is 12.66 volts. Ideal battery charging voltage should be between 13.5 and 15.1 volts. The driver/operator should normally expect to see a reading in this range with the apparatus running. For a given apparatus drivers/operators should expect voltage to normally be within a much narrower range. Whenever the voltage in a running apparatus is below 12.66 volts the battery is in the process of being discharged. If this occurs other than for a short-term, explainable reason, such as while pulling a prime, the reason should be determined and the condition corrected. At the scene, generally, some electrical items will need to be turned off (newer apparatus with load shedding will do

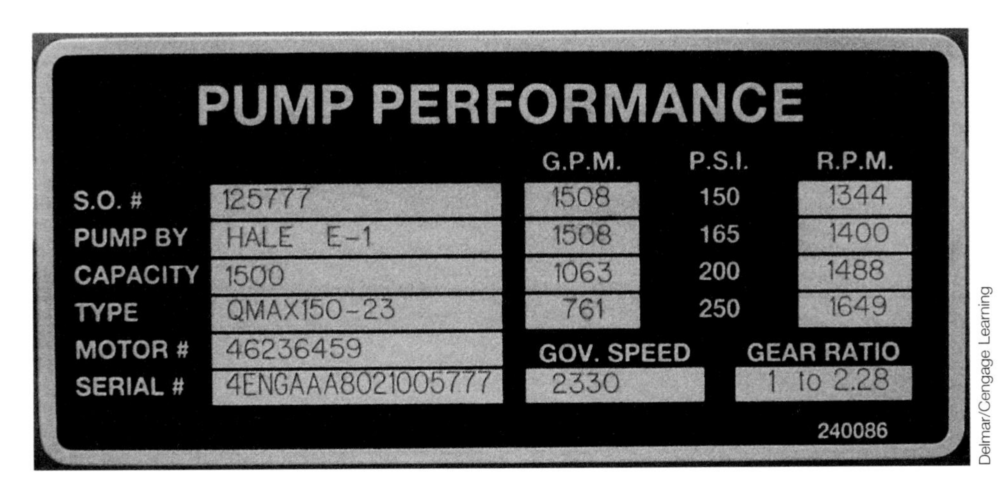

FIGURE 2-16 Never allow an apparatus to exceed the "Gov. Speed" stamped on the plate on the pump panel.

this automatically) or the generated power will need to be increased. The power produced can be increased by increasing the engine's RPM. If voltages drop below 9.6 volts, computers (electronic control units [ECUs]) on newer apparatus will produce unpredictable results. The apparatus should be taken out of service and the condition corrected by a fire mechanic before the voltage gets this low.

On the other hand, voltages too high also significantly shorten the battery's life by "cooking" the battery. If routine voltages above 15.1 volts are observed, a fire mechanic should determine the reason why. The apparatus may continue to be operated with slightly elevated voltages until the condition can be corrected. If voltages exceed 16 volts, the apparatus should be removed from service until the condition can be corrected to prevent damage to electrical components.

Some apparatus may use 24-volt, 36-volt, or 48-volt systems. For these vehicles, all of the voltages listed above should be multiplied by 2, 3, or 4, respectively.

Finally, in taking these readings, remember that gauge error also occurs. Newer digital voltmeters should be accurate; however, some of the older analog gauges could be off by a volt. An excellent way to check for gauge error is to compare the gauge in the cab with the one at the pump panel. They should read within 1 volt.

- *Speedometer.* The speedometer should read zero while parked unless in pump gear.
- *Tachometer* (Tach). The tachometer measures engine speed in revolutions per minute (RPM). This gauge is present both on the driver's gauges and at the pump panel. The tachometer allows the operator to confirm that the engine is running slow enough to engage the transmission and power take off PTO devices safely, at the ideal speed for priming, and confirm that the engine never exceeds its governed no-load RPM, which is stamped on to a plate at the pump console (see **Figure 2-16**). Inspections should confirm that both gauges provide the same readings.

Any additional gauges specific to your apparatus should be checked to make sure their readings are within an acceptable range. Operators should know what the acceptable limits for each gauge on their apparatus are and what actions they should take if a gauge moves outside of the acceptable limits. One way departments may indicate acceptable limits is to place lines on or arrows next to gauges that indicate the acceptable range.

Lights and Warning Devices

Check all lights for proper operation. This will require an individual on the outside to help. Both running lights (headlights high/low beam, brake lights, turn signals, and backup lights) and emergency warning lights should be checked. Due to NFPA emergency lighting requirements, certain lights may not activate when the parking brake is applied. The second individual can observe the emergency lighting operation while the other has the vehicle in drive with the brakes depressed. In the cab, lights and illumination devices should also be checked.

Audible warning systems should be checked for proper operations. Check to ensure that no one is in close proximity of audible warning devices prior to conducting the operations check.

Check the backup warning device, being sure to depress the brakes before placing the transmission into reverse. If the apparatus is equipped with backup cameras, they should also be checked for proper operation.

Brake and Clutch Pedals

The brake pedal should be depressed and should stop prior to reaching the floorboard.

The clutch pedal, in manual shift transmissions, should also be depressed for proper operation. While depressing the pedal, note the distance (free play) of the pedal before resistance is felt (throw-out bearing contacting the clutch plate). Insufficient or excessive free play should be reported immediately.

Consult the manufacturer's recommendations for specific guidance on testing the clutch pedal free play.

Steering Wheel

The steering wheel should be checked for proper operation. In most cases, it is not necessary, nor is it advisable, to turn the steering wheel so that the wheels turn. Doing so can cause excessive strain and damage to the steering system. Rather, turn the steering wheel until just before the wheels turn. This distance should not exceed 10 degrees in either direction. For a 20-inch (50-centimetre) steering wheel, that would be approximately 2 inches (5 centimetres) of movement. Excessive play may cause difficulty in steering and may indicate a problem with the steering system.

Other Cab Components

Any additional items in the cab including departmental added items should be inspected. Examples of common additional items include:

- Radios for proper operation
- Heating, air conditioning, and defroster for correct operation
- Computer equipment for operation and communication with base stations if applicable
- Windshield wipers and washer controls for proper operation
- Automatic snow chains when vehicles are so equipped and during appropriate seasons. (The vehicle may need to be in motion for this check to be completed.)

Outside the Vehicle

The inspection should begin as you approach the vehicle. As you walk around the vehicle once, notice its general condition and look for the following:

- Obvious damage
- Whether the vehicle is leaning to one side (not caused by surface grade)
- Liquid under the vehicle (oil, coolant, grease, water, foam, fuel, or transmission fluid)
- Missing parts or equipment
- Hazards to vehicle movement that are around the area

Next, begin a systematic inspection of the vehicle. For example, start on the driver's side toward the front of the vehicle and move in a clockwise direction until you arrive at the front of the vehicle. Examples of common components to inspect in each of the areas include the following:

Tires, Wheels, and Rims

Tire, wheel, rim, and brake inspections are important because they support the apparatus and, more important, because the tire's traction with the road allows the apparatus to turn and stop. Particularly during emergency responses, the apparatus is likely to need to stop short or make quick lane changes and therefore will be required to outperform the same components on a nonemergency vehicle. The typical full-size apparatus's tire footprint (the part that is in touch with the road) is about the size of a sheet of paper or significantly less than 1 square foot (.1 square metre), and six of these are expected to stop a vehicle going 60 MPH (100 KPH) that can weigh over 20 tons (18 tonnes) in less than 6 seconds.

In fact, the tires will not be able to meet these performance requirements if any of several problems exist. For example, if the tread is too worn, the tires will not be able to remove water from beneath the tire, resulting in hydroplaning and skidding. Tire pressure that is off even slightly greatly affects tire life and fuel economy. Larger pressure variations significantly affect traction and may even cause a catastrophic tire failure.

(For step-by-step photos of Tire, Wheel, Rim, and Brake Inspections, please refer to pages 53–54.)

- *Proper inflation.* Take a pressure reading of each tire to ensure it is properly inflated. The maximum tire pressure is listed on the sidewall of the tire. Ideal tire inflation pressure can be found using a tire-manufacturer-provided tire inflation table such as the one shown in **Table 2-4**. To use these tables, you must know the tire size, which is printed on the tire sidewall, and the actual apparatus axle weight (versus rated axle capacity). The actual axle weight can be obtained using a truck scale. Each apparatus should be weighed annually with its normal load and full tanks. In many cases, fire apparatus are loaded close to their rated weight, in which case the sidewall maximum

TABLE 2-4 Correct Cold Tire Inflation Pressures by Tire Size, Load, and Configuration

Size	Configuration	70 psi 485 kPa	75 psi 510 kPa	80 psi 550 kPa	85 psi 585 kPa	90 psi 620 kPa	95 psi 655 kPa	100 psi 690 kPa	105 psi 725 kPa	110 psi 760 kPa	115 PSI 790 kPa	120 PSI 825 kPa	125 PSI 860 kPa	130 PSI 895 kPa
11R22.5	Single (lb)	4380	4580	4760	4950	5205	5415	5625	5840	5895				
	Single (kg)	1990	2080	2160	2250	2360	2460	2550	2650	2670				
	Dual (lb)	4530	4770	4990	5220	5510	5730	5950	6175	6320				
	Dual (kg)	2050	2160	2260	2370	2500	2600	2700	2800	2870				
12R22.5	Single (lb)		10400	10900	11380	12010	12410	12810	13220	13740	14260	14780		
	Single (kg)		4720	4940	5160	5450	5630	5810	6000	6230	6470	6700		
	Dual (lb)		19960	20760	21560	22700	23140	23580	24020	25060	26100	27120		
	Dual (kg)		9050	9420	9780	10300	10500	10700	10900	11370	11840	12300		
315/80 R22.5	Single (lb)				14240	14900	15560	16220	16860	17500	18140	18760	19380	20000
	Single (kg)				6460	6760	7060	7360	7650	7940	8230	8510	8790	9070
	Dual (lb)				23540	24640	25740	26800	27880	28960	30000	31040	32040	33080
	Dual (kg)				10680	11180	11680	12160	12650	13140	13610	14080	14530	15000

Load in pounds

Load in kilograms

Do not exceed maximum pressure indicated on rim

pressure rating is the correct inflation pressure. Tire pressures listed on the driver's door of light trucks are typically too low for fire apparatus because they do not take into account the actual loading of most fire apparatus. Tire pressure should always be checked with the tires cold. When driven, tire temperature increases, causing an increased air pressure; thus, filling a hot tire will result in an underinflated tire. Dual tires should always be inflated to the same pressure and should have the same amount of tread depth. When dual tires have more than a 10-psi pressure difference or different tread depths, one of the tires carries too much of the load and becomes overloaded. Improper inflation can cause tire damage, difficulty with vehicle handling (steering, stopping, and traction), blowout and other tire failures, reduce tire life, and decrease fuel economy.

SAFETY

Any tire that has been driven at less than 80% of its recommended inflation pressure should be inspected inside and out by a trained tire professional for possible damage. Just re-inflating the tire could cause a tire explosion, resulting in serious injury or death.

NOTE

Changes have been made to NFPA 1901 to require visual indicators or tire pressure monitoring systems and to list the recommended tire pressures on a permanently mounted cab label in new apparatus. This will eliminate the need for a tire gauge check for proper fill pressures.

(A)

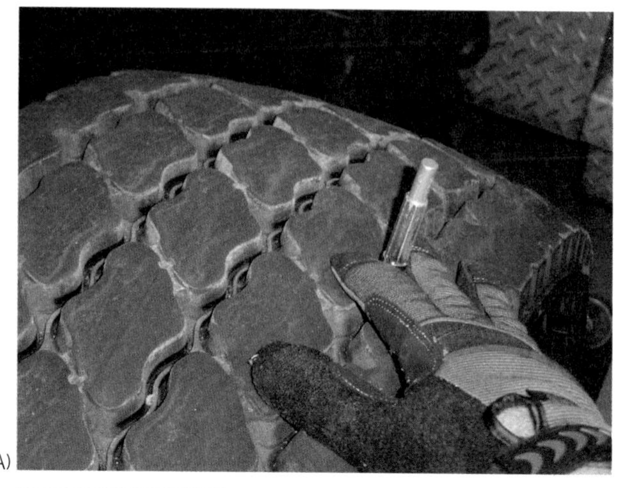

(B)

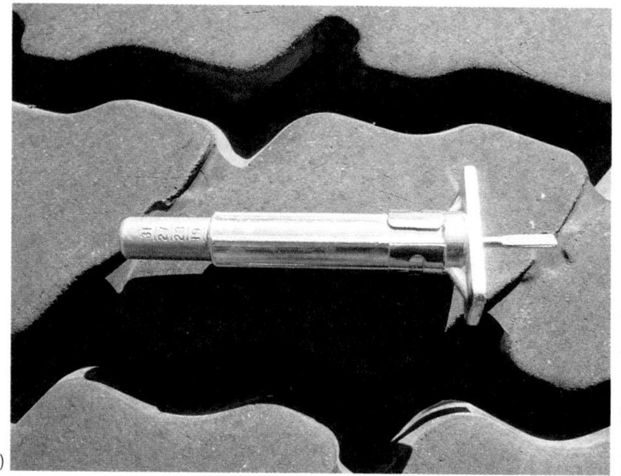

Delmar/Cengage Learning

FIGURE 2-17 (A) An extended tire tread depth gauge should be inserted in the full inner and outer tire groove to determine how much life the tire has left and if it is wearing evenly. (B) Read the tire tread depth gauge by finding the lowest number with a visible bar beneath it.

- *Tread wear*. Ensure that tire tread wear is not excessive, and look for cuts, nicks, or other damage such as impaled objects. No fabric should be visible through the tire tread or sidewall. NFPA 1911 requires tire replacement when tire wear exceeds state or federal standards. The DOT requires a tread depth of $\frac{4}{32}$ of an inch (3 mm) for major grooves in the front tire and a tread depth of $\frac{2}{32}$ of an inch (1.5 mm) for all other tires. The preferred method to determine tread depth is with an inexpensive tool called a tire tread depth gauge (**Figures 2-17A** and **2-17B**). A tread depth gauge will determine any tread depth, allowing projections on tire replacement dates and, more important, provide the earliest identification of uneven tread wear, allowing corrective action before destroying an expensive set of tires. Some tires have raised patches, called wear bars (**Figure 2-18**), in the grooves of tire treads to help determine wear. These bars are $\frac{2}{32}$ in. (1.5 mm) deep, so if any of them are even with the tread, the tire is not legal in any position. Some tires do not have wear bars. In such cases, if a tread depth gauge is unavailable a penny can be used to estimate tread wear. Place a U.S. penny into the tire tread groove with Lincoln's head down. If the tread is at or beyond the top of Lincoln's head, the tread is at or above $\frac{3}{32}$ of an inch (2 mm) of tread (**Figure 2-19**). Excessive tire wear can also affect vehicle handling.
- ***Rim condition*.** Look for damage such as dents and nicks. Also check the wheel lug nuts for tightness and for damage such as rust. Damage to a rim can cause the tire to lose pressure and even come off.

FIGURE 2-18 Tire wear bars provide quick proof of tires worn beyond their legal limit; however, even tires not worn down to the wear bars may not be legal in some positions, nor may they provide adequate control for some road conditions. Note: This tire shows uneven wear that needs to be reported and fixed.

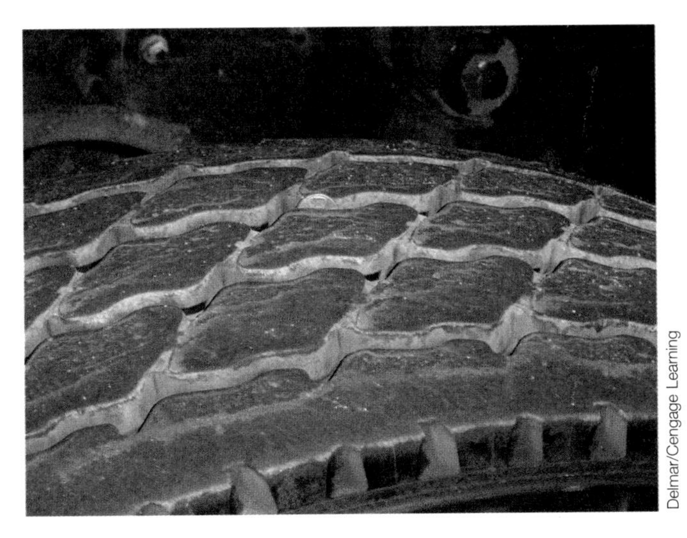

FIGURE 2-19 When a tire tread depth gauge is unavailable, tread depth can be estimated using a penny. The tire should be checked at several places in both the inside and outside groove. At each place, the tire should come to the same spot on the coin, showing even tire wear.

- *Valve stem.* Check the general condition of the valve stem to ensure that it is not loose, cracked, or otherwise damaged.
- *Tires.* Ensure that tires do not come in contact with any part of the vehicle's body. Tire sidewalls should be smooth and free of bulges or cracks. Dual tires should not be in contact with each other. Inspect the splash guards to ensure they are properly attached to the vehicle and are in good condition.

Cab Doors and Vehicle Compartments

While moving around the vehicle, check the cab and all compartments to ensure that doors open, close, and latch properly and all equipment is present and properly stowed.

As a safety feature, all vehicle doors have a double latching feature. When checking doors, make sure the doors are fully closed, not just partly closed. Doors that have only caught the first latch point can open unexpectedly in transit.

Other Outside Inspection Components

The following should also be inspected outside of the apparatus:

- *Glass.* Ensure all glass is clean, secure, and void of large cracks and nicks.
- *Steps.* All stepping surfaces, platforms, and handrails should be clean, secure, and in good working order.
- *Saddle tanks.* Check that saddle tank caps are secured, the tank is securely mounted, and that there are no leaks.
- *Suspension.* Check suspension for damage (cracks, rust, or separation) to springs, spring hangers, U-bolts, and shock absorbers.
- *Air tanks.* Air tanks should be drained to remove moisture. When the vehicle is started, the time it takes to refill the tank should be noted to make sure it fills within the manufacturer's recommendations.
- *Windshield wiper blades.* Check for any obvious defects, for proper tension, and that they are free from debris.
- *Equipment.* Equipment such as extinguishers, ladders, and hose should be free from damage and properly secured.
- *Pump intake and discharge.* Caps and connected hose should be checked for proper tightness. Connections should be tight, but not so tight that they become difficult to remove. Valves should be operated to ensure free movement and that they are in the proper position, usually closed.
- *Reflective striping.* Inspect for obvious damage. At some point, it will be important to note its reflective capability at night.

- *Hose bed.* Inspect for proper hose loading. Hose bed covers should be checked for damage and should be securely fastened.

Frequent operating of pump valves actually causes them to last longer and work easier. Minerals in the water in the pump form crystals over time, like barnacles on a boat. These crystals cut into the surface of the valves and make them harder to operate. Operating the valve while these crystals are still at a microscopic size dislodges them. If they are left to grow over time, they become firmly attached, make valves difficult to operate, and scratch the valve surfaces, causing leaks.

Pump Inspection and Test

Daily, weekly, monthly, and periodic pump inspections and tests are conducted for two primary reasons: (1) to ensure that components are in proper working order, and (2) to keep the components in working order—for example, testing the priming pump to ensure that it is operating properly while also lubricating its close-fitting parts. All valves and pump controls should be operated. Examples of inspections and tests are described next.

Inspect intake strainers to make sure they are in place, free of holes and anything that could clog the intake; inspect the prime reservoir to make sure it is filled to the recommended level with primer oil and that the top is on securely (**Figure 2-20**); inspect discharge gauges to make sure gauges respond to pressure changes and there are no cracks in the gauge glass (**Figure 2-21**); inspect the pressure relief strainer to make sure the strainer is in place and free of anything that could clog the strainer.

(For step-by-step photos of Relief Valve Strainer Inspections, please refer to page 55.)

Test the following according to the manufacturer's recommendations: priming system, transfer valve, dry vacuum, and pressure regulating device.

Delmar/Cengage Learning

FIGURE 2-20 Inspection of pump primer oil should ensure that the reservoir is at least half-filled and that the anti-siphon hole (pointed to by the yellow arrow) is not clogged. Many tanks such as this one are transparent so that the oil level can be checked without removing the cap. The red arrow points to the current oil level in this reservoir. These lines can be made more apparent by shining a flashlight near the top of the container.

Fire Pump and Related Components

The pump should be engaged following the manufacturer's procedures to ensure proper operation. Pumps can be powered in several different ways and are discussed in detail in **Chapter 4**. After engaging the pump, ensure pump panel gauges and controls are operating properly.

Check control valves to ensure smooth, unrestricted operation. Water tank levels should be visually inspected, generally by using the tank vent located on top of the vehicle and with the gauge checked to confirm that it shows the tank is full. Foam tank levels should also be inspected and filled if needed (**Figure 2-22**). When filling foam tanks, it is important not to agitate the foam, causing the liquid to bubble and making it difficult to fill the tank. Often foam is added using a funnel and tube to limit agitation as the concentrate enters the tank.

Back-flushing the Pump

Sometimes debris will get into the pump and adversely affects its performance. Debris enters the pump along with water and may enter the pump whenever water is supplied to the pump, regardless of whether the water comes from a hydrant, drafting, or other source. The likelihood of debris entering the pump is, however, more likely with some water sources. Drafting is generally considered more likely to have debris than public water systems. Debris

Delmar/Cengage Learning

FIGURE 2-21 Discharge gauges should be inspected to confirm that they are not cracked, filled with damping fluid, read zero when the pump is not running, and that all individual gauges have the same reading as the master discharge gauge with the pump running and the individual discharge capped with its valve open.

Delmar/Cengage Learning

FIGURE 2-22 Water and foam tank levels should be checked by looking through the fill tower to confirm that they are full. Gauges should also be looked at to verify that they also show the tanks are full.

entering the pump is also more likely when drafting while leaves are falling off trees than at other times of the year. Examples of debris affecting the pump include grass, leaves, sand, pebbles, rocks, and trash. Larger items can permanently damage the impeller and other pump components, resulting in very expensive and time-consuming repairs. To prevent such damage, all intakes should be equipped with strainers that prevent items too large to pass through the impeller from entering the pump. Even so, smaller items such as grass may enter the pump and wrap around the blades of the impeller, thus affecting its performance.

To ensure continued protection and performance, it is necessary to inspect intake strainers to make sure they are in place, free of any holes or damage that might mean they would not prevent larger items hitting them from getting into the pump, and that they are clear of anything that could prevent full water flow to the pump. The pump also needs to be back-flushed to remove any smaller debris that may have made it into the pump. While it is critical to back-flush the pump anytime salt water or dirty water has been pumped, many departments back-flush following any drafting operation. Many departments also check strainers after any pumping operation. A recommendation is to back-flush the pump weekly to remove sediments from the pump itself.

(For step-by-step photos of Back-flushing the Pump, please refer to pages 56–57.)

Soaping

Soaping is a periodically used process in which 32 oz. (1 L) of mild liquid dishwashing soap is added to 500 gallons (2,000 litres) of booster tank water. The water is then circulated through the pump and intake and discharge valves. Short hoselines are used to connect discharges to intakes. While circulating, valves are slowly opened and closed. (This implies that an additional intake and discharge are used so water is always flowing through the pump.) The soap acts as a lubricant and cleaning agent for the tank, pump, and valves. After soaping, valves may work more easily, and small valve leaks may even disappear.

It is interesting to note that this process works best with inexpensive soap since stronger soaps tend to remove necessary greases. The process is also often done just prior to greasing the pump. Finally, since dishwashing soap is biodegradable, disposal of the waste is not too difficult, and prior to disposal, it can be used to soap several engines. To complete the process, the booster tank should be refilled and flushed with clean water.

Detailed information on pump testing is provided in **Chapter 8** of this textbook.

Foam System Inspection and Test

The type of foam system on the apparatus may vary, so inspections and tests should be made according to the manufacturer's recommendations. This will keep the system operational and ready for use.

Tool and Equipment Inspection

All equipment carried on a given apparatus needs to be inspected. For many hand tools often this just means confirming the tool is clean and stored properly in the correct place. Particularly after use, however, a more careful inspection is necessary to confirm that no tool maintenance is needed; for example, do burrs need to be filed off striking tools? Do SCBA cylinders need to be topped off? Do batteries need recharging or replacement? Engine-operated tools need to have the engines checked for fuel, lubrication, and starting. Cutting tools need to have their edges checked to make sure they are not damaged and can cut appropriately when required.

CLEANING AND SERVICING

Cleaning and servicing a vehicle are two important tasks of a vehicle operator's responsibility. Cleaning refers to the washing, vacuuming, and waxing of a vehicle to remove dirt, grime, and debris, and to protect the finish from the elements. Servicing refers to minor maintenance activities necessary to keep the vehicle in good working order, such as topping off fluids, lubricating parts, replacing parts, and tightening connections. Sometimes these tasks are completed during preventive maintenance inspections. For example, minor corrosion around battery terminals can be quickly wiped away with a rag during the inspection. In addition, engine oil can easily be added during the inspection. These tasks can also be done when needed or at specific intervals. For example, it may be department policy to clean and wash vehicles on a specific day. In addition, a vehicle may need to be washed after an emergency run or training exercise. Regardless, all vehicle cleaning and servicing should be accomplished according to the manufacturer's recommendations.

Keeping vehicles clean at all times is important for several reasons. First, clean lens covers allow lights to project further and clean reflective striping can be seen at greater distances. This both helps you see further and be seen earlier, enhancing safety during emergency response. It helps maintain good public relations. After all, municipal apparatus usually belongs to the tax-paying public. Second, it helps ensure the pump, systems, and equipment operate as intended. Finally, maintaining clean apparatus is important to help ensure the vehicles can be inspected properly; dirt and grime could cover defects or potential problems.

How to clean a vehicle is just as important as when. Dirt should never be wiped off the surface since that tends to scratch the finish. The exterior surface should be hosed down with cool water, washed with a good automotive soap diluted in water and applied with soft clean car brushes, and then rinsed. Remaining clean rinse water should be removed with squeegees, clean chamois, or clean towels so that water spotting does not occur. Particularly in areas where road salt is used, the underbody of the apparatus needs to be washed to remove the road salt since salt is highly corrosive on metal parts.

The underbody may be cleaned with a steam cleaner or high-pressure washer, although you need to be careful not to force steam into places it does not belong and to have a catch system so that oils do not run into the groundwater system. A booster line hose can also be effective, with some care not to force water into items such as electrical connections.

(For step-by-step photos of Apparatus Cleaning, please refer to page 58.)

While cleaning, make sure not to remove lubricants essential to the long service life of apparatus components.

SAFETY

As with any activity carried out by pump operators, safety should be considered when conducting preventive maintenance inspections and tests. One way to help ensure safety is not to rush through inspections and tests. Hurrying to finish can increase the risk of an accident and increase the chance that a safety problem is overlooked. Another important safety consideration is to ensure that the work area is free from hazards such as trip and fall hazards caused by fluids that have dripped from the apparatus. This can be accomplished by walking around the apparatus, looking under and above for slippery surfaces and loose equipment. Finally, increased safety can be achieved by always keeping it in mind. Several common safety considerations for preventive maintenance inspections and tests include:

- Check for loose equipment before raising a tilt cab.
- Do not smoke around the engine compartment and fuels.
- Wear appropriate clothing (no loose jewelry; wear safety glasses, gloves, etc.).
- Consider vapor and electrical hazards.
- Always be careful when opening the radiator cap.
- Use the proper tool for the task.
- Be sure to secure equipment and close all doors prior to moving the apparatus.

SKILL 2-1

Hands-on Inspection

A When approaching the apparatus, look for problems such as leaning, physical damage, and leaking fluids.

B In the engine compartment, look for loose or missing belts, and check all fluid levels.

C In the cab, check all switches, gauge readings, and any equipment stored in the cab.

D At the pump panel, operate each valve, put the pump in gear, confirm the operation of the pressure control system, and, after taking the pump out of gear, open and then close all drain valves.

E While walking around the apparatus, systematically check all equipment in each compartment.

SKILL 2-2

Jump Starting an Apparatus

CAUTION

These procedures are for two vehicles using the same voltage and negative grounds. Some vehicles use different voltages or positive grounds; for these vehicles consult the manufacturer's owner's manual for recommended procedures. Incorrect jumping procedures can damage vehicle-based computers, requiring expensive repairs.

NOTE

Jump starting only helps for a dead or weak battery. If the vehicle cranks normally but does not start or headlights work normally when attempting to start, the problem is not a dead or weak battery. Dead or weak batteries will show a voltage of less than 9.6 volts on the voltmeter when attempting to crank.

A Park the vehicles close enough together so that the jumper cables can reach between the batteries. Shut the vehicles off and turn the main battery disconnect switches off. Open and ventilate the battery compartments of both vehicles. Confirm that corrosion does not exist on the battery terminals or cables (it may be the real source of the problem).

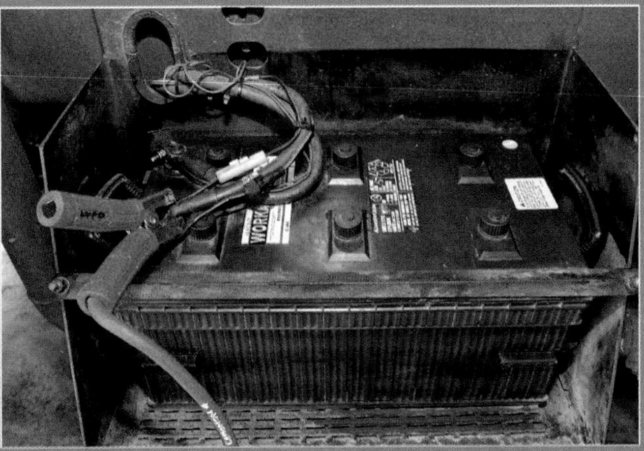

B Connect one of the red jumper terminals to the positive battery terminal of the dead vehicle.

C Connect the other red positive jumper to the positive terminal of the source vehicle battery.

D Connect the black negative jumper to the negative terminal of the source vehicle's battery.

(Continues)

SKILL 2-2

Jump Starting an Apparatus (*Continued*)

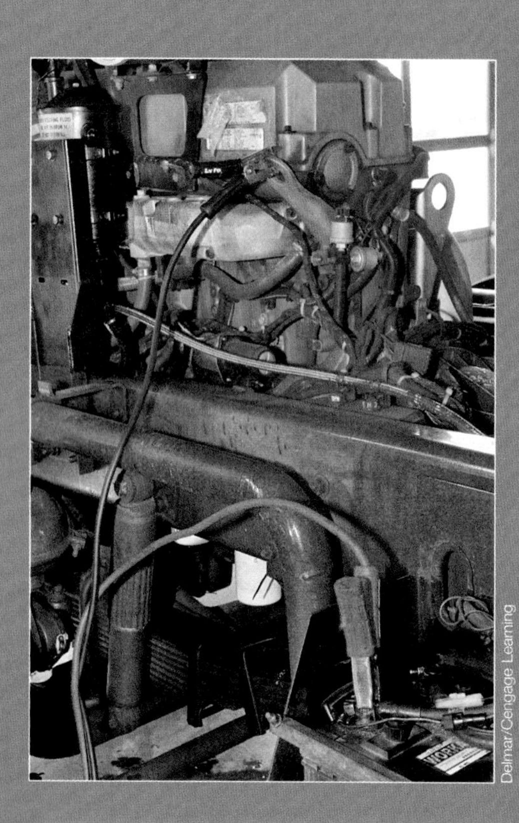

Delmar/Cengage Learning

E Connect the other black negative jumper to the dead vehicle's negatively grounded chassis or chassis terminal post, if available. Start the source vehicle and engage the high idle switch to increase alternator output.

Delmar/Cengage Learning

F Once the dead vehicle is started, disconnect the battery jumper cables in the opposite order in which they were connected.

CAUTION

Make sure energized jumper cables do not accidentally contact any unintended metal parts.

SKILL 2-3

Connecting a Battery Charger

NOTE

Connecting a battery charger should be an infrequent operation and have an explanation, such as a light left on in a vehicle without an internal charger. If unexplained charging is needed, a mechanic should be consulted to determine why.

CAUTION

Batteries produce explosive gases; therefore, good ventilation should be used to reduce their concentration, sparks (ignition source) should not be made near the battery, and personal protective gear consisting of mechanic's gloves and fully protective safety glasses should be worn.

Delmar/Cengage Learning

A For apparatus without battery jumper posts, open and ventilate the battery compartment. Confirm that corrosion does not exist on the battery terminals or cables (it may be the real source of the problem). For apparatus with battery jumper posts, remove the post covers.

Delmar/Cengage Learning

B Unplug the battery charger and set the switches to the off position (if it has an on-off switch) prior to making battery connections.

Delmar/Cengage Learning

C Connect the red positive jumper to the battery's positive terminal.

(Continues)

SKILL 2-3

Connecting a Battery Charger (*Continued*)

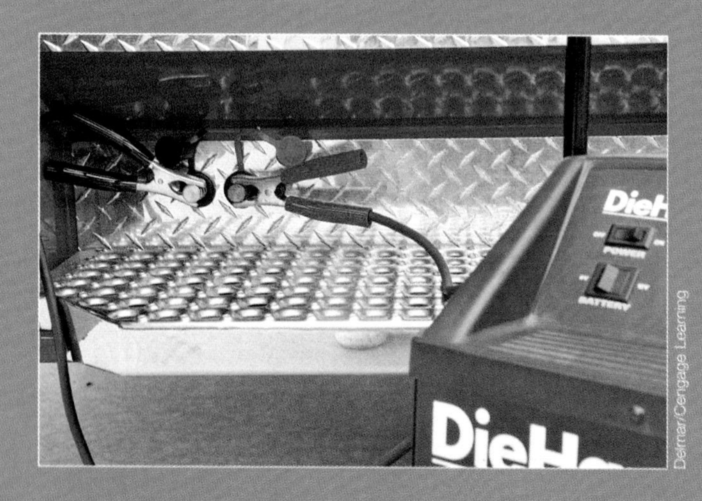

Delmar/Cengage Learning

D Connect the black negative jumper to the frame of a negatively grounded vehicle.

Delmar/Cengage Learning

E Select the correct voltage and turn the battery charger on. A slow charge is preferred if you can wait. A fast or cranking charge is used if you need to start the vehicle quickly.

Delmar/Cengage Learning

F Once the vehicle is started or the battery is charged, disconnect the negative jumper followed by the positive jumper.

SKILL 2-4

Tire, Wheel, Rim, and Brake Inspections

Delmar/Cengage Learning

A Confirm that lug nuts are not missing or loose and that rim damage does not exist. (Note: Decorative caps are not required. One has been removed for the photo to show the required lug nut below. If any lug nuts are missing or loose, the apparatus should be removed from service until corrected.)

Delmar/Cengage Learning

B No part of the tire should rub on any part of the apparatus. If a tire rubs on the apparatus, it will typically be when turning or as the apparatus moves with its suspension. Signs of such rubbing appear as shiny spots where all paint and dirt have been removed, such as where the red arrow in figure points. On dual wheels, the tire walls should not touch each other. In addition to the tire walls not touching when parked, evidence of touching, such as scuff marks on the two inner tire walls, should not exist. Underinflation is a common cause of tires rubbing, so if evidence of touching does exist, particular attention should be paid to the inflation check in step E of this skill.

Delmar/Cengage Learning

C Confirm the recommended tire inflation pressure. Ideally this comes from the tire manufacturer's chart based on the actual load on the tire. Alternatively, the pressure printed on the tire sidewall may be used since most apparatus are loaded to the tire's rating. While looking at the tire, confirm that no cuts through the cords of the tire exist. Note: Tire inflation should be checked cold.

Delmar/Cengage Learning

D Inspect the tire stem. If the apparatus is equipped with a tool-less tire pressure monitoring system (required starting with NFPA 1901, 2009 edition), confirm an acceptable reading. If the stem is equipped with a flow-thru cap, leave the cap in place. If equipped with a solid cap, it needs to be removed for the test. If the cap is missing, it should be replaced.

(Continues)

SKILL 2-4

Tire, Wheel, Rim, and Brake Inspections (*Continued*)

E Confirm that the tire is properly inflated using a tire gauge, indicating valve stem, or remote tire pressure console. If the pressure is not correct, add or remove air as required until the correct pressure is obtained. If a cap was removed, replace the cap.

F Confirm the tread depth using a tread depth gauge. Depth should be checked at several places along the complete inner and outer groove. If readings vary by more than 2⁄32 in. (1.5 mm), a mechanic should be consulted to determine why.

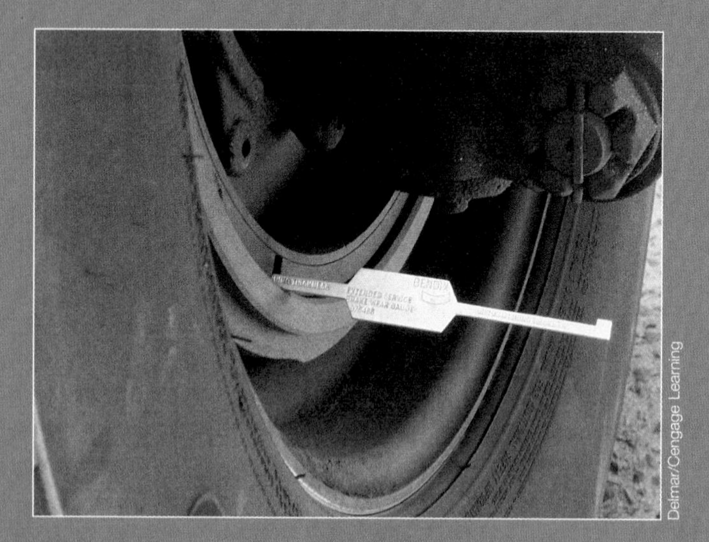

G Confirm that adequate pad depth exists on the brake shoe. Pads on full-size apparatus should be at least ¼ in. (6 mm) deep. Note that the inexpensive gauge shown here easily makes measurements.

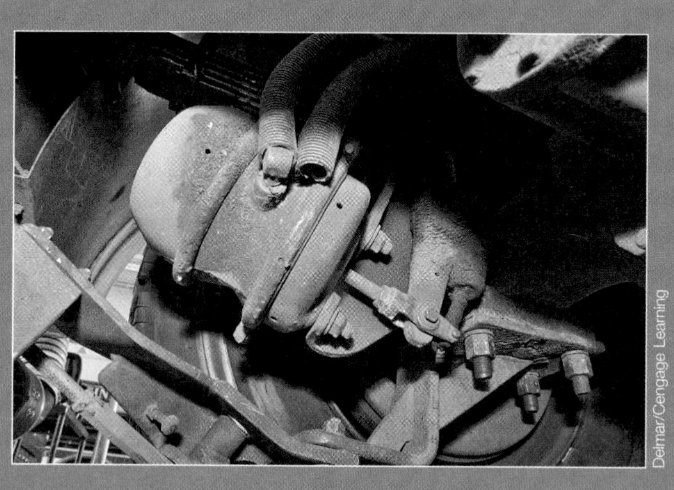

H Confirm that the air brake push rod has not extended too far (less than 90-degree angle or red showing on the rod).

SKILL 2-5

Relief Valve Strainer Inspections

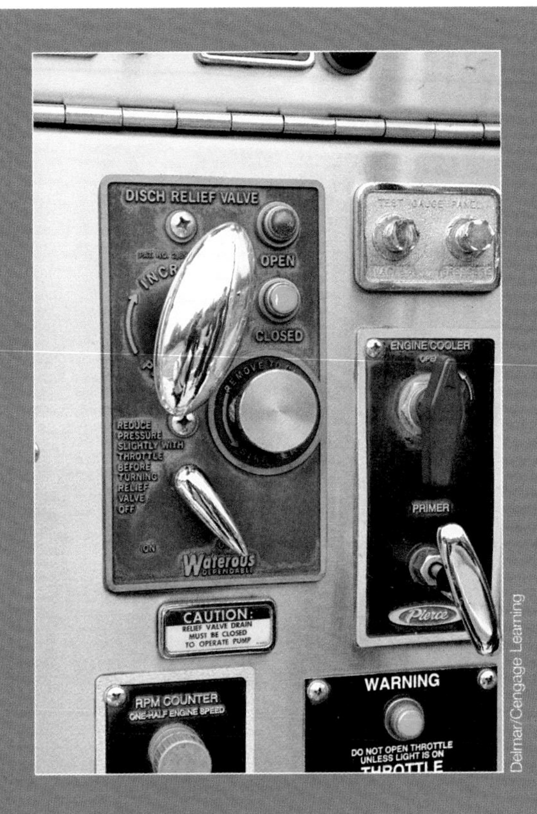

Delmar/Cengage Learning

A Some relief valves, such as Waterous, contain a strainer. These strainers should be cleaned and inspected. Start the procedure by running the pump at 150 psi (1,050 kPa) and turning the relief valve on and off. To inspect the strainer, it needs to be unscrewed.

Delmar/Cengage Learning

B With the relief valve in the off position, unscrew the strainer. Once the strainer has been removed, it should be inspected and cleaned of any debris. It can be cleaned by blowing air through it and using a wire or touch tip cleaner to remove any debris in the hose at the end of the strainer. Then screw the strainer back into its original position. Turn the T-handle control clockwise to remove tension from the valve piston. Then turn the valve on and off, being sure to follow the note on the placard about reducing pressure before turning the relief valve off.

SKILL 2-6

Back-flushing the Pump

Delmar/Cengage Learning

A With the pump *not* in gear, connect a hose from a pressurized source to a pump discharge. This will normally require the use of a double female at the male end of the hose.

Delmar/Cengage Learning

B Uncap one of the intake connections.

Delmar/Cengage Learning

C Charge the hoseline connected to the discharge and open its valve.

Delmar/Cengage Learning

D Allow water to flow out the intake until the water coming out is clean.

Delmar/Cengage Learning

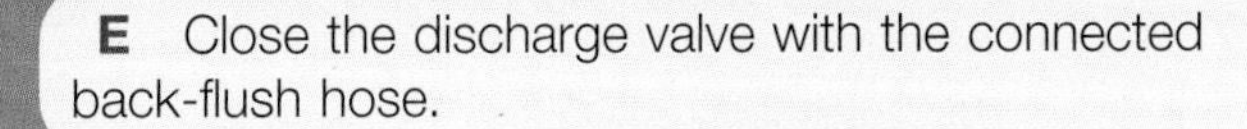

E Close the discharge valve with the connected back-flush hose.

Delmar/Cengage Learning

F Inspect the strainer to make sure it is not damaged and is clear of any obstructions such as leaves, rocks, and grass. Picking up such debris is very common when drafting but may also come through a hydrant connected to a domestic water system. Both the leaf and pebble need to be removed from the pictured strainer.

Delmar/Cengage Learning

G Recap the intake. Repeat steps B through G for each hose intake connection on the apparatus.

SKILL 2-7

Apparatus Cleaning

STREETSMART

Brushes used on heavily soiled areas such as tires and wheel wells should not be used on painted surfaces since dirt captured in these brushes can scratch the painted surfaces.

A The exterior apparatus surface should be hosed down with cool water to remove heavy dirt.

B Apply soapy water to the wetted surface with a soft brush or sponge to emulsify remaining dirt.

C Wash off the soap with clean, cool water before it has had a chance to dry.

D Remove remaining water droplets with a squeegee, clean chamois, or clean towel so that water spots do not occur.

LESSONS LEARNED

Preventive maintenance is an important pump operator duty designed to ensure that the apparatus, pump, and related components are in a safe and ready operating status. A good preventive maintenance program helps guard against mechanical failure, costly repairs, and injuries. Typically the fire department is responsible for the overall administration of the preventive maintenance program. Pump operators and qualified mechanics share the responsibility of ensuring that the apparatus is in a safe operating condition.

Preventive maintenance activities typically include inspections and pump tests. Inspections, consisting of both safety-related components and manufacturers' recommendations, verify the condition, status, and operability of the apparatus, pump, and related components. Pump tests determine the operating performance of the pump and related components. NFPA standards and manufacturers' recommendations provide a basis for what and how often components should be inspected and tested. Documentation of inspection and test results assists with tracking needed maintenance and repairs. Such documentation can also be used to show adherence to standards and manufacturers' recommendations, which can help minimize liability in a civil or criminal matter. In addition, the forms can be used to guide the pump operator through the specific inspection and test.

Preventive maintenance is an ongoing process that should occur whenever the pump operator, or others, are operating, cleaning, or training with the apparatus. As with other duties and activities conducted by the pump operator, diligence to safety should be maintained at all times.

KEY TERMS

Aerial A fire apparatus using mounted ladders and other devices for reaching areas beyond the length of ground ladders.

Aircraft rescue and firefighting (ARFF) Apparatus designed for fighting aircraft fires at or near an airport.

Inspecting Determining the condition and operational status of equipment by sight, sound, or touch.

Insurance Services Office (ISO) An organization that calculates a Public Protection Classification (PPC) based upon what a fire department does; fire insurance rates within a fire district are based upon this classification.

Manufacturer's inspection recommendations Those items recommended by the manufacturer to be included in apparatus inspections.

Preventive maintenance Proactive steps taken to ensure the operating status and readiness of the apparatus, pump, and related components.

Pump operators People who have had training, knowledge, experience, and demonstrated proof of competence in operating a fire pump as defined by NFPA 1002. For pump operators, the proof of competence is generally shown in terms of certification agencies, including departmental, state, and independent groups such as the National Board on Fire Service Professional Qualifications (PRO BOARD) and the International Fire Service Accreditation Congress (IFSAC).

Pumper The basic unit of the fire apparatus, an automotive fire apparatus that has a permanently mounted fire pump with a rated discharge rate of at least 750 gpm (3,000 Lpm), a water tank, and a hose body; also called a "triple-combination pumper." The unit is designed for sustained pumper operations during structural firefighting and is capable of supporting associated fire department operations.

Qualified mechanics People who have had training, knowledge, experience, and demonstrated proof of competence in performing mechanical repairs. For fire apparatus mechanics, the proof of competence is generally shown in terms of manufacturer certifications, Automotive Service Excellence (ASE) certifications, and Emergency Vehicle Technician (EVT) certifications.

Repair To restore or replace components that have become unserviceable, or not meeting their manufacturers' specifications for whatever reason, including damage and wear.

Safety-related components Those items that affect the safe operation of the apparatus and pump, and that should be included in apparatus inspections.

Servicing The act of performing maintenance to keep equipment working as intended.

Systematic Orderly process or following of a prescribed procedure—for example, a systematic inspection is used to make sure nothing is accidentally missed.

Testing Verifying the condition and operational status of equipment by measurement of its characteristics and comparing those measurements to the required specifications.

Tiller apparatus Aerial apparatus in which the rear wheels are steered from the back of the apparatus by a tiller operator.

Wildland Apparatus designed for fighting wildland fires; contain small pumps with pump-and-roll capabilities and limited on-board water supplies.

REVIEW QUESTIONS

Multiple Choice

Select the most appropriate answer.

1. Which of the following NFPA standards establishes responsibility for and the specific requirements of a preventive maintenance program?

- **a.** 1500
- **b.** 1901
- **c.** 1911
- **d.** All of the above are correct.

2. Which of the following NFPA standards requires that pump operators be able to conduct and document routine tests, inspections, and servicing functions to ensure that the apparatus is in a ready state?

- **a.** 1500
- **b.** 1901
- **c.** 1911
- **d.** 1002

3. Which of the following NFPA standards requires that pumps with a rated capacity of 250 gpm (1000 Lpm) or greater be performance tested at least annually?

- **a.** 1500
- **b.** 1901
- **c.** 1911
- **d.** 1002

4. A preventive maintenance inspection should begin

- **a.** with the engine compartment.
- **b.** with the pump.
- **c.** with an inventory of equipment.
- **d.** with checking previous inspection reports.

5. The NFPA standard that requires that manufacturers' certification and acceptance tests be conducted on new apparatus is

- **a.** 1500.
- **b.** 1901.
- **c.** 1911.
- **d.** 1002.

6. Ideally, transmission oil should be checked

- **a.** at room temperature.
- **b.** at normal operating temperature.
- **c.** either hot or cold, as long as the dipstick has both hot and cold levels marked.
- **d.** None of the above—checking transmission oil is not affected by temperature.

Short Answer

On a separate sheet of paper, answer/explain the following questions.

1. Explain why a preventive maintenance program is so important.

2. What are the three levels of responsibility within a preventive maintenance program?

3. List and briefly discuss five NFPA standards that contain preventive maintenance requirements related to pump operations.

4. List and explain the components to inspect during a preventive maintenance inspection.

5. List the two criteria for determining what components to include in a preventive maintenance inspection.

6. List two NFPA requirements that an apparatus braking system must meet.

ACTIVITIES

1. Conduct a preventive maintenance inspection of an apparatus.
2. Obtain three other departments' inspection forms and use them to develop a new daily preventive maintenance inspection form.
3. Demonstrate the safe use of hand tools required for routine tests, inspections, and vehicle servicing.

PRACTICE PROBLEM

1. Your department has recently experienced significant maintenance problems with several apparatus. Your chief has asked you to review the preventive maintenance program and identify any weaknesses. What will you use as the basis for evaluating the program?

ADDITIONAL RESOURCES

In addition to the owner's manuals provided with the apparatus by its manufacturer, the following are important resources.

FEMA. Guide to Developing Effective Standard Operating Procedures for Fire and EMS Departments, December 1999, http://www.usfa.dhs.gov/downloads/pdf/publications/fa-197-508.pdf

Goodyear Truck Tire Information, http://www.goodyear.com/truck/support/

Heavy Duty ProClinic. Dallas: Interstate Batteries, 1996.

Michelin Truck Tire Information, http://www.michelintruck.com/michelintruck/toolbox/reference-material.jsp

NFPA 1002, *Fire Apparatus Driver/Operator Professional Qualifications*. National Fire Protection Association, Quincy, MA, 2009.

NFPA 1071, *Emergency Vehicle Technician Professional Qualifications*. National Fire Protection Association, Quincy, MA, 2009.

NFPA 1500, *Fire Department Occupational Safety and Health Program*. National Fire Protection Association, Quincy, MA, 2009.

NFPA 1901, *Standard for Automotive Fire Apparatus*. National Fire Protection Association, Quincy, MA, 2009.

NFPA 1911, *Standard for the Inspection, Maintenance, Testing, and Retirement of In-Service Automotive Fire Apparatus*. National Fire Protection Association, Quincy, MA, 2007.

VFIS website, http://www.vfis.com/resources.htm, provides downloadable safety forms, including driver check-offs.

3 Driving Emergency Vehicles

- Learning Objectives
- Introduction
- Laws and Standards
- The Driver and Apparatus
- The Role of Officers
- Safe-driving Considerations
- Driver Operator Testing
- Emergency Response
- Lessons Learned
- Key Terms
- Review Questions
- Activities
- Additional Resources

If you drive a fire apparatus for any significant period of time, you will have some close-call stories before too long. Many of the near misses can be attributed to inattentive drivers, but some are a direct result of our own aggressive driving behaviors. This was an important lesson that I learned the first time I drove a fire engine to a fire—I almost killed four teenagers.

I was young and new at driving, and I was taking chances that I should not have been taking. As I approached a certain intersection, I could not see that a small white car was also approaching from the right—this car was in my blind spot. I had never blown through a red light with lights and siren going before; but I felt like this time, with the intersection apparently clear, it would probably be all right.

By the time I saw the car, it was too late. I tried to stop, but it was not going to do any good. Fortunately, the teenagers were completely unaware that there was a big red fire engine blasting through the intersection, so they never even hit their brakes.

This was a good thing, because if they had hit their brakes, I would have killed them for sure. I was able to walk away scared, but educated.

As I have reflected on my time as a driver/operator since then, the need for safe vehicle operations has become very clear to me. Every day we make choices that can seriously increase the chance of someone being hurt or seriously injured in an accident. It makes no difference that similar actions in the past have not had bad outcomes. It has to be a conscious decision each time we operate the apparatus to do so safely. Now, when I am the one charged with getting everyone to the emergency scene safely, I take it very seriously. We have all heard it before that we cannot help if we don't get to the scene. It may sound redundant, but it is true nonetheless.

—Street Story by J. Christopher Milne,
Apparatus Division Captain, Salt Lake City Fire Department,
Salt Lake City, Utah

LEARNING OBJECTIVES

After completing this chapter, the reader should be able to:

3-1. Explain the basic concepts contained in most state emergency driving laws.

3-2. Explain how the application of a risk management plan (as described in NFPA 1500, Section 4.2) will affect the emergency and nonemergency operation of a fire department vehicle on public roadways.

3-3. Demonstrate the use of fire apparatus passenger safety devices.

3-4. Discuss the basic attributes of a competent apparatus driver.

3-5. Describe how vehicle dimensions affect turning characteristics on fire apparatus maneuvering forwards and backwards.

3-6. Discuss how emergency vehicle characteristics (weight, width, length, height, center of gravity, loads, and liquid surge) affect safe driving.

3-7. Explain how speed, braking, load factors, road conditions, and environmental conditions affect safe driving.

3-8. Demonstrate defensive driving techniques to maintain control of a fire apparatus while driving under normal and adverse road, weather, environment, and traffic conditions.

3-9. Explain the proper procedures for engaging the vehicle transmission and shifting between gears.

3-10. Explain the considerations for using automatic braking systems in good and in bad road conditions.

3-11. Discuss how steering reactions, centrifugal force, and speed affect safe driving.

3-12. Explain the causes of vehicle skidding, how to correct a skid, and the principles of skid avoidance.

3-13. Describe the basic principles of safe vehicle operations and explain why a vehicle operator must be even more vigilant when backing a vehicle.

3-14. Explain the purpose of a spotter in communicating with the operator.

3-15. Demonstrate the use of mirrors by safely backing a fire apparatus while maintaining visual contact with spotter(s) and maintaining clearance from all obstructions.

3-16. Identify facts concerning safety in night driving.

3-17. Demonstrate the ability to safely maneuver a fire apparatus in areas with restricted horizontal and vertical clearances.

3-18. Operate a fire apparatus through various maneuvers on a public roadway so that the vehicle is operated in a safe manner and in compliance with all applicable laws and departmental policies.

3-19. Using a spotter, demonstrate safely backing a fire apparatus into restricted spaces requiring a 90-degree right or left turn.

3-20. Demonstrate the ability to safely maneuver a fire apparatus around obstructions on a roadway.

3-21. Demonstrate the ability to safely turn a fire apparatus 180 degrees within a confined space without striking obstructions.

3-22. Demonstrate the ability to drive a fire apparatus safely under emergency conditions.

3-23. Explain the importance of safe driving during emergency response.

***The driver/operator requirements, as defined by the NFPA 1002 Standard, are identified in black; additional information is identified in blue.**

INTRODUCTION

Each of the pump operator duties—preventive maintenance, driving the apparatus, and operating the pump—is important. If one of the duties is not properly executed, the ability to deliver water to the discharge point would, at best, be ineffective. Although each of the duties is important, a strong argument can be made that driving the apparatus is perhaps the most critical duty of a pump operator.

There are several reasons why driving the apparatus can be considered the most critical pump operator duty. First, a pump operator who ineffectively operates or improperly maintains a pump may still be able to provide at least some water to a discharge location; however, the best pump operator in the world and the best maintained pump will be of little value if the apparatus does not make it to the scene. Second, driving the apparatus is by far the most common duty carried out by the pump operator. More time is spent behind the wheel than behind the pump panel.

This chapter discusses the pump operator's duty of driving fire apparatus, focusing on laws and standards, driver and apparatus limitations, and driving considerations related to the safe operation of fire apparatus.

SAFETY

Apparatus drivers should be thinking of a quick yet safe response.

LAWS AND STANDARDS

Apparatus operators assume a tremendous amount of responsibility. They are responsible for the safety of the apparatus, equipment, and, most important, those riding on the apparatus. They are held responsible for the safety of the public at large by properly operating emergency vehicles.

Laws and standards assist the driver by providing a set of rules and guidelines focusing on safe operation of emergency vehicles. Laws are rules that drivers must obey, while standards provide suggested guidelines for drivers. Laws and standards assist the driver in ensuring the safety of the apparatus and of personnel riding on the apparatus, as well as the public at large. Drivers therefore must be familiar with the laws and standards that govern emergency response for their department, community, and state.

SAFETY

Laws and standards assist the driver by providing a set of rules and guidelines that focus on the safe operation of emergency vehicles.

NOTE

Drivers must be familiar with the laws and standards that govern emergency response for their department, community, and state.

Emergency Vehicle Laws

Laws enacted by state and local governments affect apparatus drivers on a daily basis. State laws, referred to as statutes, set the overall rules and standards for emergency driving within the state. All government and non-government entities involved with emergency response must obey these laws. Local laws, referred to as city codes or ordinances, set specific requirements for emergency response within the city, county, or other form of local government. Local laws may be more restrictive or provide more detail than state laws. For example, a state may allow emergency vehicles to travel through a stop light or sign after slowing down, whereas a city may require emergency vehicles to come to a complete stop prior to traveling through a stop light. Local laws are based in part on state law and, in turn, tend to use the same language and terminology as their state's law.

Although laws governing emergency response vary somewhat between states, most state laws tend to address the same basic components. Most state laws, for example, identify or define what is considered to be an **authorized emergency vehicle**. Typically this means fire department vehicles, ambulances, rescue vehicles, and police vehicles with appropriate identification and warning devices. In addition, state laws typically identify the individual empowered to designate an "authorized emergency vehicle." In most cases, a commissioner, fire chief, police chief, or county sheriff is the individual so named.

Most state laws require emergency vehicle drivers to obey the same laws as other vehicle operators unless specifically exempt. It is important to keep in mind that exemptions extended to emergency vehicle drivers are not automatic. State laws typically define several conditions that must exist for exemptions to be extended:

- Only authorized emergency vehicles are covered.
- The exemptions only apply when responding to an emergency.
- Audible and visual warning devices must be operating when taking advantage of the exemption.

Finally, state laws typically stress the duty to drive with due regard for the safety of all persons. The exemptions previously stated are not available when their use may bring harm to the public. For example, in many states, an emergency vehicle can drive through in intersection with a red light; however, they *cannot* drive through the same intersection when doing so would cause an accident because a civilian is already driving through it with a green light. If an accident occurs in this case, the fire driver, not the civilian, is at fault and may get the ticket even though the vehicle's emergency lights were on. Departments across the country are finding themselves involved in lawsuits over the disregard of these traffic laws. The fire service has been slow to remedy the misperception of what emergency lights and sirens allow the apparatus operator to do. More stringent guidelines and new standards are helping shorten the time it takes for these changes to take place.

SAFETY

State laws require emergency apparatus drivers to exercise due regard for safety as one of the prerequisites to being allowed to use the privileges.

In order to assist the driver of an emergency vehicle with an efficient response, state laws typically

Delmar/Cengage Learning

FIGURE 3-1 State laws typically exempt authorized emergency vehicles from several normal driving and parking regulations when responding to an emergency or at an emergency scene.

exempt emergency vehicle drivers from several normal driving regulations. One exemption typically found in most state laws is the ability to park emergency apparatus regardless of normal requirements (**Figure 3-1**). Other exemptions include the ability to pass through red lights or stop signs, and exceeding the maximum speed limit. Finally, emergency vehicle drivers are often exempt from normal direction or movement of travel. Again, these exemptions are typically limited to response to emergencies and are only extended when safely executed.

SAFETY

Keep in mind that it may not be safe to exceed the speed limit either due to weather conditions or the size of the apparatus, which cannot stop as quickly as a car nor turn as sharply as a car. When an accident goes to court, a key question is "was what you were doing safe?" (with the fact that an accident occurred, the first indication is that it was *not* safe). You are likely to lose the court case if the jury feels your actions differed from what a prudent person would have done.

Section 55-8-108 of the Tennessee Code Annotated is provided as an example of a state law related to emergency vehicle drivers:

a. The driver of an authorized emergency vehicle, when responding to an emergency call but not upon returning from a fire alarm, may exercise the privileges set forth in this section, but subject to the conditions herein stated.

b. The driver of an authorized emergency vehicle may:

1. Park or stand, irrespective of the provisions of this chapter;

2. Proceed past a red or stop signal or stop sign, but only after slowing down as may be necessary for safe operation;

3. Exceed the speed limits so long as life or property are not thereby endangered; and

4. Disregard regulations governing direction of movement or turning in specified directions.

c. The exemptions herein granted to an authorized emergency vehicle shall apply only when such vehicle is making use of audible and visual signals meeting the requirements of this state.

d. The foregoing provisions shall not relieve the driver of an authorized emergency vehicle from the duty to drive with due regard for the safety of all persons, nor shall such provisions protect the driver from the consequences of the driver's own reckless disregard for the safety of others.

See **Appendix D** for several other states' laws related to emergency vehicle drivers.

Standards

Two standards of most interest to apparatus operators are established by the National Fire Protection Association (NFPA) (**Figure 3-2**). NFPA 1500, *Fire Department Occupational Safety and Health Program*, identifies minimum requirements for an occupational safety and health program for the fire service. NFPA 1002, *Fire Apparatus Driver/Operator Professional Qualifications*, identifies the minimum job performance requirements for individuals responsible for driving and operating fire department vehicles. Copies of both standards should be available at fire departments.

Risk Management Plan

According to NFPA 1500, departments should have a written risk management plan covering vehicle operations. The plan should identify, evaluate, and provide methods for controlling and monitoring risks. Such a plan includes investigations of all accidents and periodic written analysis reports to the chief. From an operations perspective, this means that operators need to be familiar with the department's plan, including the identified risks and methods to control them.

NFPA 1500 (2007 Edition)

NFPA 1500, **Chapter 6**, "Fire Apparatus, Equipment, and Driver/Operator," discusses minimum safety-related requirements associated with emergency vehicles, drivers of emergency vehicles, and operation of emergency vehicles. Because minimum safety requirements are identified, emergency vehicle

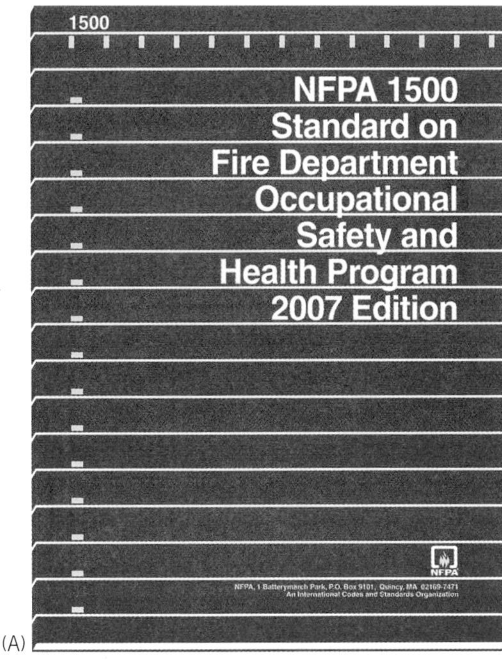

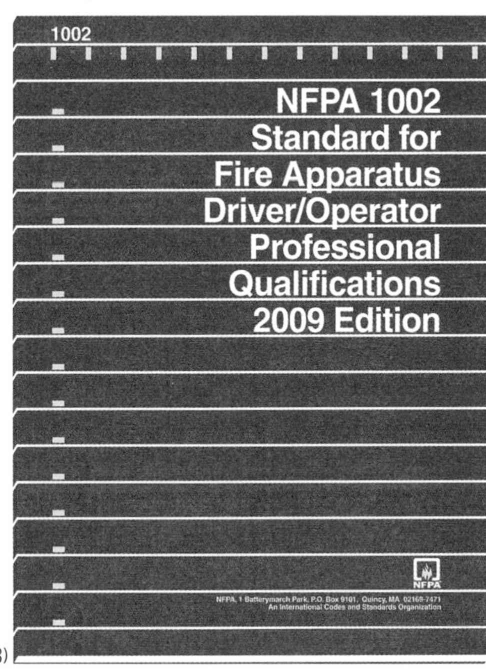

FIGURE 3-2 NFPA standards play an important role in emergency vehicle safety.

drivers should be familiar with this standard. In addition, the standard may carry the weight of law in that the standard can be viewed as the legal minimum standard in court. The standard, therefore, may be used to judge a driver's action or inaction in judicial proceedings. Finally, the standard will carry the weight of law if adopted as such by a local government.

Emergency Vehicles

According to NFPA 1500, safety and health should be considered when purchasing, maintaining, and operating emergency response vehicles. In addition, new fire apparatus, wildland fire apparatus, and marine firefighting vessels shall be specified to meet their respective NFPA standard (1901, 1906, and 1925). Refurbishing fire apparatus shall meet the applicable requirements of NFPA 1912. These standards add a number of safety items, such as reduced braking distances compared to commercial vehicles of the same size and weight.

Emergency Vehicle Drivers

NFPA 1500 requires that emergency vehicles be operated by qualified individuals. First, emergency vehicle drivers must possess a valid driver's license. Because emergency vehicle driving regulations vary, the type of license required will depend on state and local requirements. In addition, emergency vehicle drivers must meet the appropriate requirements established in NFPA 1002. Finally, emergency vehicle drivers must complete a driver training program prior to operating emergency vehicles (**Figure 3-3**).

Operating Emergency Vehicles

NFPA 1500 assigns responsibility for the safe operation of the vehicle under all conditions to the driver. The standard requires that emergency vehicle drivers obey all normal driving laws when not responding. Although this may sound too basic to be mentioned, drivers may neglect or forget this rule during

FIGURE 3-3 According to NFPA 1500, emergency vehicle drivers must complete a driver training program prior to operating emergency vehicles.

nonemergency operations. To reinforce this basic rule and to assist further in the safe operation of emergency vehicles, standard operating procedures (SOPs) should be developed for both emergency and nonemergency operations. The first priority of emergency-response-related SOPs should be the safe arrival of the fire apparatus to the incident. These SOPs should include, as a minimum, specific criteria or requirements for speed, crossing intersections, school zones, traversing railroad crossings, and use of emergency warning devices.

Safety hazards encountered during emergency response are significant.

The safety hazards encountered at intersections during emergency response are significant. Therefore, NFPA 1500 requires that emergency vehicle drivers come to a complete stop when any intersection hazard is present. Specifically, the standard requires that drivers bring the apparatus to a complete stop when any of the following exists:

- As directed by a law enforcement officer
- At red traffic lights or stop signs
- At negative right-of-way and blind intersections
- When all lanes of traffic in an intersection cannot be accounted for
- When a stopped school bus with flashing warning lights is encountered
- At any unguarded railroad grade crossing (also for nonemergency)
- When other intersection hazards are present

Riding on Apparatus

NFPA 1500 also addresses the safety of those riding on the apparatus. The standard requires that all personnel riding on the apparatus be seated and properly secured with seatbelts prior to moving the apparatus and whenever the vehicle is in motion (**Figure 3-4**). The only exceptions to this requirement are for patient treatment, the loading of hose, and tiller driver training. While the vehicle is in motion, seatbelts are not to be released or loosened at any time, including while donning self-contained breathing apparatus (SCBA) or personal protective equipment (PPE). New NFPA recommendations state "Fire helmets shall not be worn by persons riding in enclosed driving and crew areas." Not wearing fire helmets reduces strains in the event of an accident and could avoid a broken neck. Finally, personnel riding in open cab apparatus not enclosed on at least three sides and the top should wear head and eye protection.

FIGURE 3-4 Firefighters must be seated and properly secured whenever the apparatus is in motion.

Current seatbelts must be either bright orange or red and a warning device visible to the driver must indicate the status of each seat and belt. Proper use of seatbelts in all vehicles prevents injuries and deaths.

NFPA 1002

NFPA 1002 identifies the minimum requirements for fire department vehicle drivers. The standard is divided into general requirements and apparatus-specific requirements. General requirements include those items that all drivers must meet. Apparatus-specific requirements focus on those items associated with the main function of the apparatus.

General requirements focus on the safe driving of apparatus. Specifically, drivers must be licensed for the apparatus they are expected to operate. In addition, all drivers are subject to medical evaluation requirements as identified in NFPA 1500. Further, drivers must demonstrate the ability to conduct and document general vehicle inspection, testing, and servicing functions. Finally, drivers must demonstrate the ability to maneuver the apparatus safely over a predetermined route in compliance with all applicable state, local, and department rules and regulations.

THE DRIVER AND APPARATUS

The basic tools used to carry out the duty of driving apparatus are, of course, the driver and the apparatus (**Figure 3-5**). Although all drivers are not the same, there are several characteristics common among the

FIGURE 3-5 Safe drivers understand their own limits as well as the limits of their apparatus.

best drivers. Fire department apparatus also vary, yet have several common characteristics. Understanding these characteristics and the influences that affect them is vital to the safe operation of fire department apparatus.

NOTE

The publisher realizes that many good reasons exist for apparatus drivers to either wear or not wear turnout gear when driving to an emergency and that different departments have established different policies based on which of the reasons are most applicable to their department. Since a consensus standard does not exist, photographs in this book show both drivers in turnout gear and drivers in station gear. You should follow SOPs/SOGs applicable to your department even if they appear to differ from photos shown in this book.

Driver

What makes drivers react in a safe and appropriate manner? The answer to this question could be a rather long list of specific items; however, four broad characteristics will be discussed to present the important common characteristics:

- Knowledge and skill
- Attitude
- Physical and psychological fitness
- Maturity and responsibility

First, apparatus drivers must possess appropriate knowledge and skill for the position they hold. Knowledge provides the ability to think through situations and identify courses of action grounded in safety. Skill provides the ability to execute the course of action in a safe manner. Departments should make an Emergency Vehicle Operators Course part of the requirements for obtaining the position of operator. To maintain the operator's skills and knowledge at a sufficient level, this course should be repeated at predetermined intervals. For example, all apparatus drivers must understand and be knowledgeable of intersection crossing considerations to evaluate potential hazards and identify a safe course of action, as well as to be able to execute the plan skillfully by maneuvering the apparatus safely through the intersection. When drivers are knowledgeable and skillful in their jobs, they are better able to cope with situations in a safe manner.

Second, the best drivers have a good attitude toward safety. When this characteristic is present, safety becomes an integral part of every action by the driver. Safety becomes a habit with a good attitude toward safety. Drivers with a good attitude are not overconfident of their ability to drive emergency vehicles; rather, competent drivers show respect for, and knowledge of, the vehicle's limitations.

Third, drivers must be physically and psychologically fit. The stress and rigors associated with driving fire apparatus can take their toll on even the best drivers. Being physically and psychologically fit will help reduce the negative effects of stress. Being physically fit allows the driver to control the vehicle and to react quickly and appropriately. Being psychologically fit ensures that the driver is alert and can concentrate and think clearly.

Finally, emergency vehicle operators should be mature and responsible. Mature individuals have a balance of knowledge, skill, and experience and are able to utilize them in decision making and actions they take. Responsible individuals internalize their accountability for safety to themselves as well as to others. The potential risk to emergency personnel and to the public dictate that this characteristic be present in all emergency vehicle drivers.

Apparatus

In much the same way as drivers will react differently in a given situation, different apparatus will react differently in a given situation; therefore, drivers should also be familiar with their apparatus. Specifically, operators should be familiar with the physical characteristics (**Figure 3-6**) of the apparatus, such as length, height, width, and weight. Driving over bridges (**Figure 3-7**), through tunnels, and around traffic will be affected by these physical characteristics. Drivers should also be familiar with the condition of the apparatus. Across the country an

Delmar/Cengage Learning

FIGURE 3-6 The physical characteristics of individual apparatus must be considered when driving the apparatus.

Delmar/Cengage Learning

FIGURE 3-7 Drivers should know the weight of their vehicles and the weight limits of bridges within their jurisdiction.

alarming trend has emerged involving large-capacity tankers that have been involved in accidents. Many of these vehicles were found to have a record of poor maintenance or construction, were over the weight limit, and driver training had been minimal or nonexistent. Familiarization with the apparatus and adherence to NFPA standards is the key to reducing these types of accidents. Vehicles should be routinely inspected to detect and correct any deficiency that may create unsafe driving conditions. Daily inspections provide the opportunity to assist in the safe operation of the vehicle. Drivers should also be familiar with the handling characteristics of the vehicle, such as turning, braking, and shifting. For example, large-capacity tankers will handle differently than a mini-pumper when it comes to stopping and turning. Finally, drivers should also be familiar with vehicle control systems and emergency warning systems.

Vehicles should be routinely inspected to detect and correct any deficiency that may create unsafe driving conditions.

THE ROLE OF OFFICERS

Safety is everybody's responsibility: the driver's, the officer's, the crew's, and the public's, with the operator having the primary responsibility for the apparatus and the officer being responsible for the operator's actions. The relationship between the driver and officer has both a legal and practical aspect. Legally, the officer shares in the safe operation of the apparatus with the driver. Both can be held accountable for unsafe actions. In effect, the officer and driver act as a team to ensure that safety is considered in every aspect of driving. As a team, safety can be enhanced by:

- Confirming that the crew is seated and properly secured
- Identifying and communicating hazards to each other
- Allowing the officer to operate warning devices, radios, and **mobile data terminals (MDTs)** while the driver concentrates on driving
- Evaluating high-hazard areas together, such as intersections and railroad crossings

SAFE-DRIVING CONSIDERATIONS

An important requirement of all drivers is to keep the apparatus under control at all times. Even small mistakes may cause drivers to lose control of their apparatus quickly. One common mistake occurs when an apparatus is operated as if it were a personal vehicle. This happens because of habits formed while driving personal vehicles. Drivers subjected to stress or who are preoccupied will compensate by relying on habits. Another common mistake is the failure to maintain an appreciation for the unique driving requirements and characteristics of large vehicles. Fire apparatus tend to be large, heavy vehicles requiring constant vigilance toward safety to ensure that control is maintained. Finally, the failure to change driving techniques adequately to compensate for changing driving conditions is another common error.

Maintaining control of fire apparatus requires an understanding of basic safe-driving concepts and techniques, as discussed next.

Defensive Driving

Drivers must understand the need to operate their vehicles defensively. In essence, defensive driving includes a constant awareness of operating vehicles safely to avoid accidents. Key elements for defensive driving include anticipating and planning for:

- The limitations of operators and their apparatus
- Both appropriate and inappropriate actions of other drivers on the road
- The effects of speed, braking, and weather conditions
- Possible hazards while driving
- Adequate distances from other vehicles with regard to braking and turning
- The need to yield the right of way

Driving Preparations

Before driving an emergency vehicle, several important preparations must be accomplished. At the beginning of each shift, the driver should conduct a preventive maintenance inspection as discussed in **Chapter 2**. This is also a good time for the driver to adjust mirrors, seats, and restraints and look over all in-cab instrumentation and devices. In short, the driver should prepare the vehicle for immediate use. When an alarm is received, the driver should walk around the vehicle, checking for obstructions and disconnecting all ground lines such as battery chargers, heaters, and exhaust systems (unless they disconnect automatically). The vehicle should be started as soon as possible to allow the engine to warm up and in some cases build air pressure. The vehicle's operating guide and SOPs should be followed when starting the engine. After the engine starts, all dashboard instrumentation, such as oil and air pressure, engine RPM, and ammeter, should be checked for normal operation indication. The driver is responsible for ensuring everyone is seated with fastened seatbelts and that bay doors are fully open. Finally, the parking brake is disengaged, the transmission is placed in road gear, and the accelerator is slowly depressed. Automatic transmissions do not normally require the operator to shift gears while the apparatus is in motion. Automatic transmissions do, however, allow the operator to set the highest gear the transmission will use. Unusual road conditions such as steep declines should use a lower top gear, which allows the engine and transmission to aid in apparatus braking, thus reducing the chances of a runaway apparatus due to brake fade. Snow-covered roads should also use a lower top gear and reduced acceleration and deceleration to prevent loss of **traction** (friction between the tires and the road). Some vehicles also have an alternate shift schedule, which may, for example, provide a higher gear for highway driving, which is selected with the transmission mode button. Drivers should follow departmental SOPs and apparatus operators' manuals to determine when different gears should be used on automatic transmissions.

New apparatus include a seatbelt monitoring system that will notify the operator of which seatbelts are not fastened. The status is recorded in a "blackbox" that can later be reviewed by management. Drivers should expect the data will be subpoenaed for any possible legal actions resulting from an accident.

Manual transmissions require that the gears be changed as they reach the top of their bands. Operation of a vehicle with a manual transmission should be done according to the manufacturer's and department's recommendations. The operator should be trained and tested for competency prior to operating an apparatus with this type of transmission. Attention should be given to ensure the vehicle has cleared the bay doors prior to initiating a turn. Bay doors should be operated by the apparatus operator so that the doors are not activated until the correct time. Failure to do so may result in damage to the apparatus, the bay doors, or the bay opening.

The faster the apparatus travels, the greater the distance it takes to stop.

Braking Systems

Brakes are perhaps the most important safety system on an apparatus. Without them, it would not be possible to avoid an accident or stop the apparatus. To make sure that they will work, drivers must understand their capabilities and limitations.

The braking distance to stop a vehicle goes up with the square of the change in speed. This means that an apparatus meeting NFPA requirements and needing 35 ft (10 m) to stop at 20 MPH (32 KPH) will need 140 ft (43 m) to stop at 40 MPH (65 KPH) and 315 ft at 60 MPH (96 m at 100 KPH). All of these distances will increase under slippery conditions. For example, they may be nine times as long on icy roads. When the brakes are used, the friction they create converts the kinetic energy of the vehicle's motion into heat energy. If the brakes are used too much, the heat they create actually causes melting. As anyone who has stepped on a wet bathroom floor knows, if a liquid separates two solids (the floor and your feet), it is difficult to get much friction. With braking, it's called brake fade and results in the vehicle not slowing even though the brake pedal is depressed.

Two systems that help with braking are auxiliary braking systems and **anti-lock braking systems (ABSs)**. Both are typically installed on newer full-sized apparatus. Auxiliary braking systems include engine brakes or exhaust brakes such as a Jake Brake, transmission retarders, and drive line retarders, all of which reduce the amount of work the service brakes need to perform. While each auxiliary braking system has different characteristics and functions differently, they all have the same end result of reducing the RPMs of the drive shaft and wheels. The minimum stopping distance is achieved by slowing the wheels as fast as possible without leaving skid marks. Auxiliary braking systems do not change this but do help by reducing the amount of work the service brakes perform, hence keeping them cooler and lengthening the time before fade occurs. Since a skid reduces braking efficiency and vehicle control, use of an auxiliary braking system is not recommended when slick road conditions, which could cause a skid, exist. All auxiliary braking systems have a switch, reachable from the driver's position, to enable or disable them.

ABSs work in another way. They monitor wheel rotation to determine if a wheel has locked up (skid condition), which, as previously noted, reduces braking efficiency and vehicle control. When a wheel lock-up has occurred, the ABS will change from passive mode to active mode. In active mode, the ABS light on the driver's console will illuminate and the ABS will momentary release the brakes (which is also called pump the brakes) on the locked wheel. The repeated momentary brake release prevents skids, adds vehicle control, and reduces stopping distance. Many ABSs also disable the auxiliary braking system when they enter active mode to prevent the systems from working against each other.

Vehicles with ABS will contain an ABS light that will come on momentarily when the vehicle is started. If the system is working correctly, the light will go out. If a problem exists with the system, the light will stay on and normal braking will still work. When braking on vehicles without ABS, the driver should pump the brakes to avoid a skid. On vehicles with ABS, the driver should firmly press the brakes and allow the ABS system to pump the brakes when needed.

The latest systems to help are **roll stability control (RSC) systems** and **enhanced roll stability systems**. In addition to the preceding functions, these systems recognize when a vehicle is likely to roll over and can independently control power and braking to individual wheels in such a way as to maximize the chances of the apparatus responding as the operator desires while minimizing rollovers.

While each of these systems helps with vehicle control, none of them can violate the laws of physics. It still takes more space and energy to stop a truck than to stop a car. If the driver has not driven appropriately, it will not be possible to prevent an accident.

Drivers should know which of the preceding systems exist on the apparatus they drive since this will affect what they should do if the systems become active and under certain road conditions.

The only item that can absolutely prevent rollovers and accidents is for the driver to practice safe driving techniques. Items such as roll stability systems, auxiliary braking systems, and automatic traction control (ATC) all help, but you are the most important safety device!

Speed

Speed affects safe driving in two important ways: it affects the ability to stop the apparatus, and it affects the ability to steer the apparatus.

Speed and Braking

Simply stated, the faster the apparatus travels, the greater the distance it takes to stop. Doubling apparatus speed increases the stopping distance an estimated four times. Whenever possible, reduce speed to decrease stopping distance.

Total stopping distance is measured from the time a hazard is detected until the vehicle comes to a complete stop. Total stopping distance consists of perception distance, reaction distance, and braking distance (**Figure 3-8**). **Perception distance** is the distance the apparatus travels from the time the

55 MPH
Complete Stop
60' Perception Distance
60' Reaction Distance
170' Braking Distance

Delmar/Cengage Learning

FIGURE 3-8 Total stopping distance for an apparatus traveling at 55 MPH (90 KPH) on dry pavement is approximately 290 feet (88 metres).

hazard is seen until the brain recognizes it as a hazard. For an alert driver traveling at 55 MPH (90 KPH), the time it takes to perceive the need to stop is about ¾ of a second. During this time, the vehicle will have traveled approximately 60 feet (18 metres). **Reaction distance** is the distance of travel from the time the brain sends the message to depress the brakes until the brakes are depressed. During this time, the apparatus will have traveled another 60 feet (18 metres) in ¾ of a second. **Braking distance** is the distance of travel from the time the brake is depressed until the vehicle comes to a complete stop. During this time, the vehicle will have traveled approximately 170 feet (52 metres) in about 4½ seconds. Thus, the total stopping distance for a heavy vehicle with good brakes traveling 55 MPH (90 KPH) on flat, dry, and hard pavement is about 290 feet (88 metres), approximately the length of a football field (60 + 60 + 170 = 290) (18 m + 18 m + 52 m = 88 m) in 6 seconds (¾ + ¾ + 4½ = 6).

The stopping distance noted occurs when the brakes are applied properly. If the brakes are improperly applied, traction is lost. The result is both an increased stopping distance as well as loss of steering control. Road conditions that affect traction and, consequently, stopping distance and steering include:

- Just after it starts raining (water/oil mixtures)
- Standing water (potential for hydroplaning)
- Snow (packed snow and ice formed from melting snow)
- Bridges (ice will form sooner on bridges than on other road surfaces)
- Black ice (a thin clear layer of ice that makes the road appear wet)

Speed and Steering

Steering is a function of traction. When a vehicle's speed exceeds the ability of its tires to hold traction on the road, steering is virtually impossible. The road conditions previously mentioned can reduce traction, necessitating slower speeds to maintain steering control. Speed can also affect steering when entering a curve. Excessive speed in curves can cause the tires to lose traction, especially if the brakes are applied. When tires maintain traction and the speed is excessive in a curve, the vehicle may overturn.

Controlling Skids

Skidding occurs when tires lose traction against the road surface. Road conditions do not cause skidding. Rather, excessive speed, improper braking, improper tire or tread wear, or improper steering cause skids. It should be understood that skidding can occur on any road surface, including dry surfaces. When an apparatus begins to skid, it is out of control. Immediate and proper action is required to regain traction and control of the apparatus. The following can be used to help regain control:

- React quickly but do not panic or overreact.
- Take your foot off the accelerator and let the engine slow the apparatus.
- Do not disengage the clutch or slam on the brakes.
- Apply brakes slightly but at least to the extent that wheels no longer turn.
- Turn the front wheels in the direction of the skid (**Figure 3-9**).

FIGURE 3-9 Turning the wheels into a skid will help regain control of the apparatus.

Centrifugal Force and Weight Shifts

The effects of both centrifugal force and weight shifts (weight transfer) can adversely affect the ability to maintain safe control of an apparatus. Centrifugal force, the tendency to move outward from the center, occurs when apparatus navigate a curve. The key to navigating a curve safely is to maintain traction. When traction is lost, the apparatus will be out of control and will move in an outward direction (**Figure 3-10**). When apparatus change direction of travel (such as navigating curves) or when they change velocity (sudden stopping), a shift in weight occurs on the apparatus. These shifts can be severe enough to cause loss of control or even overturn the apparatus. Recall that a body in motion will tend to stay in motion. Consider a 1,000-gallon (4,000-liters) water tanker traveling at 50 MPH (80 KPH). If the brakes are suddenly applied to stop the vehicle, the on-board water will attempt to stay in motion (**Figure 3-11**). In essence, the weight of water shifts in the direction of travel as it attempts to stay in motion. The result is that it may take a substantially longer distance to stop the vehicle. In addition, the weight shift may cause the vehicle to roll over when taking a curve at too great a speed.

The key to navigating a curve safely is to maintain traction.

Backing Apparatus

Improper backing accounts for an overwhelming number of avoidable accidents. In general, backing should be avoided whenever possible. When backing an apparatus is necessary, always have a guide or **spotter** and mirrors. Some departments require at least one spotter while others require two, one in front and one in the rear. In either case, the driver

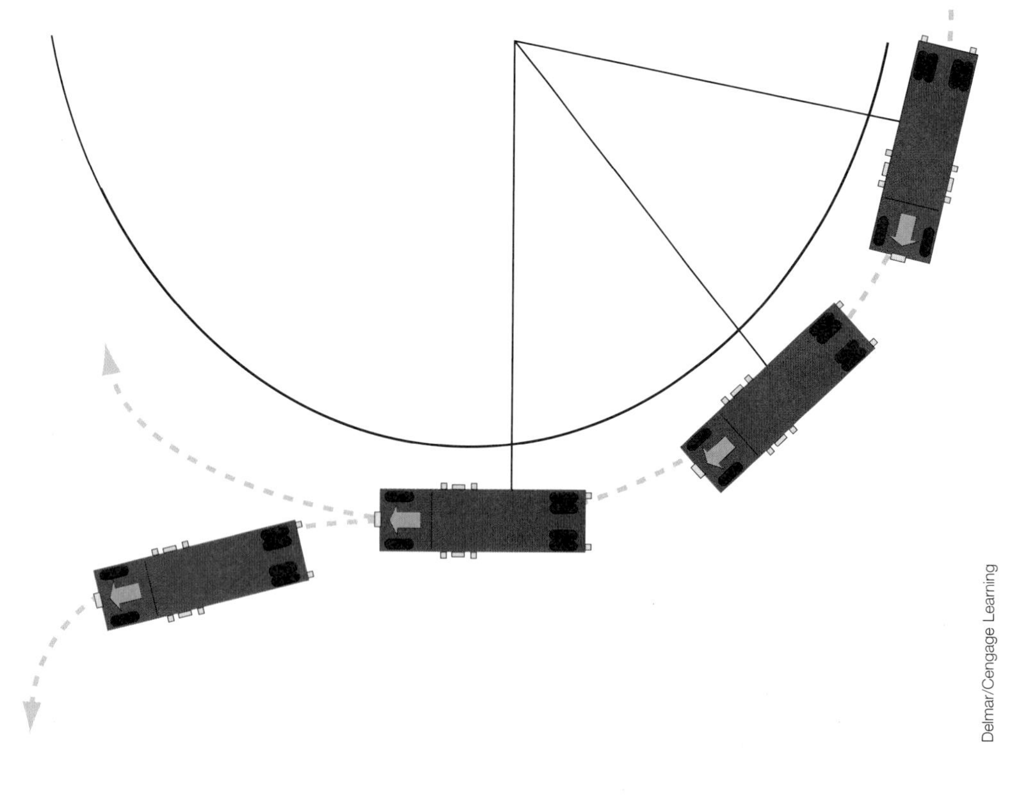

Delmar/Cengage Learning

FIGURE 3-10 When traction is lost in a curve, centrifugal force will move the apparatus in an outward direction.

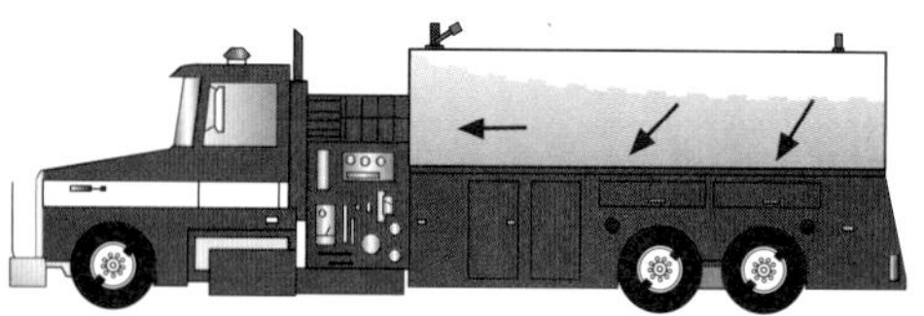

When the tanker stops abruptly, the water attempts to stay in motion.

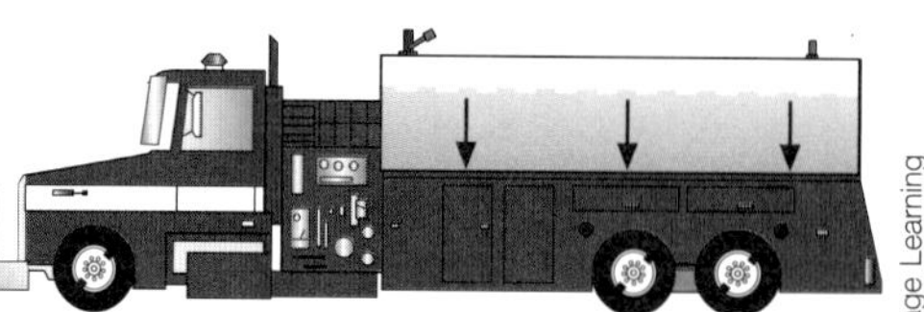

Traveling at a steady speed, the weight of water is exerted evenly on the apparatus.

Delmar/Cengage Learning

FIGURE 3-11 When the tanker stops, the on-board water attempts to stay in motion.

should maintain visual contact at all times with the spotter. The spotter should have a portable radio and the driver should also have radio contact with the spotter during the backing operation. If visual contact is lost, the driver should immediately stop the apparatus. The use of guides, however, does not relieve the driver of the responsibility for safely backing the apparatus.

Parking Apparatus

Drivers must ensure control of their apparatus when in a stationary position through proper parking techniques. Proper parking of apparatus is essential to guard against inadvertent rolling on a grade caused by loss of the parking brake or incorrect engagement of the pump. One way to maintain control of a stationary apparatus is to chock (or block) the wheels. NFPA 1901 requires two **wheel chocks** to be mounted and readily accessible on each apparatus. Apparatus should be chocked (**Figure 3-12**) when the apparatus remains in a stationary (parked) position for any length of time. Further, chocking the wheels should be the first thing taken care of when the driver, or other crew member, exits the cab. If parked on an incline, the two chocks should be placed on the downhill side of both rear wheels. NFPA 1901 requires both the chocks and the parking brake to be able to hold an apparatus on a 20% grade. When on a level grade, the two chocks can be placed on both sides of a rear tire. Chocks should be placed snug against the tire. The driver is responsible for making sure that wheel chocks are removed and properly stowed prior to moving the apparatus.

A second method used to ensure the safe control of a stationary apparatus is to align the front wheels properly. When parked next to a curb, the rotation of the front wheels should point toward the curb. When no curb is present, the front wheels should be positioned to roll the apparatus away from the road (**Figure 3-13**). When moving the apparatus from a stationary position, the driver must remember the alignment of the front tires to avoid hitting the curb.

Environmental Conditions

Laws governing the rules of the road are established for ideal conditions, that is, dry roads and good visibility. Drivers must make appropriate adjustments when adverse environmental conditions are encountered. Challenging environmental conditions affect driving by reducing visibility and traction. Limited visibility reduces the chances of early hazard recognition, while reduced traction affects the ability to steer and stop the vehicle. Examples of conditions that affect vision and/or traction are:

- Night driving
- Fog or rain
- Snow and ice
- Glare
- Wind

Night driving may be one of the more common of the difficult environmental conditions a driver will face. Difficulties include both the lack of light, reducing visibility, and newer and brighter emergency

Delmar/Cengage Learning

FIGURE 3-12 Drivers should chock the wheels when the apparatus will be stationary for any length of time.

FIGURE 3-13 To further assist with safely controlling an apparatus, drivers should properly align wheels when parked near curbs.

warning lights, which may temporarily blind motorists, resulting in unusual driving. The apparatus operator should drive with normal care at night, making sure to take into account the difference in visibility. Mechanical devices found on apparatus, such as engine, transmission, and drive line retarders used to assist brakes with slowing and stopping, can both assist and hinder traction under adverse environmental conditions. Drivers should be familiar with manufacturers' recommendations for their use and operation. Automatic chain devices provide the increased traction of snow chains on an as-needed basis. These devices are especially helpful when snow and ice conditions are encountered while driving the apparatus (**Figures 3-14A** and **3-14B**).

Limited Clearance

Limited clearances exist in some places in most jurisdictions. It may be back alleys in the old downtown business district, small single-lane roads in the outlying part of the district, or a long private driveway. (Due to the number of problems with private driveways, some departmental SOPs do not allow fire apparatus to be driven up private driveways.)

When it is necessary to drive in an area with limited clearances, the use of a spotter can significantly reduce the chances of damage to either the apparatus or private property. While the spotter may be held responsible for some accidents, remember that the driver is *always* responsible. Only the driver causes the vehicle to move or to stop; therefore, drivers and spotters must work as a team using the same signals. **Figure 3-15** provides examples of common hand signals that can be used. Radio signals may also be used. Also, the driver should never move the apparatus if the spotter(s) cannot be seen. The last thing anyone wants is to find out a spotter fell under the apparatus and was then run over.

At night, some means of making sure the spotter is seen is necessary. Either lit traffic wands (**Figure 3-15G**) can be used or a light lighting the spotter can be used. To avoid blinding the driver, lights should not be directed at the driver or apparatus mirrors.

Road Considerations

Although a driver can do little to improve the surface of a road, maintaining constant awareness of road surface conditions and reacting appropriately will help ensure the safe operation of emergency vehicles. Road conditions that require attention include:

- Sharp curves
- Severe grades
- Soft or limited road shoulders
- Unguarded railroad crossings
- Poor pavement (**Figure 3-16**)
- Center crowns and curves that are not banked
- Very narrow roads

Each of these conditions can cause the driver to lose control if appropriate action is not taken. In most cases, this means recognizing that the condition exists and reducing the speed of the vehicle.

Drivers must also become familiar with any unimproved or unpaved roads within their jurisdiction. Specifically, drivers must know which roads are safe to use given the weight, width, and length of emergency vehicles. Care must be taken while driving on these surfaces in that loose dirt, gravel, ruts, potholes, gulleys, uneven grade, and other deformities will decrease the ability to maintain safe control of emergency vehicles. The safest thing to do is not

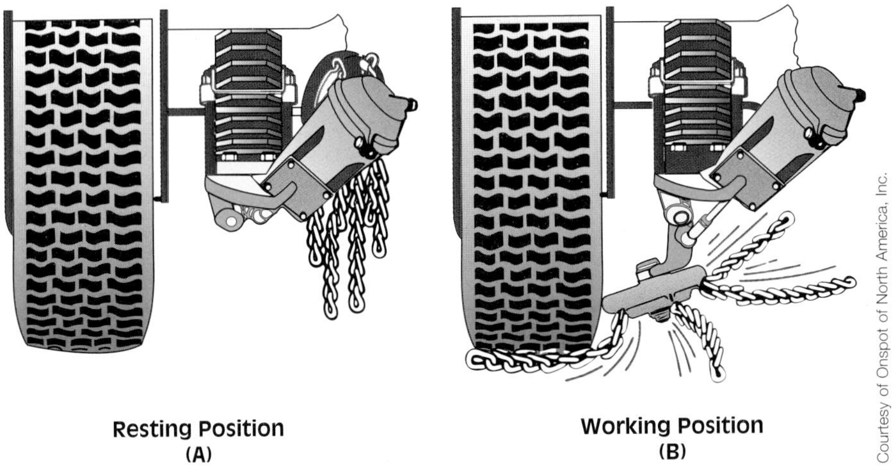

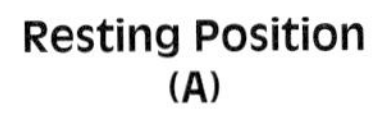

FIGURE 3-14 An automatic chain system allows the driver to engage snow chains while the apparatus is in motion. (A) In the resting position, the chains are lifted into the air and do not touch the tires or pavement. (B) In the working position, the chains swing down and the chain wheel rests on the tire sidewall. As the tire turns the chain wheel, centrifugal force causes the chain segments to be thrown in front of the tire. The tire then runs over the chain, thus increasing traction on a snow-covered surface.

to drive on unimproved road surfaces. When it is necessary to drive on these surfaces, the driver should reduce considerably the operating speed of the vehicle.

Preventing Rollovers

Rollovers are responsible for the largest proportion of apparatus-related firefighter fatalities, with three common factors: excessive speed, inexperienced drivers, and not taking appropriate corrective actions. Being prepared to make the correct maneuvers can allow you to prevent a catastrophe. For example, if one of the tires drops off the road, the driver needs to change response modes. Continuing to respond in an emergency response mode greatly increases the chances of never making it. Instead, the driver needs to change to a recovery mode in which his or her only actions are aimed at getting the apparatus back to a normal driving condition. Only after everything has returned to normal should consideration be given to resuming an emergency response mode.

A common incorrect attempted recovery from dropping a tire off the road is to attempt to quickly get back on the road by turning the wheels sharply toward the road. In this case, as the apparatus reenters the road, with its better traction, an overcorrection often occurs, with the apparatus going off the other side of the road. As another attempt to reenter the road quickly is made, apparatus tank water begins surging from side to side, making the apparatus less stable and leading to a rollover.

A preferred approach is to remove your foot from the accelerator, allowing the apparatus to gradually slow while adjusting the steering wheel slightly to ease the tires gradually back onto the road.

A less common and certainly less intuitive maneuver is for a front tire blowout. Should a tire blowout occur, the apparatus will pull to the side of the blowout. It's important to keep the apparatus within its lane, avoiding an accident with traffic in another lane. Having two hands on the steering wheel will provide better steering control than with a single hand. Though unexpected, the preferred approach is pressing the fuel pedal to the floor. When done just for a few seconds until you get the steering under control, it helps to keep the apparatus moving straight and does not result in a significant speed change. Once the steering is under control, remove your foot from the fuel pedal and let the apparatus coast to a stop and move onto the shoulder if one exists. Hard braking should be avoided since it greatly increases the chance of skidding into another lane because of the traction differences resulting from the blown-out tire.

DRIVER OPERATOR TESTING

The evaluation of drivers/operators should include both a written test and a skills assessment. In general, the written test should cover those items within NFPA 1002 listed as prerequisite knowledge, and the skills assessment should include those items listed as prerequisite skills. Several driving exercises are typically used to assess the ability to operate and control the vehicle safely. These skills typically include both closed course and road skills.

STREETSMART

Completion of an Emergency Vehicle Operators Course (EVOC) provides an excellent foundation upon which the driving skills necessary to successful master various emergency vehicle driver maneuvers can be built.

(A)

(B)

(C)

(D)

(E)

(F)

(G)

FIGURE 3-15 Typical spotter hand signals: (A) Stop is indicated by showing a clear "X." If only one hand is available, a raised fist is used. (B) "Back to my right." The wand in the left hand is moved from straight up to back at a 45-degree angle. Stop if the left-hand wand is held stationary and straight up. (To move forward and to my right, the wand in the left hand would move from straight up to a 45-degree forward.) (C) "Back to my left." The wand in the right hand is moved from straight up to back at a 45-degree angle. Stop if the left-hand wand is held stationary and straight up. (To move forward and to my left, the wand in the right hand would move from straight up to a 45-degree forward.) (D) "Slow down" is indicated by slowly moving the wands down. Without wands, this is done with palms facing down. (E) "Straighten out." Back when the wands are moving from the 45-degree angle back and move forward when the wands are moving from the 45-degree angle forward. (F) Available clearance. (G) Effective night-time signals require the driver to see the spotter's signals clearly. If lit traffic wands are not available, the spotter may be illuminated with a hand light. Lights should not be directed at the driver or the apparatus mirrors since that could reduce the driver's night vision.

FIGURE 3-16 Poor road conditions may impact the safe operation of emergency vehicles.

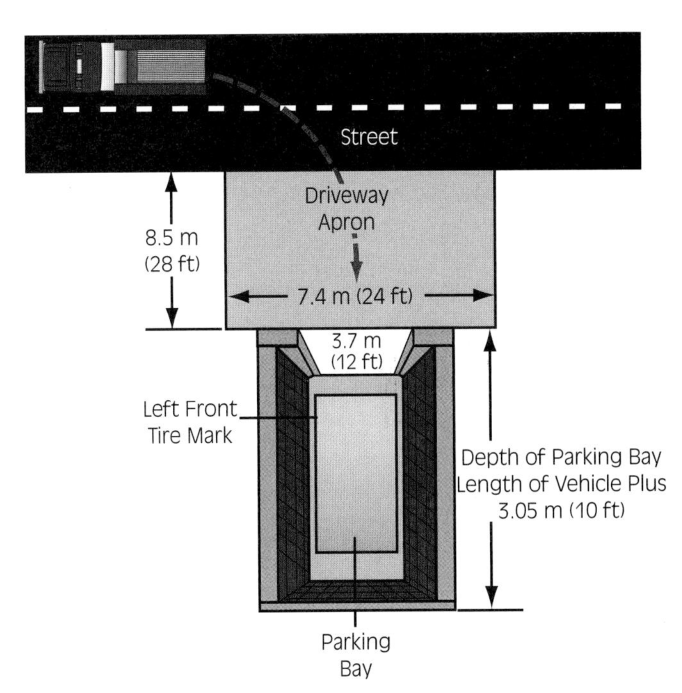

FIGURE 3-17 Station parking.

Closed Course

The closed course scenarios are designed to show the operator's ability to maneuver within a confined space with critical steering, backing, and mirror use. Before starting, drivers should therefore perform an apparatus safety check, including mirror adjustment, and make sure they understand the course route. The key to completing these scenarios successfully is to understand the apparatus position and how to get the apparatus to move along the desired track. Like many skills, start by learning the correct procedures. After learning the procedures, work on reducing the time to perform these scenarios while continuing to perform them correctly. All of these skills should be practiced in a secured area, such as a closed parking lot or drill lot, not on a public street.

These exercises include the following:

- *Station parking.* This is used to assess the ability to steer the apparatus in close limits without stopping. This maneuver (**Figure 3-17**) is designed to simulate the procedures necessary to back into a station. Like other procedures, where possible, backing should be avoided, particularly on-street backing. It is the reason why so many new stations are designed as drive-through stations. All drivers, however, should be able to perform this procedure even if they are assigned to a drive-through station.

To aid with the maneuver, drivers should establish **landmarks** and use one or more spotters where possible. Stations may also add aids such as positioning guidelines. Prior to performing this procedure, make sure no one is in your blind spot.

SAFETY

The vigilant driver should always be aware of whether anyone is in the vehicle's blind spot. This is done by constantly scanning well-adjusted mirrors, keeping track of all nearby vehicles, and noting whenever one disappears into your blind spot. Confirmation of who is in your blind spot can be made by checking the mirror on the inside of a curve when entering curves and using clues such as vehicle shadows during the day and headlight illumination at night.

Perform the procedure by pulling past the parking lane (by at least the turning radius of the apparatus). Turn the wheels and back into position. Have a landmark on the rear wall of the station to aim for. Once it has come into view, only minor steering corrections are needed.

(For step-by-step photos of Station Parking, please refer to page 85.)

- *Alley dock.* The alley dock maneuver (**Figure 3-18**) simulates backing into a restricted space, such as down an alley or into a loading dock, and is essentially the same as the station parking maneuver. The maneuver should be practiced when approaching the restricted space from both the right and the left. The driver also needs to judge the distance behind the apparatus so that it can be stopped smoothly just before a barricade, such as a loading dock.

(For step-by-step photos of Alley Dock, please refer to page 86.)

- *Serpentine.* This maneuver assesses the ability to drive around obstacles such as parked cars and tight corners. The key to maneuvering through the serpentine (**Figure 3-19**) is to know where the rear tires are relative to the cone being passed.

  Drive forward on one side of the serpentine. When backing, look in the side mirror and back until the rear tire passes a cone. Look in the other side mirror and turn the steering wheel sharply so that the apparatus will turn and clear the next cone. Since the tire has passed the cone, it is no longer possible to hit it. If you do not head for the next cone right after clearing a cone, there will not be adequate space to make the turn. As soon as the next cone comes into view, straighten the wheel and continue backing until the rear wheel passes that cone. Repeat the procedure for each successive cone.

  When driving forward, as soon as the rear tire passes a cone, turn the wheel sharply toward the next cone. Since the tire has passed the cone, it is no longer possible to hit it. If you do not head for the next cone right after clearing a cone, there will not be adequate space to make the turn. Once you can clear the next cone, straighten the wheels and continuing driving forward until the rear tire clears the cone. Repeat the procedure for each additional cone.

(For step-by-step photos of Serpentine, please refer to pages 87–88.)

- *Confined space turnaround.* This maneuver assesses the ability to turn the vehicle around within

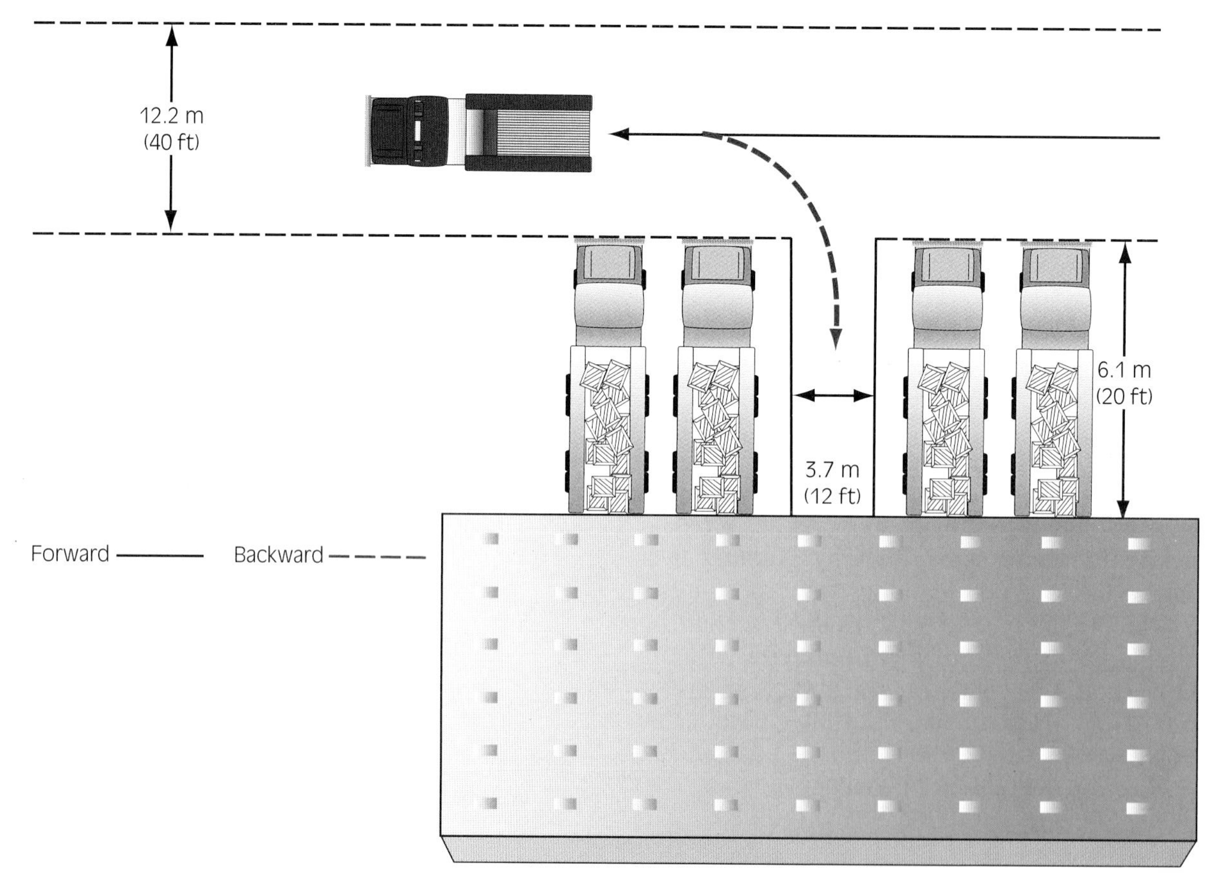

FIGURE 3-18 Alley dock.

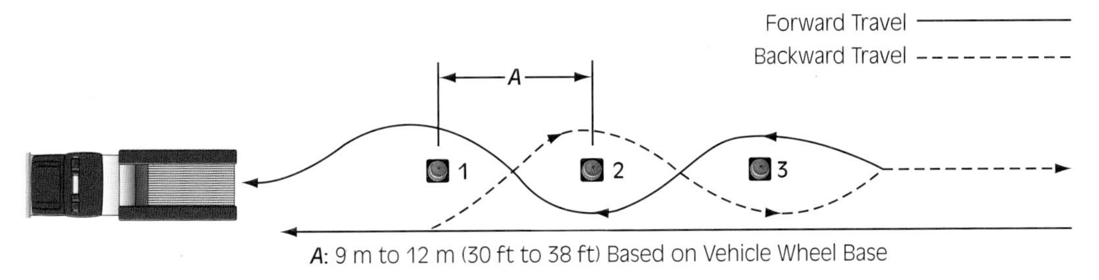

FIGURE 3-19 Serpentine.

a confined space such as a narrow street or driveway. The confined space turnaround is similar to an automotive three-point turn, with the significant difference of a longer vehicle. This maneuver may be necessary to get out of a subdivision with no outlet, but like all operations involving backing, it should be avoided where possible. Preferred methods include going around the block. Drivers should avoid allowing the tires to leave the pavement since other surfaces, including some shoulders, are soft and residential driveways may not be able to support the weight of a fire apparatus. For areas with no outlet, it may be possible to back into a side street to turn around.

In a confined space such as the one shown in **Figure 3-20**, turn the wheel sharply to the right as the rear wheels enter the marked area. (Turning the wheels before this can result in hitting the cone on the right side.) As the vehicle approaches the right side, turn the wheels sharply to the left and proceed toward the left side. As the front left side of the apparatus approaches the left side, turn the wheels sharply to the right and back until the left rear of the apparatus approaches the right side of the enclosed space. Now turn the wheels sharply to the right and exit the confined space.

(For step-by-step photos of Confined Space Turnaround, please refer to pages 89–90.)

- *Diminishing clearance.* This maneuver (**Figure 3-21**) assesses the ability to drive the vehicle in a straight line, such as on a narrow street or road. It also judges the distances both on the sides of the apparatus and in front of the apparatus. Key techniques to use to perform this successfully include looking far ahead at your destination, not right in front of the apparatus, and making only small steering changes (at most turning the steering wheel a few degrees or a couple inches off of center). Not following these techniques frequently results in following an "S" pattern, where overcorrections are made first in one direction and then in the other direction.

(For step-by-step photos of Diminishing Clearance, please refer to page 91.)

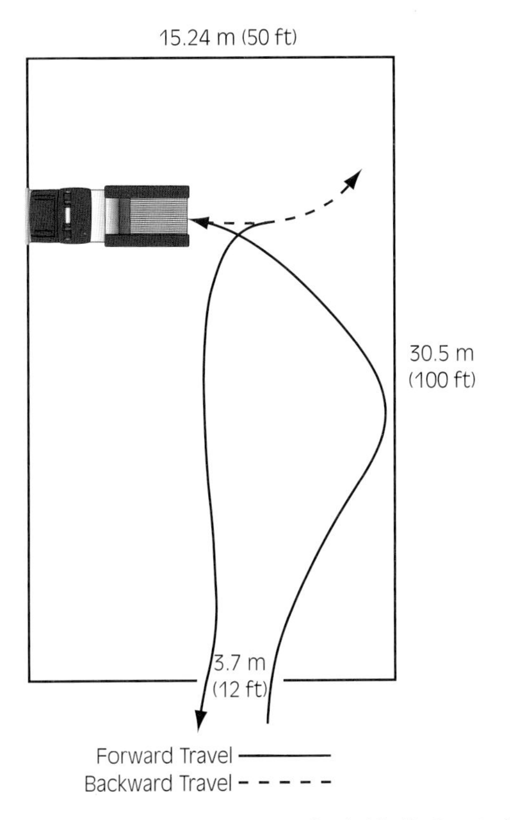

FIGURE 3-20 Confined space turnaround.

Road Skills

Prior to emergency response driving, drivers should practice maneuvers that they may need to perform. These include: right- and left-hand turns; rural, urban, and multilane roads; lane changes; up and down

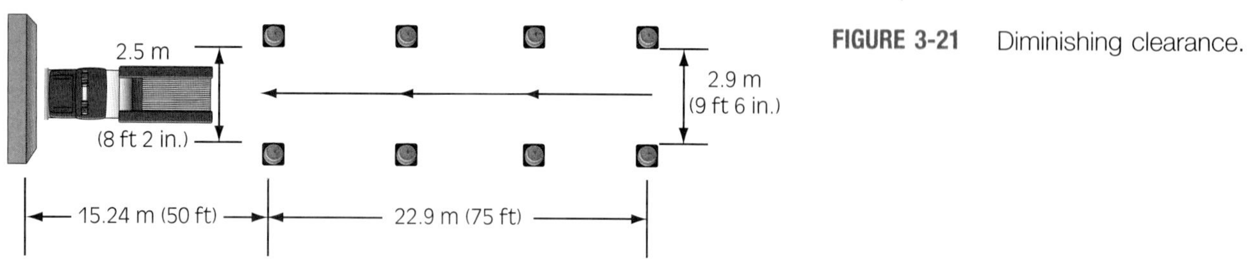

FIGURE 3-21 Diminishing clearance.

steep grades; day and night driving; curves; railroad crossings; and a bridge or underpass. Practicing of road skills should be done with another experienced driver who can instruct the new driver about various best practices.

EMERGENCY RESPONSE

Emergency response begins well before an alarm is ever received because drivers must ensure that apparatus are ready for immediate response. Drivers must be familiar with their response districts. Hazards such as congested areas, road construction, limited height and weight factors of bridges and tunnels, and school zones must be identified and avoided if possible. Knowledge and skill in safe-driving techniques must be mastered. During emergency responses, all safe-driving techniques must be followed. Exemptions from driving regulations do not relieve the driver of the responsibility to maintain control of the apparatus at all times. Although safe-driving techniques must be followed for both nonemergency and emergency driving, several additional factors must be considered when responding to an emergency. Due to this fact, some departments have reviewed response policies for larger vehicles such as tankers. This includes responding nonemergency on calls due to the increased hazards of operating this vehicle in an emergency mode.

When the alarm comes in, a physiological response is initiated that increases the pulse rate, causes the pupils to dilate, and pumps adrenaline into the bloodstream. This type of stress is often called the "fight or flight" syndrome. The purpose is to prepare the body for a rapid burst of strength and quickness with a heightened alertness. This condition can be both beneficial for and detrimental to the emergency driver. On the detrimental side is the tendency of the stress to encourage speed over safety. Perhaps this is caused by the "flight" aspect. On the beneficial side is the driver's heightened alertness and reaction. Drivers must learn to control the negative side of stress while capitalizing on the positive side of stress.

Prior to leaving the station, several activities must be completed. Most commercial trucking companies require their trucks to idle before being driven, a policy that significantly reduces engine wear. While not reasonable to delay an emergency response, this is a good practice for nonemergency response. Starting the apparatus as early as possible to allow some idling while getting ready is advantageous. First, primary and alternate response routes must be selected and agreed upon between the driver and the officer. Knowledge of response routes allows the driver to concentrate on safety rather than on street signs. Second, the driver must ensure that the apparatus is ready for the response; everyone must be seated and secured, exhaust ducts and chargers must be disconnected, bay doors must be fully open, and emergency warning devices must be activated.

During the response, the driver must aggressively avoid being distracted from safely operating the apparatus. Distractions can be especially dangerous when first leaving the station, when confronted with intersections or other hazardous locations, and when approaching the emergency scene. Distractions are often insidious in that drivers do not realize that they are being distracted. These distracters can easily accumulate to the point at which the driver experiences sensory overload. When this occurs, the driver may not process and react to information correctly. Several potential distracters include:

- The initial excitement of the response
- Thinking about response routes
- Thinking about the nature of the emergency
- Focusing on the emergency on approach
- Emergency lights and sirens

As previously mentioned, the driver and officer work as a team to ensure the safe arrival of the apparatus and crew. Both the driver and officer must attempt to stay focused on safety during the response; however, the officer may be required to gather information on the particular incident. For example, the officer may need to review pre-fire plans, material safety data sheets (MSDSs), and the emergency response guidebook. Although this activity is necessary, looking up information can take time and shifts the officer's attention away from response hazard recognition. The use of on-board computers and software allows officers to access required information quickly (**Figure 3-22**). The less time it takes to gather information, the more time the officer has to assist the driver in a safe response.

Drivers and officers must also take into consideration other responding vehicles. Extreme care should be exercised when two responding vehicles approach an intersection at the same time. The driver and officer should be familiar with SOPs that prescribe actions to follow when such an incident occurs. Other considerations include police officers, ambulances, and mutual aid companies that may be encountered while responding to an incident.

(For step-by-step photos of Pre-trip Activities, please refer to pages 92–93.)

Limitations of Emergency Warning Devices

Most state laws require warning devices to be in operation whenever the apparatus is responding to an emergency. Drivers should understand, however, the limitations of both visual and audible warning devices. The use of visual warning devices does not guarantee that all drivers will see the vehicle during daylight hours and when approaching vehicles from behind. Studies have shown that visual warning devices mounted on top of apparatus are not as effective as those mounted lower on the apparatus. Other limitations to visual warning devices include:

- Environmental factors (glare, fog, rain)
- Visibility from within the vehicle (tinted, dirty, or blocked windows)
- Geographical factors (buildings, hills, trees)
- Other vehicles (especially large tractor trailers)
- Drivers (sunglasses, visual ability)
- Distracters such as cell phones

Audible warning devices are most effective directly ahead of the apparatus. Vehicles approaching emergency apparatus from the side may not hear audible signals as soon as those directly ahead. Other limitations to audible warning devices include:

- Noise inside vehicles (radio, heater/air conditioner, passengers)
- Soundproofing of newer vehicles (some commercials stress the quiet ride of new models)
- Noise caused by environmental conditions (rain, wind, hail)

FIGURE 3-22 The use of on-board computers and software provides officers with the ability to quickly gather needed information on an incident. In turn, the officer can spend more time assisting the driver in a safe response.

- Geographical factors (buildings, hills, trees)
- Other vehicles (especially large tractor trailers)
- Drivers (hearing ability)
- Distracters such as cell phones

Drivers should understand that even when emergency warning devices are heard or seen, other drivers may not react appropriately. Some drivers simply disregard warning devices. Others may become nervous or confused and slam on their brakes or turn in the wrong direction. Still others may speed up, attempting to cross an intersection or to make a turn to avoid having to slow down or stop. A driver should make no assumptions about what another driver will do and should be prepared to react at all times to the vehicle in front of the apparatus.

On-scene Considerations

Perhaps the most significant consideration when approaching the incident is that the emergency scene will most likely be congested. The driver must exercise extreme care when approaching the emergency scene. Emergency scene congestion can occur from several sources. First, congestion can occur from:

- Preexisting, sources such as parked cars
- Bystanders and narrow roads
- Utility poles and trees

Second, the source of on-scene congestion can be the result of the incident, which could include:

- Debris from fallen buildings
- Wrecked vehicles
- Downed electrical wires
- Backed-up traffic

Finally, on-scene congestion can be increased by:

- Emergency operations such as other responding vehicles
- Emergency personnel
- Emergency equipment such as hoselines, rehab locations, and staging locations

After arriving on the scene, the driver must safely position the apparatus. In most cases, this means following the department's SOPs. Several general considerations for positioning the vehicle include the following:

- Position to reduce the likelihood of the vehicle being struck by traffic.
- Use the vehicle to shield emergency personnel from traffic.
- Position the vehicle to enhance emergency operations, such as water supply, suppression, and rescue activities.
- Position the vehicle away from hazards such as liquid spills, overhead electrical lines, and poor surface conditions (soft surface or severe grade).
- Consider wind direction, exposure, and emergency escape routes when positioning emergency vehicles.
- Do not part in the collapse zone.
- Never park the apparatus on railroad tracks.
- Park the vehicle on the side of the incident.
- Park so as not to impede later-arriving apparatus, particularly aerial apparatus, which have the most restrictive positioning requirements for safe and effective operations.

Returning from Emergencies

Returning from an emergency response can be dangerous. The stress of the incident coupled with fatigue can cause drivers to be less alert and have sluggish reactions. The desire to get back to the station should not overshadow safety. Finally, drivers should inspect the apparatus to ensure that equipment is properly stowed and personnel are seated for the return trip.

SKILL 3-1

Station Parking

Delmar/Cengage Learning

Delmar/Cengage Learning

A Pull past the station so that the apparatus can back directly into its bay. While pulling past the station, visually confirm that no obstructions exist in the path that the apparatus will take because while you are backing, this area will be in the vehicle's blind spot.

B When safe to do so, enter reverse and turn the wheel so that the vehicle can enter the station. The primary focus should be on the mirror on the inside of the curve and ideally on the driver's side of the apparatus.

Delmar/Cengage Learning

C As the turning apparatus aligns with its parking space, straighten out the wheels. Ideally, this is done with a single arc, avoiding "S" curves as multiple adjustments are made to position the apparatus. Once the wheels are straightened out, only very minor deflections of the wheel should be necessary. To simplify parking, many stations have aids to help with positioning. These aids include positioning lines, wall-mounted mirrors, and tire positioning humps.

SKILL 3-2

Alley Dock

Delmar/Cengage Learning

A Pull past the alley so that the apparatus can back directly into it. While pulling past the alley, visually confirm that no obstructions exist in the path that the apparatus will take because while you are backing, this area will be in the vehicle's blind spot.

Delmar/Cengage Learning

B When safe to do so, enter reverse and turn the wheel so that the vehicle can enter the alley. The primary focus should be on the mirror on the inside of the curve. While better vision exists when this is on the driver's side of the apparatus, it should be performed both backing from the right and from the left.

Delmar/Cengage Learning

C As the turning apparatus aligns with its parking space, straighten out the wheels. Ideally, this is done with a single arc, avoiding "S" curves as multiple adjustments are made to position the apparatus. Once the wheels are straightened out, only very minor deflections of the wheel should be necessary. Ideally, the operator should stop within 6 inches (15 cm) of the dock and without hitting the dock. For practice, traffic cones or a chalk line are typically used, preventing any possible damage.

SKILL 3-3

Serpentine

A Drive forward along the side of the serpentine cones.

B Start backing and cutting the wheel sharply until the cone becomes visible in the side-view mirror.

C As soon as the cone becomes visible, straighten the wheels and continue backing until the rear wheel passes the cone. Repeat steps B and C, alternating side-view mirrors based on which side of the apparatus will pass the cone for each successive cone.

(Continues)

SKILL 3-3

Serpentine (*Continued*)

Delmar/Cengage Learning

D Pull forward, cutting the wheel sharply until the apparatus will be able to pass the next cone on the correct side.

Delmar/Cengage Learning

E Continue driving forward until the rear tire clears the next cone. Repeat steps D and E for each additional cone.

SKILL 3-4

Confined Space Turnaround

A As the rear wheels of the apparatus enter the confined space, turn sharply to the right. (This will cause the rear wheels to track to the right and provide more room for maneuvering.)

B As the apparatus approaches the right-hand boundary, the wheels should be turned sharply to the left. This will allow the front of the apparatus to just miss the right-hand boundary. The turned front wheels should be parallel to the boundary and the apparatus should be at a 45-degree angle to the boundary.

C Proceed until the front of the apparatus approaches the left boundary. (The apparatus should be at an angle of at least 45 degrees.)

(Continues)

SKILL 3-4

Confined Space Turnaround (*Continued*)

Delmar/Cengage Learning

D Once the apparatus approaches the left boundary, turn the wheels sharply to the right, and then back to the right boundary. (The apparatus should be at an angle of at least 45 degrees.)

Delmar/Cengage Learning

E Turn the wheels sharply to the left and pull forward and straighten the wheels so that you leave the confined space through the center gate.

SKILL 3-5

Diminishing Clearance

A The operator should enter at the center of the course and focus on the final destination. Proceed at a constant speed to the destination, stop just before the destination, and don't hit the marker.

B Back to the starting position. Remember to focus on the destination and avoid over-steering.

SKILL 3-6

Pre-trip Activities

A Start the vehicle and allow it to idle while the other steps are completed. Prior to cranking, make sure the parking brake is set, the battery switch(es) are on, the vehicle is in neutral (or park), and that power-up diagnostics, including glow plug warm-up, have completed.

B Open apparatus bay doors.

C Walk completely around the vehicle, ensuring doors are completely closed (double latched) and wheel chocks are stowed. (While you are doing this, the crew is typically suiting up and the officer is confirming the route.)

D Remove any loose items that may have been left on the apparatus.

E Make sure everyone is belted in and the route to the call is known.

F After exiting the station, close the bay door. (For stations with vehicles parked two deep, make sure the vehicle behind you is not also exiting.)

LESSONS LEARNED

Apparatus drivers must ensure the safe arrival of the apparatus, equipment, and personnel at the scene. To accomplish this important duty, drivers must be familiar with the laws and standards related to safe driving. In addition, drivers must understand basic driving factors such as speed and environmental conditions. Finally, drivers must understand the unique requirements for operating their apparatus during emergency response.

KEY TERMS

Anti-lock braking system (ABS) A safety system that monitors wheel lock-up (wheels that stop turning or skid while the vehicle is still in motion) and, when wheel lock-up occurs, automatically pumps (quickly releases and then reapplies) the brakes. The pumping action under this condition increases vehicle control and reduces stopping distance.

Authorized emergency vehicles Legal terminology for vehicles used for emergency response, such as fire department apparatus, ambulances, rescue vehicles, and police vehicles equipped with appropriate identification and warning devices.

Braking distance The distance of travel from the time the brake is depressed until the vehicle comes to a complete stop.

Enhanced roll stability systems Systems that recognize when a vehicle is likely to roll over and can independently control power and braking to individual wheels in such a way as to maximize the chances of the apparatus responding as the operator desires while minimizing rollovers.

Landmark An aid to positioning an apparatus for maneuvers; it could be marking tape on a wall, floor pavement marking, or a natural item such as a tree.

Mobile data terminals (MDTs) An apparatus-mounted computer providing access to 9-1-1 dispatch information and perhaps other functions such as fire preplans, Internet, maps, and reporting software; may also be called a mobile communications terminal (MCT).

Perception distance The distance the apparatus travels from the time a hazard is seen until the brain recognizes it as a hazard.

Reaction distance The distance of travel from the time the brain sends the message to depress the brakes until the brakes are actually depressed.

Roll stability control (RSC) systems Systems that recognize when a vehicle is likely to roll over and can independently control power and braking to individual wheels in such a way as to maximize the chances of the apparatus responding as the operator desires while minimizing rollovers.

Spotter An individual used to assist in backing up an apparatus.

Total stopping distance Measured from the time a hazard is detected until the vehicle comes to a complete stop.

Traction Friction between the tires and road surface.

Wheel chock Device placed in front of or behind wheels to guard against inadvertent movement of the apparatus.

REVIEW QUESTIONS

Multiple Choice

Select the most appropriate answer.

1. Which of the following NFPA standards discusses minimum safety-related requirements associated with emergency vehicles, drivers of emergency vehicles, and operation of emergency vehicles?

a. 1001
b. 1002
c. 1500
d. 1901

2. The requirement that emergency vehicle drivers bring the apparatus to a complete stop at red traffic lights or stop signs is contained in which of the following NFPA standards?

a. 1001
b. 1002
c. 1500
d. 1901

3. Which of the following NFPA standards requires drivers to demonstrate the ability to safely maneuver the apparatus over a predetermined route in compliance with all applicable state, local, and department rules and regulations?
   a. 1001
   b. 1002
   c. 1500
   d. 1901
4. Which of the following should *not* be used to help regain control while in a skid?
   a. take foot off the accelerator
   b. depress accelerator slightly
   c. apply brakes slightly
   d. turn front wheels in the direction of the skid
5. The distance the apparatus travels from the time a hazard is seen until the brain recognizes it as a hazard is known as
   a. braking distance.
   b. total stopping distance.
   c. reaction distance.
   d. perception distance.
6. The distance measured from the time a hazard is detected until the vehicle comes to a complete stop is known as
   a. braking distance.
   b. total stopping distance.
   c. reaction distance.
   d. perception distance.
7. The distance of travel from the time the brake is depressed until the vehicle comes to a complete stop is known as
   a. braking distance.
   b. total stopping distance.
   c. reaction distance.
   d. perception distance.
8. The distance of travel from the time the brain sends the message to depress the brakes until the brakes are depressed is known as
   a. braking distance.
   b. total stopping distance.
   c. reaction distance.
   d. perception distance.
9. To regain control during a skid,
   a. slightly increase speed.
   b. disengage the clutch.
   c. turn the wheels in the direction of the skid.
   d. turn the wheels in the direction of the skid and depress the brake aggressively.

Short Answer

On a separate sheet of paper, answer/explain the following questions.

1. Explain why driving an apparatus is such a critically important duty.
2. List and discuss four common factors associated with apparatus accidents.
3. Explain the purpose of laws and standards for the safe operation of emergency vehicles.
4. Briefly discuss the role of NFPA 1500 and NFPA 1002 in connection with emergency apparatus.
5. List at least four instances where responding apparatus must come to a complete stop.
6. Discuss the attributes of a safe driver.
7. List several examples of defensive driving techniques.
8. Explain the relationship of speed to both steering and stopping apparatus.
9. Discuss how environmental conditions affect the safe handling of apparatus.
10. Explain how stress can both help and hinder a driver while responding.
11. List three general considerations for positioning vehicles on-scene.
12. List three basic components contained in most state driving laws.
13. Explain the limitations (conditions) of exemptions extended to emergency vehicle drivers.
14. State laws typically exempt emergency vehicle drivers from several normal driving regulations. List four of these exemptions.
15. List four key elements of defensive driving.
16. What do you need to maintain in order to safely navigate a curve?
17. List four weather-related conditions that affect traction and, consequently, stopping distance and steering.
18. List four environmental conditions that may affect a driver's vision.

19. List five activities that should be accomplished prior to leaving the station for an emergency.
20. List four limitations to visual warning devices and four limitations to audible warning devices.
21. Explain the considerations for using automatic braking systems in good and bad road conditions.

ACTIVITIES

1. Obtain a copy of your state law, local law or ordinance, and department SOPs or policies related to emergency response.
2. Evaluate your department's driving-related SOPs or policies in comparison to state and local laws as well as in relation to NFPA 1500 and NFPA 1002.
3. Collect and evaluate data on emergency vehicle accidents at the local or state level.
4. Develop SOPs for (1) emergency driving through intersections and (2) the use of emergency signaling devices during emergency response.

ADDITIONAL RESOURCES

Most states produce a commercial driver's license (CDL) manual. Regardless of whether your state requires a CDL for driving emergency apparatus, this manual provides valuable information on safe-driving techniques for large vehicles.

Allison Transmission Publications, http://www.allisontransmission.com/publications/

Emergency Response Guidebook. Washington, DC: United States Government Printing Office, 2008. Available online at http://www.phmsa.dot.gov/hazmat/library/erg

FEMA. *Safe Operation of Fire Tankers.* U.S. Fire Administration Publication # FA 248. Available online at http://www.usfa.dhs.gov/downloads/pdf/publications/fa-248.pdf

FEMA. *Emergency Vehicle Driver Training Program.* U.S. Fire Administration Publication # FA 110. Available online at http://www.usfa.dhs.gov/downloads/pdf/publications/fa-110.pdf

NFPA 1500, *Standard on Fire Department Occupational Safety and Health Program.* National Fire Protection Association, Quincy, MA, 2007.

NIOSH. *Fire Fighter Fatality Investigation and Prevention Program* Available online at http://www.cdc.gov/niosh/fire/reports/face200905.html

Pump Construction and Peripherals

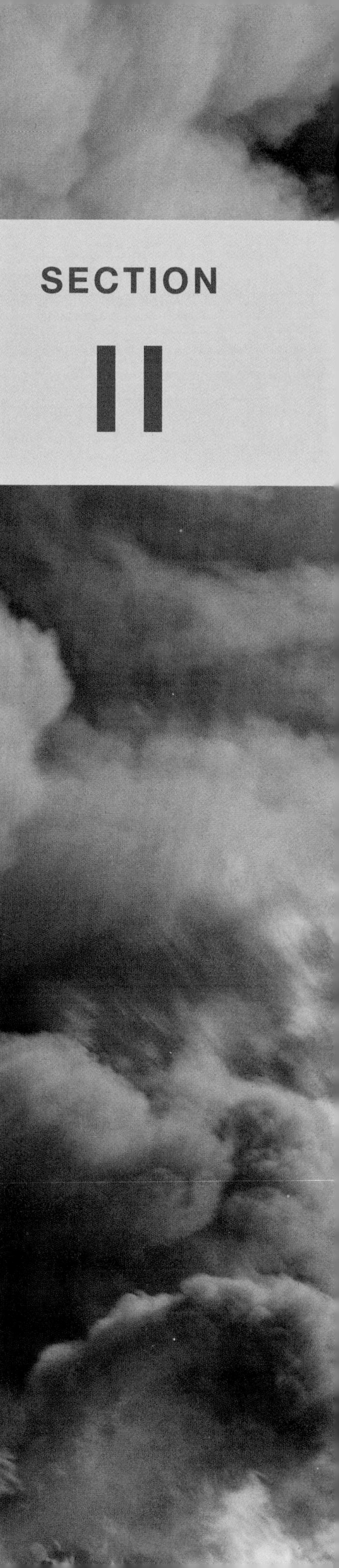

The first section of this book discusses basic requirements and concepts for both pump operators and emergency vehicles. This second section of the book covers the requirements of NFPA 1002 presented in sections 5.2.1, 5.2.2, and 5.2.3. It discusses the operating principles, theories, and construction of pumps as well as the systems and components typically used in conjunction with fire pumps. **Chapter 4** discusses the need for pumps, types of pumps, and basic pump concepts. The many components used to control and monitor the pump are presented in **Chapter 5**. **Chapter 6** discusses the different types of hose, nozzles, and appliances used in pump operations.

Section III continues the discussion of the three duties of the pump operator and the utilization of the tools of the trade in more detail.

4 Pump Operating Principles and Construction

L × W ÷ % involved

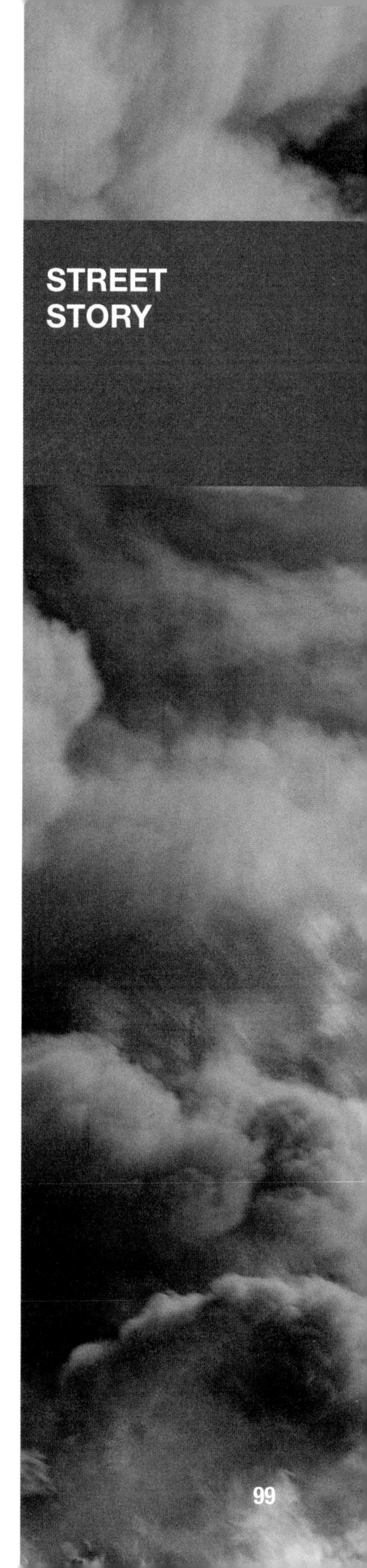

STREET STORY

It's really important to know how the pump itself works. About 2 years ago, we got called to a fire in a truck yard just outside the city limits at 3 AM. There was no hydrant there, so we took the tanker (along with the engine, and I was driving the tank). It turned out that four dump trucks positioned in a row were on fire. The captain wanted Class B foam because there was fuel involved. So I put the pump into gear, got water flowing, and turned on the foam, radioing "how does the foam look?" to the crew. But they responded that they had no foam.

On this tanker, the pump panel is located on the outside. So I scrolled through the digital display—you can see the total foam flowed, the percentage of foam, or the gallons (liters) per minute. But all of them were zero. I worried that the display might be broken.

So I went back to the other side of the truck and opened the side panel to take a look at the pump with my flashlight. I found the lines involved inside the pump compartment and then found their path of travel from the source to the hoseline. I looked to see if there was foam in the line and there wasn't, so I went back to the Class B tank, and there was foam there. So I checked the valves along the Class B foam line, and it turned out that one of them was shut off. Apparently the crew on the previous shift had closed the line manually because the foam valve wasn't closing all the way with electronic control, and they didn't want foam being drawn into the pump when trying to flow just water. But they hadn't told anyone else. Fortunately I was able to open the valve and get foam to my crew.

Knowing how the pump operated, I was able to fix this issue in under a minute. It's important to understand that the pump panel is just there for ease of use: You could operate the pump and never touch the panel if you understand its physical makeup. So many pump operators put up pump charts on their windows to have a quick reference, but you can't get too reliant on these or the panel. You have to know why and how the pump works so that if there's a problem, you can fix it quickly.

—Street Story by Shannon Askew,
Firefighter (Apparatus Operator), Calgary Fire Department,
Engine Company #14, Calgary, Alberta, Canada

LEARNING OBJECTIVES

After completing this chapter, the reader should be able to:

4-1. Define the term *pump*.

4-2. Explain why pumps are needed in the fire service.

4-3. Discuss the following pump terms and describe how they affect pumping operations: intake and discharge side of pumps, pump efficiency (flow, pressure, speed, slippage), rated capacity, single stage, multiple stage, pressure or series mode, volume or capacity mode, priming.

4-4. Identify the two types of pumps used in the fire service and the significant characteristics of each of these pumps.

4-5. Identify the major subsystems of a fire pump, as well as the major components of each subsystem and their purpose.

*The driver/operator requirements, as defined by the NFPA 1002 Standard are identified in black, additional information is identified in blue.

14.7lbs

INTRODUCTION

Pumps are defined as mechanical devices that raise and transfer liquids from one point to another. Note the similarity of this definition to the definition of fire pump operations presented in **Chapter 1**, which was "the systematic movement of water from a supply source through a pump to a discharge point."

The ability of a pump to accomplish the task of moving water rests, in part, with the pump itself as well as with the hose and appliances on the intake and discharge sides of the pump. To some extent the hoselines and appliances can be changed to meet the varying needs of different pumping situations. For example, if higher pressures are needed, smaller hoselines can be used. For greater water flows, larger hoselines can be used. One of the major limiting factors of water movement, however, is the type and size of the pump. Different pumps have different capabilities and characteristics for moving water.

Knowledge of pump operating principles and construction will help ensure the pump is operated in a safe manner.

Understanding the different types of pumps and their operating principles and construction will assist the pump operator in completing the duties of driving the apparatus, operating the pump, and conducting preventive maintenance. Understanding the limits of a pump will affect the driving and positioning of apparatus for the most effective use of water supplies. In addition, knowledge of the pump will help ensure that the pump is safely operated within its limits to deliver appropriate water quantity and pressure. Providing proper maintenance for safe and efficient operating performance of the pump requires a basic knowledge of pump construction and principles.

NEED FOR PUMPS

Water has several characteristics that make it an excellent extinguishing agent for a variety of sizes and types of fires. It is relatively stable at normal temperatures and has a tremendous capacity to absorb heat. Water is either readily abundant or at least accessible in most areas. Water is also relatively easy to move from one place to another. Because of these and other characteristics, the predominant extinguishing agent used in the fire service is water. It makes sense that the fire service would utilize pumps as a means to move water. Fire pumps have played an important role in the history and success of the fire service. From human-powered piston pumps to high-capacity pumps to foam systems on modern apparatus, the use of and reliance on pumps in the fire service have steadily increased.

Pumps continue to play an important role in the fire service today for several reasons (**Figure 4-1**). First, pumps on apparatus play an important role in a municipal water distribution system in that they boost pressure and flow to move water from hydrants to the scene. Second, pumps provide some degree of safety from pressure surges. Fire suppression and other emergency activities often require large quantities of water and/or pressure. Either equipment failures or operator mistakes can cause dangerous pressure surges during pumping operations. Pressure relief devices protect both personnel and equipment from these powerful pressure surges. Third, pumps are often required to move water supplies maintained on the apparatus to the fire scene. As previously stated, most fire apparatus have at

FIGURE 4-1 Emergency vehicles equipped with pumps continue to play an important role in the fire service.

least a minimum water supply on board. Fourth, in areas lacking a municipal water distribution system, pumps are required to move water from static sources such as ponds, lakes, and rivers to the scene. Finally, pumps can provide a continuous supply of water at the proper pressure and quantity to the emergency scene.

BASIC PUMP CONCEPTS AND TERMS

Several concepts and terms are common among all types of pumps. The first concept concerns the terms used to identify the intake and discharge sides of the pump. The **intake** side of a pump is the point at which water enters the pump, also referred to as the "supply side" and the "suction side" of the pump. The use of the term *suction side* is misleading in that water is not actually sucked into the pump, but rather flows into the pump under pressure. The **discharge** side of the pump, also referred to as the "pressure side" of the pump, is the location where water leaves the pump. For the remainder of this text, *intake* and *discharge* are used to identify the two main sides of the pump.

The second concept concerns the terms used to describe the efficiency of a pump: flow, pressure, speed, and slippage.

- **Flow** refers to the rate and quantity of water delivered by the pump and is expressed in gallons per minute (gpm) or litres per minute (Lpm).
- **Pressure** refers to the amount of force generated by the pump or the resistance encountered on the discharge side of the pump. Pressure is typically expressed in pounds per square inch (psi) or kilopascals (kPa). Note: Chapter 10 of this textbook discusses flow and pressure in greater detail.
- **Speed** or **pump speed** refers to the rate at which the pump is operating and is typically expressed in revolutions per minute (RPM).
- **Slippage** is the term used to describe the leaking of water between the surfaces of the internal moving parts of the pump.

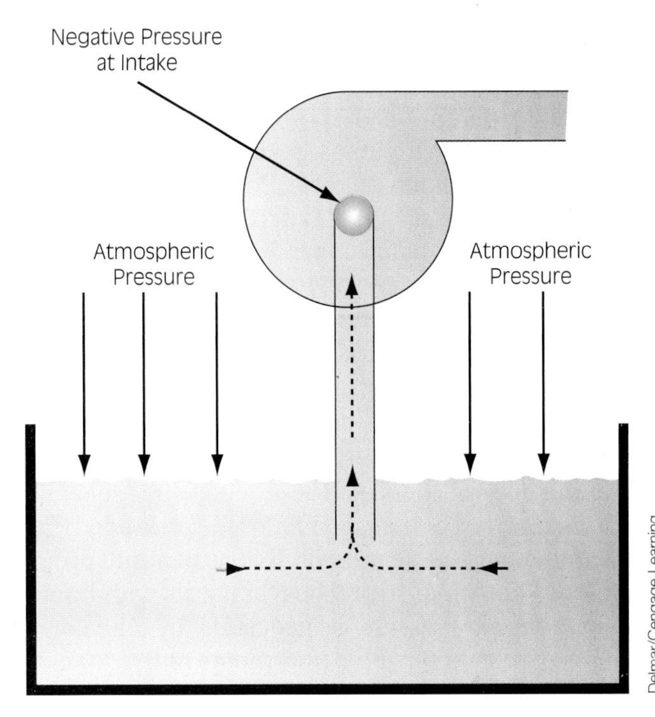

FIGURE 4-2 A negative pressure allows atmospheric pressure to force water into the pump.

The third concept concerns the **priming** of pumps. All pumps must be primed to operate. Priming involves removing the air inside the pump and replacing it with water. Positive displacement pumps are typically self-priming while centrifugal pumps require separate primers. Often priming is misunderstood as a suction process, that is, that water is sucked into the pump. The process is actually the creation of a lower pressure on the intake side of the pump that allows atmospheric pressure to force water into the pump (**Figure 4-2**) through a noncollapsible waterway.

TYPES OF PUMPS

Pump can be grouped in two types: by their principle of operation and by their intended use in the fire service. The two broad categories of pumps based on operating principles are positive displacement and dynamic. Modern-day pumping apparatus utilize pumps of both categories.

- Positive displacement pumps are typically used as priming pumps, high-pressure auxiliary pumps, or portable pumps, and for use in **pump-and-roll** situations. These pumps generally provide higher pressures with low flows. Positive displacement pumps are based on **hydrostatic** principles (a branch of hydraulics that deals with liquids at rest and the pressures they exert or transmit). Pressure is generated within the pump by the application of mechanical force.
- Dynamic or non-displacement pumps are the **main pumps** on modern pumping apparatus. They may also be used for auxiliary, portable, and pump-and-roll pumps. These pumps are generally used for large flows with lower pressures. Dynamic pumps are based on **hydrodynamic** principles (a branch of hydraulics that deals with liquids in motion). Pressure is generated by movement and momentum within the pump rather than by force. The common term for dynamic pumps in the fire service is *centrifugal pumps*. This term is therefore used in the remainder of this textbook.

The three broad categories of pumps based on their intended use are main, priming, and auxiliary. Note that this categorization may differ depending on regional preferences.

The three broad categories of pumps based on their intended use are main, priming, and auxiliary.

- The main pump is the primary working pump permanently mounted on the apparatus. The size and type of the pump may vary based on the intended purpose of the apparatus. According to NFPA 1901, *Standard for Automotive Fire Apparatus*, fire apparatus with the primary purpose of combating structural and associated fires must have a permanently mounted fire pump with a **rated capacity** of at least 750 gpm (3,000 Lpm). Fire apparatus with the primary purpose of initiating fire suppression efforts and supporting associated fire department operations must have a permanently mounted fire pump with a rated capacity of at least 250 gpm (1,000 Lpm). Specialized vehicles such as tankers, also called water tenders, and high-pressure apparatus may have either centrifugal or positive displacement pumps of various sizes. In most cases, however, the term *main pump* identifies a centrifugal pump.
- **Priming pumps** are positive displacement pumps permanently mounted on an apparatus. Apparatus equipped with a centrifugal pump as the main pump will have a priming pump. The sole purpose of these pumps is to prime the main pump when needed.
- **Auxiliary pumps** refer to pumps other than the main pump or priming pump that are either permanently mounted or carried on an apparatus. The most common of these is the booster pump. Booster pumps do not actually boost the main pump. Rather, they are a smaller pump used to supply small hoselines. Booster pumps can be either positive displacement or centrifugal pumps. Pumps used for pump-and-roll operations are included in this category. The pump-and-roll operation is the process of discharging water while the vehicle is in motion.

  Wildland firefighting apparatus and air crash fire and rescue vehicles often have this capability. High-pressure foam pumps are also included in this category, as are portable pumps that can be either positive displacement or centrifugal.
- **Floating pumps** are used when drafting with a high total pressure loss in the intake hose. High pressure losses occur both from long intake hoses and static water supplies significantly lower than the engine. Engines have a maximum lift of 20 to 25 feet (6 to 7.5 metres) or intake vacuum of 20 to 22 inches of Hg (500 to 550 mm), and at their maximum lift, they have a significantly reduced capacity. When significant lifts are required to get water to the engine, a high-capacity, low-pressure floating pump is placed at the water source to pump the water to the engine's intake.

POSITIVE DISPLACEMENT PUMPS

Positive displacement pumps are pumps that discharge a fixed quantity of water during each cycle of the pump's operation. Although no longer used as the primary pump on engines, they are commonly used in the fire service since they have capabilities, such as being self-priming, that do not exist in other types of pumps. Since they are the easiest type of pump to understand, our discussions will start with these pumps.

Principles of Operation

Two important principles are fundamental to the operation of positive displacement pumps.

Water is virtually noncompressible under normal conditions.

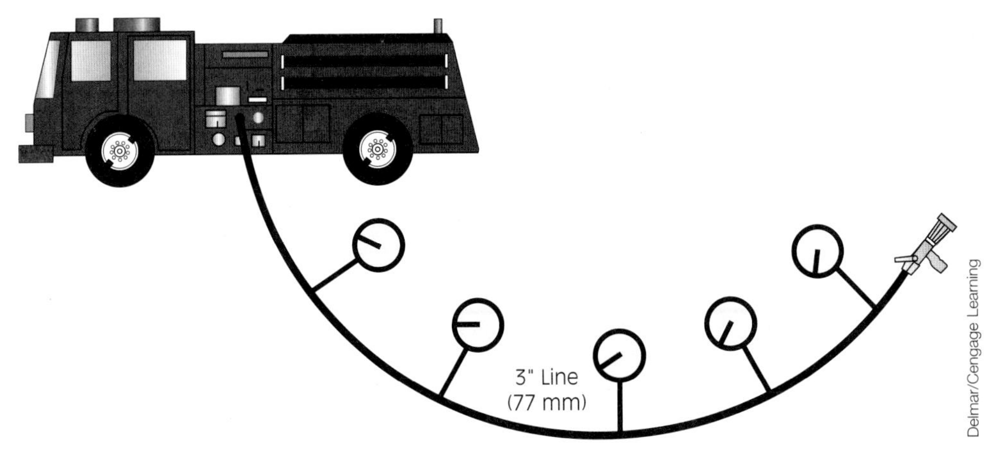

FIGURE 4-3 When no water is flowing, the pressure at any point in the line will be the same.

Principle 1: Water is virtually noncompressible under normal conditions. This means that forces applied to water will tend to push or move water rather than compress it. This principle can be demonstrated by filling a balloon with water. Pressure exerted on the balloon tends to move the water. The elastic nature of the balloon provides the freedom of the water to move.

Principle 2: When pressure is applied to a confined liquid, the same pressure is transmitted within the liquid equally. Take, for example, a pump supplying an attack line (**Figure 4-3**). If the nozzle is closed, the pressure at any point in the line will be the same. Positive displacement pumps utilize these principles to move water.

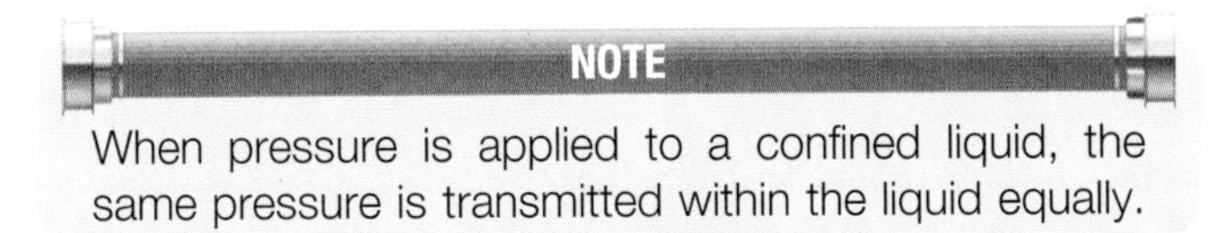

When pressure is applied to a confined liquid, the same pressure is transmitted within the liquid equally.

Operating Characteristics

Positive displacement pumps theoretically discharge (displace) a specific quantity of water for each revolution or cycle of the pump. In reality, slippage occurs that reduces the quantity of water discharged. The greater the slippage, the less efficient the positive displacement pump. Increasing the flow of positive displacement pumps is a simple matter of speeding up the pump. The result of increased pump speed is the same quantity of water per revolution delivered at a faster rate, thus increasing the flow rate.

The relationship between pump intake, discharge, and pressure in positive displacement pumps is, for the most part, straightforward. Each revolution of the pump yields (discharges) a specific quantity of water. Consequently, the pressure on the intake side of the pump is irrelevant. No matter what the intake pressure is, the pump will only discharge its specific quantity per revolution. Increases in pressures may, however, help the pump-driving mechanism function more efficiently in that the drive mechanism will not have to work as hard to pump the water if intake pressure is increased.

Similarly, the water discharged from positive displacement pumps is independent of pressure. Another way of looking at it is that positive displacement pumps simply transfer water. No matter how big the hose or how long the lay on the discharge side, positive displacement pumps will yield the same quantity of water per revolution. Pressure buildup is simply the result of confinement on the discharge side of the pump and does not affect pump discharge. The larger the hose, the longer the pump will take to fill the hose. If the water leaves the hose in the same quantity as the pump delivers it, pressure buildup will be minimal. If the hose or associated appliances, however, restrict the amount of water being discharged, pressure increases will occur. If the water is completely restricted as in the case of closing a nozzle, pressure will continue to build up until the pressure forces the pump to stop, the water is released (the nozzle is opened), or something breaks. Because of the potential serious buildup of pressure, some method of relieving excessive pressure on the discharge side of the pump is used. Pressure relief devices are discussed further in **Chapter 5**.

Positive displacement pumps are typically designed to operate at specific speeds with predetermined pressures to deliver a specific quantity of water. These variables are typically designed into the operation of the positive displacement pump. Priming pumps, for example, often have a simple on-and-off switch to operate the pump.

Wear and tear in positive displacement pumps can increase slippage, which significantly reduces performance.

Construction

The two main types of positive displacement pumps are piston (also referred to as reciprocating) and rotary. Piston pumps were once the main pump on fire apparatus. The first piston pumps used for firefighting were hand operated (see **Figure 4-4**). Although rarely used as a main pump today, piston pumps are making a comeback as high-pressure units. Rotary pumps replaced piston pumps and today they are typically found on modern apparatus in support of the main pump (a centrifugal pump) as either auxiliary or priming pumps. Rotary pumps are further divided into rotary gear, lobe, and vane. Rotary pumps are differentiated from piston pumps based on their circular motion rather than the up-and-down action of piston pumps.

Common among positive displacement pumps is the reliance on closely fitting moving parts. Because the parts fit so closely, various liquids and sometimes air can be pumped. The close-fitting moving parts are used to force water or air from the intake side of the pump to the discharge side. The efficiency of positive displacement pumps is related to the close fitting of parts. Consequently, wear and tear of these pumps can increase slippage, which significantly reduces their performance. Improper speeds and debris such as sand in the pump can dramatically increase wear and tear and consequently increase slippage.

Piston (Reciprocating) Pumps

Piston pumps have a minimum of three moving elements: the piston, the intake valve, and the discharge valve (**Figure 4-5**). The piston is contained in a cylinder and moves up and down by means of the piston rod. The intake and discharge valves open and close to direct the water.

Figure 4-6 illustrates the operation of a single-action piston pump. When the piston moves upward, the volume between the piston and the cylinder increases.

This action creates a slight vacuum in the chamber that causes the intake valve to open, allowing water to enter the chamber. Water will continue to enter until the piston stops. When the piston moves down, the pressure exerted on the water closes the intake valve and opens the discharge valve.

Note that water is discharged only during the downward movement of the piston. This action creates a pulsating effect. A second piston can be added to both reduce the pulsating effect as well as to double the flow of water (**Figure 4-7**). The pistons simply operate in reverse of each other. While piston A is filling the void space with water, piston B is discharging water. During the next cycle, piston A discharges its water while piston B is filling its void space. Even with two pistons, the pulsating effect still exists. To further reduce the pulsating effect, multiple pistons can be used. Each action of the piston is slightly offset from the others. Air domes have also been used to reduce the pulsating effect. Instead of pumping directly to discharge lines, water is pumped into the air dome. The air dome has a pocket of air that acts to cushion the pulsating effect. When the air dome fills to capacity, the internal pressure

Courtesy National Museum of American History, Kenneth E. Behring Center, Smithsonian Institution.

FIGURE 4-4 Example of an early hand-operated piston pump.

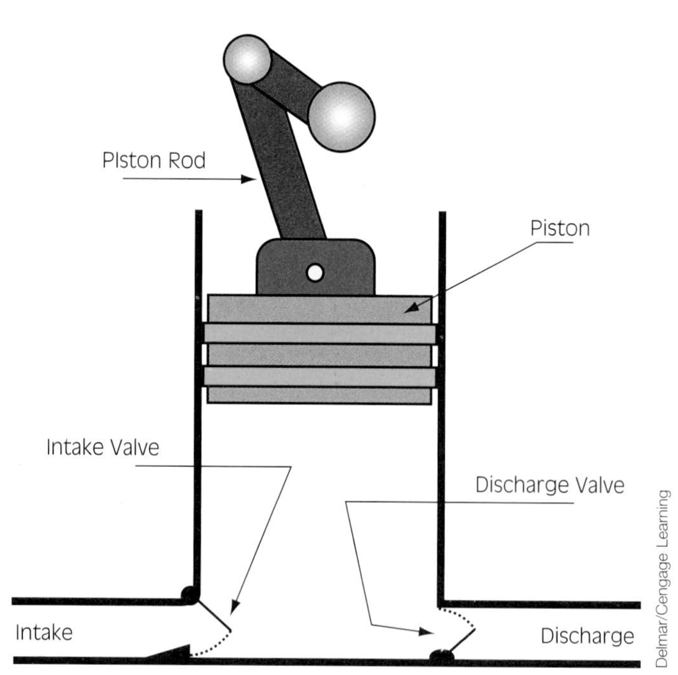

Delmar/Cengage Learning

FIGURE 4-5 The basic components of a piston pump.

First Stroke
(Intake Stroke)

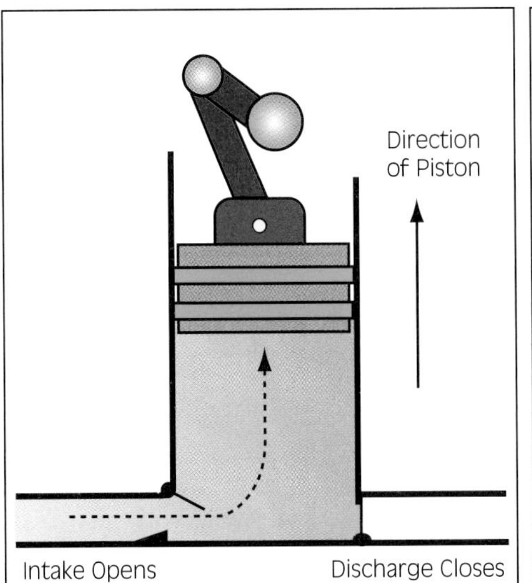

Second Stroke
Discharge Stroke

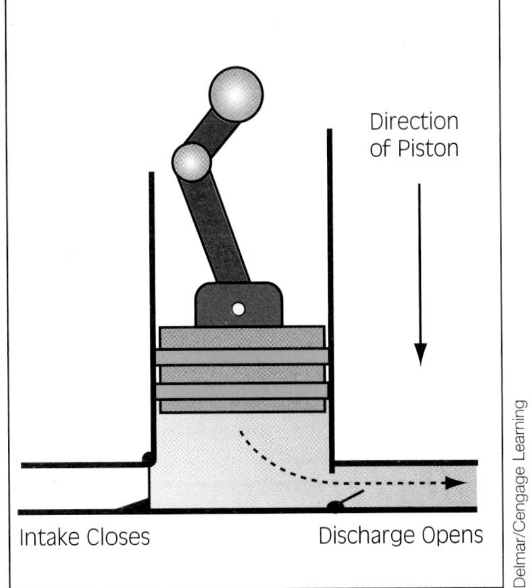

FIGURE 4-6 Two-stroke working cycle of a piston pump.

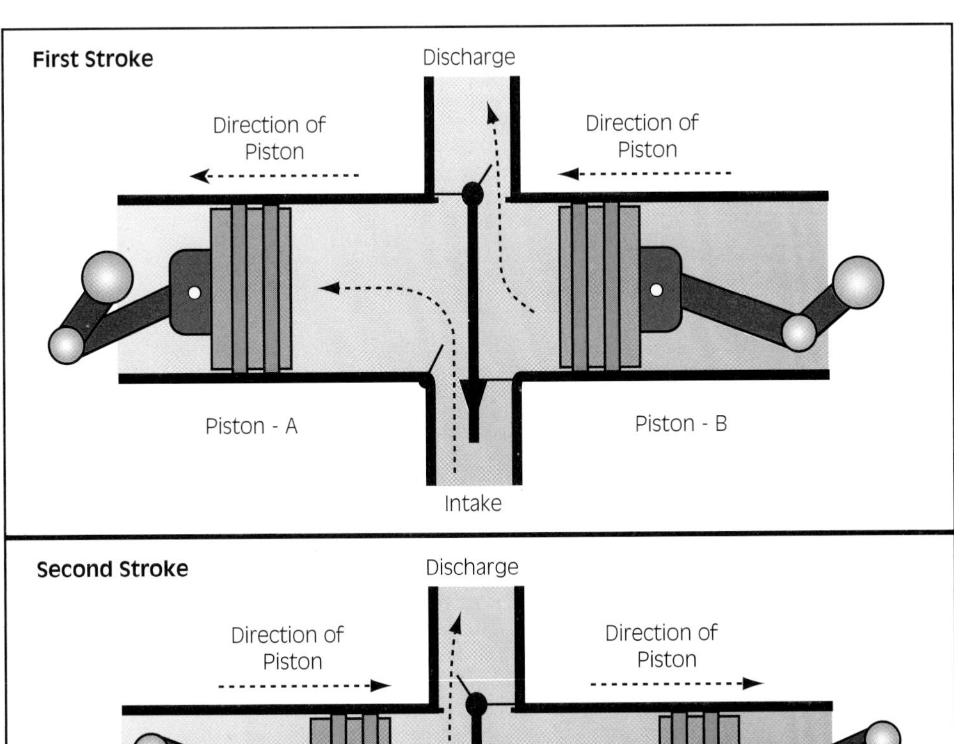

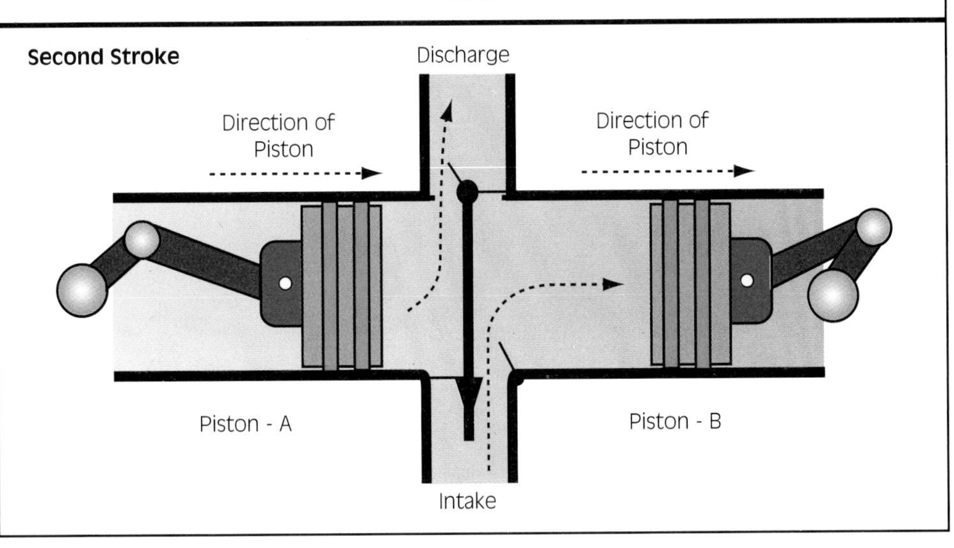

FIGURE 4-7 Two-stroke working cycle of a dual piston pump.

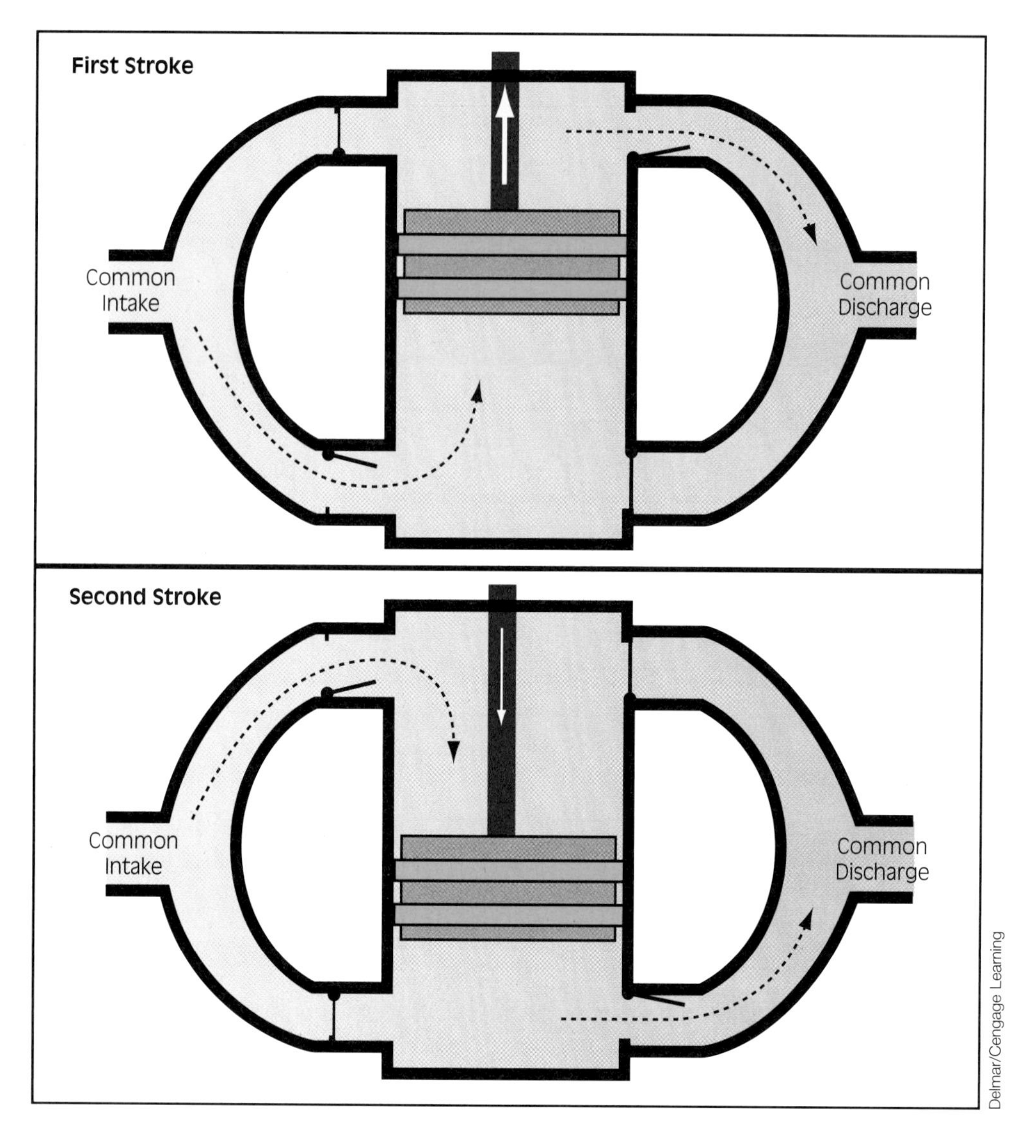

FIGURE 4-8 Two-stroke working cycle of a double-acting piston pump.

created from both the water and compressed air forces water out of the dome at a constant pressure.

Piston pumps can also be double acting (**Figure 4-8**). Double-acting piston pumps discharge water during both the upward and downward motion of the piston. As the piston moves up, water is both drawn in and discharged at the same time. The same is true when the piston is moving down. This, however, still does not solve the problem of pulsating since water flow stops each time the piston needs to change direction.

Rotary Lobe

The moving elements in rotary lobe pumps are two lobes (**Figure 4-9**). These lobes rotate in opposite directions within a pump casing. The volume between the pump casing and lobe increases alternately between the two lobes. Water enters these cavities

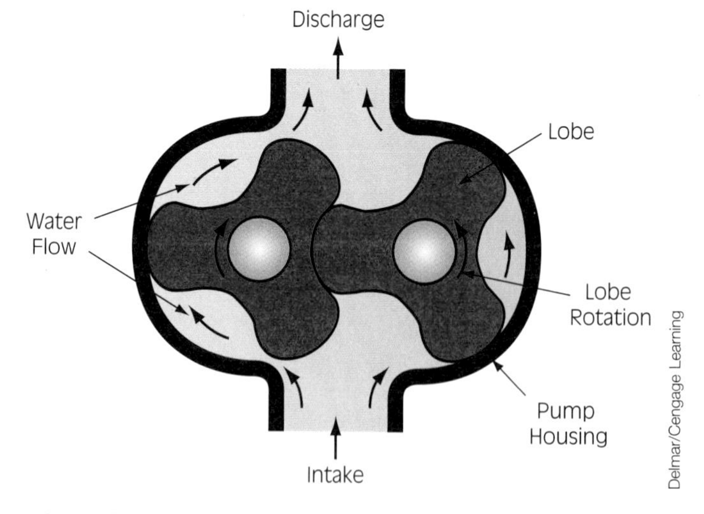

FIGURE 4-9 Illustration of a rotary lobe pump showing the flow of water from the intake to the discharge side of the pump.

and is trapped between the lobes and the pump casing. The lobes move the water along the outside until it reaches the discharge side of the pump. Note that the close mesh of the lobes prevents water from returning to the intake side of the pump.

Rotary Gear

Rotary gear pumps operate in the same manner as rotary lobes. Instead of lobes, however, closely meshed gears are used to trap water and move it to the discharge side of the pump (**Figure 4-10**). In some cases, the shape of the rotary gears allows one gear to be powered and in turn power the second gear.

Common drawbacks of rotary gear and rotary lobe pumps include the need for lubrication to prevent wear. (Environmental concerns now restrict the use of designs that discharge oil, which were common in prior years.) Since they require tight tolerances they will also fail from wear if the water they pump contains suspended debris such as silt or sand.

Rotary Vane

The moving elements in a rotary vane pump are the rotor and its vanes (**Figure 4-11**). The rotor is offset (eccentric) from the pump center. This eccentric alignment of the rotor creates a void space of increasing and decreasing volume. The vanes are located in the rotor and are able to move in and out within their slot.

The vanes maintain contact with the pump lining by spring tension, centrifugal force, or both. The vanes are typically made of a softer metal than the inside of the pump casing. Although the vanes wear down with use, they are able to maintain contact

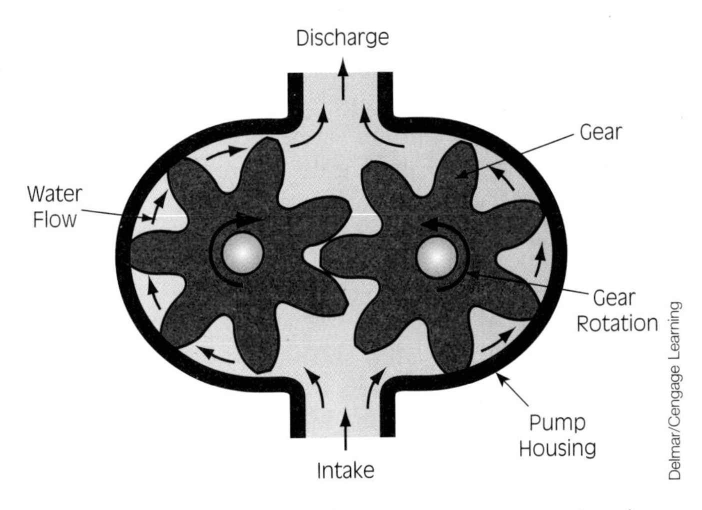

FIGURE 4-10 Illustration of a rotary gear pump showing the flow of water from the intake to the discharge side of the pump.

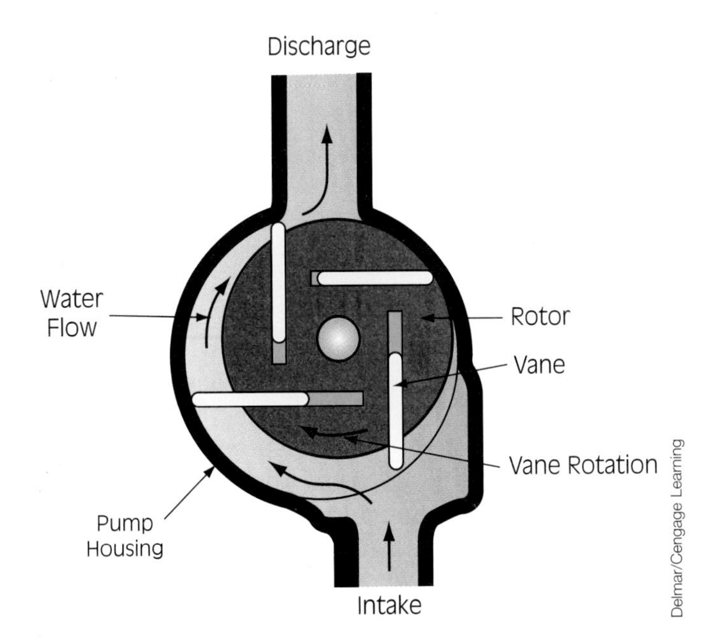

FIGURE 4-11 Illustration of a rotary vane pump showing the flow of water from the intake to the discharge side of the pump.

with the inside of the pump casing. When the vanes wear down to a certain point, they are replaced at a lower cost than replacing a worn pump casing. As the rotor turns, the volume (space) between the pump lining, the vane, and the rotor increases near the intake side of the pump. Water enters this increasing void space and is trapped. The trapped water is forced to the discharge side of the pump along the pump lining. At the discharge side of the pump, the volume (space) decreases in size, and the water is forced out. The close fit of the rotor and pump lining at the discharge side of the pump helps prevent water from returning to the intake. Older rotary vane primers used oil as both a lubricant and to improve the seal around the vanes, thus improving the pump's efficiency. Rotary vane designs also exist that do not require priming oil, thus eliminating environmental concerns about discharging oil.

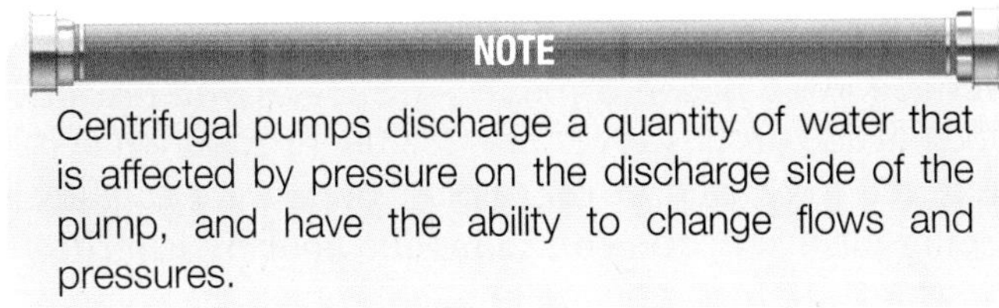

NOTE

Centrifugal pumps discharge a quantity of water that is affected by pressure on the discharge side of the pump, and have the ability to change flows and pressures.

CENTRIFUGAL PUMPS

Centrifugal pumps have replaced positive displacement pumps as the main pump on modern apparatus. The principles of operation and construction of centrifugal pumps are a little more complicated and confusing than those of positive displacement pumps.

Recall that positive displacement pumps discharge a specific quantity of water per revolution independent of pressure on the discharge side of the pump. Centrifugal pumps, however, do not discharge a specific quantity of water for each revolution. The quantity of water discharged is affected by pressure on the discharge side of the pump. With centrifugal pumps, flow, pressure, and the speed of the pump (RPM) are interrelated in that a change in one of these factors will change the others. This interrelationship is one of the useful aspects of centrifugal pumps. The ability to change flows and pressures makes centrifugal pumps very useful for the changing demands at an emergency scene. Calculations for centrifugal pump flows and pressures are discussed further later in this textbook.

Principles of Operation

The basic operating principle of centrifugal pumps is centrifugal force. **Centrifugal force** is often defined as the tendency of a body to move away from the center when rotating in a circular motion. Technically, there is no centrifugal force that acts on a body to move it away from the center. Rather, the term is used to explain the effective force or behavior of objects in a circular or rotating motion. Centrifugal force is perhaps better thought of as a term used to describe the outward force associated with rotational motion.

Newton's first and third law of motion can be used to help explain the concept of centrifugal force. Newton's first law of motion indicates that a moving body travels in a straight line with constant speed (velocity) unless affected by an outside force. Imagine that you are riding in the passenger seat of a car traveling down the road in a straight line and at a constant speed. As you travel down the road you do not experience or feel any forces acting upon you. If the speed is held constant but the car travels in a circle, you feel or experience what can be described as an outward force as it presses you toward the outside of the curve. This feeling or experience is referred to as centrifugal force. You are not actually being pushed or forced against the door by centrifugal force. Rather, you are pressed against the passenger door by the inertia of your body as it attempts to keep you moving in a straight line. Another way of thinking about this phenomenon is that the car turns in front of you and your body runs into the passenger door. According to Newton's third law of motion, for every action there is an equal and opposite reaction. You change direction following the path of the car because the passenger door pushes you toward the center of the circle or rotation. This is known as centripetal force, a Latin term meaning central (*centrum*) seeking (*petere*). As you travel in a circle, the force you feel is often described as centrifugal force, a Latin term meaning central (*centrum*) fleeing (*fugere*). As indicated earlier, the force you feel is actually caused by inertia. If the door suddenly opened, you would be thrown out of the car and would once again follow a straight line. Centrifugal force can also be thought of as the lack of centripetal force.

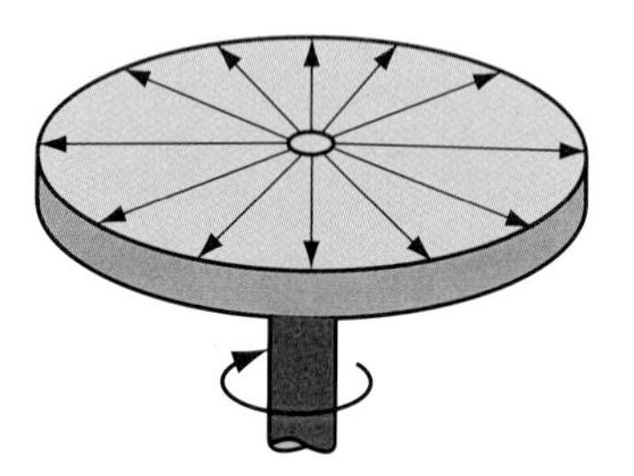

FIGURE 4-12 The centrifugal force (outward motion) generated by a spinning disk.

Figure 4-12 illustrates this concept as it relates to centrifugal pumps. Water placed on the spinning disk would tend to move away from the center in all directions because centripetal force is not exerted on the water. The distance water would travel is related to the speed of rotation and the diameter of the disk. If the water is trapped or contained, pressure will increase relative to the speed of the turning disk. The operation of centrifugal pumps is based on this principle

Operating Characteristics

The operating characteristics of centrifugal pumps are quite different than those of positive displacement pumps. Centrifugal pump performance is contingent on three interrelated factors: flow (quantity of water discharged), pressure of the water discharged, and speed of the pump. With one factor remaining constant, a change in one of the remaining factors will change the other factor, as follows:

- *Speed.* If the speed of the pump is held constant and the flow of water is increased, pressure will drop. Remember, pressure is the result of resistance on the discharge side of the pump. The less resistance, the less pressure. Specifically, if the speed of the pump impeller is doubled and the flow in gpm (Lpm) is kept constant, the pressure will increase four times. If the speed of the pump is doubled with the pressure kept constant, the flow in gpm (Lpm) will also double, which is to say the flow is proportional to the speed of the pump. Thus an increase in speed results in more pressure or more flow or some combination of the two. Additionally, the horsepower required to operate the pump varies with the cube of the change in speed or the impeller diameter.

- *Flow*. If the flow of water is held constant and the speed of the pump is increased, pressure will increase. In this case, the same amount of water is being discharged yet the pump is attempting to discharge more water. The result is an increase in pressure.
- *Pressure*. If the pressure is held constant and the speed of the pump is increased, flow will increase. In this case, the increased speed of the pump will increase the flow. The pressure is maintained constant by increasing or reducing the resistance on the discharge side of the pump. If the discharges are closed, the pump reaches something known as the cut-out or cut-off pressure. This is the point at which the amount of water returning due to slippage is equal to the amount of water going through the eye of the **impeller**. A new pump's cut-off pressure will be the highest due to its small clearances and it will decrease as the pump wears and clearances through which slippage occurs increase.

Three other important operating characteristics of centrifugal pumps are the use of intake pressure, slippage, and the need for priming. Unlike positive displacement pumps, centrifugal pumps take advantage of positive pressure on the intake side of the pump. So, if all other items are kept constant, any increase in intake pressure will result in the same amount of an increase in discharge pressure. For example, if the desired discharge pressure is 150 psi (1,050 kPa) and the intake pressure is 50 psi (350 kPa), the pump will only have to produce 100 psi (700 kPa), which is called **net pump discharge pressure (NPDP)**.

NOTE

Three important operating characteristics of centrifugal pumps are the use of intake pressure, slippage, and the need for priming.

Since centrifugal pumps are rated at 150 NPDP (1,050 kPa) and are able to discharge more water at lower NPDPs, pumps are able to discharge more than the rated gpms at 150 psi (1,050 kPa) **pump discharge pressure (PDP)**, when supplied with a positive pressure intake such as a fire hydrant.

Two types of slippage occur in centrifugal pumps. One is water not going through the eye of the impeller because of back pressure on the discharge, which is not a factor in the efficiency of a centrifugal pump. Centrifugal pumps have an open path from the intake to the discharge side of the pump. If the discharge side of the pump is closed, water simply spins within the pump. When this occurs, the pump is considered to be experiencing 100% slippage.

Another type of slippage is water moving between the impeller and the wear ring, thus returning from the discharge (pressure) side of the pump to the intake (suction) side of the pump. This slippage limits both how much water can pumped and its maximum pressure since it increases as the NPDP increases. To reduce this slippage, pump manufacturers leave only a few thousandths of an inch clearance between the impeller and the wear ring. Over time, wear can increase this clearance to the point that the pump will no longer be able to meet its rated capacity. When this occurs, the wear rings must be replaced with oversized wear rings that will restore the correct tolerance. Replacing wear rings, while much less costly than purchasing a new pump, is still an expensive, time-consuming operation, which with proper care should not normally be needed over the life of a fire pump.

One common cause of wear ring wear is overheating the pump. As the impeller heats up, it expands, eventually scraping the wear ring. This type of wear can be avoided by always circulating water through the pump to keep it cool.

Another cause of wear is pumping water that contains suspended debris such as silt and sand, which effectively sandblasts the interior of the pump. Procedures to minimize this wear include flushing hydrants prior to connecting lines, periodically replacing the water in booster tanks, draining sediment from pump drains, and, whenever possible, avoid pumping contaminated water.

Because centrifugal pumps rely on the movement of water to operate, the lack of water creates a serious problem. Centrifugal pumps must be primed in order to operate from a static water source. Centrifugal pumps are primed using a device called a primer, which is outside of the centrifugal pump and can reduce the pressure inside of the pump. This allows higher atmospheric pressure to push water into the pump.

Construction

Centrifugal pumps essentially have one moving part: the impeller (**Figure 4-13**). The impeller is a disk mounted on a shaft that spins within the pump casing (**Figure 4-14**). Water enters the impeller through its

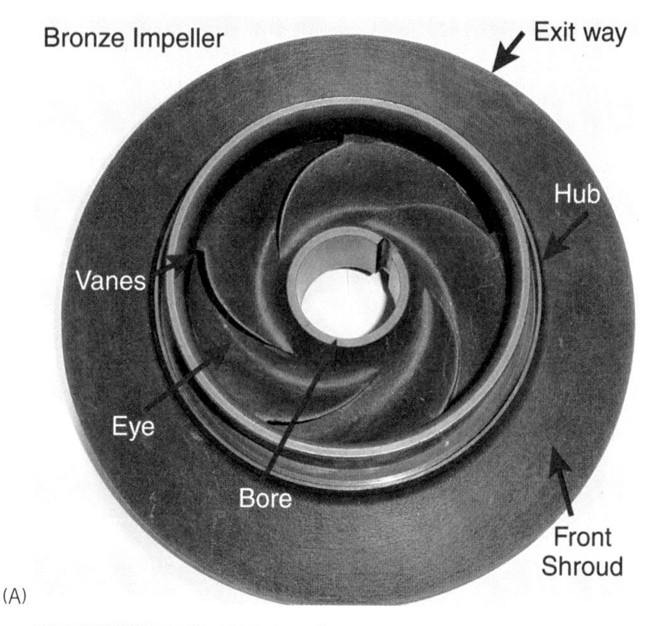

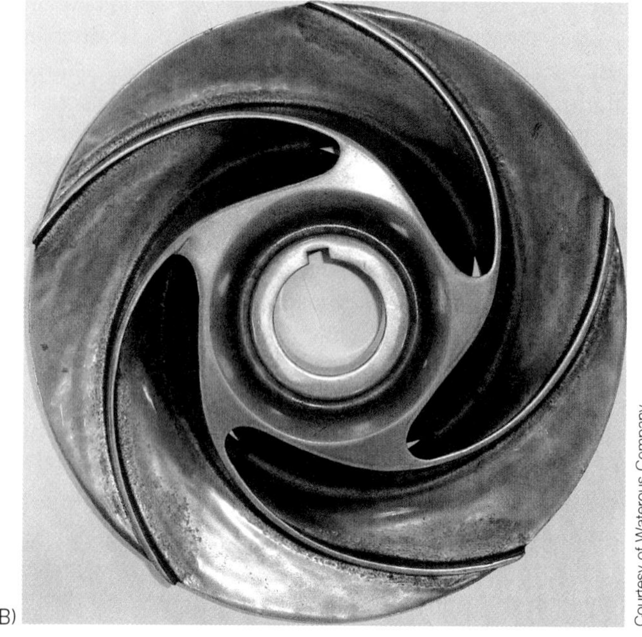

FIGURE 4-13 Centrifugal pump impellers. (A) The front of the impeller along with the names of components. (B) The back of a single-intake impeller with the rear shroud machined off, revealing the continuation of the vanes between the two shrouds. The vanes are the part of the impeller that imparts velocity to water as the impeller rotates.

eye. Some impellers have an opening on one side and are called a single-intake impeller while others have openings on both sides. The latter is typically called a double-intake or dual-intake impeller. After entering the eye, water enters the void space between the curved impeller vanes (sometimes called blades) and the sides of the impeller, called the shroud. When the impeller spins, water is forced to the outer edge of the impeller vanes by centrifugal force, creating velocity. The vanes also guide the water toward the discharge.

The impeller is typically mounted eccentric to the pump casing or housing wall. The void space created by this design, which increases toward the pump's discharge, is called the **volute** (**Figure 4-15**). The volute has several functions in a centrifugal pump. First, it converts the velocity created by the impeller's vanes into pressure. Second, the increasing void space in the volute allows for increases in water flow created by the impeller. Third, the volute directs water from the impeller to the discharge. Fourth, the volute helps to streamline water, which reduces pressure loss due to turbulence.

The path of a single drop of water through a centrifugal pump is illustrated in **Figure 4-16**. The drop enters the eye of the spinning impeller. The centrifugal force generated by the impeller forces the water to the outer edge of the impeller vane. As the drop moves away from the eye toward the volute, it picks up velocity or speed from the impeller. The velocity increases, in part, because the velocity of the impeller increases with the increase in distance from the eye. When the drop leaves the impeller, it is channeled by the volute and directed to the discharge.

Centrifugal pumps can be either single-stage or multistage in construction. Each impeller is referred to as a stage; thus, a single impeller within a pump casing is called a single-stage centrifugal pump. The speed of the impeller controls the performance of single-stage pumps. Multistage pumps have two or more impellers (stages) enclosed with individual volutes. These pumps provide greater flexibility of flows and pressures. Multistage pumps operate in either volume (parallel) or pressure (series) modes (see **Figure 4-17**). A transfer valve enables the pump operator to switch between the two modes. Most multistage pumps have two stages (meaning they have two impellers) and are called two-stage centrifugal pumps or series-parallel pumps.

STREETSMART

Historically, most fire pumps were two-stage pumps. Today, however, most pumps are high-volume single-stage pumps, although each of the major pump manufacturers still makes two-stage pumps. In the past, when most pumps had two stages, gasoline engines were common in fire engines. These gasoline engines would last much longer if run at lower RPMs, which was possible with two-stage pumps in pressure mode pumping a fraction of their rated capacity, as was the norm.

Today, diesel engines are used and they do not have problems if run at high RPMs for prolonged periods of time. Pump advances such as dual-intake impellers, which cancel out unbalanced forces existing

FIGURE 4-14 Cutaway of a single-stage centrifugal pump showing the impeller on a shaft, with intake water flow shown with blue arrows and discharge water flow shown with yellow arrows.

FIGURE 4-15 Centrifugal pump impeller and volute.

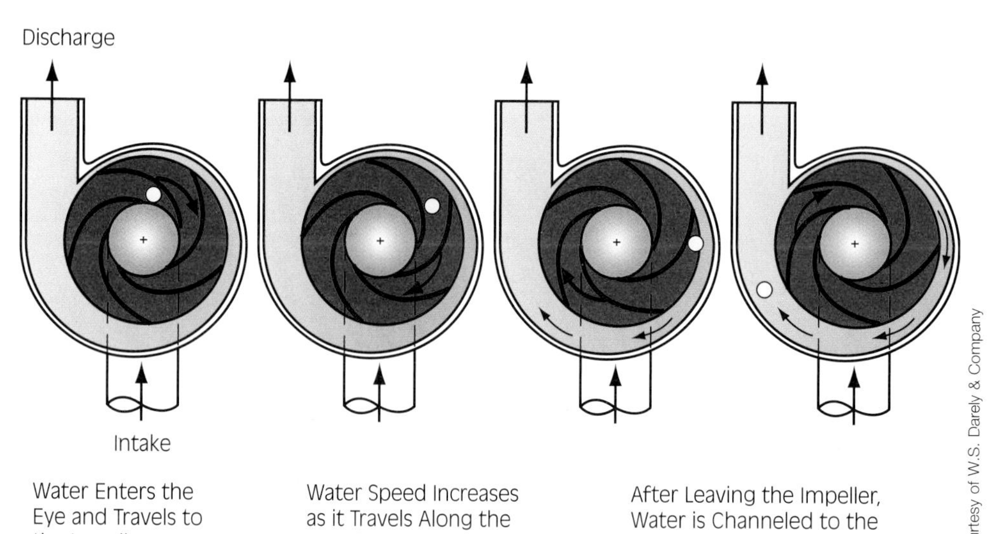

FIGURE 4-16 Path that water follows in a centrifugal pump.

with single-intake impellers and limited pump size due to impeller deflections the unbalanced forces caused. (Remember that only a few thousandths of an inch of clearance exists between the impeller and the wear ring.) Multi-stage pumps are still preferred when high pressures are needed for a sustained operation such as supporting a standpipe or sprinkler in a skyscraper.

Operating in Volume Mode

Operating in volume mode is similar to that of a single-stage operation. The main difference is that water enters both impellers directly from a common intake and leaves directly to a common discharge (**Figure 4-18**). Because both impellers are on the same drive shaft, they spin at the same speed and hence provide the same quantity of water. This means that half the total discharge is produced by each of the impellers. If the total discharge of a pump is 1,000 gpm (4,000 Lpm) at 100 psi (700 kPa), each impeller is discharging 500 gpm at 100 psi (2,000 Lpm at 700 kPa). Multistage pumps are operated in volume mode when large quantities of water are required for maximum flow rates at lower pressures, typically when exceeding 50% of the rated pump capacity for longer operations.

FIGURE 4-17 Cutaway of a dual-stage centrifugal pump showing the impellers on a shaft. The yellow arrows indicate where water enters the eye of each impeller.

Operating in Pressure Mode

When operating in pressure mode, the transfer valve redirects the discharge from one impeller to the intake of the second impeller (**Figure 4-19**). The second impeller can only discharge the quantity given to it by the first. The pressure generated by the first impeller, however, can be used by the second. Again, because the impellers are rotating on the same drive shaft, they will each produce a specific quantity and pressure. The second impeller takes the pressure given to it by the first impeller and adds its pressure. For example, if the total discharge of the pump is 500 gpm at 250 psi (2,000 Lpm at 1,750 kPa), then the first impeller is discharging 500 gpm at 125 psi (2,000 Lpm at 875 kPa). The discharge of the first impeller is then directed to the intake of the second impeller. The second impeller is only given 500 gpm (2,000 Lpm) to pump; however, it takes the 125-psi (875-kPa) intake pressure and adds its 125 psi (875 kPa) for a total pressure of 250 psi (1,750 kPa). Multistage pumps are operated in pressure mode when higher pressures with lower volumes are required.

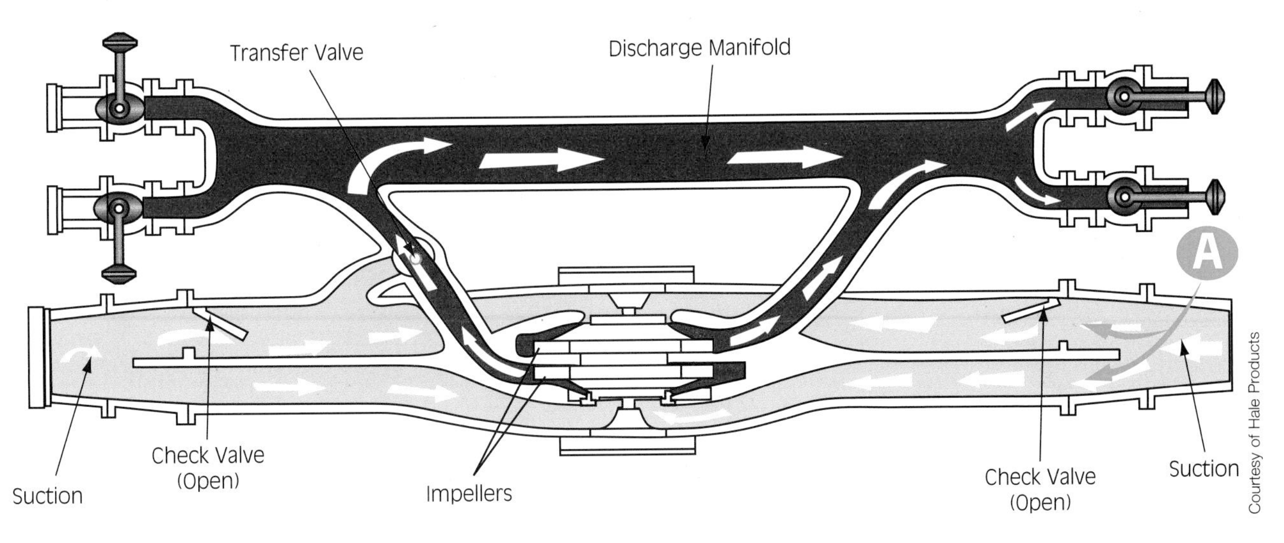

FIGURE 4-18 Water flow through a two-stage pump operating in volume mode.

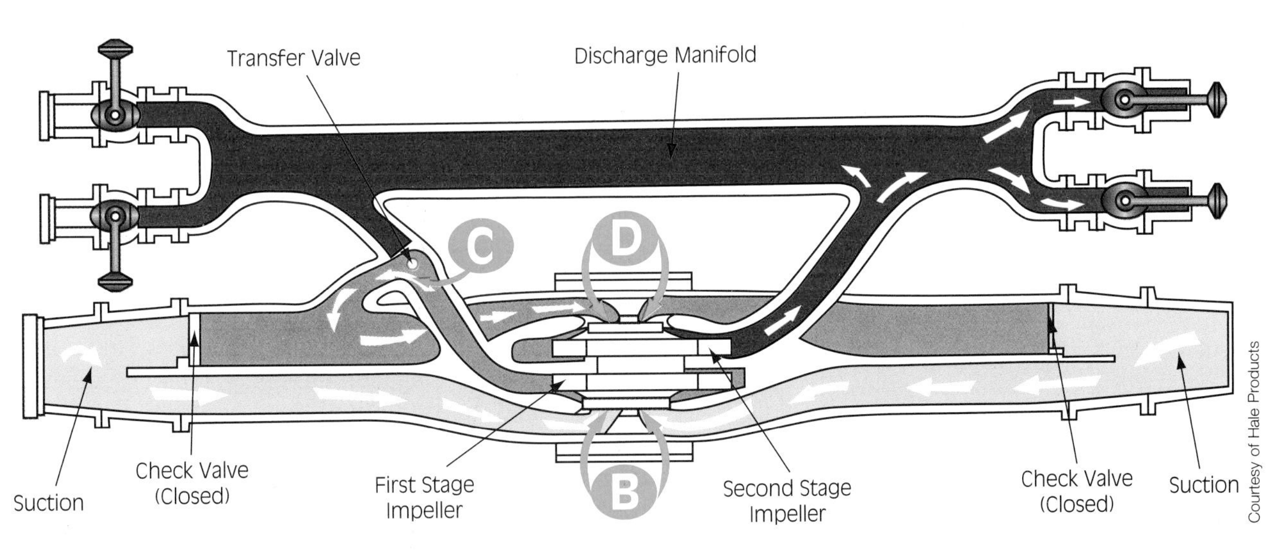

FIGURE 4-19 Water flow through a two-stage pump operating in pressure mode.

Rated Capacity and Performance

According to NFPA 1901, *Automotive Fire Apparatus*, pumps used for extended emergency operations must have standard pump capacities of 750, 1,000, 1,250, 1,500, 1,750, 2,000, 2,250, 2,500, or 3,000 gpm (3,000, 4,000, 5,000, 6,000, 7,000, 8,000, 9,000, 10,000, 12,000 Lpm). In addition, they must pump their rated capacity as follows: 100% at 150 psi (1,050 kPa), 70% at 200 psi (1,400 kPa), and 50% at 250 psi (1,750 kPa). If the pump is rated at 1,000 gpm (4,000 Lpm), then it should deliver 1,000 gpm at 150 psi (4,000 Lpm at 1,050 kPa), 700 gpm at 200 psi (2,800 Lpm at 1,400 kPa), and 500 gpm at 250 psi (2,000 Lpm at 1,750 kPa). See **Table 4-1** for the rated capacity of several pumps. Pumps must also have discharges for their rated capacity. For rating purposes, each 2½-in. (65-mm) or larger discharge is assigned a capacity based on the data in **Table 4-2**; the sum of all of these capacities is the largest rating that can be assigned to the pump. In reality these discharges are able to flow much more water than their rated capacities; however, this method helps to make sure that a pump has an adequate number of discharges for its size.

Fire Pump Subsystems

The four major components of a centrifugal fire pump are (1) the *intake manifold*, which previously was sometimes called the pump suction; (2) the

TABLE 4-1 Sample Rated Capacities in gpm (Lpm) for Several Common Sizes of Fire Pumps

Customary			Metric		
100% @ 150 psi	70% @ 200 psi	50% @ 250 psi	100% @ 1,050 kPa	70% @ 1,400 kPa	50% @ 1,750 kPa
500	350	250	2,000	1,400	1,000
750	525	375	3,000	2,100	1,500
1,000	700	500	4,000	2,800	2,000
1,250	825	625	5,000	3,500	2,500
1,500	1,050	750	6,000	4,200	3,000
1,750	1,225	825	7,000	4,900	3,500
2,000	1,400	1,000	8,000	5,600	4,000

TABLE 4-2 NFPA Pump Discharge Allowances for Rating Purposes

Outlet Size in Inches	Rated Flow Capacity in gpm
2.5	250
3	375
4	625
5	1000
6	1440

Reproduced with permission from NFPA 1901–2009, Standard for Automotive Fire Apparatus, Copyright © 2008, National Fire Protection Association, Quincy, MA. This is not the complete and official position of the NFPA on the referenced subject, which is represented only by the standard in its entirety.

Source: NFPA 1901 2009 ed. Table 16.7.1

discharge manifold, which previously was sometimes called the pressure manifold; (3) the *pump casing*; and (4) the *pump transmission* (**Figure 4-20**). Depending on how the pump casing is made, these components may be difficult to separate physically; however, each still exists and performs the functions described next. Understanding the function of each of these components and where they are physically on your assigned apparatus will allow quicker diagnostics and faster resolution of problems that will occur during operations.

Intake Manifold

The intake manifold collects water from various intakes and directs it to the pump casing. Therefore, all water entering the pump goes through the intake manifold. In a two-stage pump, the swing valve separates the intake manifold from the pump casing. It includes the intake relief valve, which prevents excessive pressures from reaching the pump; the return for the pressure relief valve; all intakes, including both gated and ungated auxiliary intakes, steamer intakes, and tank-to-pump intakes; connections to the master intake pressure gauge; and the auxiliary cooling return line. It typically also changes the round casings of the intake to a rectangular connection to the pump casing, which reduces problems with water spinning in circles.

Discharge Manifold

The discharge manifold, also called the discharge header, is located above the intake manifold and pump casing. All water leaving the pump goes through the discharge manifold. In a two-stage pump, it includes the transfer valve, connections to the master discharge gauge and the relief valve, all discharge valves, tank fill connection, and water to the engine cooler.

Pump Casing

The pump casing houses the impeller, where velocity is added to incoming water and then changed into pressure by the restriction of the volute, which is the part of the casing that surrounds the impeller and increases in size as it get closer to the discharge

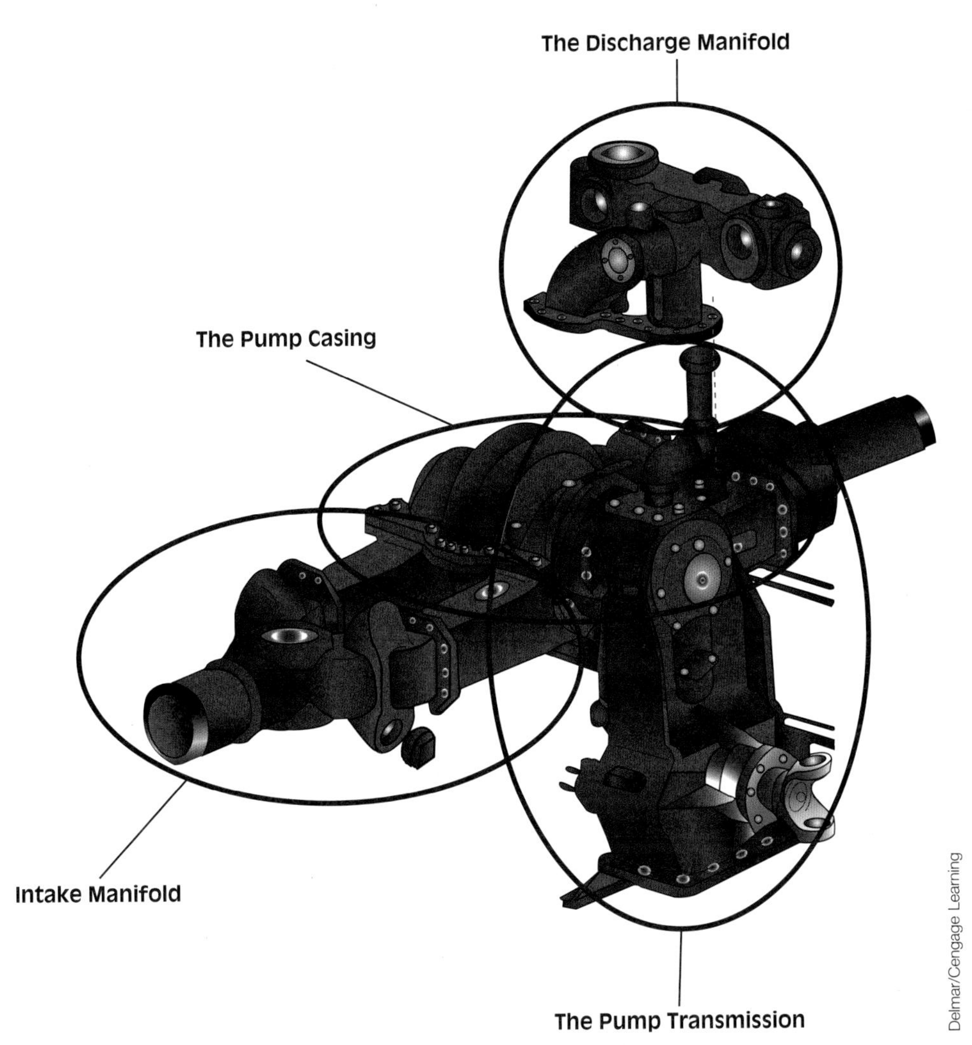

FIGURE 4-20 Major pump components.

to allow the increasing amount of water to flow. At the end of the volute a striping edge directs water from the volute into the discharge header. While the pump casing is hydrostatic tested to 500 psi (3,500 kPa), the entire pump including seals and valves is only tested to 250 psi (1,750 kPa), unless the fire department specifies a higher rating for its specific needs. While manufacturers do allow a safety margin in their numbers, operators should not exceed 250 psi (1,750 kPa) PDP to avoid blowing out a pump's seals. The seals will not blow out at 250 psi (1,750 kPa), but if pressures are exceeded to the point that they do blow, it is a sudden, catastrophic failure occurring without warning and will require expensive, time-consuming repairs.

The pump casing contains a stationary fin near the eye of the impeller called the Francis vane. It prevents induced cavitation by preventing the rotation of the impeller to cause the intake water to rotate with the impeller. To rotate the impeller(s), a shaft runs through the center of the impeller(s), with the shaft being supported by ball bearings. Since the shaft needs to rotate and go from atmospheric pressure outside the pump to the high pressures that can exist inside the pump, it needs to pass through either a mechanical seal or packing that prevents water from leaving the pump. Water is used both to lubricate and cool the packing, so a constant slow drip when the pump is under pressure is normal. If a pump continues to drip when not under pressure, the packing needs adjustment. The adjustment should be made by someone trained in packing adjustments to avoid any potential of damaging the pump shaft. In the case of a two-stage pump, a diaphragm, which is a disk usually made of bronze, is placed between the first-stage and the second-stage volutes of the pump. It prevents water at the higher pressure in the second stage from moving back to the first stage.

Pump Transmission

The pump transmission contains a shaft taking power from the engine and another shaft providing power to the pump. Within the transmission, either a chain (some Waterous models) or gears increase the speed of the impeller shaft so that the impeller rotates between 2,000 and 4,000 RPM. On split-shaft pumps, the transmission also contains a shaft that transfers power to the drive wheels when in road gear. To do this, the transmission needs to be able to transfer incoming power from the engine to either the drive wheels or pump based on options the operator has selected. Finally, to make all of these things work, the transmission contains gears, bearings, gear oil, and possibly filters.

LESSONS LEARNED

The reliance on water as the main extinguishing agent for fire suppression efforts dictates the reliance on pumps in the fire service. To operate pumps in a safe and efficient manner, pump operators must understand the basic operating principles and construction of pumps. The two main types of pumps based on operating principles are positive displacement and centrifugal pumps. The three categories of pumps based on intended use are main, priming, and auxiliary pumps. Pumps can be mounted almost anywhere on the apparatus and can be powered by a separate engine or the drive engine through a direct connection to the engine crankshaft, through a PTO from the transmitter, or through a split-shaft transmission.

KEY TERMS

Auxiliary pumps Pumps other than the main pump or priming pump that are either permanently mounted or carried on an apparatus.

Centrifugal force The tendency of a body to move away from the center when rotating in a circular motion.

Discharge The point at which water leaves the pump; also called the pressure side.

Floating pumps Pumps placed in a static water source to pump water to the apparatus intake, avoiding restrictions the apparatus would have with maximum lift.

Flow The rate and quantity of water delivered by a pump, typically expressed in gallons per minute (gpm) or litres per minute (Lpm).

Hydrostatics The branch of hydraulics that deals with the principles and laws of fluids at rest and the pressures they exert or transmit.

Hydrodynamics The branch of hydraulics that deals with the principles and laws of fluids in motion.

Intake The point at which water enters the pump; also called the supply side or suction.

Impeller A disk mounted on a shaft that spins within the pump casing.

Main pump Primary working pump permanently mounted on an apparatus.

Net pump discharge pressure (NPDP) The difference between the intake pressure and the discharge pressure. It is the amount of pressure the pump adds.

Positive displacement pump Moves a specified quantity of water through the pump chamber with each stroke or cycle; it is capable of pumping air, and therefore is self-priming, but must have a pressure relief provision if pumping or hoses have shutoff nozzles or valves; examples of positive displacement pumps are the rotary gear pump, the piston pump, the rotary lobe pump; and the rotary vane pump.

Pressure The force exerted by a substance in units of weight per area; the amount of force generated by a pump or the resistance encountered on the discharge side of a pump; typically expressed in pounds per square inch (psi) or kilopascals (kPa).

Priming The process of replacing air in a pump with water.

Priming pump Positive displacement pump permanently mounted on an apparatus and used to prime the main pump.

Pump A mechanical device that raises and transfers liquids from one point to another.

Pump-and-roll An operation where water is discharged while the apparatus is in motion.

Pump discharge pressure (PDP) The amount of pressure on the discharge side of the pump. It is the pressure read on the pump discharge gauge.

Pump speed The rate at which a pump is operating, typically expressed in revolutions per minute (RPM).

Rated capacity The flow of water at specific pressures that a pump is expected to provide.

Slippage The leaking of water between the surfaces of the internal moving parts of a pump.

Speed The rate at which a pump is operating, typically expressed in revolutions per minute (RPM).

Volute An increasing void space in a pump that converts velocity into pressure and directs water from the impeller to the discharge.

REVIEW QUESTIONS

Multiple Choice

Select the most appropriate answer.

1. Positive displacement pumps generally provide

a. lower pressures with higher flows.

b. lower pressures with lower flows.

c. higher pressures with lower flows.

d. higher pressures with higher flows.

2. Centrifugal pumps are based on principles from the branch of hydraulics known as

a. hydrodynamics.

b. hydrostatics.

c. hydromotion.

d. hydrophysics.

3. According to NFPA 1901, an apparatus intended for sustained operations at structural fires must have a centrifugal pump rated at least

a. 250 gpm (1,000 Lpm).

b. 500 gpm (2,000 Lpm).

c. 750 gpm (3,000 Lpm).

d. 1,000 gpm (4,000 Lpm).

4. Which of the following is incorrect concerning positive displacement pumps?

a. Each revolution of the pump yields a specific quantity of water.

b. The pressure on the intake side of the pump does not affect the discharge.

c. The water being pumped is independent of discharge pressure.

d. The quantity of water discharged per revolution will increase as pump speed increases.

5. Each of the following is an example of a positive displacement rotary pump, *except*

a. gear.

b. vane.

c. lobe.

d. piston.

6. If the speed of a centrifugal pump is held constant and the flow of water is increased (say, for example, a larger line is attached), the pressure will

a. decrease.

b. increase.

c. not change.

d. increase at first but then decrease.

7. If the flow of water in a centrifugal pump is held constant (same size hose and nozzle) and the speed of the pump is increased, pressure will

a. decrease.

b. increase.

c. not change.

d. increase at first but then decrease.

8. The void space created by the impeller being mounted eccentric to the pump casing is called

a. centrifuge.

b. countervailance.

c. hydro-symmetry.

d. volute.

9. According to NFPA 1901, pumps must deliver their rated capacity for each of the following, *except*

a. 100% at 150 psi (1,050 kPa).

b. 70% at 200 psi (1,400 kPa).

c. 50% at 250 psi (1,750 kPa).

d. 25% at 300 psi (2,100 kPa).

Short Answer

On a separate sheet of paper, answer/explain the following questions.

1. Explain why it is important to understand pump operating principles and construction.

2. Explain the difference between a positive displacement pump and a centrifugal pump.

3. Explain the uses of positive displacement and centrifugal pumps in the fire service.

4. What is slippage and why is it an important factor in positive displacement and centrifugal pumps? Explain.

5. List and discuss four reasons why pumps are needed in the fire service.
6. List and discuss the terms used to describe the efficiency of a pump.
7. Explain the priming process.
8. Explain why positive displacement pump discharge is independent of discharge pressure.
9. Draw a simple piston pump and label the three moving parts.
10. What keeps the vanes in a rotary vane pump in contact with the pump lining?
11. If the pressure of a centrifugal pump is held constant and the speed of the pump is increased, flow will __________. Why does this occur?
12. Explain what happens when water enters the eye of the impeller.
13. If the total discharge of a two-stage pump operating in the pressure mode is 1,000 gpm at 150 psi (4,000 Lpm at 1,050 kPa), what is the flow and pressure for each of the impellers?

ACTIVITIES

1. For each of the following pumps, provide a rough drawing of the pump, identify major components, and indicate the path of water flow in each: rotary (either gear or vane), piston, and centrifugal.
2. Develop a rated capacity table for all standard pump capacities. Use the following headings across the top of the table: @ 150 psi (1,050 kPa), @ 200 psi (1,400 kPa), and @ 250 psi (1,750 kPa).

PRACTICE PROBLEMS

1. You are operating in the volume mode of a two-stage 1,250-gpm (5,000-Lpm) centrifugal pump flowing capacity @ 150 psi (1,050 kPa). Diagram the water flow, indicating pressure and gpm (Lpm). Be sure to indicate pressure and gpm (Lpm) for each stage of the pump as well as total flow and pressure.
2. You are operating in the pressure mode of a two-stage 1,250-gpm (5,000-Lpm) centrifugal pump flowing 50% of rated capacity @ 250 psi (1,750 kPa). Diagram the water flow, indicating pressure and gpm (Lpm). Be sure to indicate pressure and gpm (Lpm) for each stage of the pump as well as for the total flow and pressure at the intake and discharge of each stage.

ADDITIONAL RESOURCES

Darley website, http://www.darley.com/pumps.html

Hale Pumps website, http://www.haleproducts.com

Waterous website, http://www.haleproducts.com

5 Pump and Apparatus Peripherals

- Learning Objectives
- Introduction
- Pump Panel
- Pump Instrumentation
- Control Valves
- Pumping Systems
- Miscellaneous Pump Panel Items
- Auxiliary Apparatus Equipment and Fixed Systems
- Lessons Learned
- Key Terms
- Review Questions
- Activities
- Practice Problem
- Additional Resources

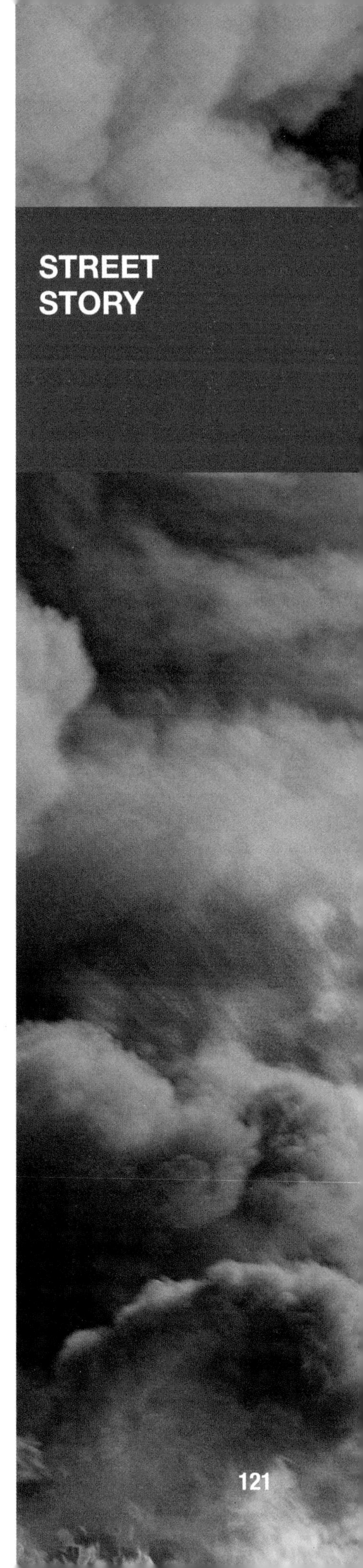

Not long ago we had a fire in a neighboring township, and the city companies were called in. It was a residential fire in a 2½-story wood-frame home. When we arrived, we parked right behind the ladder company, and by the time we got there, the command staff was setting up a defensive attack and had made the decision to attack from the platform of the ladder company using master streams.

It was our job to supply the water to the ladder company, and instead of just giving water, it occurred to me that we could use a solution of water mixed with foam.

Our fire engine is an unusual vehicle—it's the only one in the city with a single-agent foam system that can apply Class A or Class B foam. Our system is set up to supply foam to seven different discharges, and the apparatus is equipped with a separate 30-gallon (100 litre) foam tank.

From my training, I know that Class A foam is attracted to carbon; it wants to coat materials that are burning. Thus it breaks down the surface tension and enables the water to absorb into the material. I also know from my training that foam allows more fire to be extinguished than the same amount of water. Using Class A foam effectively doubles our water supply, so our 750 gallons becomes as effective as 1,500 gallons (6.000 litres) because the water is used more efficiently. If we are using one attack line flowing 175 gallons per minute (700 litres per minute), our tank would last approximately 4 minutes—with the addition of Class A foam that stretches to 8 minutes.

With a fire of this size, I was worried that we might not have enough water. So I told the operator, "I'll make the hookups, and send you some foam."

In this situation, knowing the full capacity of the truck and the capabilities of the foam pump allowed me to think outside the box, and come up with a more efficient solution as far as overall extinguishment. Using the foam made the master streams more effective, so we got a quicker knockdown. And the quicker we can get the fire out, the better it is for everyone involved: The guys inside didn't have to attack the fire for as long, and the structure was less compromised.

—Street Story by Steve Saksa,
Firefighter/Paramedic, Engine Co. 14, 2 Unit,
Columbus Division of Fire, Columbus, Ohio

LEARNING OBJECTIVES

After completing this chapter, the reader should be able to:

5-1. Identify the pump peripherals typically found on pump panels.

5-2. Describe the advantages and disadvantages of different pump panel locations.

5-3. Discuss the importance of continuously monitoring instrumentation.

5-4. Explain the two methods of developing discharge flows using flow meters and discharge pressure gauges.

5-5. Explain how flow meters simplify discharge flow development.

5-6. Describe where to find foam generation and application information.

5-7. Identify the common types of foam systems found on pumping apparatus.

5-8. List the components of a foam eduction system and describe how they function.

5-9. List the components of a foam injection system and describe how they function.

5-10. Identify the basic components of a foam-producing system with an eductor.

5-11. Identify the advantages and limitations of eductor and proportioner foam systems.

5-12. Verify the operational status of the foam system (s) found on a pumping apparatus by performing the necessary inspections and tests.

5-13. Identify the basic components of a foam-producing system with a proportioner, and explain the differences in operation compared to an eductor system.

5-14. Produce a properly proportioned foam fire stream given foam-producing equipment.

5-15. Explain how foam concentrations and proportioning rates affect foam generation capabilities and foam qualities.

5-16. Demonstrate operating foam-producing equipment after correctly setting the foam metering valve for the concentrate being used.

5-17. Properly connect a foam eductor, hose, and foam nozzle to a pumping apparatus.

5-18. Describe the procedures for assembling the equipment to produce foam fire streams.

5-19. Demonstrate correctly operating all fixed systems and equipment on a fire department vehicle.

5-20. Identify the manufacturer's operating instructions and local policies and procedures applicable to operating auxiliary air, electrically, and hydraulically operated equipment and fixed systems on a fire apparatus.

5-21. Demonstrate the safe operation of auxiliary air, electrically, and hydraulically operated systems on a fire apparatus.

5-22. Monitor the operation of auxiliary equipment and fixed systems on a fire apparatus.

5-23. Demonstrate the appropriate actions necessary to assure reliable operation of auxiliary systems and equipment contained on a fire apparatus.

***The driver/operator requirements, as defined by the NFPA 1002 Standard, are identified in black; additional information is identified in blue.**

INTRODUCTION

The ability to move water from one location to another is a function of the pump and the peripheral components associated with the pump. As stated earlier, understanding the construction and operating principles of a pump will enable the pump operator to:

- Position the pump for best use of available water supply
- Safely operate the pump to deliver proper flows and pressures
- Conduct proper preventive maintenance and inspections on the pump

Understanding the construction and operating principles of pump peripherals will likewise assist the pump operator in pumping activities.

Pump peripherals are those components directly or indirectly attached to the pump that are used to control and monitor the pump and the engine. Pump peripherals can be grouped into three categories: instrumentation, control valves, and systems. Instrumentation, such as gauges, flow meters, and indicators, is used to ensure that the pump is operating safely and efficiently while providing appropriate pressures and flows. Control valves include devices used to initiate, restrict, or direct water flow. Systems are components that operate as a support function to the pump. Examples include priming systems, pressure relief systems, cooling systems, and foam systems. Most pump peripherals are located on, or operated from, the pump control panel.

In **Chapter 4**, pump construction and operating principles were discussed. In this chapter we look at the pump peripherals found on modern pumping apparatus, their construction, and operating principles. Pump peripherals are important for several reasons. First, they allow the pump operator to control the pump by, for example, increasing pump pressure, opening discharge and intake valves, and priming the pump. Second, they help the pump operator ensure that the pump is operating in a safe and efficient manner through the use of gauges and indicator lights and by operating components such as transfer valves and cooling systems. Finally, pump peripherals provide a margin of safety to both the pump and personnel with such items as intake and discharge pressure relief systems.

Pump peripherals can be grouped into three categories: instrumentation, control valves, and systems.

Delmar/Cengage Learning

FIGURE 5-1 Pump operators control and monitor the pump from the pump panel.

PUMP PANEL

The **pump panel** is the mission-control center for pumping operations. It is the one location where the pump operator can control and monitor the pump as well as evaluate its efficiency (**Figure 5-1**). Pump control, monitoring, and evaluation are accomplished through **pump peripherals** (instrumentation, **control valves**, and systems) located on the pump panel. **Figure 5-2** identifies typical pump peripherals found on pump panels. Instrumentation components provide the ability to monitor and evaluate pressures, flows, and pump configuration. Engine operating parameters can also be monitored from **instrumentation** located on the control panel. Operators should periodically also check other instrumentation to confirm consistent readings on duplicated gauges and monitor non-duplicated gauges. For example, the apparatus fuel level gauge is normally only readable from the driver's seat; nevertheless, the pump operator is responsible for communicating up the chain of command a low-fuel situation as soon as it becomes clear that the on-board fuel may not be adequate to complete the call. This allows necessary actions to be implemented to ensure a successful incident outcome.

Control valves are necessary to physically change the pump configuration, such as operating discharge valves, transfer valves, and pump-to-tank valves. Systems on the pump are those components that operate in conjunction with the pump. For example, while the pump is operating, pressure control systems and cooling systems can be used to help ensure pressures and temperatures remain within safe operating limits. Without a central location for pump peripherals, pump operators would have a more difficult time operating the pump.

Physical Characteristics

All pump panels are not created equal. The number, type, and location of pump peripherals on a pump panel are affected by several factors. First, the size and type of pump will affect the number of peripherals on a pump panel. Large-capacity pumps will often have more instrumentation and control valves, while smaller pumps have fewer peripherals (**Figure 5-3**). Second, the intended use of the pump will affect the types of peripherals on a pump panel. For

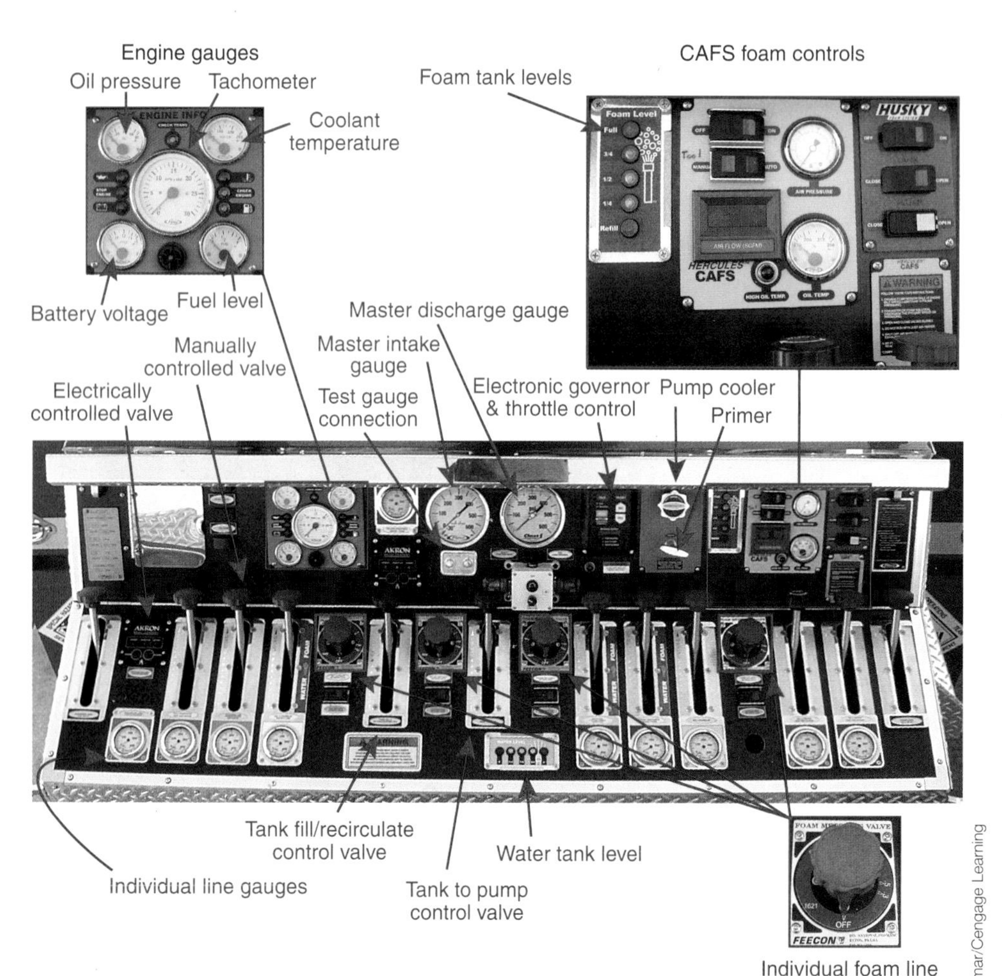

FIGURE 5-2 Typical components found on a pump panel.

example, pumps intended to supply foam will have a foam system on the pump panel (see the foam system on the larger pump panel in **Figure 5-3**). Third, types and locations of peripheral components on a pump panel will be different based on manufacturer preferences. Finally, NFPA 1901, *Standard for Automotive Fire Apparatus*, can affect pump panels. For example, according to NFPA 1901, only 2½-inch (65-mm) or smaller discharges can be located on the operator's panel. In addition, NFPA 1901 requires the following warning on pump panels:

"WARNING: Death or serious injury might occur if proper operating procedures are not followed. The pump operator as well as individuals connecting supply or discharge hoses to the apparatus must be familiar with water hydraulics hazards and component limitations."

Compare the pump panels shown in **Figure 5-3** with the pump panels in **Figure 5-4**. Note the different sizes and configurations of the panels.

The traditional appearance of pump panels is in a state of change. One reason for this change is the use of electronic and microprocessor-driven components. Increasingly, **flow meters** are augmenting pressure devices, and small electric actuating motors are installed on pump panels. **Figure 5-5** illustrates the dramatic change in the appearance of a modern electronic pump panel over traditional panels. A second reason that pump panels are in a state of change is the requirements of NFPA standards. In the case of the pump panel, the change is toward standardization. For example, NFPA 1901 sets minimum gauge sizes, requires intake relief valves, and suggests a standard color code to match discharges with their gauges (see **Table 5-1**). It even specifies the wording for some labels. For example, master gauges on some pumps had been labeled "suction" and "pressure"; today they are labeled as "intake" and "discharge." Finally, pump panels are changing to provide a wider margin of safety to pump operators, for example, by the removal of intake and discharge hose connections from the pump panel (**Figure 5-6**). Because of this change, the pump operator is less likely to be injured from lines rupturing or from simply tripping over them.

(A)

(B)

FIGURE 5-3 Pump panels differ, based in part on the size of the pump.

Although pump panels can be different in a number of ways, they are often more alike than different. These common items simplify the job of operators who may need to operate different apparatus by allowing them to understand the panel quickly with a recognition of common elements they should look for.

- NFPA 1901 specifies basic peripheral components that should be on the pump panel. These components include: engine throttle, discharge valves, gauges, master intake and discharge gauges, a tank level gauge, and tank-to-pump and pump-to-tank valves.
- Pump discharge control valves and their gauges are typically grouped together on most pump panels (see **Figure 5-7**).
- The general location of peripherals is similarly located on most pump panels. For example, the master intake and master discharge gauges are larger than other intake and discharge gauges and are located close together in the upper section of the panel with the master intake gauge on the left and the master discharge gauge on the right. Pump intakes are typically located in the lower section of the pump panel with auxiliary intakes being below the center of the steamer intake and discharges being above them.
- The size of pump panels tends to vary based on the size of the pump, in part because larger pumps tend to have more discharges and because the pumps themselves are larger, both of which require more space. Larger pumps are also more likely to have optional systems, such as a compressed-air foam (CAF) system, which requires space on the pump panel. Pump panels provide for illumination and ease of access to components.
- Pump panels are typically constructed of stainless steel, with some having a vinyl coating. Paint is not used on pump panels because the damping fluid in gauges will damage paint.

Compare the pump panels in the previous figures for their similarities.

Intakes and Discharges

Intakes (with the exception of the steamer connection) will be equipped with female threads while discharges are equipped with male threads. All intakes should also contain strainers that will not permit anything to enter the pump that will not fit through the impeller.

(A)

(B)

(C)

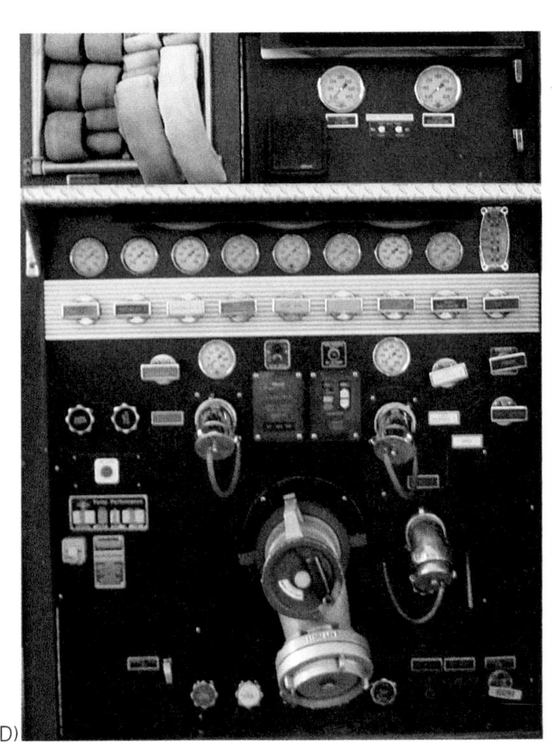
(D)

(E)

Delmar/Cengage Learning

FIGURE 5-4 Various pump panel configurations exist.

Delmar/Cengage Learning

FIGURE 5-5 A modern electronic pump panel.

Delmar/Cengage Learning

FIGURE 5-6 The apparatus has no intake or discharge connections on the pump panel.

TABLE 5-1 NFPA 1901 Suggested Color Coding for Matching Discharge Gates with Their Respective Gauges

Discharge Number	Color
Preconnect #1 (front bumper jump line)	Orange
Preconnect #2	Red
Preconnect or discharge #3	Yellow
Preconnect or discharge #4	White
Discharge #5	Blue
Discharge #6	Black
Discharge #7	Green
Deluge/deck gun	Silver
Water tower	Purple
Large-diameter hose	Yellow with white border
Foam line(s)	Red with white border
Booster reel(s)	Grey
Inlets	Burgundy

Reproduced with permission from NFPA 1901–2009, Standard for Automotive Fire Apparatus, Copyright © 2008, National Fire Protection Association, Quincy, MA. This is not the complete and official position of the NFPA on the referenced subject, which is represented only by the standard in its entirety.

Delmar/Cengage Learning

FIGURE 5-7 Discharge control valves and their gauges are typically located next to each other.

Pump Panel Location

The location of pump panels on the apparatus has been restricted, in part, by the constraints of mechanical operating components and by tradition. However, the introduction of electronic automation allows pump panels to be located almost anywhere on the apparatus. In addition, tradition appears to be giving way to safety and efficiency issues related to pump configuration and location.

One safety issue centers on the connecting of intake and discharge lines to the pump panel. The concern is the possibility of tripping over, or the rupturing of, intake and discharge lines. Another safety issue focuses on the location of the pump operator while at the pump panel. The concern is the possibility of the operator being hit by moving vehicles. An issue related to efficiency includes the ability to get pump operations initiated quickly while at the same time staying clear of equipment being removed from the apparatus. Other issues include the ability to observe emergency scene operations, noise levels, and the ability to communicate.

Different locations of pump panels on apparatus have both advantages and disadvantages. In some cases, trade-offs are made in moving the pump panel from one location to another.

Traditional Location

The traditional location of the pump panel (**Figure 5-8**) is just behind the crew cab entrance on the driver's side. One advantage of this location is that pump operators are practically at the pump panel when they exit the cab. In addition to quick access to the pump panel, this position keeps the pump operator out of the way while equipment is removed from the apparatus.

Delmar/Cengage Learning

FIGURE 5-8 The traditional location of the pump panel is just behind the crew cab on the driver's side.

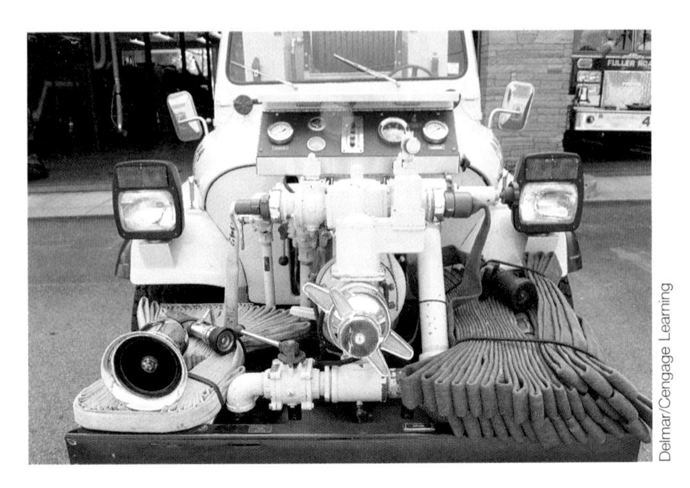

Delmar/Cengage Learning

FIGURE 5-9 Apparatus with the front-mounted pump panel located next to the pump.

Safety appears to be the greatest disadvantage of this location. First, intake and discharge lines typically connected to the pump panel subject the pump operator to hazards from tripping, as well as the potential rupture of a hose at the panel. Second, this position typically exposes the pump operator to the possibility of being hit by passing vehicles. Finally, this position offers pump operators limited visibility of the emergency scene.

Front-Mounted Pump Panels

Pump panels mounted on the front of apparatus are not common. When pump panels are mounted on the front of the apparatus, the pump is typically mounted there as well (**Figure 5-9**). Front-mounted pumps are usually smaller, resulting in fewer peripherals. Consequently, the pump panel is relatively small with only basic components and a lower price than other configurations (**Figure 5-10**). One advantage to this

Delmar/Cengage Learning

FIGURE 5-10 Close-up view of a front-mounted pump panel.

position is that pump operators have a relatively good view of activities on the emergency scene. Another advantage is positioning for drafting evolutions. The overriding concern with this location is the potential for inadvertent movement of the apparatus while the pump operator is standing in front of it. Tripping hazards and breakage of lines are also problems at this location. Disadvantages of front mounts include less protection from freezing weather and front-end collisions.

Top-Mounted Pump Panels

Pump panels found on the top of the apparatus are typically mounted across the apparatus so the pump operator faces the rear while operating the pump (**Figure 5-11**). The major reason for having the pump panel on top of the apparatus is to reduce the likelihood of pump operators being hit by traffic. In addition to being safe from moving vehicles, this position often provides a good view of fireground activities. Another advantage is that pump operators have fewer obstructions to deal with while working at the pump panel since intakes and discharges are typically located elsewhere on the apparatus.

There are two main disadvantages to a top-mounted pump panel: the potential for leaving the pump panel unattended while connecting or changing hoselines, and the increased risk of slipping and falling when the pump operator ascends and descends the apparatus. Other disadvantages include a high price resulting from the additional linkages needed for valve controls and more complex pump transmission, and an increased apparatus length, which can affect maneuverability in tight places and even fitting into some stations.

Rear-Mounted Pump Panels

Rear-mounted pump panels (**Figure 5-12**) can be located on the end or toward the end on the sides of the apparatus. One advantage of pump panels located on the back of the apparatus is that discharge lines can be preconnected to the panel. In addition, other hoselines are readily accessible to the pump operator. This position also allows the pump operator to assist with pulling lines. One disadvantage of this location is that the pump operator may get in the way of equipment being taken off the apparatus. In addition, the distance of the pump panel from the cab requires further travel by the pump operator, and, occasionally, emergency apparatus are struck in the rear by vehicles while operating on the emergency scene.

NOTE

Instrumentation on pump panels can be grouped into three categories: pressure gauges, flow meters, and indicators.

Inside Location

Pump panels are also located inside the cab/crew compartment (**Figure 5-13**). This position has several advantages. First, the pump operator is off the street, thus reducing the likelihood of being hit by traffic. Second, the working conditions are more favorable. Heating and air conditioning as well as reduced noise levels allow the pump operator to

Delmar/Cengage Learning

FIGURE 5-11 Apparatus with a top-mounted pump panel.

Delmar/Cengage Learning

FIGURE 5-12 Rear-mounted pump panel showing close-up and position on apparatus.

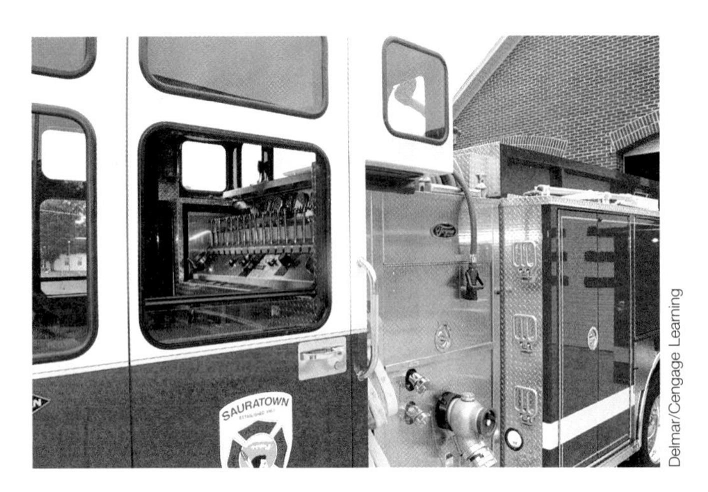

FIGURE 5-13 Apparatus with the pump panel inside the crew cab.

concentrate on operating and monitoring. Inside pump panels are most common in areas where serve cold weather is common. One disadvantage of this location is the limited view provided inside the cab. Small windows, fog, and equipment may obscure the view of the fireground. In addition, the pump operator must go in and out of the cab to hook up hoselines, resulting in increased slipping potential, leaving the pump unattended, and loss of contact with the fireground.

PUMP INSTRUMENTATION

All modern pumping apparatus have instrumentation located on their pump panels. Instrumentation is vital for setting up and monitoring pumping operations as well as for evaluating the efficiency of the pump and its engine. In order to operate and monitor the pump and its peripherals safely, the pump operator must be familiar with all instrumentation located on the pump panel. In addition, continuous monitoring of this instrumentation should take place whenever the pump is used. Instrumentation can provide warnings prior to the onset of many problems that affect pumps, but only if the operator is alert to the changes and understands the significance of the readings. According to NFPA 1901, the following instrumentation must be located on the pump panel:

- Master pump intake and discharge gauges
- Engine tachometer
- Pumping engine coolant temperature and oil pressure indicators
- Voltmeter
- Water tank level indicator

Instrumentation on pump panels can be grouped into three categories: **pressure gauges**, flow meters, and **indicators**.

Pressure Gauges

Obviously, the purpose of a pressure gauge is to measure pressure. The purpose of pressure gauges on a pump panel is to provide a central location for the collection of information on pressures during pumping operations. On the fireground, pump operators use this information in hydraulic calculations to determine flow. Additional information on calculated flow is presented in Section IV of this text. Pressure gauges on pump panels may measure positive and negative pressure. Positive pressure gauges measure pressure in pounds-per-square-inch (psi) (kilopascals or kPa) above atmospheric pressure. Negative pressure gauges measure pressure in inches of mercury (in. Hg, kilopascals or millimetres [kPa or mm]). A **compound gauge** reads both positive pressure (psi) (kPa) and negative pressure (in. Hg) (kPa or mm).

The **bourdon tube gauge** is the most common analog pressure gauge found on pump panels. It consists of a small curved tube filled with liquid (**Figure 5-14**). One end of the tube is connected to the indicating needle through a linkage system. When the liquid in the tube is under pressure, the tube straightens itself, causing the indicating needle to turn. When the pressure of the liquid decreases, the tube begins to return to its original curved position, again causing the indicating needle to turn.

NOTE

All gauges should be calibrated at regular intervals to ensure that they provide accurate readings.

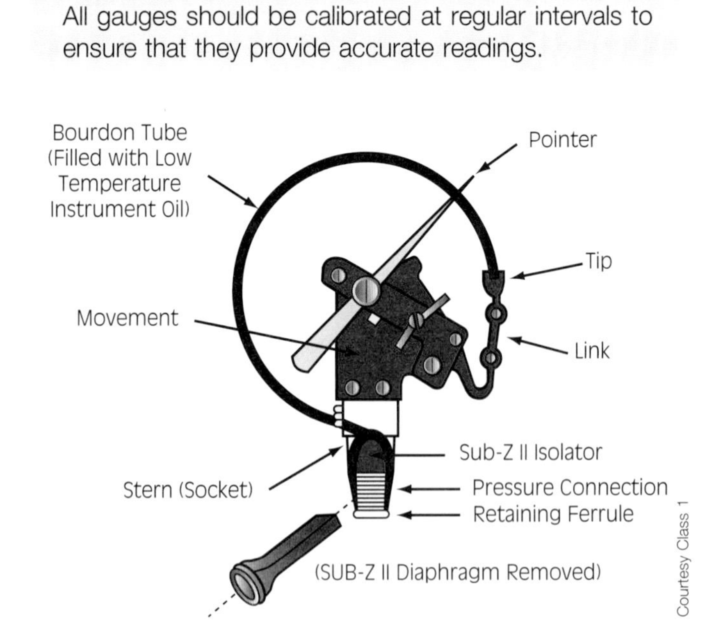

FIGURE 5-14 Diagram of a bourdon tube gauge.

Vibrations of the engine and pump and varying water pressures can cause the indicating needle to fluctuate or bounce. To compensate for this movement, some manufacturers fill the gauge with a heavy clear liquid, such as silicone, to help keep the needle from excessive fluctuation. Another means used to reduce needle bounce is a dampening device. These devices place pressure on the needle to reduce the fluctuation; however, if too much pressure is placed on the needle, the gauge will not function correctly.

Bourdon tube gauges are used because they provide the degree of accuracy and durability required on mobile apparatus. However, the often harsh working conditions and environments these gauges operate in can, over time, affect their accuracy. Therefore, the calibration of all gauges should be checked at regular intervals to ensure that they provide accurate readings. As per NFPA 1911, calibration should be checked as part of an annual pump service test. Unfortunately, calibration of pump gauges is often not checked. All of the efforts of a pump operator to provide adequate pressures and flows will be negated if pump panel gauges do not provide accurate readings. Pump operators should take the initiative to ensure that pump panel gauges are in calibration or recalibrated or replaced by qualified personnel.

Digital pressure gauges are increasingly finding their way onto pump panels. A sensing device measures the pressure and sends an electrical signal to the pump panel. The signal is received, translated, and digitally displayed on the pump panel (**Figure 5-15**). When digital pressure gauges are used, NFPA 1901 requires the device to display pressure in increments of 10 psi (70 kPa) or less.

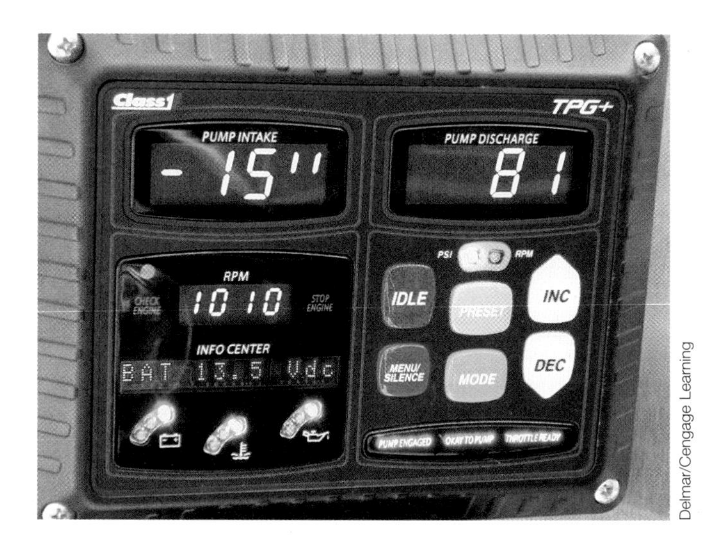

Delmar/Cengage Learning

FIGURE 5-15 The pressure transducer sends an electronic signal to the digital pressure gauge. On this particular apparatus, the digital pressure gauges incorporated into the pressure governor replace master pump intake and discharge gauges.

Pressure gauges found on most pump panels can be grouped into main pump gauges and individual line discharge gauges. The main pump gauges are located close to each other on the pump panel and are larger than the individual discharge gauges.

Main Pump Gauges

Most pump panels have two main pump gauges, sometimes referred to as master gauges. According to NFPA 1901, these gauges are to be labeled "Pump Intake" and "Pump Discharge." The pump intake gauge is a compound gauge attached to the intake side of the pump. This gauge measures negative and positive pressures from 30 in. Hg. (760 mm) to at least 300 psi (2,100 kPa), with 600 psi (4,200 kPa) being the most common and 300 psi (2,100 kPa) and 400 psi (2,800 kPa) also used. Negative pressure readings are important for monitoring drafting (a process for priming the pump) operations. Positive pressure readings are important for monitoring pressurized water supplies, such as from hydrants or relay operations. The other main gauge, called the pump discharge gauge, is a positive pressure gauge mounted on the discharge side of the pump. The main pump discharge gauge measures the highest positive pressure generated on the discharge side of the pump. NFPA 1901 requires the master intake and discharge gauges to be mounted within 8 inches of each other, with the intake gauge installed to the left or below the discharge gauge.

Individual Discharge Gauges

Individual gauges are commonly provided for each 1½-inch (38-mm) and larger discharge outlet located on the pump. These gauges measure positive pressure in psi (kPa) and are smaller than the main gauges. Individual discharge gauges measure positive pressure for the specific outlet to which they are attached and are usually located next to their respective discharge control valves.

Knowing the discharge pressure of each outlet is important to ensure that excessive pressures are not supplied to attached discharge lines. In addition, by knowing the pressure of the discharge outlet, the pump operator can calculate the amount of water flowing through the line. This is a critical point in that the effectiveness of extinguishing efforts rests, in part, with the quantity of water flowing. (More information on the relationship between water flow and extinguishment and the calculation of water flow is provided in Section IV.)

Flow Meters

Flow meters are placed on pump panels to augment or replace pressure gauges (**Figure 5-16**). The purpose of flow meters is to measure the quantity and rate of water flow in gallons per minute (gpm) or litres per minute (Lpm). Unlike pressure gauges that require hydraulic calculations to determine gpm (Lpm), flow meters measure gpm (Lpm) directly. The result is reduced hydraulic calculations and increased accuracy of flow rates.

Flow meters are comprised of several components. A sensor mounted on the discharge pipe provides a reading of the flow. In most cases, a paddlewheel is used to measure flow (see **Figure 5-17**). The sensor transmits the signal to the pump panel, where a microprocessor chip translates the signal into the flow rate that is displayed on the digital readout. Because of the "smart" electronics of flow meters, calibration is a simple process that does not require special instruments. Flow meters can provide current flow and flow accumulation for an operation, and some flow meters provide pressure readings as well. Discharge flow meters, according to NFPA 1901, must have display digits of at least ¼ inch (6 mm) and display flow in increments of 10 gpm (40 Lpm) or less. The current trend is to include both pressure and flow information on pump panels (see **Figure 5-18**).

Delmar/Cengage Learning

FIGURE 5-16 Flow meters are increasingly being installed on pump panels.

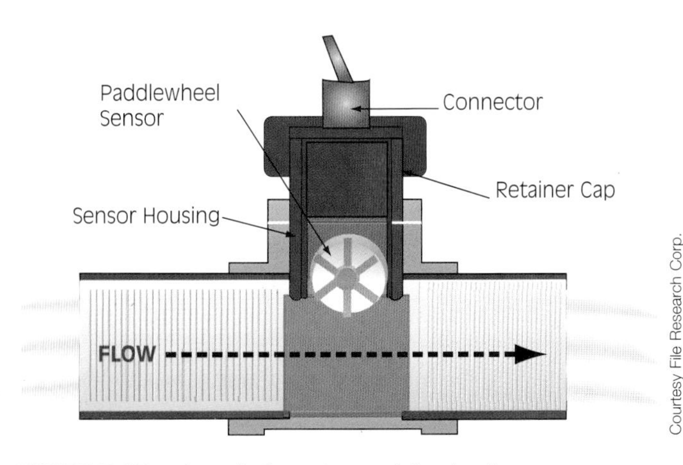

Courtesy File Research Corp.

FIGURE 5-17 Installation of a paddlewheel sensor.

While flow meters simplify hydraulic calculations, operators must be cautious not to increase pressures to unacceptable levels in an attempt to maintain a desired flow. For example, a kinked hoseline will require an abnormally high pressure for a desired flow and is one of the reasons why larger-diameter discharges require a pressure gauge even if they have a flow gauge. Furthermore, large diameter hoses (LDH) and appliances have a lower maximum operating pressure.

Indicators

Indicators are loosely categorized as all instrumentation on the pump panel other than pressure gauges or flow meters. They can be simple or complex devices that provide information on a component. Several indicators can be found on the pump panel to provide information on the pump engine. The required engine indicators include the tachometer, battery voltage, oil pressure, and coolant temperature. Departments may add additional gauges such as transmission temperature. Digital readouts that combine engine information into a single display are becoming more common on pump panels (**Figure 5-19**). The on-board tank has an indicator on the pump panel that provides the status of the water level contained in the tank. A pump engagement indicator is provided on the pump panel to let the pump operator know when the pump is properly engaged. A pressure control indicator is typically found on pump panels; this indicator lets the pump operator know when the pressure control system has been activated.

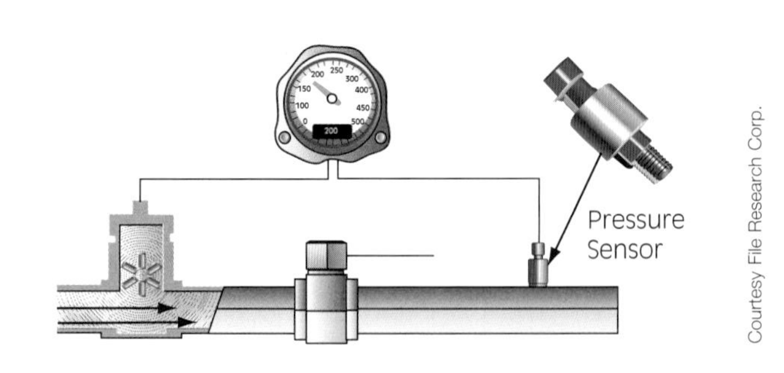

Courtesy File Research Corp.

FIGURE 5-18 Both a flow meter and pressure sensor sending electronic signals to a combination flow/pressure instrument.

Delmar/Cengage Learning

FIGURE 5-19 Digital displays provide information on engine RPM, oil pressure, engine temperature, voltage, and can display engine alert messages.

Courtesy of Waterous Company

FIGURE 5-20 Example of a control mechanism and valve.

CONTROL VALVES

Control valves are used by the pump operator to open, close, and direct water flow. Control valves consist of a valve and an operating mechanism (**Figure 5-20**). The valve is the component that physically directs water flow. The most common types of valves found on pump panels are ball valves and butterfly valves. They are the most common because when open these valves have the fewest waterway obstructions and the least changes to the direction of water flow resulting in the smallest valve-created friction loss. Piston and gated valves can also be found on the pump panel or on other intake/discharge locations on the apparatus.

The operating mechanism is the device that controls the valve. These can be manual mechanisms such as push-pull (commonly referred to as "T-handles"), quarter turn, and crank control. Electric, pneumatic, and hydraulic actuators are also used. Common among the different types of operating mechanisms is the requirement to hold the valve in position, when set, so that it will not change while operating at the maximum flows and pressures of the pump. In addition, control valves are designed to operate smoothly within normal operating pressures of the pump. According to NFPA 1901, discharges of 3 inches (77 mm) or larger must have slow-acting (cannot go from full close to full open in less than 3 seconds) control mechanisms (**Figure 5-21**).

Electric Control Valves

Electrically activated control valves are used on pump panels. **Figure 5-22** is a schematic of an Elkhart electric control valve. The pump operator controls the valve through the switch/indicator that sends information through the wiring harness to and from the control valve. The switch activates the motor to open and close the valve. The indicator lights (the left one is green, the middle one is amber, and the right one is red) provide information on the position of the valve (**Table 5-2**).

Delmar/Cengage Learning

FIGURE 5-21 Example of a slow-acting control mechanism in compliance with NFPA 1901.

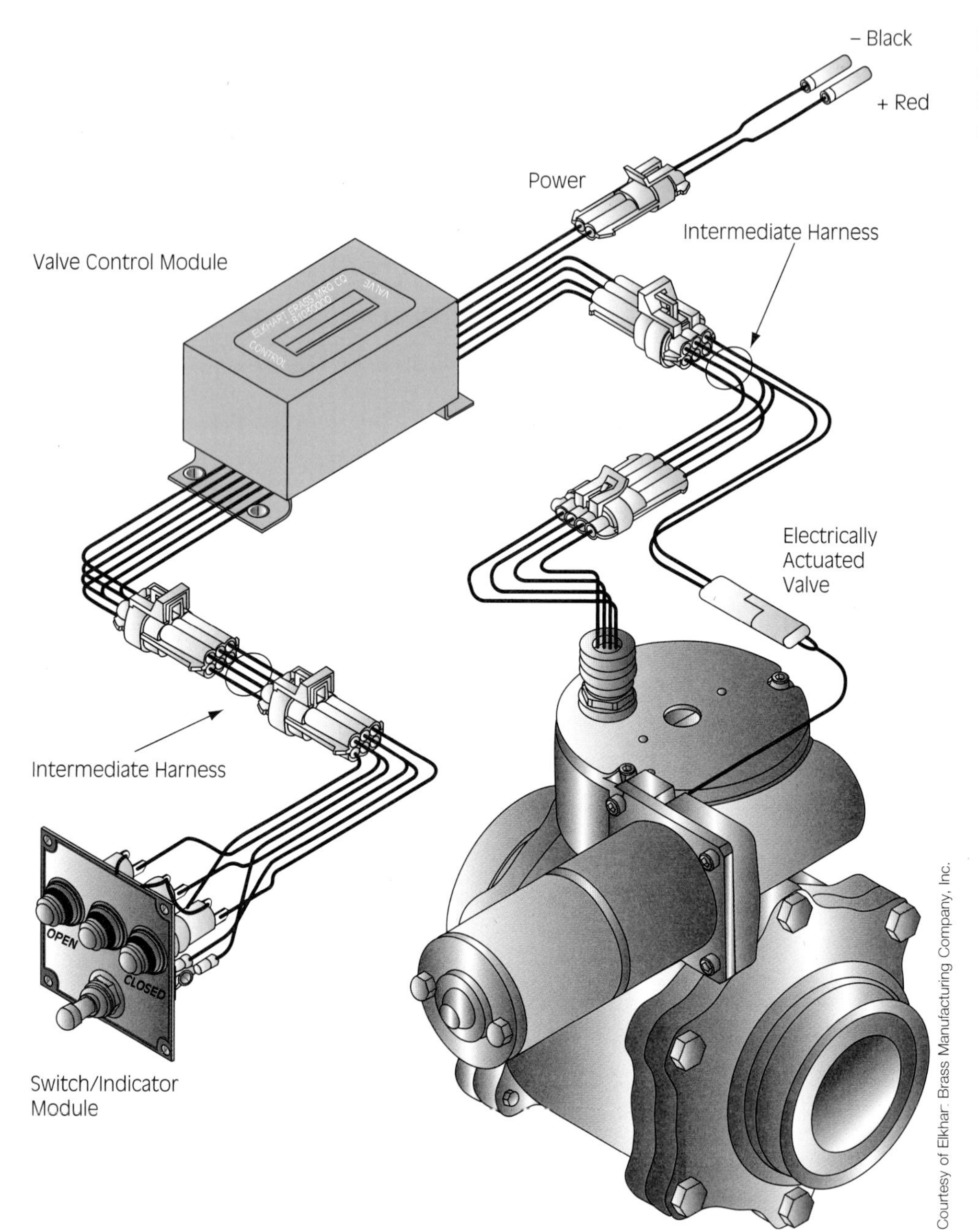

FIGURE 5-22 Schematic of an electrically actuated valve. No change other than adding color is required.

Intake Control Valves

Intakes provide for the connection of external supply sources to the pump. Intake control valves allow the pump operator to regulate external supply sources to the pump. Pumps usually have several intakes located in various positions on the apparatus. NFPA 1901 has several requirements for intakes. **Table 5-3** indicates the minimum number and size of intakes per rated capacity of the pump. Not all intakes have control valves; however, NFPA 1901 requires that at least one auxiliary gated 2½-inch (65-mm) intake be operable from the pump panel. All intakes are required to have a removable or accessible strainer to keep large debris from entering the pump. All 3½-inch (90-mm) or larger intakes with a control valve must also have an automatic pressure relief device. **Figure 5-23** shows the intake relief valve in operation. Note the water being discharged under the apparatus.

Discharge Control Valves

Unlike intakes, each individual discharge requires a control valve. Pumps typically have at least two, 2½-inch (65-mm) discharges. Furthermore, the pump usually has as many 2½-inch (65-mm) or larger discharges as are needed to discharge the rated capacity of the pump. As a rule of thumb, a pump will have one 2½-inch (65-mm) discharge for each 250 gpm (1,000 Lpm) of rated capacity. For example, if the

TABLE 5-2 Indicator Signal Meanings on Electric Control Valves

Indicating Signal	Meaning
Steady burning red	Valve ball in fully closed position
Steady burning green	Valve ball in fully open position
Steady burning amber	Valve ball in an intermediate position
Steady burning amber with flashing green	Valve ball moving toward open position
Steady burning amber with flashing red	Valve ball moving toward closed position

Delmar/Cengage Learning

FIGURE 5-23 Example of an intake pressure relief valve in operation. Note the water discharging to the ground.

TABLE 5-3 Suction Hose Size, Number of Suction Lines, and Lift for Fire Pumps

Rated Capacity		Maximum Suction Hose Size		Maximum Number of Suction Lines[1]	Maximum Lift	
gpm	Lpm	in.	mm		ft	m
250	1,000	3	77	1	10	3
300	1,200	3	77	1	10	3
350	1,400	4	100	1	10	3
500	2,000	4	100	1	10	3
750	3,000	4½	110	1	10	3
1,000	4,000	6	150	1	10	3
1,250	5,000	6	150	1	10	3
1,500	6,000	6	150	2	10	3
1,750	7,000	6	150	2	8	2.4
2,000	8,000	6	150	2	8	2.4
2,000	8,000	8	200	1	6	1.8
2,250	9,000	6	150	3	6	1.8
2,250	9,000	8	200	1	6	1.8
2,500	10,000	6	150	3	6	1.8
2,500	10,000	8	200	1	6	1.8
3,000	12,000	6	150	4	6	1.8
3,000	12,000	8	200	2	6	1.8

[1] *Where more than one suction line is used, all suction lines do not have to be the same hose size.*

Reproduced with permission from NFPA 1901–2009, Standard for Automotive Fire Apparatus, Copyright © 2008, National Fire Protection Association, Quincy, MA. This is not the complete and official position of the NFPA on the referenced subject, which is represented only by the standard in its entirety.

pump is rated at 1,250 gpm (5,000 Lpm), then most likely the pump will have a minimum of five 2½-inch (65-mm) discharges. Pumps typically have at least two, 1½-inch (38-mm) preconnect discharges as well. Discharge control valves provide a great deal of flexibility over individual hoselines, allowing multiple hoselines of different pressures and flows to be deployed at the emergency scene.

Other Control Valves

Several other control valves are located on the pump panel. One control valve, called the tank-to-pump, allows water to flow from the on-board water supply to the intake side of the pump. Another control valve located on the pump panel is called the pump-to-tank or tank fill valve. This valve allows water to flow from the discharge side of the pump to the tank. These two control valves can be used together to help keep the pump from overheating by circulating water from the pump to the tank and then back to the pump again. The transfer valve is a control valve found on multistage pumps that redirects the water flow within the pump between the pressure mode and volume mode.

PUMPING SYSTEMS

Pump peripheral systems are those components that directly or indirectly assist in the operation of the pump. The common systems found on pump panels include priming, pressure regulating, cooling, and compressed-air foam systems.

Priming Systems

Priming systems are used by the pump operator to prime the pump. Recall that priming a centrifugal pump is the process of removing air from the intake side of the pump and replacing it with water. Priming systems accomplish this task. Any type of positive displacement pump can be used as a primer. Rotary gear pumps were commonly used for this purpose in the past, but recently have declined in use due to current environmental requirements not to discharge the oil they need to operate. Rotary vane primers are the most common primers on current Class A pumpers with some new primer types, such as air primers, coming on the market. Small brush trucks often use exhaust primers.

The Waterous priming system is used to illustrate a typical priming system. This system is comprised of a priming pump, an oil reservoir, and a priming valve activated on the pump panel (**Figure 5-24**).

The priming process is initiated when the pump operator activates the priming valve. Referring to **Figure 5-25**, when this occurs, a sliding plunger (A) in the priming valve accomplishes two tasks. One task is to open a passageway between the intake

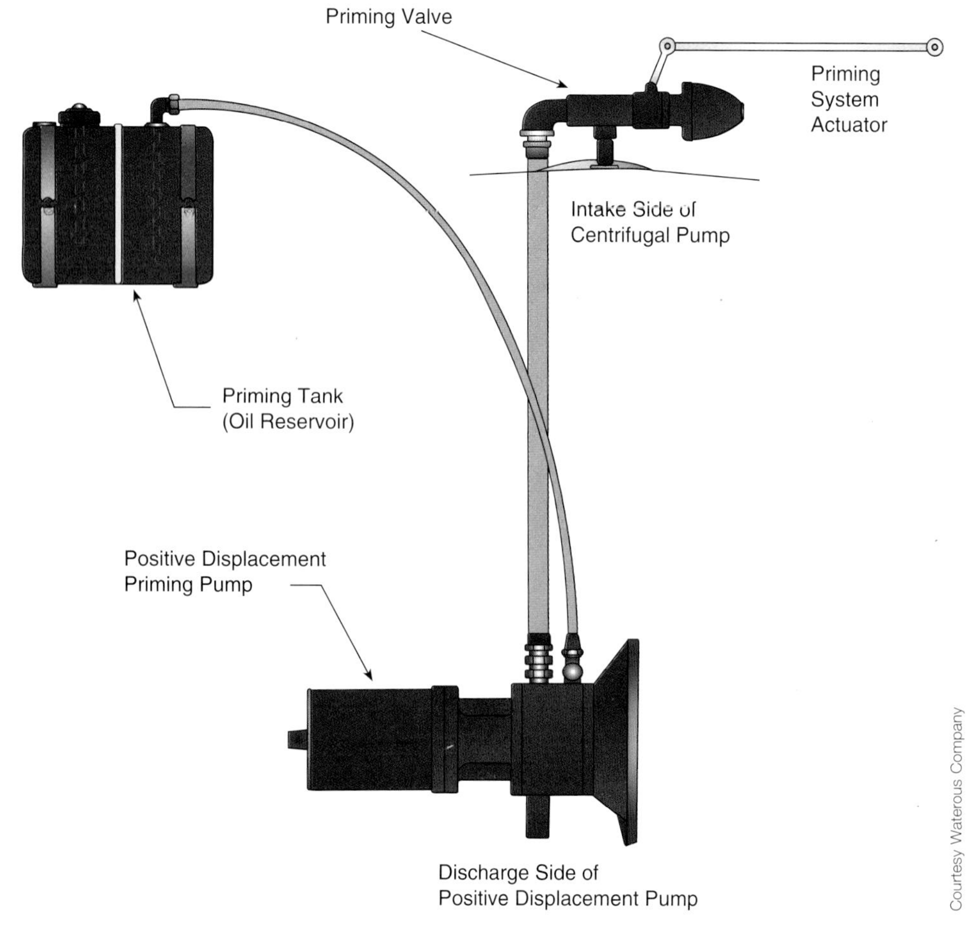

FIGURE 5-24 Components of a typical priming system.

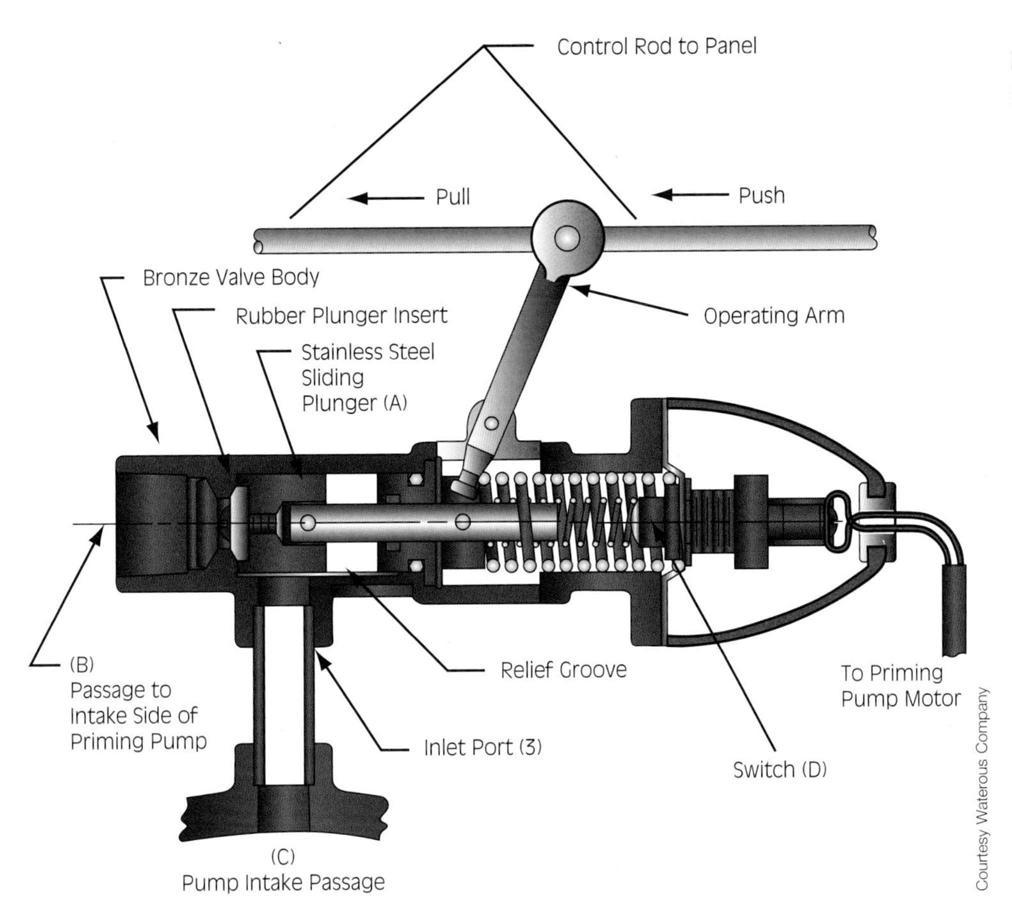

FIGURE 5-25 Example of a priming valve.

side of the priming pump (B) and the intake side of the pump being primed (C). The second task is to depress the priming pump motor switch (D).

With the priming motor activated, the priming pump begins to pump air from the intake side of the main pump and oil from the priming oil tank reservoir (refer back to **Figure 5-24**). Oil is included in the priming process to help lubricate the priming pump as well as to help provide a tighter seal between the moving parts. Recall that the ability to move air and water rests, in part, with the close fit of moving parts in the positive displacement priming pump. The air and oil mixture is typically discharged under the apparatus. Because this water/oil mixture is discharged, an environmentally friendly priming fluid is now being used with some priming devices while other systems have been redesigned to be oil-less.

STREETSMART

Since oil-less primers have some design differences, some differences in check-out procedures are recommended to ensure trouble-free operations. Specifically, it is important to allow water to run through these primers. Operating them for just a second during each day's apparatus check and/or long operations without water flowing through them actually increases the likelihood of failure when needed. Use of the primer without water flowing through it creates dust as the vanes wear. If water does not flow through the primer to flush the dust, it causes sticking, increasing failures. Lack of lubricating and cooling water increases vane sticking and breakage.

As the air is removed by the priming system, a negative pressure is created on the intake side of the pump. When this happens, atmospheric pressure forces the water into the pump (refer back to **Figure 4-2**). When the discharge from the priming pump changes from air to water, the priming system has accomplished its task of removing air from the pump and replacing it with water. Other indicators that the pump is primed are:

- There is a positive reading on the discharge pressure gauge.
- The priming motor sounds as if it is slowing.
- The main pump sounds as if it is under load (begins to pump water).

Priming devices on modern pumping apparatus are typically rotary vane positive displacement pumps. Older apparatus may use exhaust or vacuum primers. Exhaust primers redirect exhaust gases

FIGURE 5-26 Electric starter motor used to drive an oil-less rotary vane primer motor.

through a chamber connected to the intake side of the centrifugal pump. Current standards, however, do not allow their use on Class A pumpers because they need more maintenance. Vacuum primers use the vacuum created in the manifold of gasoline-powered engines. A float is typically used to help ensure water does not travel from the pump into the engine manifold.

The following sections provide more detailed information on various common types of primers.

Rotary Primers

When rotary gear, lobe, or vane pumps (previously described) are used as primers, they are typically driven by a heavy-duty electric starter motor (**Figure 5-26**). Because of the duty cycle of the starter motor, these primers cannot be run for more than a minute and should generally be run for much less than a minute before being given several minutes to cool down. Running one of these primers for too long creates excessive heat that either destroys the priming motor or greatly shortens its life. Use of rotary gear and rotary lobe primers has decreased due to their need to use priming oil.

Air Primers

Air primers have recently been approved for use as primers on pumpers. They use air from the apparatus brake system and a Venturi effect to create the partial vacuum needed to prime the pump. Their advantages include not requiring the large power consumption that is required to run electric-motor-based primers and not having the problems of burning out these motors. They also may be less expensive to purchase. They do, however, use large quantities of air. While air is not required for the air brakes when the apparatus is parked for pumping, an air primer may still require increases in the size of the apparatus air compressor and/or air tanks. Furthermore, since it will use air faster than the compressor can replace it, if the pump is not successfully primed before the air tank is depleted, the operator will need to wait while the compressor recharges the system.

Exhaust Primers

Exhaust primers use the engine exhaust to create a Venturi effect similar to what an air primer does with air. It allows an inexpensive exhaust primer to be used on small gasoline-driven pump engines without adding much additional weight. When used, exhaust is redirected through a smaller Venturi discharge, increasing its speed and the amount of vacuum created. Higher engine RPMs are needed to create enough exhaust gases for the rapid movement needed for this device to work efficiently. On smaller engines with separate controls to divert the exhaust gases and to connect the pump to the primer, the primer valve must be closed before the exhaust diverter or air will leak back into the pump, causing the pump to lose its prime (**Figure 5-27**).

The disadvantage with exhaust primers is that they tend to need a lot of maintenance because of the soot in the exhaust gases.

STREETSMART

Class A pumpers are pumpers that will deliver their rated capacity at 150 psi (1,050 kPa) net pump discharge pressure at a lift of not more than 10 feet (3 m) and with an engine load of not more than 80% of the certified peak on the brake horsepower curve; will deliver 70% of rated capacity at 200 psi (1,400 kPa); and will deliver 50% of rated capacity at 250 psi (1,750 kPa). The Class A pumper is also called an engine.

Class B pumpers are pumpers that will deliver their rated capacity of 120 psi (840 kPa) net pump discharge pressure at a lift of not more than 10 feet (3 m) and with a engine load not exceeding 80% of the certified peak of the brake horsepower curve. They will deliver 50% of rated capacity at 200 psi (1,400 kPa) and 33⅓% of rated capacity at 250 psi (1,750 kPa). Class B pumpers have not been manufactured since the mid-1950s.

Today the term *Class A pumper* is still used to refer to a primary engine as opposed to a secondary unit such as a brush truck, which does not meet all of the requirements of pumper fire apparatus.

FIGURE 5-27 To use the exhaust primer on this pump, hold the exhaust cover (A) over the exhaust; this reroutes the exhaust past the Venturi. While this is occurring, open valve (B) to allow the air in the pump to be drawn out by the pressure reduction the Venturi creates.

Pressure Regulating Systems

All modern pumping apparatus have a pressure regulating system attached to the centrifugal pump. **Pressure regulating systems** provide safety to personnel and equipment and assist with the efficient delivery of discharge flows. Safety is attained by controlling sudden and excessive pressure buildup during pumping operations. Efficiency of discharge flows is assisted by compensating for fluctuating pressures, which helps keep the pump and engine under a constant load. Without pressure regulating devices, pump operators would have an impossible task of maintaining safe and efficient pressures and flows.

Pressure regulators are needed because pressures can fluctuate during any pumping operation. These fluctuating pressures can be severe enough to damage equipment and injure personnel. In addition, they can occur rapidly and without warning. Even the most astute pump operator cannot adequately correct and control these pressure fluctuations without the aid of a pressure regulator.

Pressure increases can result from simply closing a discharge line. Recall that pressure buildup results, in part, from restrictions on the discharge side of the pump. Take, for example, a pump flowing a total of 500 gpm (2,000 Lpm) through two discharge lines. Each discharge line flows half of the volume, or 250 gpm (1,000 Lpm). If one line is closed, the pump attempts to flow the 500 gpm (2,000 Lpm) through the one line. A sudden pressure increase occurs due to the restrictions of the one remaining open discharge. Pressure regulators can be set to compensate automatically for such an increase in pressure.

The two pressure regulators commonly found on pumps are pressure relief devices or valves and **pressure governors**. Pressure relief devices control pressure buildup by sending excess water pressure back to the intake side of the pump. Governors control pressure buildup by controlling the speed of the engine, which in turn controls the speed of the pump. Pump speed is increased and decreased to handle pressure fluctuations. According to NFPA 1901, the pressure regulating system must limit discharge pressure increases to less than 30 psi (210 kPa) when valves are closed within 3 to 10 seconds and an NPDP of at least 90 psi (630 kPa) exists. In addition, the pressure regulating system must be operational from the pump panel and must indicate when the system is in operation.

Pressure Relief Devices

To protect against excessive pressure buildup by diverting excess water flow from the discharge side of the pump back to the intake side of the pump, pressure relief devices are used (see **Figure 5-28**). In the previous example, the excess 250 gpm (1,000 Lpm) caused by closing one line would be pumped back into the intake side of the pump. The pump, in essence, would continue to flow 500 gpm (2,000 Lpm)—250 gpm (1,000 Lpm) through the open line and 250 gpm (1,000 Lpm) back to the intake side of the pump.

Various pressure relief devices are installed on pumps; however, the operating principle behind each of them is virtually the same. Pressure relief devices operate, in part, based on a hydraulic principle stating that the pressure exerted by a liquid on a surface is proportional to the area of the surface. In essence, a small quantity of water can be used to force a piston against higher pressures. Relief valves that operate automatically utilize this principle.

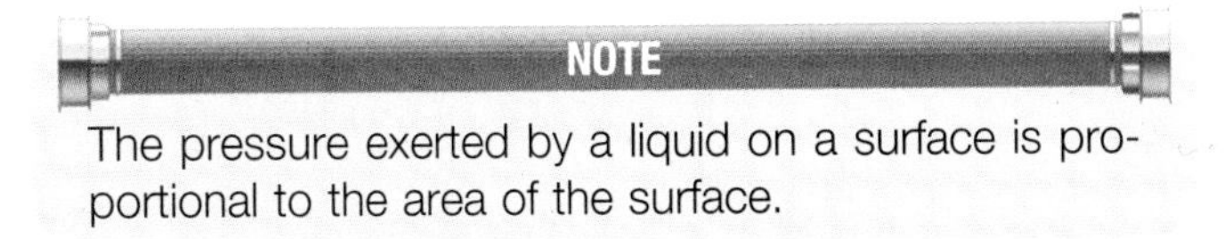

The pressure exerted by a liquid on a surface is proportional to the area of the surface.

Pressure relief devices consist of two main components: a control mechanism and a relief valve. The control mechanism is used to set the relief device to

FIGURE 5-28 Pressure relief device.

operate automatically at specific pressures, allowing excess pressure to be diverted back to the intake side of the pump.

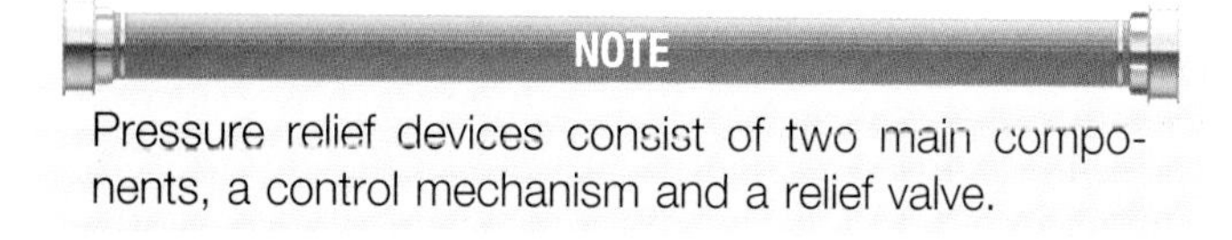

Pressure relief devices consist of two main components, a control mechanism and a relief valve.

The relief valve is a piston contained in a cylinder that opens and closes a passage between the discharge side of the pump and the intake side of the pump (**Figure 5-29**). Both sides of the piston receive pressure from the discharge side of the pump; however, the smaller side of the valve receives pressure directly from the discharge side of the pump while the larger side receives pressure that is regulated from the control mechanism. When the pressure from the control mechanism is equal to or greater than the discharge pressure, the relief valve is closed. When pressure increases on the discharge side of the pump, the piston is forced open and excess pressure is diverted to the intake side of the pump. When discharge pressure decreases, the relief valve is forced to close by the regulated pressure.

Two operating characteristics affect the efficiency of pressure relief devices. First, because excess pressure is diverted to the intake side of the pump, high intake pressures can negate the operation of the relief valve. Increasingly, intake relief devices are also used to ensure that intake pressure is maintained at a safe level. Second, pressure relief devices decrease pressure by diverting excess water flow back to the intake. They are not designed to increase pressure. Consequently, when discharge pressure is reduced, the engine speed must be increased to maintain desired discharge pressures.

Governors

To protect against excessive pressure buildup by controlling the speed of the pump engine to maintain a steady pump pressure, pressure governors are used. When discharge pressures increase, the governor reduces engine speed and the desired pressure is maintained. When pump discharge pressures decrease, the governor increases engine speed to maintain the desired pressure.

Older pressure governors utilize a mechanical linkage to control engine speed. When discharge

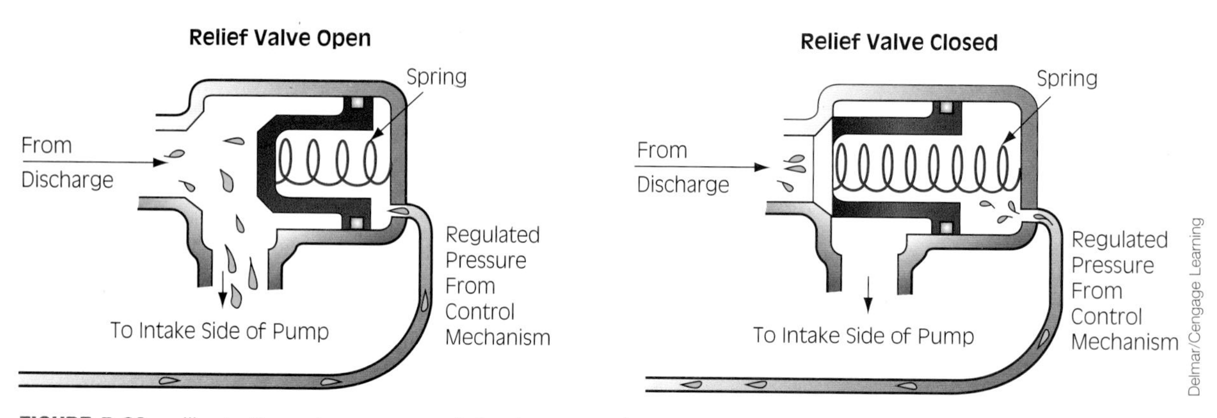

FIGURE 5-29 Illustration of pressure relief valve operation.

pressures rise above the governor setting, a device connected to the engine accelerator by mechanical linkage reduces engine speed. When discharge pressures fall below the governor setting, the device increases engine speed. Two problems are associated with these older governors. First, if intake pressure is reduced or lost, the discharge pressure would decrease, causing the governor to attempt to increase pressure by increasing engine speed. Because the discharge pressure cannot attain the governor pressure setting, the governor continues to increase engine speed to its maximum RPM. Second, problems with the mechanical linkages could cause a slow response by the governor, resulting in a continuous increase and decrease of the engine speed by the governor as it attempts to reach its pressure setting.

Newer pressure governors utilize microprocessors to control engine speed. Information on intake and discharge pressures is sent to the microprocessor (see **Figure 5-30**). The processor analyzes the information and increases or decreases engine speed as needed. Utilizing intake and discharge pressure information guards against high engine speed as the result of lost intake pressure or cavitation. In addition, electronic devices tend to react quickly, reducing the increase and decrease of engine speed as the governor attempts to maintain the pressure setting.

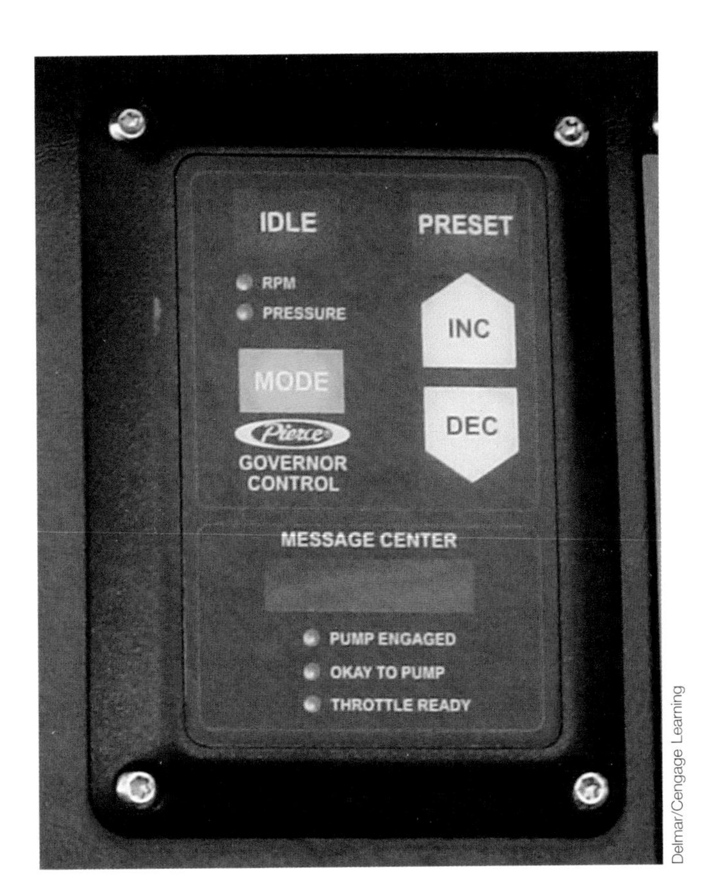

FIGURE 5-30 Electronic automatic pressure governors may also monitor essential engine vital signs.

Limitations of Pressure Regulating Systems

In addition to setting the pump's pressure regulating system, operators need to understand the limitations of both relief valves and pressure governors. They also need to know the proper procedures to use when affected by these limitations. Relief valves, or, more formally, discharge relief valves, allow water to flow from the pump's discharge manifold to the pump's intake manifold. Pressure governors control the throttle of the pump's engine. Neither reduces the intake pressure nor prevents the pump from adding a pressure increase as the intake water passes through the impeller, resulting in a higher discharge pressure. Thus, in cases where the desired pump discharge pressure is below or only slightly above the pump's intake pressure, neither the relief valve nor pressure governor will be able to provide the desired discharge pressure.

In situations with a high intake pressure, a number of operators have attempted to control the pump discharge pressure by throttling back on the intake valve. While this maneuver appears to work, it should be avoided. Throttling an intake can quickly result in the pump becoming starved for water if flow rates change, as is likely to occur when multiple lines are operating. Instead, operators should throttle back the discharge valves to obtain the desired line pressures in the same way they would when the line is a second line operating at a lower pressure than some other line. The pressure regulating system can also be set. It is set for the pressure of the imaginary higher pressure line but also protects the pressure on the gated line just as it does when flowing multiple lines of different pressures.

Intake Relief Valves

Intake relief valves protect the apparatus from excessively high intake pressures. Many in-service apparatus do not contain these valves; however, recent changes to NFPA 1901 added intake relief valve requirements for all new pumps containing 3-inch (77-mm) or larger intakes. On valve-controlled intakes, they exist on the intake side of the valve with an additional relief valve required on pumps containing any 3-inch (77-mm) or larger valve-less intakes. Intake relief valves are adjustable and may or may not require tools for their adjustments. Those requiring tools generally are preset to 125 psi (875 kPa) and should be reset based on the highest expected intake pressure within the apparatus's service area.

The piping on intake relief valves will generally discharge the water away from the operator and end with a hose thread. These threads are to allow the connection of hoses that will discharge water in a safe location, such as down a storm drain instead of on a street in freezing weather where ice would create a slip hazard. Intake relief discharges should not generally be capped since doing so would inhibit the intended safety functions provided by an intake relief valve.

Cooling Systems

All combustion engines are equipped with a cooling system. This cooling system maintains the engine within designed operating temperatures under normal environmental and working conditions. Engines used to drive pumps are often subjected to intense working conditions as well as harsh environments. In addition, automotive engines are designed to maintain operating temperatures while in motion. Engines used to drive pumps work hard while in a stationary position. Because of this, additional cooling systems are required to maintain engine temperatures.

Auxiliary Cooling Systems

Sometimes called heat exchange systems or engine coolers, auxiliary cooling systems are used to maintain engine temperatures within operating limits during pumping operations. Such a system consists of tubing running from the discharge side of the pump through a heat exchange system and back to the intake side of the pump. Water flows from the pump's discharge header through the heat exchanger and back to the pump's intake header due to the differences in the pump's intake and discharge pressures. The heat exchange system allows the transfer of heat from the engine coolant to the pump water in the auxiliary cooling system. The engine coolant and the **auxiliary cooling system** water never actually mix. Since the fluids do not mix, pressures are not affected in either the pump or engine/radiator. This system helps keep the engine coolant system cool enough to maintain engine temperatures within acceptable operating limits. The auxiliary cooling system is controlled by a valve on the pump panel.

Pump Cooler

The pump cooler is a small valve on the pump panel that allows water to flow from the discharge side of the pump to the booster tank. The discharged water is replaced by cool water entering the pump through the pump intake; thus, the pump cooler provides an easy way to make sure the pump does not overheat by circulating a small amount of water through the pump. Alternatively, the valve may be labeled a "recirculate valve," or instead of providing a separate small-diameter valve, the tank-to-pump valve may be opened slightly. Since it is essential that water is always flowing through the pump to keep it cool, all pumps provide a method of circulating water through the pump. Of course if hoselines are actually flowing, the pump cooler is unnecessary.

A few cases exist where it may be undesirable to have the pump cooler open. Since it discharges water into the booster tank, if the pump's water is being supplied by an external source (e.g., a hydrant instead of the booster tank) once the booster tank becomes full, it will overflow, which could create problems if the temperature is below freezing and the ice this makes creates a slip hazard. When water is supplied via a tanker shuttle, it is also undesirable to waste it. More important, it can affect priming. If the booster tank connection is above the booster tank water level, and this valve is open when priming the pump, it will allow air to enter the pump, preventing successful priming.

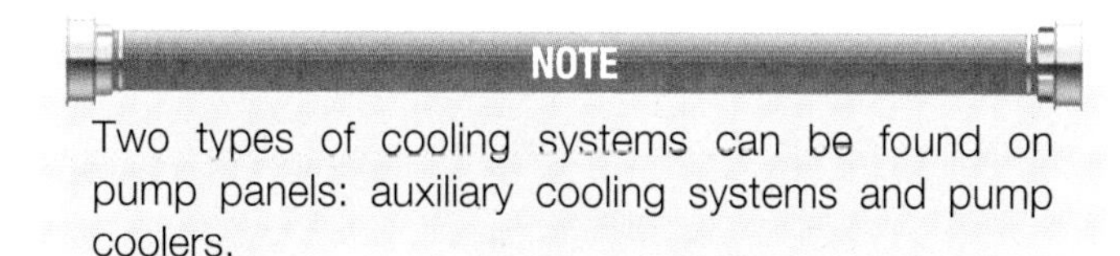

Two types of cooling systems can be found on pump panels: auxiliary cooling systems and pump coolers.

Thermal Relief Valve

Thermal relief valves are an option that can be added to any pump either at the time of manufacture or in the field. Most apparatus, however, do not have thermal relief valves since, if the pump operator is doing his or her job correctly and always circulating water through the pump, thermal relief valves are unnecessary. Departments that do purchase thermal relief valves are providing an inexpensive additional backup system that can prevent very expensive, time-consuming repairs should the water in the pump become too hot. Thermal relief valves (**Figure 5-31A**) are connected to an unused port on the discharge side of the pump. They contain a wax that will melt at a specified temperature and allow a small amount of water to leave the pump discharge and be replaced by cooler water coming into the pump's intake.

(A)

(B)

Delmar/Cengage Learning

FIGURE 5-31 (A) An example of a thermal relief valve that opens automatically if the pump is allowed to get too hot. This allows hot water to exit the pump and be replaced with cooler intake water. (B) When the thermal relief valve opens an indicator is displayed at the pump panel so the operator can take corrective action.

They also have an indicator on the pump panel (**Figure 5-31B**) to notify the operator of the problem. Thermal relief valves typically discharge their water into the booster tank and will automatically reset when the pump is cool enough.

Foam Systems

The use of foam as an extinguishing agent has increased in popularity because it appears to increase the efficiency and safety of fire suppression efforts. Systems that produce foam (both Class A and B foams) are being added to pumps, with the controls typically located on the pump panel. Such systems include premixed systems, in-line eductor systems, around-the-pump proportioning systems, balanced pressure systems, and direct injection/compressed-air systems. The basic differences between the systems lie in how and where the foam and water come together. Foam generation and foam application data can be found from the manufacturers of the equipment and **foam concentrate** used.

Efficient foam operations require appropriate application techniques by the firefighters, selection of a suitable foam type based on the type of fire, and the use of the correct pressure and proportion by the pump operator. A number of foam containers list a range of proportioning rates. Which rate should be chosen depends on the type of fuel being extinguished and the desired characteristics of the **finished foam**. For example, Aqueous Film Forming Foam/Alcohol Resistant AFFF/AR, a common Class B foam, often lists application rates of 3% and 6% with the correct application rate depending upon the fuel being extinguished. Polar solvents, or fuels that can mix with water, such as alcohol, use the 6% rate, and fuels that cannot mix with water, such as diesel, jet fuel, and home heating oil use the 3% rate. A fuel that is often incorrectly categorized is gasoline. Gasoline is really a blend of multiple ingredients, with many of today's gasolines including ethanol containing significant amounts of alcohol. Therefore, the 6% rate should be used to prevent alcohol in the gasoline from destroying the foam blanket.

NOTE

Some newer Class B foams are designed for use with an application rate of 1%. Lower application rates allow a given quantity of foam concentrate to produce a greater quantity of finished foam or an apparatus to carry less foam concentrate and still produce the same amount of finished foam.

The concentrates known as Class A foam are technically "wetting agents," which means they lower the surface tension of water so that it sheets over a surface instead of forming droplets, the way water does on a newly waxed car. When water forms droplets, it tends to run off the surface, limiting its extinguishment efficiency. When it sheets over a surface, it covers more surface area and better absorbs heat from the surface, allowing extinguishment with less water and less water damage.

Class A foams often have a wide proportioning rate, such as 0.1% to 1%. At the higher foam percentage, a dry foam is produced that looks like shaving cream and sits on top of the surface, separating the air supply above the foam from the fuel supply below the foam. At the lower foam percentage, a wet foam is produced that coats the fuel, adding moisture to the fuel and cooling it. Wet foams will also flow into voids, displacing air in these voids while coating surfaces along the way. This helps with subsurface fires such as dumpster fires.

Premixed System

Premixed systems or batch mixing consists of a tank in which foam concentrate and water are added at appropriate proportions. Often, the on-board water tank is used, and the foam concentrate is simply added to the tank, with the mixing time dependent on both the viscosity and solubility of the foam concentrate. When the tank-to-pump control valve is opened, the foam/water mixture enters the intake side and is pumped to the discharge lines.

Advantages: No special components, such as metering valves or special gauges, are needed and the correct amount of foam can be added for effective foam/water streams.

Disadvantages: A limited foam supply is available (when the tank is empty, the foam supply is exhausted), and there is the potential for harm caused by the foam to the tank, valves, and fittings.

It is also not efficient with foams that do not readily mix with water. Generally, Class B foams are not premixed for this reason.

In-line Eductor Systems

In-line eductor systems utilize eductors to add foam to water in appropriate proportions. Water passes through the eductor, which causes foam to mix with the water (**Figure 5-32**). In addition to the eductor, a foam tank (or supply) is required. This system can be either external or internal to the apparatus. Internal systems have a foam concentrate valve and metering control installed on the pump panel. The eductors are typically installed in-line between the pump discharge and the discharge outlet on the pump panel. In addition, a foam tank is usually installed on the apparatus. External systems place the eductor, with built-in metering control, in-line between the discharge outlet on the pump panel and the nozzle. The foam is typically supplied in small containers and drawn into the eductor through a pickup hose connecting the eductor to a pick up tube that is placed in the foam concentrate bucket. (See **Figure 5-33**)

To provide correct proportioning, the manufacturer's instructions must be followed, and eductors must have the correct inlet pressure, which is generally 180 to 200 psi (1,260 to 1,400 kPa); they cannot have too much back pressure and the limitations on the amount of foam concentrate lift must be followed, which is generally less than 6 feet. Correct back pressure is achieved by using a nozzle capable

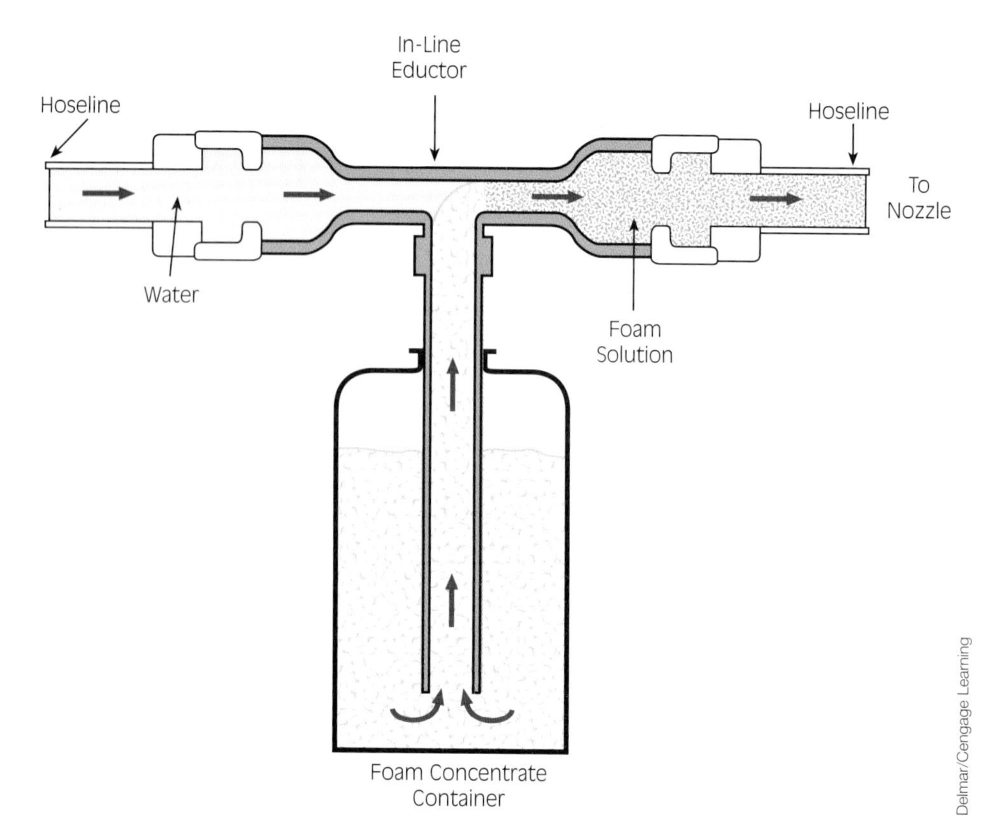

FIGURE 5-32 Foam eductor system.

FIGURE 5-33 Foam concentrate can be drawn out of a bucket via a pickup tube that is inserted into the bucket during foam operations.

of flowing eductors rated gpm (Lpm), avoiding kinks in the foam line, and keeping foam lines short enough, which is generally less than 200 feet (60 m). With the correct configuration, pump operators may either set the foam line by pressure 200 psi (1,400 kPa) unless the manufacturer of their eductor has specified a different inlet pressure or by setting for the eductor's specified flow rate when using a pump with flow meters.

Advantages: Ease of operation, and foam can be resupplied to this system for extended foam operations. Their ease of operation combined with relative low cost have made these the most common systems for municipal fire departments.

Disadvantages: The narrow operating pressure requirements of the eductor (because eductors require a specific pressure for maximum operating efficiency, hoseline length and size can be limited) and extremely cold operating temperatures may limit the operating efficiency of eductors.

(For step-by-step photos of Using an Externally Mounted Foam Eductor, please refer to pages 152–153.)

(For step-by-step photos of Using an Apparatus-mounted Foam System, please refer to page 154.)

The remaining systems (around-the-pump and balanced pressure systems), although more expensive, have the advantage of working over a large range of flows.

Around-the-pump Proportioning Systems

An around-the-pump proportioning system also uses an eductor to mix foam with water. In this system, the eductor is located between the discharge and intake sides of the pump. When the system is activated, water from the discharge side of the pump is allowed to flow through the eductor. After picking up foam, the solution then travels to the intake side of the pump and can be pumped through any and all discharge outlets attached to the pump (**Figure 5-34**).

Advantages: Includes the ability to provide foam to any discharge.

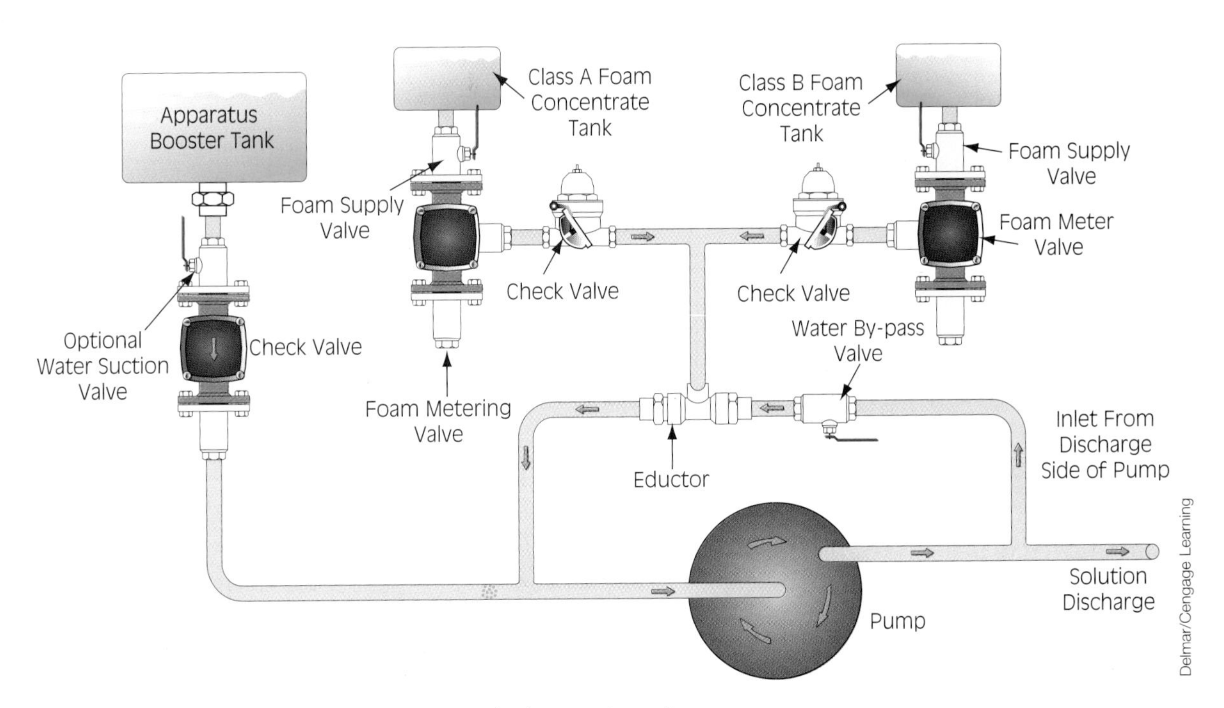

FIGURE 5-34 Around-the-pump proportioning system diagram.

Disadvantages: Water and foam cannot be pumped at the same time. The system also places limitations on the maximum intake pressure, often requiring that intake pressure may not exceed 10 psi, although some newer designs have eliminated this restriction. Finally, concerns exist since the foam, which can react with pump components, actually goes through the pump, therefore allowing these reactions. Since the reactions take time to occur, most of these problems can be resolved by promptly and properly flushing the pump after foam operations.

Balanced Pressure Systems

Balanced pressure systems mix foam with water by means of pressure. Two types of balanced pressure systems are used on pumps. One system utilizes discharge pressure to force foam from a bladder contained in a vessel. As discharge water enters the vessel, the pressure buildup forces foam from the bladder. This type of system is also known as a pressure proportioning system. The balancing part of the system comes into play when foam is forced from the bladder in proportion to the pressure under which water enters the vessel. The foam, under pressure, passes through a metering system and is then mixed with a water stream. One advantage of this system is that foam enters the system under pressure; therefore, this system is efficient over a variety of pressures and flows.

The second balanced pressure system consists of a separate foam pump and foam tank. Foam from the tank is pumped under pressure through a metering system and is mixed with a water stream. The two common types are bypass and demand systems. In the former, foam pressure is controlled through a bypass system that diverts excess foam concentrate back to the foam tank. With the demand system (**Figure 5-35**), the discharge pressure of the main pump is measured and used to regulate the pressure at which the foam pump will discharge foam.

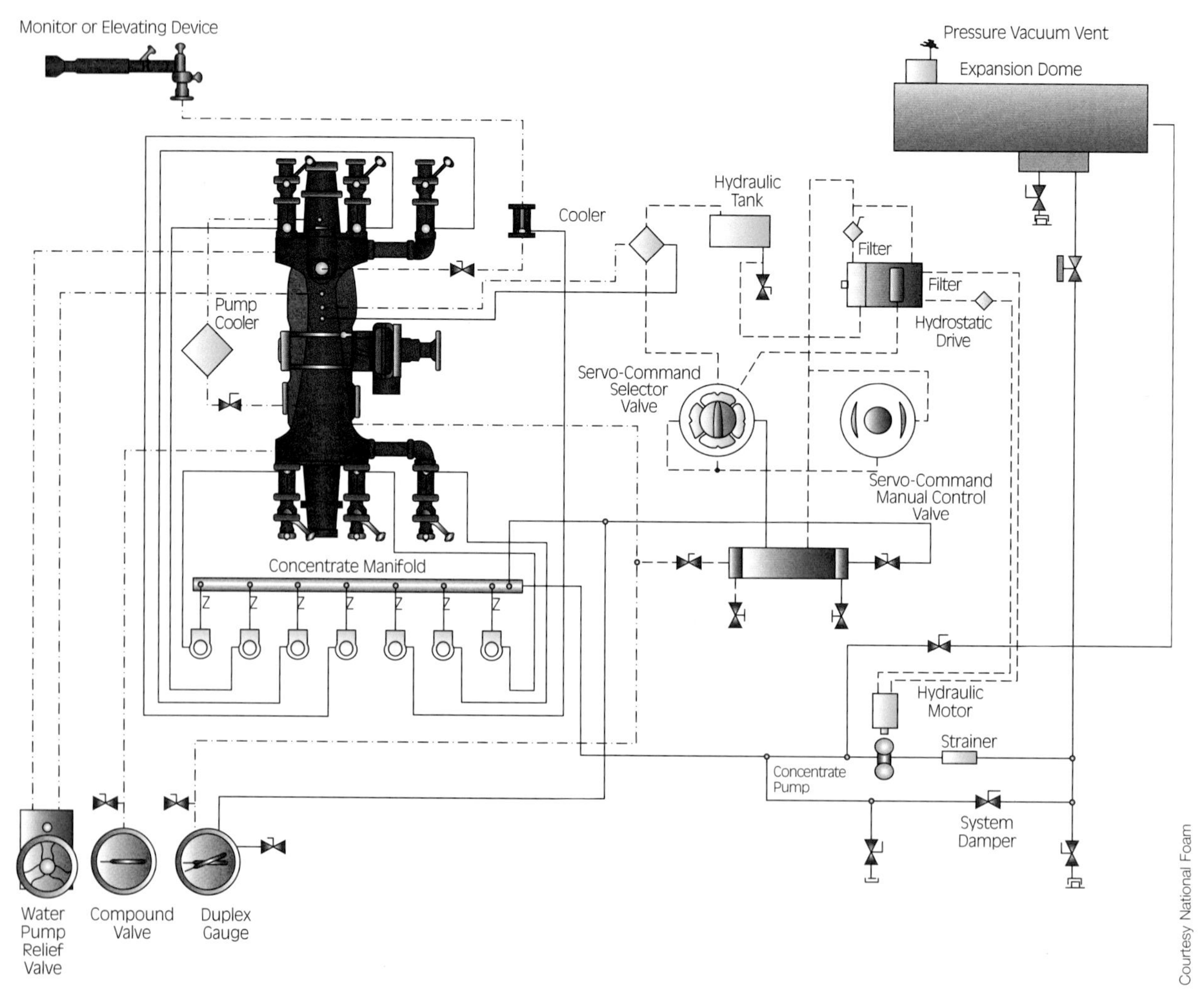

FIGURE 5-35 Balanced pressure demand-type foam proportioning system.

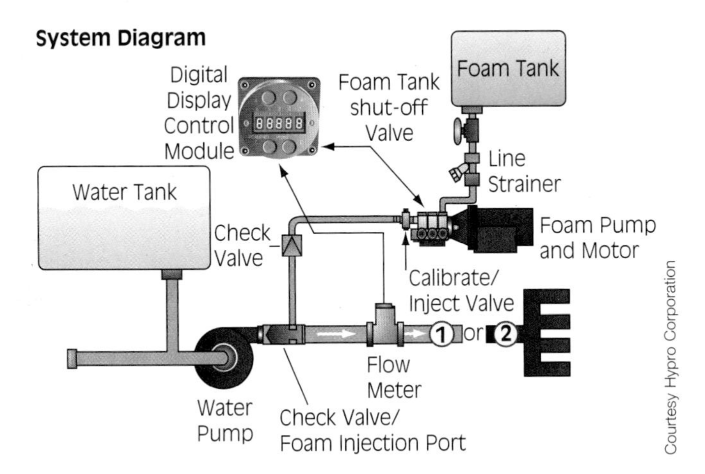

FIGURE 5-36 Diagram of a direct injection foam system with digital control and display.

Direct Injection/Compressed-air Foam Systems

Direct injection systems consist of a separate foam pump and foam tank. Foam from the tank is pumped directly into discharge lines. The rate at which foam is injected into discharge lines is controlled by a microprocessor that receives a signal from a discharge flow meter (see **Figure 5-36**). Compressed-air foam systems (CAFSs) add one more step to the process. After the foam is injected into the discharge line, and prior to leaving the discharge outlet, compressed air is added to the foam/water (**Figure 5-37**). This process creates a unique lightweight foam of fluffy consistency. CAFSs have become a common sight on apparatus due to their efficiency and ease of use. The ability to vary the type of foam from dry for vapor suppression to wet foam for optimum heat absorption gives these systems incredible flexibility; but, like many other systems on the apparatus, they requires expertise and regular training.

Advantages: The system is not affected by intake or discharge pressures, and it keeps foam out of the main pump.

Disadvantages: Price is the major disadvantage to direct injection systems. Because the PDP must be balanced with the air pressure, it may not be possible to use the same engine to concurrently supply non-foam lines with water that needs to be at a higher PDP.

(For step-by-step photos of Using a Compressed-air Foam System, please refer to page 155.)

Foam System Maintenance

Foam systems will not work correctly if certain maintenance procedures are not followed. After the application of foam, all areas that came in contact with **foam solution** need to be thoroughly flushed with fresh water, otherwise the foam concentrate will dry, preventing the proper operation of items such as eductors the next time foam is required. Foam tanks should also be kept full since the concentrate agitation that would occur as the apparatus moves would result in frothing of the foam. Such mixing with air reduces the shelf life of foam concentrate. Even partially filled 5-gallon (20-litre) buckets should not be carried.

Foams from different manufacturers or of different types should not be mixed unless the manufacturer has indicated the foams are compatible and may be mixed. Mixing of other foams, even of the same type, such as AFFF/AR, can result in interactions between different additives, which results in an unusable foam concentrate.

MISCELLANEOUS PUMP PANEL ITEMS

There are a number of miscellaneous components typically found on pump panels. Common to most pump panels is the pump engine hand throttle. This hand throttle controls the speed of the pump engine and can set and hold the speed necessary for appropriate flows. Several auxiliary outlets can also be found on pump panels. One auxiliary outlet allows the engine speed to be measured independent of the tachometer mounted on the pump panel. Two other outlets commonly found on pump panels allow pressure test gauges to be connected, one to the intake and the other to the discharge (**Figure 5-38**). Drain valves are commonly found on pump panels (**Figure 5-39**). Drain valves allow pressure to be relieved from the hose connected to the pump or to drain the pump and its associated piping. Finally, water tank level indicators can be found on pump panels. Several examples of modern tank level indicators are provided in **Figure 5-40**.

AUXILIARY APPARATUS EQUIPMENT AND FIXED SYSTEMS

Generators, extrication equipment, and air systems were traditionally part of truck company operations. Today, where equipment and functions are located is changing to the point that quints exist with some of everything on them and many new aerial apparatus include their own pump. The use of powered equipment has also become more common, adding the

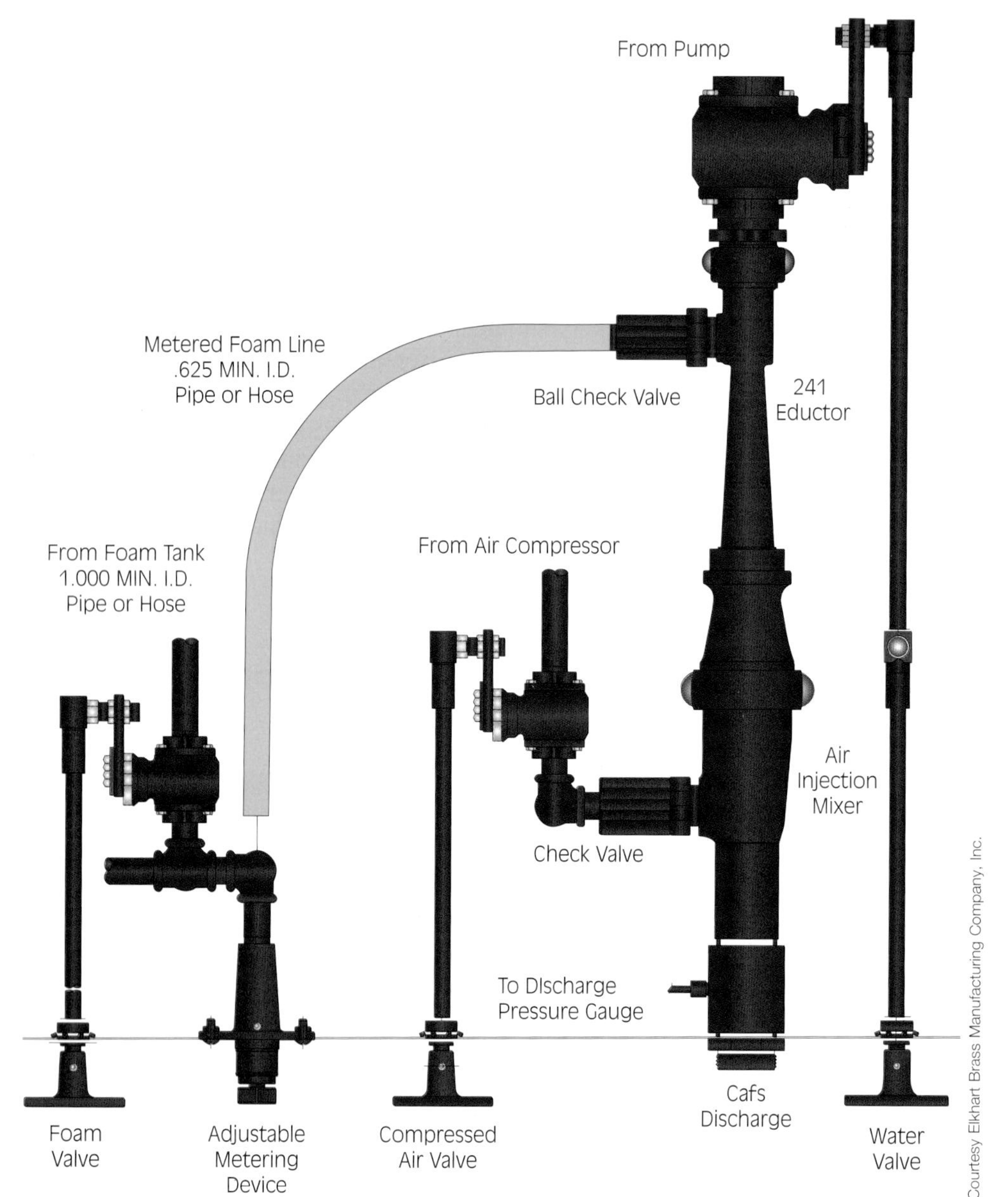

FIGURE 5-37 Typical compressed-air foam system (CAFS).

requirement for the fixed systems needed to power that equipment. Operators should be familiar with the use and operation of all equipment on their assigned apparatus.

The primary source of information on the use of equipment is the manufacturer's operating manuals, which are supplied with the apparatus. Many manufacturers also have copies of their operating instructions on their websites for easy access. In addition, the department may have SOPs that include various manufacturer recommendations along with specific department requirements.

In some cases, departments may have copies of operating instructions on the apparatus or on an apparatus-mounted MDT; nevertheless, an understanding of the operating procedures should be known prior to their need at an emergency scene.

Line Voltage Equipment

Line voltage equipment uses household power instead of the lower voltages used by automotive equipment. Creating line voltages requires either an inverter, which changes the automotive power into a line voltage, or a generator. Inverters are limited in terms of the amount of power they can produce and therefore are used only when limited power requirements exist. Generators may be sized for whatever power requirements the department requests. They are available as gasoline-driven, diesel-driven, and PTO-driven generators.

Operators should know the rated capacity of generators installed on their assigned apparatus. Generators are rated in watts or more commonly kilowatts, which are thousands of watts. Each appliance connected to a

FIGURE 5-38 Test gauge outlets.

FIGURE 5-39 Drain valve.

generator consumes some number of watts. Operators should know the maximum number of watts or maximum amperage, for generators provided with an amp meter, that their generator can provide. This number is a percentage of the generator's rated capacity and may vary by type of load, with resistive loads, such as lighting, being allowed a larger capacity of the generator's output than inductive loads such as motors. As generator capacity is approached, some devices will need to be removed before adding others to prevent generator problems.

Light Towers

Light towers like the one shown in **Figure 5-41** are included on some engines. On such apparatus, the operator needs to know the procedures for deploying and stowing the tower, including the operation of the generator. Many towers may be controlled by a handheld controller such as the one shown in **Figure 5-42**.

(A)

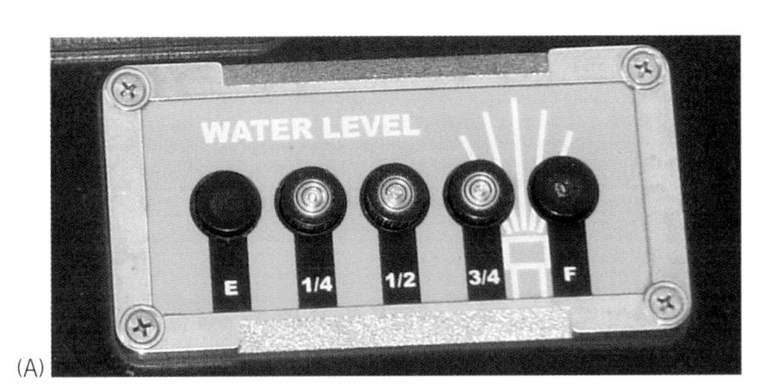

(B)

FIGURE 5-40 Examples of tank level indicators.

FIGURE 5-41 Light towers may be quicker to deploy than handheld lights, provide more lighting, and are positioned at an angle that can be less distracting, particularly for highway incidents.

Hydraulically Operated Equipment

Hydraulically operated equipment typically refers to extrication tools but also includes hydraulic ladder racks, hydraulic dump tank racks, and hydraulic cab tilt units. Each of these requires a hydraulic pump and some type of hydraulic oil. Different hydraulic units may use different types of hydraulic oil, including oils with labels such as power steering oil, automatic transmission oil, and mineral oil. Only the type of oil specified by the equipment manufacturer should be used in any particular unit. The oil must be clean and the reservoir must be filled to the correct level if the hydraulic tool is to work correctly. When filling the reservoir make sure to follow the manufacturer's instructions on the position of hydraulic tools. Many manufacturers indicate that hydraulic cylinders should be in the retracted position when checking fluid levels. This is because the actual position of the cylinder will affect the level of fluid in the reservoir.

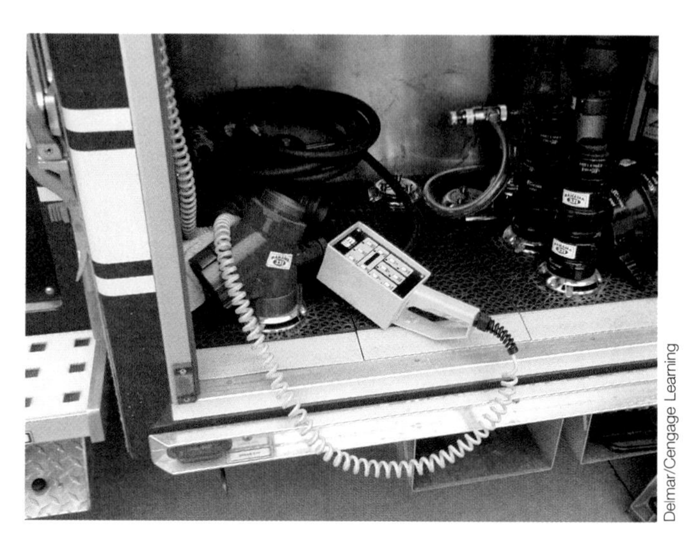

FIGURE 5-42 Portable handheld light tower controllers such as the yellow one shown here allow operators to back away from the apparatus so they can see what is illuminated, even if they are illuminating something on the other side of the apparatus. The controller provides raise, lower, tilt, and rotate functions.

Most extrication tools are connected to the hydraulic lines via quick-connect fittings. Connecting or disconnecting of these fittings should only be done when the hydraulic lines are *not* pressurized. So that lines are only pressurized when needed for actual tool operation, a firefighter should be stationed at the control valve that pressurizes the hydraulic line whenever the hydraulic pump is in operation. Ideally this firefighter is in visual contact with the firefighter using the extrication tool.

SAFETY

All firefighters operating near charged hydraulic lines should be wearing full PPE, including gloves and eye protection. Hydraulic fluid is an irritant and once used tends to become contaminated with materials that make it even more hazardous. Firefighters have been hospitalized when a couple drops not under pressure have entered a firefighter's eyes under a helmet's shield that inadequately protected the eye. With operating pressures in excess of 10,000 psi (70,000 kPa) for some hydraulic tools, a pinhole leak can produce a nearly invisible stream that at close range can penetrate turnout gear and skin, as a firefighter who was seriously hospitalized after such an accident can testify. Again, safety dictates that lines only be charged when necessary and all equipment be properly maintained and used.

Pneumatically Operated Equipment

Pneumatic or air-operated equipment may receive its air from one of three types of compressors that may exist on an engine. Most engines use air brakes and contain an engine-mounted compressor for the air brakes. In addition to supplying the air brakes, this compressor may also supply convenience items such as air-operated seats and steps, to warning and safety devices such as air-operated horns and air-operated tire chains, to on-scene requirements such as air-operated primer pumps. Apparatus with a CAFS will have a pump-mounted compressor to provide air for the CAFS. CAFS compressors are covered in the foam section of this chapter and with service testing in **Chapter 8**. Some apparatus is capable of filling SCBA bottles. These apparatus will have high-pressure storage bottles and may have their own high-pressure-breathing air compressor.

STREETSMART

In all cases, the compressors have a design operating pressure rated in pounds per square inch (psi) and a duty cycle—the number of minutes on within a given minute period. Exceeding either of these limits will significantly shorten the compressor life. Compressors also have a maximum designed flow rated in cubic feet per minute (cfm). Numerous devices, including air-operated horns and air primers, exceed these flow rates. When used briefly and intermittently, the compressor recovers normally. When the devices are used excessively, the compressor's duty cycle is exceeded and its life is shortened.

This is the reason why a small air brake leak may cause the compressor to fail suddenly even though the brakes appear to be operating normally. It is why equipment with a problem that may not appear significant is taken out of service until a qualified individual can look at it. A small leak may be fixed in minutes. If unfixed, a compressor failure may result, requiring a tow, and the apparatus could be out of service for days waiting for replacement parts. The repair will take hours and will result in a much higher cost.

Air Cylinder Fill Systems

Some departments have mobile SCBA fill systems. These systems may be mounted on separate support vehicles, specialty vehicles such as Harmat vehicles, or engines. The system may be part of a cascade system that fills bottles from previously compressed air stored in cylinders or may include a breathing air compressor to create its own compressed air. Care should be exercised in parking apparatus with breathing air compressors to make sure that the air intake for the compressor is positioned in clean air free from hazards and containments from either the incident or exhausts of other apparatus.

(For step-by-step photos of Using an Apparatus-mounted SCBA Fill System, please refer to pages 156–157.)

SKILL 5-1

Using an Externally Mounted Foam Eductor

A Select a discharge for use with the eductor. Often eductors are mounted directly on the apparatus discharge, which can eliminate hauling foam to another location. The area next to the apparatus often also has good working conditions, with lighting and a hard, stable surface. If the discharge point is going to be some distance from the apparatus, a 2½-in. (65-mm) line can be used between the apparatus and the foam eductor.

B Connect a 1½-in. or 1¾-in. (38-mm or 45-mm) attack line to the eductor discharge. Attack line length should not exceed 200 feet (60 m). Longer lines create too much back pressure, which prevents the eductor from mixing foam solution at the correct ratio.

C Place the foam concentrate pickup tube into the foam container. Successful operations are dependent on the amount of lift between the foam container and the eductor.

D Connect a fog nozzle to the attack line. If an aspirator is available, it should be used since it will create superior finished foam.

E Confirm the foam proportioning rate, move the eductor bypass valve to the foam position, open the appropriate pump discharge valve, increase the pump to the appropriate discharge pressure, and set the pressure control device (relief valve or pressure governor).

F Once foam operations are completed, move the pickup tube to a bucket of clean water and run the foam line to flush any remaining foam concentrate, which could dry and cause future foam operations to fail.

SKILL 5-2

Using an Apparatus-mounted Foam System

Delmar/Cengage Learning

Delmar/Cengage Learning

A For internal eductors, select a discharge supporting foam. Typically, only a subset of the discharges support foam operations, often including some with preconnected lines and some without. Discharges supporting foam should be labeled.

B Connect a fog nozzle to the attack line. If an aspirator is available, it should be used since it will create superior finished foam.

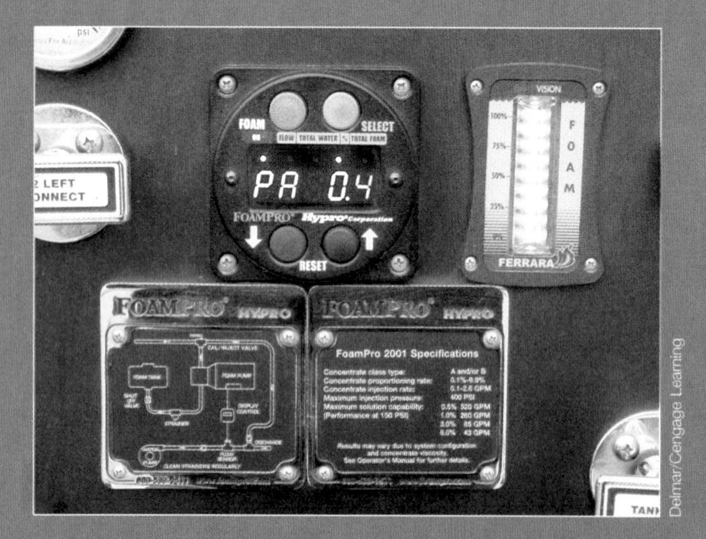

Delmar/Cengage Learning

Delmar/Cengage Learning

C Set the foam proportioning equipment to provide the correct ratio based on the type of foam concentrate being used and the desired behavior of the finished foam. Open the appropriate discharge valve.

D During foam operations, monitor gauges for proper operations and to make sure adequate foam concentrate exists. Apparatus with foam tanks will have a foam tank level gauge that should be monitored. The foam system may also provide information. The one pictured provides flow in gpm (Lpm) (like a flow meter), total water consumed in gallons (litres), proportioning percentage, and gallons (litres) of foam concentrate used.

SKILL 5-3

Using a Compressed-air Foam System

Delmar/Cengage Learning

A For a compressed-air foam system select a discharge supporting foam. Typically, only a subset of the discharges support foam operations, often including some with preconnected lines and some without. Discharges supporting foam should be labeled.

Delmar/Cengage Learning

B Connect a smooth-bore nozzle to the attack line.

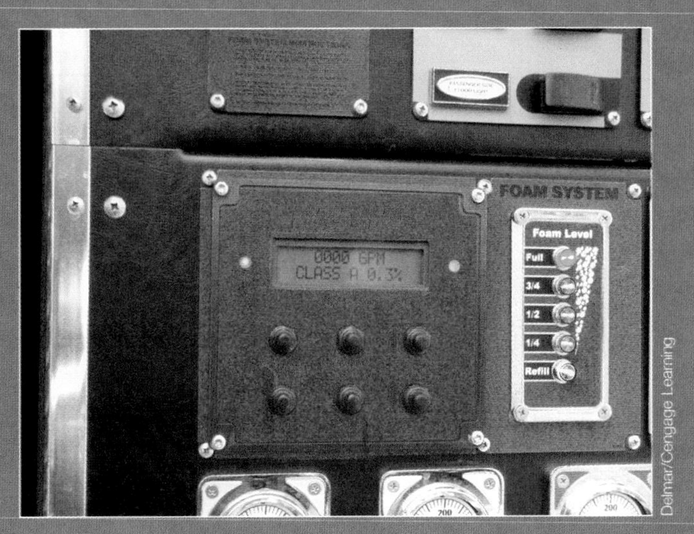

Delmar/Cengage Learning

C Set the foam proportioning equipment to provide the correct ratio based on the type of foam concentrate being used and the desired behavior of the finished foam. Turn on the foam injection switch for the selected discharge line.

Delmar/Cengage Learning

D Turn on the CAFS air compressor. The CAFS air compressor pressure should be the same as the PDP, typically around 100 psi (700 kPa). Open the appropriate hoseline discharge valve. During foam operations, monitor gauges for proper operations and to make sure adequate foam concentrate exists. Apparatus with foam tanks will have a foam tank level gauge which should be monitored. The foam system may also provide information. The one pictured, provides flow in gpm (Lpm) (like a flow meter), total water consumed in gallons (litres), proportioning percentage and gallons (litres) of foam concentrate used. CAFS oil temperature and CAFS air pressure need to be monitored with CAFS is in operation. Just as fire pumps should only be operated with water flowing, CAFS air compressors should only be operated with air flowing.

SKILL 5-4

Using an Apparatus-mounted SCBA Fill System

Delmar/Cengage Learning

A Visually inspect the cylinder to confirm that it is within the hydrostatic test date and does not have physical damage; then place the cylinders into the containment chambers.

Delmar/Cengage Learning

B Connect the lines to the SCBA bottles. Lines need to be fully threaded to avoid leaks and O-ring damage; however, excessive tightening force should *not* be used.

Delmar/Cengage Learning

C Close the fill-station bleed valve, then open the SCBA bottle valves.

Delmar/Cengage Learning

D Close the fill-station door(s). Confirm that the fill-station regulated pressure equals the SBCA's full pressure. If it does not, adjust the fill-station regulator. Open the fill-station fill valve. SCBA bottles should not be filled faster than 500 psi (345 kPa) per minute. This may require feathering of the fill valve if not all positions contain a bottle.

Delmar/Cengage Learning

E Close the fill-station fill valve. Open the fill-station door(s) and close the valves on the SCBA bottles.

Delmar/Cengage Learning

F Close the SCBA bottle valve. Open the fill-station bleed valve. After the fill-line pressure has dropped to zero, disconnect the fill lines from the SCBAs.

Delmar/Cengage Learning

G Remove the SCBAs from the fill station and return them to the full bottle area. If additional bottles need filling, repeat the procedure; otherwise, secure the fill lines.

LESSONS LEARNED

The ability to move water from one location to another is a function of the pump and its attached peripheral components. The pump panel is the mission-control center of pump operations, allowing the pump operator to control and monitor pump efficiency from one location. The peripheral components on the pump panel help control the pump, ensure the efficiency of pump operations, and provide safety to personnel and equipment during pump operations.

Pump peripherals can be grouped into three categories: instrumentation, control valves, and systems. Instrumentation allows the pump operator to gather information on pressures, flows, and components. Control valves operate to initiate, restrict, and direct water flow through the pump and associated piping. Systems support the function of the pump and include priming, pressure regulating, cooling, and foam systems. Miscellaneous components typically found on pump panels include the pump engine hand throttle, auxiliary outlets, and drain valves.

KEY TERMS

Auxiliary cooling system A system used to maintain the engine temperature within operating limits during pumping operations.

Bourdon tube gauge The most common pressure gauge found on an apparatus, consisting of a small curved tube linked to an indicating needle.

Compound gauge A pressure gauge that reads both positive pressure (psi) (kPa) above atmospheric pressure (psi) (kPa) and negative pressure below atmospheric pressure (in. or mm of Hg).

Control valves Devices used by a pump operator to open, close, and direct water flow.

Finished foam The final foam that is applied after the foam solution has been aerated.

Flow meter A device used to measure the quantity and rate of water flow in gallons (Litres) per minute (gpm) (Lpm).

Foam concentrate The material purchased from the manufacturer prior to dilution.

Foam solution Foam concentrate that has been mixed with water but not yet mixed with air; what is in a hoseline after the eductor and before an aerating nozzle.

Indicators Devices other than pressure gauges and flow meters (such as tachometer, oil pressure, pressure regulator, and on-board water level) used to monitor and evaluate a pump and related components.

Instrumentation Devices such as pressure gauges, flow meters, and indicators used to monitor and evaluate the pump and related components.

Pressure gauge Device used to measure positive pressure in pounds per square inch (psi) (kilopascals or kPa).

Pressure governor A pressure regulating system that protects against excessive pressure buildup by controlling the speed of the pump engine to maintain a steady pump pressure.

Pressure regulating systems Devices used to control sudden and excessive pressure buildup during pumping operations.

Pressure relief device A pressure regulating system component or valve that protects against excessive pressure buildup by diverting excess water flow from an area of excessive pressure such as the discharge side of the pump or intake valve back to the intake side of the pump or to the atmosphere.

Pump panel The central location for controlling and monitoring the pump and related components.

Pump peripherals Those components directly or indirectly attached to the pump that are used to control and monitor the pump and related components.

REVIEW QUESTIONS

Multiple Choice

Select the most appropriate answer.

1. Most pump peripherals are located
   - **a.** in the cab of the apparatus.
   - **b.** on the pump panel.
   - **c.** at various locations on the apparatus.
   - **d.** under the apparatus.
2. Which NFPA standard suggests a standard color code to match discharges with their gauges?
   - **a.** 1500
   - **b.** 1901
   - **c.** 1911
   - **d.** 1961
3. Instrumentation on pump panels can be grouped into three categories. Which of the following is *not* one of the categories?
   - **a.** emergency warning lights
   - **b.** pressure gauges
   - **c.** flow meters
   - **d.** indicators
4. Which of the following is *not* considered an indicator on the pump panel?
   - **a.** tachometer
   - **b.** water level instrument
   - **c.** pump engagement light
   - **d.** flow meter
5. Each of the following is an example of a valve, *except*
   - **a.** ball.
   - **b.** butterfly.
   - **c.** ram.
   - **d.** gated.

Short Answer

On a separate sheet of paper, answer/explain the following questions.

1. Explain how pump panels are both similar and different.
2. List the different locations where pump panels can be found on apparatus and explain the advantages and disadvantages of each location.
3. How can the fluctuations or bounces of indicating needles be controlled?
4. What is the function of the two main gauges and individual discharge gauges located on pump panels?
5. What is the main advantage of using flow meters on pumps?
6. As a rule of thumb, what is the rated capacity for a pump with one 2½-inch (65-mm) discharge?
7. Explain the function for each of the following:

   Tank-to-pump valve

   Pump-to-tank valve (recirculating valve)

   Transfer valve
8. What is the purpose of oil in the priming system?
9. How can you tell when the priming process has actually primed the pump?
10. List the different types of foam systems and identify the advantages and disadvantages of each.
11. Explain the difference between pressure gauges and flow meters.
12. What are control valves? Provide several examples of common control valves found on pump panels.
13. Why are pressure regulators so important to pump operations?
14. Explain how air is replaced with water during the priming process.

ACTIVITIES

1. Draw a schematic, or take a picture, of two different pump panels. First, identify all of the components found on the pump panels. Next, discuss the differences and similarities between the two pump panels.
2. After careful investigation and testing, your department has decided to take the plunge and begin using compressed-air foam during suppression efforts. The next big question is, what system should your department use to generate foam streams? Because of your excellent research talents, your chief has asked you to contact manufacturers and vendors to determine what types of compressed-air foam-generating systems are available. Contact the manufacturers and vendors and prepare a report to your chief highlighting the different systems available. Be sure to indicate which system you think is the best for your department and why. Also, include the information you collected during your research.
3. A new firefighter has been assigned to your station. During the morning apparatus check he tells you he is unsure how to check the foam system's readiness on your station apparatus. Explain the required steps and provide a hands-on lesson by performing the necessary inspections and tests that verify the operational status of the foam system.

PRACTICE PROBLEM

1. You are the lead apparatus operator for your department. Because of your experience and knowledge of pump operations, your chief has asked you for your advice on pump panel location and configuration. Respond to your chief's request in the form of a memo. Be sure to provide justification for each of your recommendations.

ADDITIONAL RESOURCES

Pump manufacturers' literature is an excellent source for information on pumps and pump peripherals construction, operation, and maintenance.

A Firefighter's Guide for Foam, Kidde Fire Fighting, http://www.kidde-fire.com

Compressed Air Foam for Structural Fire Fighting: A Field Test, FEMA, USFA-TR-074/January 1994, http://www.usfa.dhs.gov/downloads/pdf/publications/tr-074.pdf

Questions About Class A Foam in Municipal Fire Operations, Hale Products, http://www.haleproducts.com/_Downloads/hale/articles/Questions%20about%20Class%20A%20foam%20in%20Municipal%20Fire%20.PDF

Class A Foam and CAFS briefing—Structural Firefighting, http://cafsinstitute.org/pdf/CAFS_Briefing.pdf

FoamPro Presentations, http://www.foampro.com/en-us/Presentations/Presentations.htm

Security Blanket, Ansul Incorporated, http://www.ansul.com/AnsulGetDoc.asp?FileID=17930

Mike Richards, Compressed Air Foam Systems in Structural Firefighting, November 27, 2003, http://www.atu.edu/lfa/Brucker/Engl2053/Samples/bmr15.pdf

6 Hose, Appliances, and Nozzles

- Learning Objectives
- Introduction
- Hose
- Appliances
- Nozzles
- Lessons Learned
- Key Terms
- Review Questions
- Activities
- Practice Problem
- Additional Resources

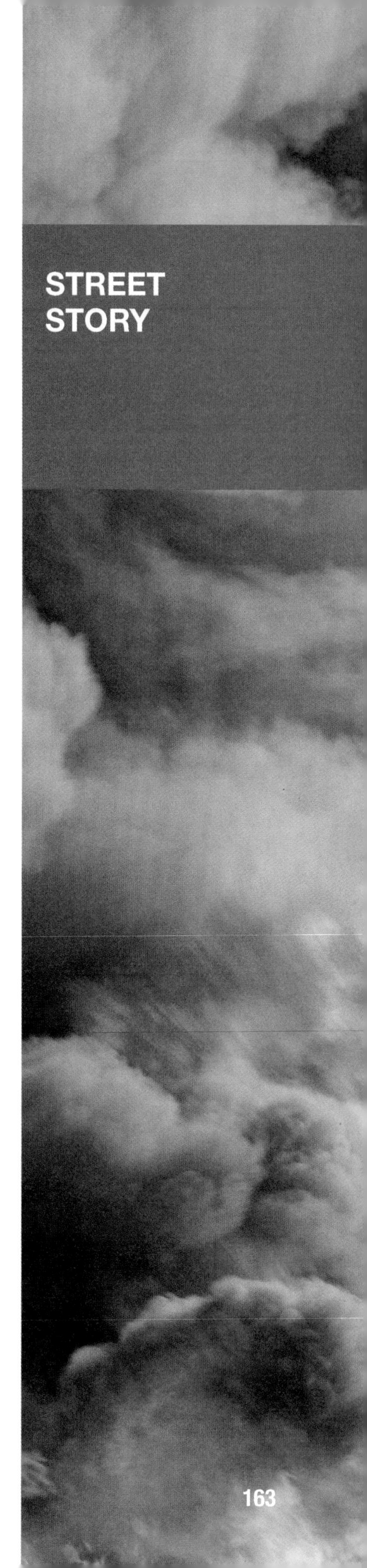

STREET STORY

I remember my first big fire call as an apparatus engineer. I was filling in at a station on the other side of town. Around 9:30 at night, a call came in for an eight-unit apartment complex fire. Arriving on the scene, I stopped at a hydrant, and a firefighter got out and hooked it. Once I was able to supply water to my first attack line, I thought "Now what can I do?"

I went over the course of my training, and began looking at the building to try to see where I could help. The main body of the fire was on the B and C side of the building. I was on the completely opposite side, and I could still see a lot of fire above the roof line. A ladder truck, which was the first on-scene, had a 1¾-inch (45-mm) hose coming off it, and my crew took a 1¾-inch (45-mm) line too. I thought, "Wow, that's a lot of fire for those hoses." In my training I'd learned the calculations you can do to help determine how much water you need to knock the fire down. A 1¾-inch (45-mm) hoseline throws about 180 gallons (680 litres) per minute. Those two lines were not putting the fire down. They were not delivering enough water to extinguish the fire. So I said to the guy driving the ladder, "Do you think they'll go defensive?"

We figured that they would, unless they could get ahead of it. So between the two of us, we began preplanning what we could do from our side. I knew from my training that good pump operators anticipate and prepare for problems and incident commander (IC) requests so that they can quickly handle any situation that arises. I figured I could take down the deck gun, a master stream that can deliver 1,000 gallons (4,000 litres) per minute when mounted on the engine and 800 gallons (3,200 litres) per minute when used as a ground monitor. I got the deck gun down, and put it on the ground to hit the C side. Just as I was getting the last hose connected, the IC called for "everybody out" over the radio and directed my truck company engineer to sound his air horn.

Soon after, my lieutenant was coming toward me. He looked at the ground monitor, looked at me, and said "fire it up!" I made a decision based on what I was seeing and my training—I had been able to read the building and think ahead. It was really cool to know I had done the right thing."

—Street Story by Tracy Burrus,
Apparatus Engineer, Engine Company #3,
City of Madison Fire Department, Madison, Wisconsin

LEARNING OBJECTIVES

After completing this chapter, the reader should be able to:

6-1. Identify the NFPA standards that focus on hose, appliances, and nozzles.

6-2. Explain the role hose plays in pump operations.

6-3. List the basic parts of a hose and explain the three main types of hose construction.

6-4. Discuss the two broad categories of hose and identify several tools used when working with hose.

6-5. List and discuss several appliances used during pump operations.

6-6. Describe how foam solution is drawn into the water stream when using an eductor.

6-7. Explain the role nozzles play in pump operations.

6-8. Explain the relationship of flow, pressure, shape, and nozzle reaction in the design and operation of a nozzle.

6-9. Demonstrate the various procedures for initiating, extending, and removing discharge lines.

6-10. Demonstrate the ability to assemble hoselines, nozzles, valves, and appliances.

*The driver/operator requirements, as defined by the NFPA 1002 Standard, are identified in black; additional information is identified in blue.

INTRODUCTION

Why study such items as hose, appliances, and nozzles in a textbook on pump operations? First, keep in mind that pump operations is the process of moving water from a supply source through a pump to a discharge point. Getting water from a supply source to the pump and from the pump to a discharge point is the job of hose, appliances, and nozzles. Second, misuse or inappropriate use of hose, appliances, and nozzles can reduce the effectiveness of a pump. Too small a hose on the intake side of the pump may not be able to supply the required flows on the discharge side. Finally, and perhaps most important, lack of knowledge can create hazardous situations for both personnel and equipment. Providing too much pressure can rupture hose or overpower personnel operating nozzles. Understanding the operating principles and construction of hose, appliances, and nozzles will help ensure that pump operations are conducted in a safe, efficient, and effective manner while providing appropriate flows and pressures.

Pump operators should be familiar with the NFPA standards that focus on hose, appliances, and nozzles. The following standards are referenced in this chapter:

- NFPA 1961, *Standard on Fire Hose*. This standard establishes requirements for the design, construction, inspection, and testing of new fire hose.
- NFPA 1962, *Standard for the Inspection, Care, and Use of Fire Hose, Couplings, and Nozzles and the Service Testing of Hose*. This standard applies to all fire hose, couplings, and nozzles currently in operation or that will be placed in operation.
- NFPA 1963, *Standard for Fire Hose Connections*. This standard defines the minimum requirements for new fire hose connections.
- NFPA 1964, *Standard for Spray Nozzles*. This standard provides the minimum performance requirements for new adjustable-pattern spray nozzles for the following firefighting applications: general use, marine and offshore platform use, standpipe system use.
- NFPA 1965, *Standard for Fire Hose Appliances*. This standard provides the minimum performance, operation, and testing requirements for appliances with up to 6-inch (150-mm) connections.

Lack of knowledge can create hazardous situations for both personnel and equipment.

HOSE

Hose is an essential, yet often misused and abused, component in fire pump operations. The ability to move water from a supply source to the pump as well as from the pump to a discharge point would be impossible without hose. NFPA 1961 defines fire hose as a "flexible conduit used to convey water." Because of its importance in transporting water, pump operators should be familiar with the different classifications of hose, hose tools, hose construction, and hose care principles.

Classification of Hose

The two broad classifications of hose are intake or **supply hose** and discharge or **attack hose**. Intake hose moves water from a supply source to the pump, while discharge hose moves water from a pump to the discharge point. In some cases, the same type of hose can be used for either intake or discharge water movement. NFPA 1961 identifies specific requirements for supply and **suction hose** (intake hose) and attack hose (discharge hose). This standard also identifies **large-diameter hose** (LDH) as any hose with a diameter of 3½ inches (90 mm) or larger.

The two broad classifications of hose are intake hose and discharge hose.

Intake Hose

Intake hose, often referred to as a supply line, moves water from a source to the pump. Typically, this means getting as much water to the pump as possible. Depending on the water supply and the discharge flow requirements, several types of hose can be used. Supply hose, according to NFPA 1961, is designed to move water from a pressurized water source such as a hydrant to a pump. Supply hose has a minimum trade size diameter of 3½ inches (90 mm), which means it is also classified as LDH. NFPA 1962 requires that supply hose not be operated at pressures exceeding 185 psi (1,295 kPa). Supply hose can transfer water over longer distances with minimum loss of pressure due to friction because of the large-diameter size of the hose. Two other types of intake hose are used to transfer water over shorter distances. **Soft sleeve** or soft suction is typically a shorter section of supply hose with female couplings on both ends used to ease installation when the pump is close to a pressurized water supply such as a hydrant. Soft sleeve hose ranges in size from 2½ inches to 6 inches (65 mm to 150 mm) in diameter. Suction hose, also referred to as hard suction hose, is defined by NFPA 1961 as a "hose designed to prevent collapse under vacuum conditions so that it can be used for drafting water from below the pump (lakes, rivers, wells, etc.)." The size of suction hose ranges from 2½ inches to 6 inches (65 mm to 150 mm) in diameter and is typically 10 feet (3 metres) in length. Note the confusing term "hard suction" given to this hose. Recall that drafting operations raise water to the pump by atmospheric pressure rather than by a suction process.

Conventional hard suction hose looks like automotive radiator hose except it is larger in diameter. It is constructed from rubber, black in color, and it contains an imbedded helix wire to make it hard. Over time, the inner section of the hose may separate and collapse when drafting, which significantly reduces the flow of water. Operators should be aware that this internal collapse is not externally visible; confirm that it has not collapsed by using a transparent hose cap, operating the primer to create a vacuum, and looking through the cap with a flashlight to confirm that the hose interior remains intact. A newer alternative is a PVC hard suction hose. It is formed from clear PVC with an external helix reinforcement rib and a smooth interior. Its advantages include being able to see into the hose, which aids in leak detection and correction, lighter weight, and greater flexibility. Its disadvantages include greater damage from sustained ultraviolet (UV) light, and its restriction to suction use (i.e., no positive pressure).

Discharge Hose

Discharge hose, referred to as attack hose in NFPA 1961 and NFPA 1962, is used to move water from the pump to a discharge point. The discharge point could be a nozzle used for suppression efforts or for covering exposures. In addition, the discharge point could be a different pump that uses the discharge hose as a supply line. Often, hoses used to supply a nozzle are called attack lines while those used to supply other pumps or devices are called supply lines. NFPA 1961 defines *attack hose* as "hose designed to combat fires beyond the incipient stage." Attack hose can be either woven-jacket or rubber-covered, ranges in size from 1 inch to 3 inches (25 mm to 77 mm), and has a normal highest operating pressure of 250 psi (1,750 kPa). The typical sizes used for attack hose include 1½ inches (38 mm), 1¾ inches (45 mm), 2 inches (50 mm), 2½ inches (65 mm), and 3 inches (77 mm), with 1¾-in. and 2-in. (45-mm and 50-mm) hose typically having 1½-in. (38-mm) couplings and 3-in. hose (77-mm) available with either 2½-in. or 3-in. (65-mm or 77-mm) couplings. In this book, 3 in. (77 mm) is assumed to mean 3-in. (77-mm) hose with 2½-in. (65-mm) couplings unless otherwise specified.

Construction

The basic parts of a hose include a reinforced inner liner and an outer protective shell with couplings attached at both ends. The inner liner, sometimes referred to as the inner tube, keeps the water contained in the hose. According to NFPA 1961, liners are made from one of the following materials:

- Rubber compound
- Thermoplastic material
- Blends of rubber compounds and thermoplastic material
- Natural rubber-latex-coated fabric

An extrusion method, a process of forcing heated rubber or plastic through a die to create a uniform component, is typically used to create a continuous and seamless liner. In addition to keeping the hose leak-proof, the liner provides a smooth surface to reduce the friction of water as it passes over the surface. Hose liners by themselves cannot withstand the pressures often required during pump operations. Therefore, liners are reinforced with natural fiber, synthetic fiber, or a combination of the two. The purpose of the outer shell is to protect the reinforced inner liner from the many hazards that may damage or destroy the integrity of the hose, such as abrasion, cuts, chemicals, and heat. The outer shell or cover can be made of the same selection of material as is available to the liner.

Although NFPA 1961 no longer specifies the required length of hose, a section of hose is commonly 50 feet (15 m), sometimes 100 feet (30 m), in length with couplings attached to both ends. Hose and coupling diameters range from 1 inch to 6 inches (25 mm to 150 mm). The two main types of hose construction are woven-jacket and rubber-covered hose. The two main types of couplings are threaded and Storz. **Storz couplings** are also called quarter-turn and sexless couplings.

The basic parts of a hose include a reinforced inner liner and an outer protective shell with couplings attached at both ends.

Woven-Jacket Hose

Woven-jacket hose has been the hose of choice in the fire service for many years. The inner liner is usually made from a synthetic rubber compound or special thermoplastic. The reinforcement material is typically a synthetic fiber weave called a jacket. Synthetic fiber has, for the most part, replaced cotton in the construction of woven-jacket hose. The inner liner and reinforcement jacket are bonded together by a number of different processes. A liner with a reinforced woven jacket is commonly referred to as a single-jacketed lined hose. Single-jacketed hose is commonly used for building occupant standpipe connections and for forestry hose.

Most hose used for structural firefighting is double-jacketed. Double-jacketed hose has a second jacket, or outer shell, added to provide additional strength and resistance to the harsh working environments of the hose. **Figure 6-1** shows both a single- and double-jacketed hose. To increase the durability of the hose, various treatments are used on the outer shell to increase resistance to abrasion, heat, and chemicals. To increase water flow with less friction loss and lower pressure, numerous special inner coatings are available. These inner coatings are smoother (i.e., they don't have the peaks and valleys of the jacket weave in them) and slicker, which offers less resistance to water movement.

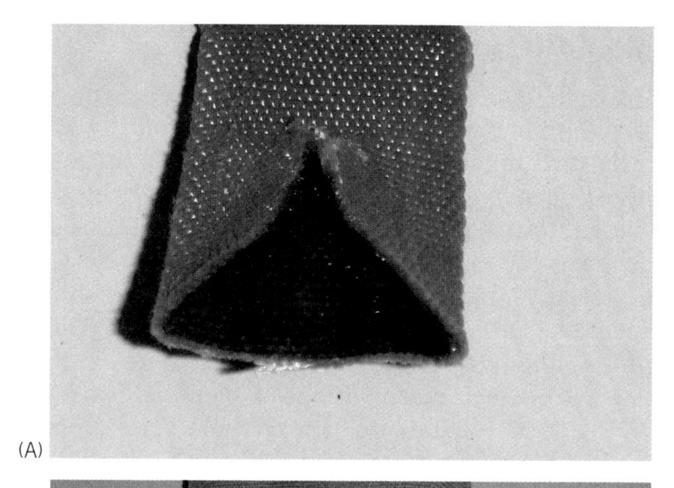
(A)

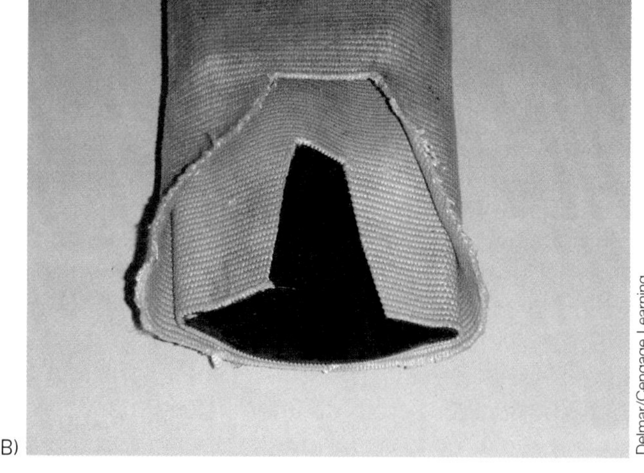
(B)

FIGURE 6-1 Photos of a single-jacketed and double-jacketed hose.

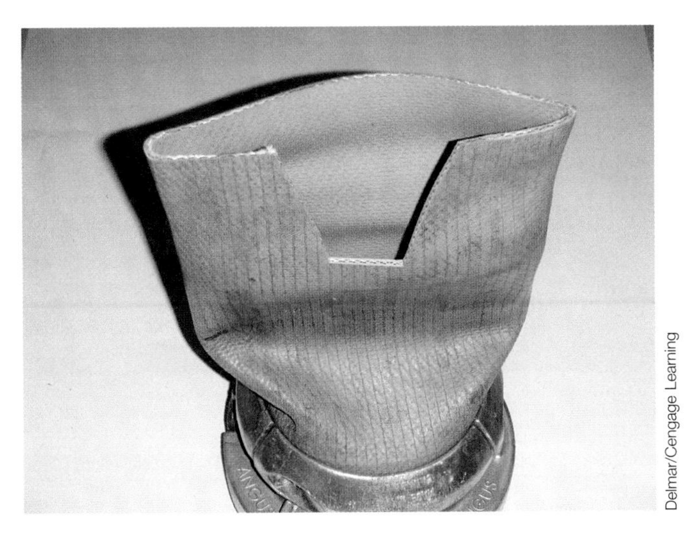

FIGURE 6-2 Cutaway of a rubber-covered hose. Note that the weave, which provides strength, can be seen encased by the rubber jacket.

Rubber-Covered Hose

Today, larger-diameter rubber-covered hose is popular in the fire service because it is lightweight and durable. As the term implies, the hose has a rubber outer cover (**Figure 6-2**). Several methods are used to construct rubber-covered hose. In one method the liner is attached to a reinforcing synthetic fiber jacket, as is the case with woven-jacket hose; however, instead of using a second woven jacket as the outer shell, a rubber outer cover is extruded over the reinforced liner. In another method, the rubber liner and outer shell are extruded over the reinforced woven fiber in one process. Both processes produce a unified hose appearance.

The two main types of hose construction are woven jacket and rubber covered.

Couplings

NFPA 1963 defines couplings as a "set or pair of connection devices attached to a fire hose that allow the hose to be interconnected to additional lengths of hose or adapters and other firefighting appliances." The two most popular materials used to make couplings are brass and lightweight aluminum alloys. Couplings used on hose, appliances, and nozzles are typically either threaded or sexless. Couplings must be easy to connect yet able to withstand high pressures and the demands of fireground operations. NFPA 1963 requires that couplings pass a variety of rigorous tests, including being dropped from 6 feet

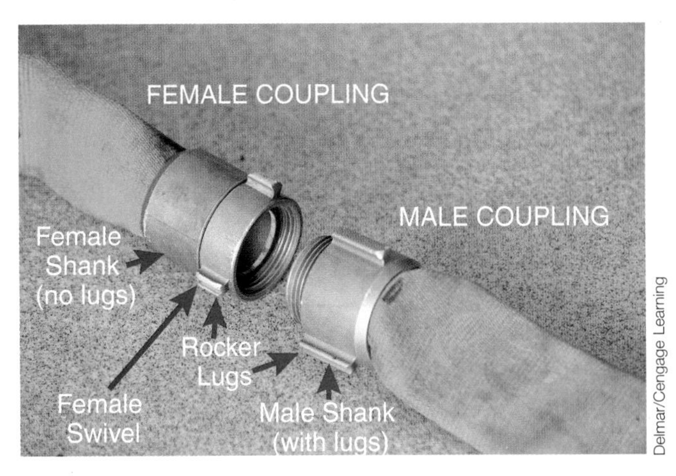

FIGURE 6-3 Example of a male and female threaded coupling.

(2 m) without damage and being able to withstand the service test pressure of the hose without leakage or failure after being connected 3,000 times.

Threaded couplings utilize two different couplings, both having a common thread. Of the two different couplings, one is called a male and the other a female (**Figure 6-3**). The thread used in nearly all threaded coupling construction is referred to in NFPA 1963 as "American National Fire Hose Connection Screw Threads" and is abbreviated with the thread symbol NH. Sometimes this thread type is referred to as national standard thread (NST). Some municipalities, however, still utilize nonstandard threads. In such cases, adapters should be readily available to connect NH couplings and appliances. Threaded couplings have lugs or handles to assist with connecting and disconnecting the couplings. The most common lug on threaded couplings is the rocker lug (see **Figure 6-3**).

Sexless couplings, sometimes referred to as nonthreaded couplings or quarter-turn couplings, are so named because the two connecting couplings are identical to each other. Storz couplings (**Figure 6-4**) are the most common sexless couplings used in the fire service. Storz couplings are connected by joining the two identical couplings and twisting (one-quarter turn) to lock them.

The two main types of couplings are Storz (sexless) and threaded.

Hose Tools

Pump operators have a variety of hose tools from which to choose when working with hose.

FIGURE 6-4 Storz couplings are sexless in that the couplings are identical; that is, there are no male and female couplings.

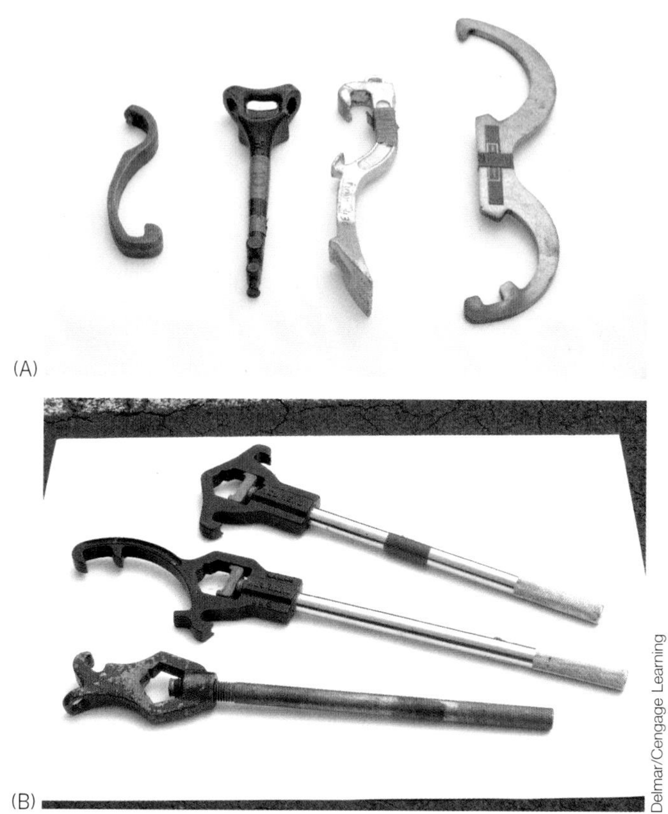

FIGURE 6-5 (A) Spanner wrenches are available in a variety of sizes and styles to fit the different types of hose couplings. (B) Many hydrant wrenches also include a spanner wrench. These are also available in a variety of styles to fit the different types of hose couplings.

Spanner wrenches come in many sizes and shapes to help connect and disconnect couplings of various types, sizes, and lug construction (**Figure 6-5A**). Many hydrant wrenches also include a spanner wrench as part of their design (**Figure 6-5B**). Hose clamps are critical for pump operators in that they control the flow of water in the supply hose. The three main types of hose clamps are: press, screw, and hydraulic. **Figure 6-6** shows a press hose clamp in action. Care should be exercised when using clamps, as they can cause damage to hose and/or injury to personnel. Hose bridges (**Figure 6-7**) are used to allow vehicles to move across hose without causing damage to the hose. Hose jackets provide a quick method of minimizing flow loss from a leaking hose (**Figure 6-8A**); however, it is more common to replace the leaking section of hose with two replacement sections. More important, hose jackets can be used to connect two hoses with different tread types (**Figure 6-8B**). Although less frequent, they can also be used to connect to a hose with a damaged tread or in place of a double male or female. There are many more components used in conjunction with hose. Some are discussed in the following section, "Appliances." Others are discussed later in the text when presenting specific pump operations.

FIGURE 6-6 Example of a press hose clamp.

Delmar/Cengage Learning

FIGURE 6-7 Hose bridges help protect hose from damage when the hose is driven over.

(A)

(B)

Delmar/Cengage Learning

FIGURE 6-8 (A) Hose jackets can be used to temporarily stop the flow from a leaking hose. (B) When hose jackets are used to connect mismatched or damaged couplings, or instead of using a double male or female, both unconnected couplings are placed inside the hose jacket.

NOTE

The three main types of hose clamps are press, screw, and hydraulic.

STREETSMART

Prior to the industrial revolution with its improvements to mass production and easy transportation, most items were made near their point of use. As cities and improved water systems developed, locally made fire hydrants were installed. With no accepted standards, many designs and thread types were created. While hydrants were consistent within a city, they would differ from those in other cities. Then in 1904 a great fire occurred in Baltimore, destroying over 1,500 buildings and much of the downtown business district, exceeding the capabilities of Baltimore's fire department. Many pumpers came to assist from neighboring cities but they were hampered in their ability to help due to their inability to connect to Baltimore's incompatible hydrant threads. With general agreement that the incompatibility caused much greater fire losses than would have existed if the threads were compatible, insurance companies covering the replacement buildings wanted a resolution to the problem, and the NFPA was formed. NFPA developed a standardized hydrant thread; however, many cities with existing hydrants did not replace them, citing the expense. Since all threads within a city (hydrant, hose, engines, adapters, etc.) were compatible, all would need to be changed to adopt the national standard.

Today, with more than 1,000 different hydrant threads in use, ISO requires two items to circumvent incompatible thread problems. Every engine with nonstandard threads should carry adapters to standard threads. All engines should carry hose jackets (**Figure 6-8B**), which may be used to connect two hoses of different threads.

SAFETY

Care should be exercised when using hose clamps, as they can cause damage to hose and/or injury to personnel.

Hose Care

Because of the important role hose plays in pump operations, appropriate attention should be given to its care. Care of hose begins with preventing it from getting damaged. This includes proper use as well as the recognition of situations that may help prevent damage to hose, such as

- Minimizing chafing
- Reloading so that hose folds are in different locations each time
- Keeping vehicles from running over hose
- Minimizing heat and cold exposure
- Allowing air circulation in hose loads to reduce mildew growth and rust/corrosion

Care of hose continues with three important activities: inspecting, cleaning, and record keeping.

Inspecting

All hose should be inspected prior to being placed in service and after each use. According to NFPA 1962, hose should be inspected for the following when placed in service:

- Not vandalized and free of debris
- No evidence of mildew or rot
- No damage by chemicals, burns, cuts, abrasions, or vermin
- Service test of hose is current
- Liner shows no sign of delamination

In addition, hose should be tested annually and after any repairs have been made, or if it has been subjected to freezing. The inspection of in-service hose after it is used should include the following components, as a minimum:

- Hose outer cover (inspecting for its condition, discoloration, abrasions, cuts, fraying)
- Coupling (looking for damage, proper operation, condition of threads, slippage from hose)
- Gasket (checking its presence and condition)
- Liner (no sign of delaminating)
- Hydrostatic test date (ensuring hose is within annual test date)

Should any section of hose fail this inspection, it should be removed from service and either repaired or condemned.

Hose should be tested annually and after any repairs have been made or if it has been subjected to freezing.

Cleaning

Hose should be properly cleaned after each use. For woven-jacket hose this means a good cleaning and thorough drying. The main reason for cleaning and drying is that dirt, grime, and chemicals can work their way into the woven jacket, attacking and deteriorating the fibers. For rubber-covered hose, this means a good wipe down depending on the amount of dirt and grime on the hose. The cleaning process can be manually accomplished by laying hose section out and then scrubbing with a mild soap or detergent. Some departments purchase or build a mechanical hose washer. In most cases, the inspection of hose can be accomplished during this cleaning process.

Record Keeping

According to NFPA 1962, fire departments must establish and maintain an accurate record of each hose section. This requires that each hose section be assigned an identification number. The identification number must be stenciled on the jacket/cover or stamped on the shank or swivel of the coupling. In addition, NFPA 1961 requires that each hose section be indelibly marked on both ends using 1-inch (25-mm) letters with the following information:

- Manufacturer's identification
- Month and year of manufacture
- The statement "Service Test to xxx psi per NFPA 1962" or "Service Test to xxx bar per NFPA 1962"

LDH must use 2-inch (50-mm) letters and be marked as either "supply hose" or "attack hose." In the past, most hose record-keeping systems were paper based. Today, many departments develop or purchase electronic hose record-keeping systems. Typical information maintained includes the following:

- Identification number
- Manufacturer and vendor
- Hose size and length
- Type and construction of hose
- Date placed in service
- Hose location (station and/or apparatus)
- Date of each service test
- History of damage and repairs

The collection and analysis of hose information/ data is important for the safe use of the hose. Analysis of hose data can help determine the cost-effectiveness of hose and how well or poor a hose performs over time. For example, a trend in high repair rates for a hose may be due to either poor hose construction or improper use. In either case, corrective action can be taken to ensure the safe use of hose.

Hose Testing

Most new attack hose has a service test of 250 psi (1,750 kPa), while supply hose has a service test pressure of 200 psi (1,400 kPa). Hose that complies with NFPA 1962 will have the service test pressure stenciled on each section of hose. The testing of fire hose requires an appropriate location with a water source. When possible, the location should be level with a smooth surface and adequate space to lay out the hose being tested. The water supply can be a hydrant, tank, or static source, such as a pond or lake. The quantity of water sufficient to fill hose being tested is the primary consideration for a water supply. In general, the fire hose is laid out, filled with water, and the pump discharge pressure is increased to the service test pressure. NFPA 1962 provides excellent guidance on conducting fire hose service tests and includes important safety considerations and sample forms to document the test. Several considerations for conducting annual service tests of fire hose are discussed next.

Preparation (Prior to Starting the Test)

Identify the correct test pressure for the hose. More than one hose size and section can be tested at one time, but all hose should have the same test pressure. Conduct an inspection of the hose, making sure to check gaskets, threads, and obvious damage. A maximum of 300-foot (90-m) of hose per discharge should be tested in a single evolution. Ensure testing instrumentation calibration is within 12 months.

Conducting the Test

Basic Steps for Conducting Hose Tests

1. Connect the water supply and the test hose sections.
2. Open the discharge control valves to which the test hose sections are connected.
3. Slowly increase the discharge pressure to 45 psi (310 kPa).
4. Remove all air from within the hose, check for leaks, and mark the hose near the coupling to determine coupling slippage.
5. Increase pressure slowly; NFPA 1962 suggests a rate no faster than 15 psi (103 kPa) per second, to the service test pressure.
6. Conduct the test for 3 minutes, periodically checking for leaks. Keep all nonessential personnel away from testing area.
7. Record the results of the test per fire department requirements.

After the Test

Any hose that fails the test should be removed from service and repaired or condemned. Repaired hose must pass a retest prior to being placed back in service. All hose passing the pressure test should be cleaned and dried as per normal operating procedures.

APPLIANCES

During pump operations, various hose configurations are used to move water from a supply to a discharge point. The accessories and components used to support these varying hose configurations are called appliances. Appliances can be used on either the intake or discharge side of hose configurations. In most cases, friction loss occurs when water flows through an appliance. Pump operators must include appliance friction loss factors when performing hydraulic calculations. The most common appliances used in pump operations include **wyes** and **Siamese**, adapters, and several specialized devices such as manifolds, **water thieves**, and **eductors**. NFPA 1965 establishes requirements for appliances up to 6 inches (150 mm) in diameter. According to NFPA 1965, appliances must have a maximum operating pressure of 200 psi (13.8 bar) or greater, and the maximum operating pressure along with the manufacturer name and product/model identification must be permanently marked on the appliance.

The most common appliances used in pump operations include wyes and Siamese, adapters, and several specialized devices.

Wyes and Siamese

A wye (**Figure 6-9**) is used to divide one hoseline into two or more lines. The inlet side of the wye has a female coupling and the discharge outlets are male couplings. A plain wye cannot control flow in the divided lines as it has no control or clapper valves.

Delmar/Cengage Learning

FIGURE 6-9 This plain wye has one 2½-inch (65-mm) female NH inlet and two 1½-inch (65-mm) male NH outlets.

Most wyes, however, have control valves, typically a ball valve, and are called gated wyes. According to NFPA 1965, appliances with lever-operated handles must indicate a closed position when the handle is perpendicular to the hoseline. A reducing wye has a larger coupling on the intake side and smaller couplings on the discharge outlets; for example, a 2½-inch (65-mm) inlet and several 1½-inch (38-mm) outlets. Wyes come in a variety of sizes and configurations (**Figure 6-10**).

(A)

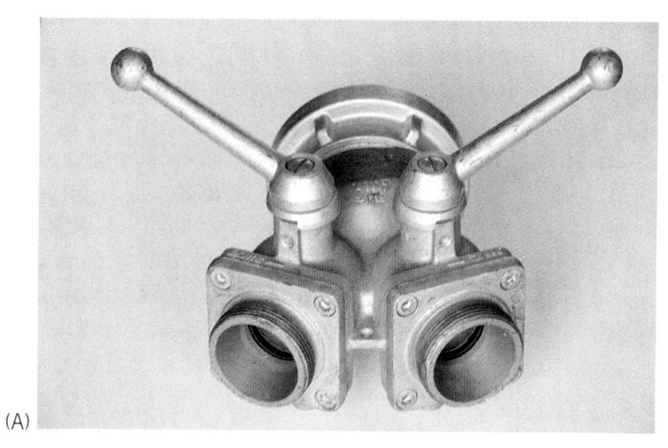

(B)

Delmar/Cengage Learning

FIGURE 6-10 Various wye sizes and configurations are used in the fire service.

Delmar/Cengage Learning

FIGURE 6-11 This clappered Siamese has two 2½-inch (65-mm) female NH inlets and one 2½-inch (65-mm) male NH outlet.

A Siamese is used to combine two or more lines into a single line. The inlet sides of a Siamese have female couplings while the discharge outlet has a male coupling. As with wyes, plain Siamese have no control or clapper valves (**Figure 6-11**). Some Siamese have clapper valves. Clapper valves on a Siamese prevent higher pressure from one of the intake hoselines from entering other connected intake hose of lower pressure. The majority of Siamese will have either the same size couplings on both the inlets and the outlet, or a larger outlet than the inlets (**Figure 6-12**). Like wyes, Siamese come in a variety of sizes and configurations.

Delmar/Cengage Learning

FIGURE 6-12 A variety of Siamese sizes and configurations are used in the fire service.

FIGURE 6-13 Double male and double female adapters are commonly used by pump operators.

Adapters

When laying out hose configurations for pump operations, mismatched hose (2½-inch to a 1½-inch line or 65-mm to a 38-mm line) and couplings (two female couplings) may occur. To connect these mismatched lines, various adapters are available. Double male and double female adapters (**Figure 6-13**) are used to connect threaded couplings of the same size and sex. Increasing and decreasing adapters allow couplings of different size to be connected. **Figure 6-14** shows a common decreaser, 2½ inches (65 mm) (female) to 1½ inches (38 mm) (male), and a 2½ inches (65 mm) (female) to 3 inches (77 mm) (male) increaser. Decreasers are also called reducers. Increasers can be made by adding an adapter to a decreaser. For example, an increaser can be made by adding a double 1½-inch (38-mm) female to the 1½-inch (38-mm) male side of the decreaser and a double 2½-inch (65-mm) male added to the female side of the decreaser.

STREETSMART

Increasers are far less common than decreasers because unless joined by a Siamese, the smaller-diameter lines cannot provide enough water to operate devices that connect to larger-diameter lines. Larger-diameter lines can easily provide enough water to operate devices that connect to smaller-diameter lines, so reducers are common.

Specialized Devices

Numerous specialized devices are used to support and enhance pumping operations. **Four-way hydrant valves** are used to increase pressure without

FIGURE 6-14 An example of a smooth decreaser and smooth increaser. The smooth transition of the increaser decreases turbulence, thereby reducing friction losses.

FIGURE 6-15 A four-way hydrant valve.

FIGURE 6-17 A distribution manifold.

FIGURE 6-16 A water thief.

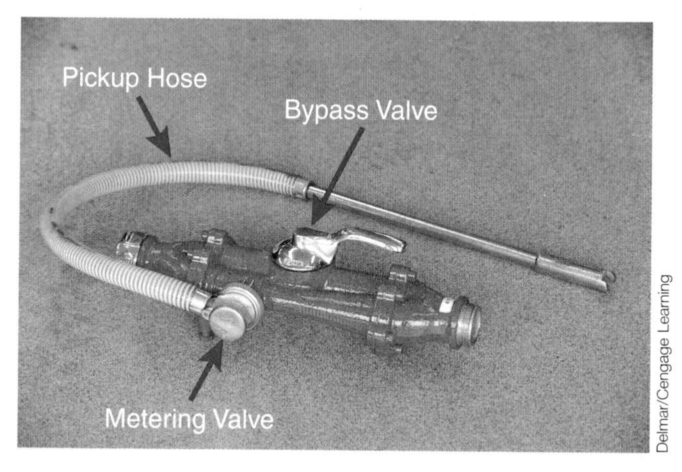

FIGURE 6-18 Eductors are used to pick up and mix foam during foam operations.

interrupting flow (**Figure 6-15**). Water thieves are similar to gated wyes in that they are used to connect additional smaller lines from an existing larger line (**Figure 6-16**). **Manifolds** are larger devices that provide the ability to connect numerous smaller lines from a large supply line (**Figure 6-17**). Pressure relief devices and gauges are often mounted on manifolds. Manifolds provide a greater degree of flexibility when used for both forward and reverse hose lays. Sharp curves and twisting in the hoselines connected to manifolds should be avoided when possible, especially with longer hose lays. Excessive curves and twisting can cause the manifold to move and even flip. Use of manifolds and hose lays is discussed further in **Chapter 7**. Care must be exercised when using manifolds.

Eductors are specialized devices used in foam operations. The basic components of an eductor include a metering valve and pickup hose (**Figure 6-18**). The metering valve controls the percentage of foam drawn into the eductor. The pickup hose is a noncollapsible tube used to move the foam to the eductor. The eductor utilizes the **Venturi principle** to draw foam into the water stream (**Figure 6-19**). Water pressure increases as the eductor tube decreases in size. The pressure is converted to velocity as it enters the induction chamber and then travels out of the eductor. This process creates a low-pressure area in the induction chamber that allows foam to be drawn into the eductor and mixed with the water stream.

(For step-by-step photos of Initiating a Discharge Line, please refer to page 180.)

(For step-by-step photos of Extending a Discharge Line with a Break-apart Nozzle, please refer to page 181.)

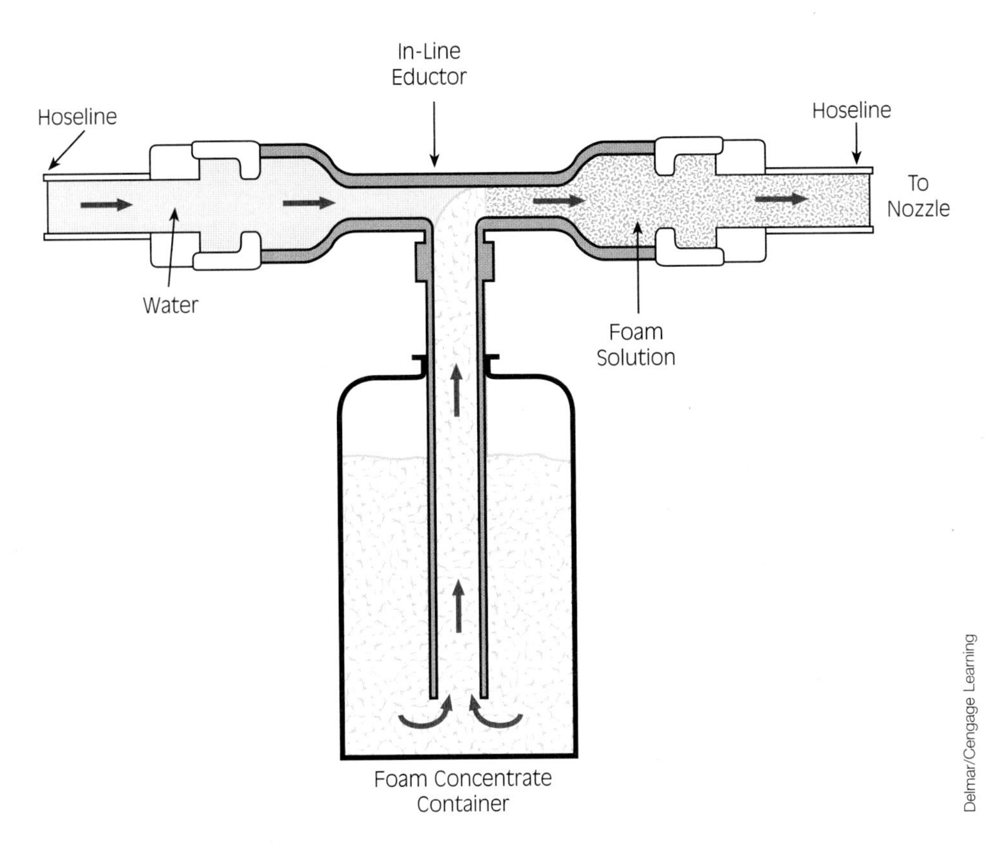

FIGURE 6-19 Eductors use the Venturi principle to draw foam into water streams.

(For step-by-step photos of Removing a Discharge Line, please refer to page 182.)

(For step-by-step photos of Extending a Discharge Line Without a Break-apart Nozzle, please refer to pages 183–184.)

When proper flow and pressure are provided, the three nozzle functions of controlling flow, reach, and stream shape will be the most effective for suppression efforts.

NOZZLES

Nozzles are an important component in fire suppression efforts. Without nozzles, fire suppression activities would be minimal at best. In addition, nozzles that are supplied with inefficient flows and pressures can reduce extinguishment effectiveness as well as increase hazards to personnel. You might argue, then, that the primary purpose of pump operations is to supply nozzles with appropriate flows and pressures so that efficient and effective suppression efforts can be made. Therefore, pump operators should have a good working knowledge of the operating principles and classification of nozzles.

Nozzles that are supplied with inefficient flows and pressures can reduce extinguishment effectiveness as well as increase hazards to personnel.

Nozzle Classification

Although a wide variety of nozzles exists for fire suppression activities, the vast majority of nozzles can be grouped into either **smooth-bore** or **combination nozzles**.

Smooth-Bore Nozzles

Also referred to as solid-stream nozzles, smooth-bore nozzles are the least complex of nozzles. The nozzle illustrated in **Figure 6-20** is an example of a smooth-bore nozzle. These nozzles are designed to produce a compact, solid stream of water with extended reach. The nozzle diameter determines the quantity of water the nozzle is designed to flow. Smooth-bore nozzle diameters, also called tips, range from ¼ inch to 2½ inches (6 mm to 65 mm). Some smooth-bore nozzles have stacked or interchangeable tips that provide the ability to change

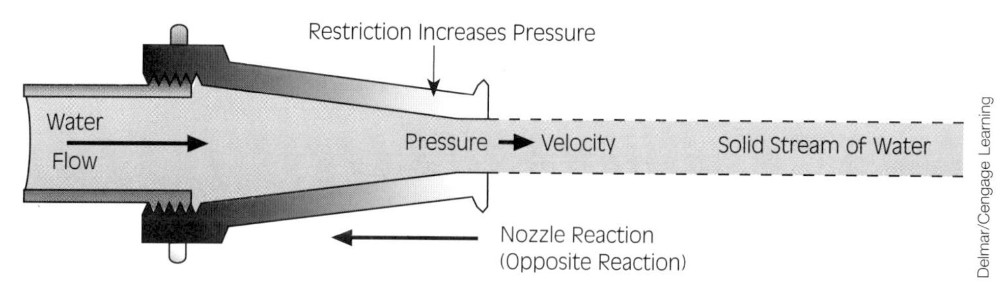

FIGURE 6-20 Illustration of a smooth-bore nozzle. The nozzle acts to restrict water flow, which increases pressure. Pressure is converted to velocity as it leaves the nozzle. Nozzle reaction is the tendency of the nozzle to move in the opposite direction of water flow.

the quantity of water depending on flow requirements. Smooth-bore nozzles are used on handlines (**Figure 6-21A**) and master stream applications (**Figure 6-21B**). When used as handlines, tip sizes up to 1⅛ inch (29 mm) are typically used with a **nozzle pressure** of 50 psi (350 kPa) Actual nozzle pressures can be varied within a range of 25 psi to 65 psi (175 kPa to 450 kPa), depending on reach and flow requirements, with good results and a proper stream. When used as master stream devices, tip sizes of 1¼ inches (32 mm) or larger are typically used with a nozzle pressure of 80 psi (560 kPa). Solid master stream devices may be used with a pressure range of 25 psi (175 kPa) with a 60-ft (18-m) nozzle reach to 100 psi (700 kPa) with a 120-ft (36-m) nozzle reach with good results and a proper stream.

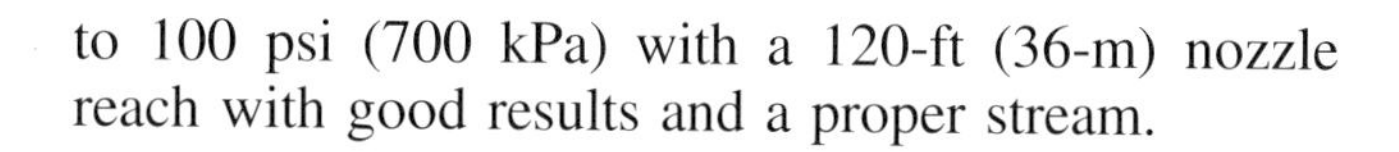

STREETSMART

It is essential that smooth-bore nozzles not be larger than one-half the diameter of the hoseline in order to increase the pressure, as shown in **Figure 6-20**, and produce an effective fire stream.

Combination Nozzles

More complex than smooth-bore nozzles in both design and operation, combination nozzles (**Figure 6-22**) are designed to provide both a straight stream and a wide fog pattern. They are currently the type of nozzle most widely used in the fire service. Most combination nozzles are designed to provide their rated flow at 100-psi (700-kPa) nozzle pressure. However, some combination nozzles, called low-pressure nozzles, provide their rated flow with nozzle pressures between 50 and 90 psi (350 and 630 kPa), with 75 psi (525 kPa) being a common low-pressure nozzle.

Combination nozzles can be divided into three groups: fixed-flow, selectable-flow, and automatic-flow nozzles. Fixed-flow nozzles, also referred to as constant-flow or fixed-gallonage nozzles, are

(A)

(B)

FIGURE 6-21 (A) Handline smooth-bore nozzle with tip. (B) Smooth-bore master stream stacked tips.

FIGURE 6-22 Example of a combination nozzle.

designed to provide a specific flow at 100-psi (700-kPa) nozzle pressure. These nozzles deliver the same flow or gpm (Lpm) regardless of the stream pattern. For example, a nozzle designed to deliver 125 gpm (500 Lpm) will continue to deliver 125 gpm (500 Lpm) when flowing a fog stream or a straight stream as long as the nozzle pressure remains at 100 psi (700 kPa). Some special application fixed-flow nozzles are designed to deliver their rated flow at 75 psi (525 kPa). It is important to note that the pump operator must maintain a 100-psi (700-kPa) nozzle pressure to obtain maximum flow rates and stream patterns. Nozzle pressures less than 100 psi (700 kPa) will result in less flow and reach as well as poor pattern development. Nozzle pressures higher than 100 psi (700 kPa) will result in poor pattern development and excessive **nozzle reaction**. Attempting to control excessive nozzle pressure at the nozzle by partially closing the nozzle will result in decreased flow as well as ineffective patterns.

Selectable-flow nozzles, also referred to as manual-adjustable nozzles or selectable-gallonage nozzles, provide the ability to change the flow at the nozzle. These nozzles provide some flexibility on the firegrouind by allowing the flow to be increased or decreased as needed. Typically, a series of flow settings is available on these nozzles that will increase or decrease the nozzle discharge opening to deliver the selected flow at 100-psi (700-kPa) nozzle pressure (see **Figure 6-23**); however, when changes are made at the nozzle, adjustments must also be made by the pump operator to maintain the designed operating nozzle pressure of 100 psi (700 kPa). If changes are made at the nozzle without the pump operator's knowledge, the nozzle will not flow the selected rate.

Delmar/Cengage Learning

FIGURE 6-23 Example of a selectable-flow nozzle. The red arrow points to the gpm (Lpm) setting. At 100 psi (700 kPa) the nozzle will flow this gpm (Lpm).

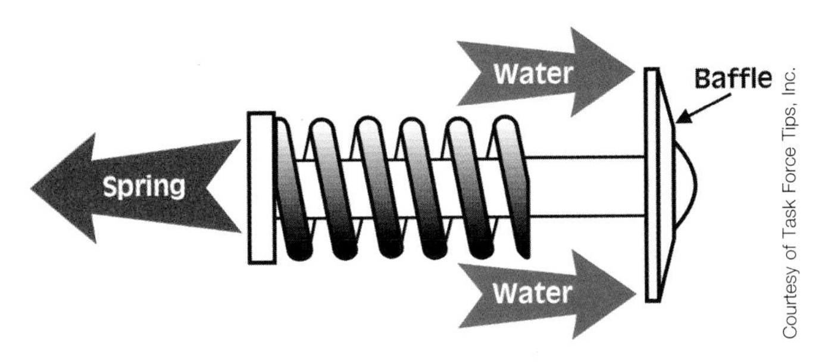

Courtesy of Task Force Tips, Inc.

FIGURE 6-24 Automatic nozzle-sensing device that increases and decreases the nozzle's discharge opening.

Automatic nozzles, also referred to as constant-pressure nozzles, are designed to maintain a constant nozzle pressure over a wide range of flows. These nozzles have a built-in sensing device (spring) that automatically increases or decreases the discharge opening (see **Figure 6-24**). As pressure increases, the baffle moves forward, increasing the nozzle discharge orifice. When pressure decreases, the spring pulls the baffle back, decreasing the nozzle discharge orifice. The result is an increase or decrease in flow while maintaining 100-psi (700-kPa) nozzle pressure (see **Figure 6-25**). In addition to maintaining a constant nozzle pressure, automatic nozzles also maintain effective flow patterns over **nozzle flow** range. Because of this, patterns tend to "look good" at both the low and high end of the nozzle flow range. The pump operator must ensure that "acceptable" flows are provided when using automatic nozzles to ensure firefighter safety during fire suppression activities. A newer feature on some automatic nozzles is the ability to switch between two operating pressures. These dual-pressure nozzles can be switched between 100-psi (700-kPa) and approximately 60-psi (420-kPa) nozzle pressure, allowing for great flows.

Operating Principles

All nozzles are designed to provide a given flow, or range of flows in the case of automatic nozzles, at a specific pressure. Recall that flow is the quantity of water delivered at a specific rate, typically expressed in gallons per minute (gpm) (litres per minute [Lpm]). The specific pressure, called *nozzle pressure*, is the designed operating pressure for a particular nozzle. Most nozzles are designed for an operating nozzle pressure of either 50, 75, 80, or 100 psi (350, 525, 560, 700 kPa). When proper flow and pressure are provided, the three nozzle functions of controlling flow, reach, and **stream shape** will be the most effective for suppression efforts. In addition, the nozzle reaction will most likely be within acceptable limits for those operating the

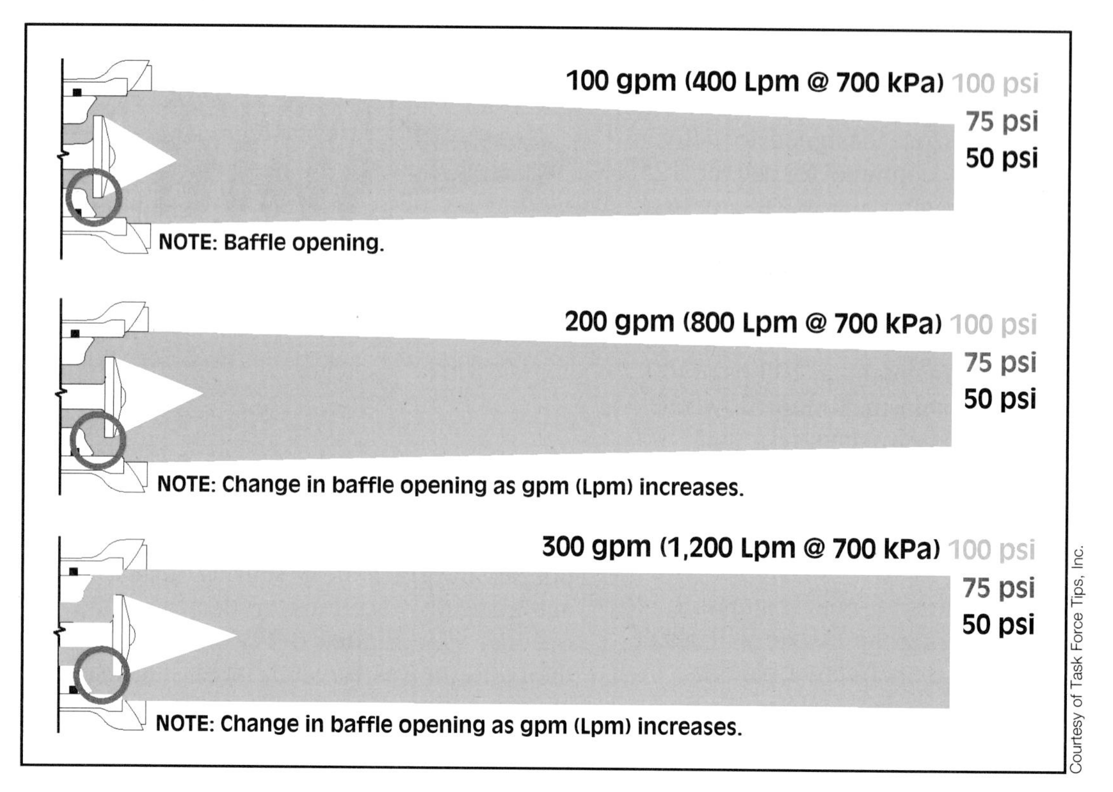

FIGURE 6-25 Automatic nozzles maintain 100-psi (700-kPa) nozzle pressure while flows increase or decrease.

nozzle. Conversely, if the specific flow and pressure are not provided to the nozzle, the nozzle will not operate as designed, resulting in poor stream quality and reduced suppression capability, as well as increased risks to those operating the nozzle.

Nozzle Flow

The amount of water flowing from a nozzle (*nozzle flow*) is a function of the discharge orifice of the nozzle. The larger the orifice, the more water the nozzle is capable of flowing. Some nozzles have a fixed orifice size in which a specific flow is provided. Others have the ability to select from several orifice sizes to provide the ability to vary the flow. Finally, some nozzles can automatically change the orifice size while maintaining proper nozzle pressure.

Reach

A nozzle uses a constriction to speed up and direct the flow of water. Note the terms *constriction* and *speed*. Restrictions in a hose will cause pressure to increase. One inherent design feature of a nozzle is to restrict the flow of water which, in turn, increases pressure (see **Figure 6-20**). This increased pressure is converted to velocity, or speed, as water leaves the nozzle. The end result is an increase in the distance water travels when it leaves the nozzle. The distance water travels after leaving a nozzle is called **nozzle reach**.

The concept of restriction and reach can be illustrated using a garden hose. The distance water will travel when flowing from the open end of a garden hose without a nozzle is minimal. However, when a thumb is placed over the opening, causing a restriction, the distance water will travel is significantly increased. Nozzles used for fire suppression utilize this principle to provide maximum usable reach of the nozzle at a specific pressure. Gravity, wind, and the stream's shape all affect the distance water travels.

Stream Shape

The shape of a stream of water (stream shape) after it leaves a nozzle is determined by the type and function of the nozzle. Nozzles produce a variety of stream shapes ranging from a compact water stream, referred to as a solid or straight stream, to a disperse stream consisting of small water drops, referred to as a fog stream. A solid stream is a compact column of water used to attain greater reach. A fog stream is a wide pattern of water used to provide a large quantity of water droplets over a wide area immediately in front of the nozzle. The most effective shape the nozzle is capable of producing will be attained when the correct flow and pressure are provided.

STREETSMART

Confusion sometimes exists between a straight stream and a solid stream. Part of the confusion exists because the terms have at times been incorrectly used interchangeably. A combination or fog nozzle produces a straight stream when the pattern selector is rotated to the right (clockwise) and produces a narrow, long-reaching fire stream. While all of the water droplets in the stream are all heading in the same direction, and water droplets may combine into larger droplets, there are still many separate water droplets created by the actions of the fog nozzle. By contrast, a solid stream, produced by a smooth-bore nozzle, is a single solid cylindrical mass of water. As a result of the these different physical properties, a straight stream can more readily absorb heat due to its increased surface area in contact with air and is preferred for these applications. A solid stream has a longer reach and is less affected by wind and is preferred for these applications.

Nozzle Reaction

In addition to providing the correct flow and pressure required by a nozzle, pump operators must also consider nozzle reaction. Newton's third law of motion states that "for every action there is an equal and opposite reaction." Relating this law of motion to that of a modern fire suppression nozzle flowing water, the first action would be water leaving the nozzle in a forward direction, and the second action would be the tendency of the nozzle to move in the reverse direction (see **Figure 6-20**). This tendency to move in the opposite direction of water flow is called nozzle reaction.

The person operating the nozzle constantly battles the nozzle's opposite reaction to the water being discharged. Note that this opposite reaction is equal to the first action. Consequently, when an increase in flow or nozzle pressure is experienced at the nozzle, a corresponding increased nozzle reaction will occur. The concern to pump operators, and especially those operating the nozzle, is to provide the required pressures and flows, which in turn will help maintain manageable nozzle reaction.

(For step-by-step photos of Assembling Hoselines, Nozzles, Valves, and Appliances, please refer to page 185.)

Nozzle Maintenance

According to NFPA 1962, nozzles should be inspected after each use and annually. At a minimum, the inspection should include the following:

- Waterway is clear of obstructions
- Tip is not damaged
- Controls and adjustments operate properly
- No missing or broken parts
- Gasket is in good condition

It is also a good practice to clean nozzles. Nozzles may be effectively cleaned with a mild soap and water.

According to NFPA 1964, nozzles must be permanently marked with the following information:

- Manufacturer name
- Product or model
- Rated nozzle pressure and flow(s)
- Minimum and maximum discharge (automatic nozzles)
- FLUSH operating position if so equipped
- Straight stream and fog pattern indications

SKILL 6-1

Initiating a Discharge Line

NOTE

Additional information on operating discharge lines is covered in **Chapter 9**.

Delmar/Cengage Learning

A Select a discharge, remove its cap, and connect the hoseline to the discharge.

Delmar/Cengage Learning

B Advance the hoseline, connecting the other end to a nozzle, appliance, or intake of another apparatus as required.

Delmar/Cengage Learning

C When notified that it is safe to do so, the pump operator should slowly open the discharge valve, avoiding any possible water hammer, observe the line gauge, and adjust to the appropriate pressure.

SKILL 6-2

Extending a Discharge Line with a Break-apart Nozzle

SAFETY

Tie a small piece of rope to the bail so that it cannot accidentally close on the nozzle crew as the line is dragged.

NOTE

Due to the increased friction loss of the longer hoseline, increase the pump discharge pressure to maintain the same nozzle pressure and flow rate. Determining these pressures and friction loss is covered in Section IV.

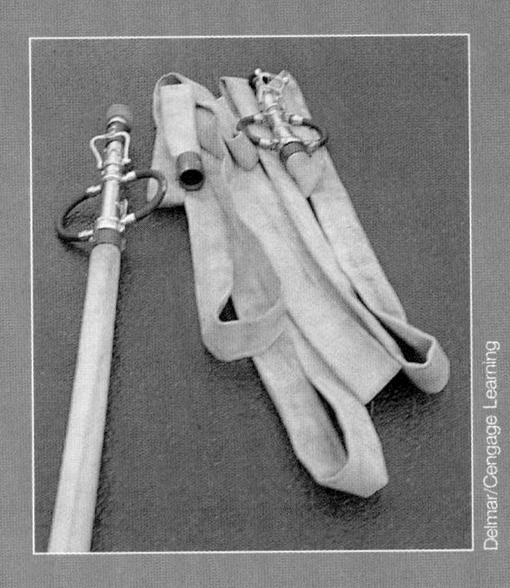

Delmar/Cengage Learning

A The preferred method of extending a hoseline is to use a break-apart nozzle. The additional hose is brought to the nozzle end of the hoseline and the line is shut down with the control handle.

Delmar/Cengage Learning

B The nozzle tip is removed and the additional hose is added.

Delmar/Cengage Learning

C The additional line with nozzle tip is advanced and charged from the control valve.

SKILL 6-3

Removing a Discharge Line

Delmar/Cengage Learning

A The pump operator should shut down the line by closing its valve.

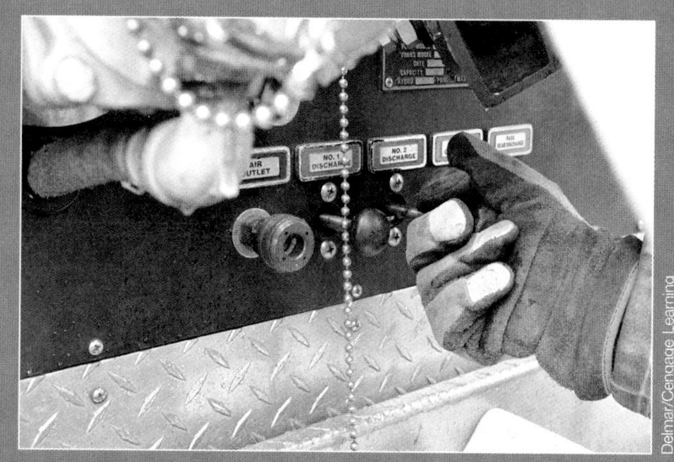

Delmar/Cengage Learning

B Open the line's bleeder valve.

Delmar/Cengage Learning

C Confirm that the line pressure drops to zero by observing the line gauge.

Delmar/Cengage Learning

D Disconnect the discharge line.

SKILL 6-4

Extending a Discharge Line Without a Break-apart Nozzle

Delmar/Cengage Learning

A The pump operator should (a) shut down the line, (b) open the line's bleeder, and (c) confirm that the line pressure drops to zero.

Delmar/Cengage Learning

B Additional hose is brought to the nozzle end of the hose, where the nozzle crew confirms that the line has been shut down, opens the nozzle, and removes the nozzle.

Delmar/Cengage Learning

C The nozzle is attached to the end of the added hose and advanced to remove kinks.

(Continues)

SKILL 6-4

Extending a Discharge Line Without a Break-apart Nozzle (*Continued*)

Delmar/Cengage Learning

D When the pump operator is notified to recharge the line, the line bleeder is closed, the line valve is slowly opened, and the line gauge pressure is confirmed.

Delmar/Cengage Learning

E The nozzle crew bleeds, checks the pattern, and continues the operations.

SKILL 6-5

Assembling Hoselines, Nozzles, Valves, and Appliances

Delmar/Cengage Learning

A Select the hoseline and valve, nozzle, or appliance. Confirm that the gasket is in place. Align the higbee indicators if present and rotate the swivels in a clockwise direction. With good, clean gaskets, swivels that are hand tight should not leak.

Delmar/Cengage Learning

B Make sure valves and shut-offs are in the correct position and, when ready, have the line charged.

LESSONS LEARNED

Hose, appliances, and nozzles are used by pump operators to accomplish the task of moving water from a supply to a discharge point. Hose provides the ability to move water over a distance and is used for different tasks depending on its size and construction. Nozzles provide flow, reach, and shape to extinguish fires. For optimal and safe performance, nozzles must be supplied with the proper flow and pressure. Pump operators utilize the pump and pump peripherals to control flows and pressures to nozzles. Pump operators utilize appliances to provide the variety of hose configurations required to move water to the point of discharge. Hose, appliances, and nozzles have certain flow, pressure, and friction loss characteristics. Hydraulic calculations for providing proper flows and pressures must include these characteristics. (**Section IV** of this text discusses hydraulic calculations in detail.)

KEY TERMS

Adapter Appliance used to connect mismatched couplings.

Appliances Accessories and components used to support varying hose configurations.

Attack hose 1½-in. to 3-in. (38-mm to 77-mm) hose used to combat fires beyond the incipient stage.

Combination nozzle A nozzle designed to provide both a straight stream and a wide fog pattern; the type most widely used in the fire service.

Couplings A set or pair of connection devices attached to a fire hose that allow the hose to be connected to additional lengths of hoses, adapters, and other firefighting appliances.

Eductor A specialized device used in foam operations that utilizes the Venturi principle to draw the foam into the water stream.

Four-way hydrant valve Appliance used to increase hydrant pressure without interrupting the flow.

Hose Flexible conduit used to convey water; also see **Attack hose**, **Supply hose**, and **Suction hose**.

Hose bridge Device used to allow vehicles to move across a hose without damaging the hose.

Hose clamp Device used to control the flow of water in a hose.

Hose jacket Device used to temporarily minimize flow loss from a leaking hose or coupling.

Large-diameter hose Hose with a diameter of 3½ inches (77 mm) or larger.

Manifolds Devices that provide the ability to connect numerous smaller lines from a large supply line.

NH A common thread used in the fire service to attach hose couplings and appliances.

Nozzle flow The amount of water flowing from a nozzle; also used to indicate the rated flow or flows of a nozzle.

Nozzle pressure The designed operating pressure for a particular nozzle.

Nozzle reach The distance water travels after leaving a nozzle.

Nozzle reaction The tendency of the nozzle to move in the opposite direction of water flow.

Siamese Appliance used to combine two or more hoselines into a single hoseline.

Smooth-bore nozzle Nozzle designed to produce a compact, solid stream of water with extended reach.

Soft sleeve Shorter section of hose used when the pump is close to a pressurized water source such as a hydrant.

Spanner wrench Tool used to connect and disconnect hose and appliance couplings.

Storz couplings The most popular of the nonthreaded hose couplings; they are also quarter-turn and sexless couplings.

Stream shape The configuration of water droplets (shape of the stream) after leaving a nozzle.

Suction hose Special noncollapsible hose used for drafting operations; also called hard suction hose.

Supply hose Used with pressurized water sources and operated at a maximum pressure of 185 psi (1,295 kPa).

Venturi principle Process that creates a low-pressure area in the induction chamber of an eductor to allow foam to be drawn into and mixed with the water stream.

Water thieves Similar to gated wyes, water thieves are used to connect additional smaller hoselines from an existing larger hoseline.

Wye Appliance used to divide one hoseline into two or more hoselines. The wye lines may be the same size or a smaller size, and the wye may or may not have gate control valves to control the water flow.

REVIEW QUESTIONS

Multiple Choice

Select the most appropriate answer.

1. The basic parts of a hose include each of the following, *except*

a. reinforced inner liner.

b. outer protective shell.

c. couplings.

d. safety relief ring.

2. The most common length of a section of hose is

a. 50 feet (15 m).

b. 100 feet (30 m).

c. 150 feet (45 m).

d. 200 feet (60 m).

3. Which of the following is considered the most common sexless coupling used in the fire service?

a. rocker **c.** Storz

b. lug **d.** NST

4. Each of the following can be used for intake hose, *except*

a. LDH. **c.** hard suction.

b. soft sleeve. **d.** SDH.

5. Which of the following NFPA standards provides guidelines for the care, use, and maintenance of fire hose?

a. 1961 **c.** 1963

b. 1962 **d.** 1964

6. A ________________ is used to divide one hoseline into two or more lines.

a. wye **c.** adapter

b. Siamese **d.** increaser

7. A ____________________ is used to combine two or more lines into one line.

a. wye **c.** adapter

b. Siamese **d.** increaser

8. The correct nozzle pressure for a smooth-bore handline is:

a. 50 psi (350 kPa).

b. 100 psi (700 kPa).

c. 150 psi (1,050 kPa).

d. depends on the hose length.

9. Which of the following NFPA standards defines minimum requirements for new fire hose connections?

a. 1961 **c.** 1963

b. 1962 **d.** 1964

10. Which of the following NFPA standards focuses on fire hose appliances?

a. 1961 **c.** 1963

b. 1962 **d.** 1964

11. According to NFPA, supply hose have a maximum operating pressure of _________ while attack hose have a maximum operating pressure of ________.

a. 185 and 285 psi (1,295 and 1,995 kPa)

b. 185 and 275 psi (1,295 and 1,925 kPa)

c. 150 and 250 psi (1,050 and 1,750 kPa)

d. 95 and 150 psi (665 and 1,050 kPa)

12. The thread used in nearly all fire hose threaded coupling construction is referred to as

a. GHT. **c.** HN.

b. NH. **d.** STORZ.

Short Answer

On a separate sheet of paper, answer/explain the following questions.

1. Explain the importance to pump operations of studying hose, appliances, and nozzles.
2. Discuss the basic types of hose construction. Which do you feel is better? Why?
3. Identify four appliances commonly used during pump operations.
4. What are the critical points associated with nozzles operating as designed?
5. List the standard nozzle pressures most often used by fire service nozzles.
6. Explain the operation of a nozzle. Be sure to discuss reach, shape, flow, and nozzle reaction.
7. What are the differences between wye and Siamese appliances?
8. List and discuss the purpose of NFPA standards that focus on hose, appliances, and nozzles.
9. List four common types of hose tools used by pump operators.
10. What should you look for when conducting a hose inspection?
11. Explain why proper flow and pressure must be provided for effective suppression efforts and list the three nozzle functions controlled by flow and pressure.
12. Explain the difference between smooth-bore nozzles and combination nozzles.
13. List and explain the difference between the three major types of combination nozzles.
14. List several types of wyes commonly used in the fire service.
15. List several adapters commonly used in the fire service.

ACTIVITIES

1. Identify and categorize the hose, appliances, and nozzles used by your department.
2. Using NFPA 1962 as a guide, develop a detailed procedure for testing fire hose, including safety considerations and records documentation.

PRACTICE PROBLEM

1. You have been asked to identify all of the pump-related items needed for a new 1,250-gpm (5,000 Lpm) pumper that the chief will propose in the new budget. List the items you feel should be provided with the new pumper.

ADDITIONAL RESOURCES

NFPA 1961, *Standard on Fire Hose, National Fire Protection Association*. Quincy, MA, 2007.

NFPA 1962, *Standard for the Inspection, Care, and Use of Fire Hose, Couplings, and Nozzles and the Service Testing of Fire Hose*. National Fire Protection Association, Quincy, MA, 2008.

NFPA 1963, *Standard for Fire Hose Connections, National Fire Protection Association*. Quincy, MA, 2009.

NFPA 1964, *Standard for Spray Nozzles, National Fire Protection Association*. Quincy, MA, 2008.

NFPA 1965, *Standard for Fire Hose Appliances, National Fire Protection Association*. Quincy, MA, 2009.

Pump Procedures

The first section of this text discusses basic requirements and concepts for both the pump operator and emergency vehicles. The second section discusses pumps and the many components connected to, or used with, pump operations.

This third section discusses the concepts and procedures for the three interrelated fire pump operation activities: securing a water supply, operating the pump, and maintaining discharge pressures referencing NFPA 1002, *Standard for Fire Apparatus Driver/Operator Professional Qualifications*, 2009 edition, section 5.2. **Chapter 7** of this text presents the various water supply sources and considerations for their use. **Chapter 8** discusses various procedures for operating the pump. **Chapter 9** discusses how to initiate and maintain discharge hoselines.

Section IV presents detailed explanations of hydraulic theories and principles as well as fireground flow considerations and pump discharge calculations.

7 Water Supplies

- Learning Objectives
- Introduction
- Basic Water Supply System Considerations
- On-board Water Supplies
- Municipal Systems
- Static Sources
- High-capacity Rural Water Supplies
- Relay Operations
- Tanker Shuttle Operations
- Nurse Feeding
- Lessons Learned
- Key Terms
- Chapter Formulas
- Review Questions
- Activities
- Practice Problems
- Additional Resources

We had a residential structure fire involving a 3,000-square-foot (900-square-metre) two-story house, and I was a newly promoted driver assigned to driving a quint. My apparatus was the first on the scene. We immediately deployed our 200-foot, 2-inch (60-metre, 50 mm) handline to the front of the house. When the battalion chief arrived on-scene, he requested the aerial ladder be set for a potential defensive attack. Once the ladder was set, I was able to operate it from the ground with a remote control, instead of on the turnstile, and be at the pump panel getting the lines flowing. Knowing this quint has a 2,000-gallon-per-minute (8,000-litre-per-minute) capacity, I was able to pump to near maximum capacity with approximately 1,250 gallons per minute (5,000 litres per minute) from the aerial, 225 gallons per minute (900 litres per minute) from the 200 feet (60 metre) of 2-inch (50 mm) hose, 150 gallons (550 litres) per minute from 150 feet (45 metre) of 1¾-inch (45 mm) hose, and 300 gallons (1,200 litres) per minute from 200 feet (60 metre) of 3-inch (77 mm) hose on the Blitzfire (portable master stream).

When the second-in apparatus arrived, they hooked me in to the hydrant at the end of the cul-de-sac. I knew that I could maintain this output as long as my intake from the hydrant wasn't compromised. I just needed to watch my intake to make sure that I was maintaining a 20-psi (140 kPa) residual pressure from the hydrant. Within a few minutes, our CAFS engine arrived on-scene and hooked up to a different hydrant at the beginning of the street. Strangely, once they started flowing from their pump, my pressures started fluctuating. I could see this on the gauges and could hear the changes in the engine sound. (The quint is equipped with a pressure governor; when this is in "pressure mode" the engine speed is constantly adjusted to maintain the desired pump discharge pressure.) I realized my intake was affected.

It turns out that because both hydrants were on the same dead-end main, they shared the same water line. Most good water systems have a circulating or looped system, but this one didn't. Realizing this, I knew I had to prioritize my discharges, since there was no other water supply available. I dropped the gallons-per-minute (litres-per-minute) output from the aerial, because the safety of the guys on the ground with the handlines and the Blitzfire was more important. By doing so, I was able to maintain safe fire streams for the firefighters on the handlines.

Knowledge and understanding of the capacity of my quint, and the flow capabilities of all the hoses and appliances that my apparatus carries, helps me identify my abilities and limitations in providing required fire flows. Since this incident, our fire crew has also utilized the city water main maps to identify dead-end hydrants, as not all cul-de-sac hydrants are on dead-end mains.

—Street Story by Traci McGill,
Driver/Engineer and Paramedic,
Grapevine Fire Department, Grapevine, Texas

LEARNING OBJECTIVES

After completing this chapter, the reader should be able to:

7-1. List the three basic steps to efficient and effective pump operations.

7-2. List and explain basic considerations for preplanning and selecting water supplies.

7-3. Define and discuss the concept of available water.

7-4. Explain each of the water supply sources typically used and list the considerations for preplanning and selecting.

7-5. Identify the potential problems associated with small-diameter and dead-end mains.

7-6. Describe the hydrant coding system suggested by NFPA 291.

7-7. Identify the potential problems associated with low-pressure systems, and private water supply systems.

7-8. Determine available water when given the static and residual pressures of a hydrant used for water supply.

7-9. List the considerations used to determine a reliable static water source.

7-10. Describe the options a relay pump operator may consider in order to maintain the minimum supply pressure.

7-11. Explain what determines the maximum supply pressure used in a relay pumping operation.

7-12. Describe what the relay pump operator must do to ensure the maximum supply pressure is not exceeded.

7-13. Define the terms *open relay* and *closed relay*, and give an example where each would be used in a water supply relay scenario.

7-14. Explain why a minimum residual supply pressure is important in relay pumping operations.

7-15. Identify the factors that limit the distance water can be pumped in supply lines.

7-16. Describe three methods of increasing the distance water can be pumped through supply lines.

7-17. Given the length and size of the supply line and the desired pressure and flow rate in a relay pumping evolution, pump a supply line so the correct pressure and flow are provided to the next pumper in the relay.

***The driver/operator requirements, as defined by the NFPA 1002 Standard, are identified in black; additional information is identified in blue.**

INTRODUCTION

When an alarm is received, the first duty of a pump operator is to ensure the safe arrival of the apparatus, equipment, and personnel at the scene. The next duty is to initiate fire pump operations. As stated earlier, fire pump operations consist of three interrelated activities:

1. Securing a water supply
2. Operating the pump
3. Maintaining discharge pressures

The critical first step of a pumping operation is to secure a water supply. Securing a water supply is the first step in the process of moving water from a source to the intake side of a pump. This particular activity is perhaps the most challenging and difficult of the three, in that controlling the pressure and flow of a supply is typically limited.

It is also common for the pump operator to need to change water sources. Often the initial water source is the water carried in the pump's booster tank, with the requirement to change to an external water source prior to depleting the on-board water. Such transitions need to be performed by the pump operator transparently to the line crews—that is, without a change in discharge pressure or an interruption in fire flow. Procedures for making these transitions will be covered in the next chapter.

This chapter presents information as though the pump operator makes the decision concerning which water supply to secure. In reality, this is not always the case. Depending on a department's standard operating procedures (SOPs), the determination of which water supply to secure can rest with any of the following:

- Preplans and departmental SOPs
- Pump operator
- Assigned officer
- Pump operator and the officer (they share the responsibility)
- Incident commander, water supply officer, or other officer within the command system

Regardless of who actually makes the decision, the pump operator should be capable of evaluating and securing water supplies. Therefore, pump operators should be familiar with the strengths and limitations of available water supplies.

Recall from **Chapter 1** that water supplies come from three kinds of sources: the apparatus, static sources, and pressurized sources. Apparatus water supplies include water tanks carried on the apparatus and are secured through on-board tank operations. Static sources include ponds, lakes, and rivers and are secured through drafting operations. Pressurized sources include elevated towers and municipal water supplies. Municipal water supplies are the most common pressurized source used in the fire service and are secured through hydrant operations. Private water supplies, common in large industrial complexes, are similar in nature to municipal systems. All three kinds of sources can be used in combination through relay operations and tanker shuttle operations to provide initial and sustained water supply to the scene.

Obviously, when only one source is available, pump operators have no choices. When more than one source is available, pump operators should be able to choose, or recommend, the water supply that maximizes the efficient and effective use of both the supply source and the apparatus to meet the flow requirements of an incident.

This chapter discusses the pump operator's task of securing a water supply, starting with basic water supply considerations. Then common supply systems typically used in fire pump operations are presented, including on-board supplies, municipal private supplies, static supplies, relay operations, and tanker shuttle operations.

BASIC WATER SUPPLY SYSTEM CONSIDERATIONS

The first few minutes of an emergency typically have many tasks that must be performed, accompanied by stress, anxiety, and confusion. In most cases, preplanning water supplies will help reduce this stressful time of an incident. In addition, preplanning will help the pump operator complete the task of securing a water supply in an efficient and effective manner. The following considerations should be included in the preplanning and selection of a water supply:

- *Required flow*. Determining the **required flow**, an estimated flow of water needed for a specific incident, is one of the first considerations when selecting a water supply. Considerations for calculating required flow are discussed in **Section IV**. In some cases, multiple pumpers are needed to provide the required flow for an incident. In other cases, the required flow may not be known. In either case, a water supply should be selected that can provide enough water to flow the capacity of the pump. In doing so, the pump operator is in the best position to assist with providing the required flow. The bottom line is that when securing a water supply, the amount of water the pump operator is expected to provide should be known.

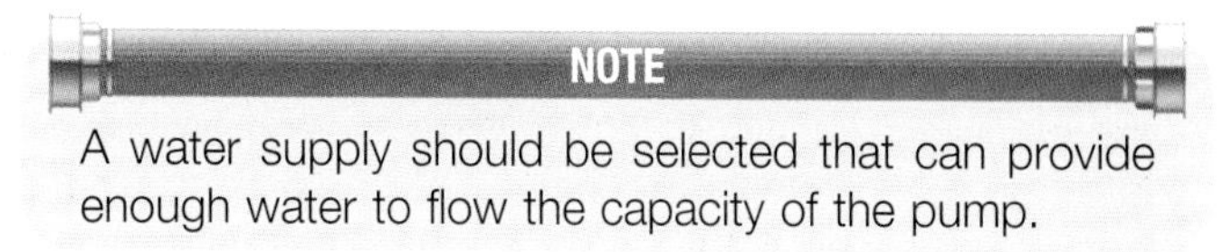

A water supply should be selected that can provide enough water to flow the capacity of the pump.

- *Pump capacity*. Another important consideration when securing a water supply is the capacity of the pump (**Figure 7-1**). In general, supplies should be chosen that are capable of providing enough water to allow the pump to flow its capacity. Larger-capacity pumps may rule out some water supplies, while smaller-capacity pumps will have a wider selection of supplies to choose from. Keep in mind that pumps, at draft, are expected to flow 100% capacity at 150 psi (1,050 kPa) with 10 ft (3 m) of lift. When pump pressures in the operation reach 200 psi (1,400 kPa), the pump can only be expected to deliver 70% of its rated capacity. When pressures reach 250 psi (1,750 kPa), only 50% of the pump's rated capacity should be expected. In essence, increased pump pressures reduce the quantity of water the pump is able to flow. Capacity also decreases with increases in lift. (See "Rated Capacity and Performance" in **Chapter 4**.)
- *Supply hose capacity*. The size and length of supply hose is another consideration when selecting a supply source. A 2½-inch (65-mm) supply hose will not be able to flow as much water over as long a distance as large-diameter hose (LDH). In general, larger-diameter supply hose will provide less friction loss over longer distances. Where high flows are required and LDH is not available, multiple hoselines should be used.

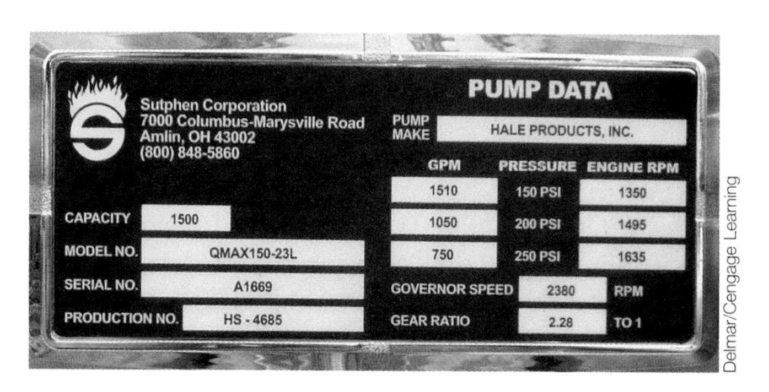

FIGURE 7-1 The rated capacity and performance of a pump is usually identified on or near the pump panel.

STREETSMART

Until recently, carrying long lengths of LDH was uncommon. Instead, hose beds often carried a split load of 3-in. (77-mm) hose. The split load allowed a forward lay, a reverse lay, or, since the 3-in. (77-mm) hose could not carry enough water to meet the pump's capacity, a lay known as "dual 3's." This involved dropping two 3-in. (77-mm) lines in a single lay. The advantage of two 3-in. (77-mm) lines is that either the same amount of water could be moved as with one 3-in. (77-mm) line but with ¼ of the friction loss, or twice as much water could be moved as with one 3-in. (77-mm) line with the same friction loss.

The key advantage of LDH is being able to move even more water with less friction loss than with dual 3's. Unlike dual 3's, it allows larger pumps to operate at capacity with long lays. It requires less storage space than dual 3's and adds less weight to the apparatus. Since LDH often uses sexless Storz couplings, it also eliminates the need for double male and double female adapters. Laying and connecting a single LDH hose also requires less manpower and time than connecting two 3-in. (77-mm) hoses.

Disadvantages of LDH include the near impossibility of moving it once filled due to its weight, and blocking streets since even with hose bridges it can be too high for cars to clear. In areas without hydrants the amount of water required to fill LDH hose can be a problem. A 5-in. (125-mm) hose requires about a gallon a foot to fill, versus a quart a foot to fill 2½-in. hose (12.5 litres to fill versus 3.5 litres per metre to fill 65-mm hose). Since the medium diameter hose MDH can flow all the water many water shuttles can move, having less water in the hose means getting more water on the fire. If dual hoses are laid, the water shuttle can start with one hose, and if the water shuttle capacity grows to exceed its capacity, the second hose can be charged. This allows a minimum amount of water to be in the hoseline to start, growth in capacity, and, even after both hoselines are used, significantly less total water in hoselines than with LDH.

- *Water availability.* Another important consideration is the availability of the source water. Several factors contribute to the availability of water. One is the quantity of water available. For example, a swimming pool will not provide as much water as a large lake. Other factors are the flow (gpm or Lpm) and pressure (psi or kPa) at which water is available from the supply. **Municipal supplies**, for example, may provide water at a variety of flows and pressures. Finally, the physical location of the water is a factor. A marginal hydrant distant from the incident may not be the best water supply if a good **static source** such as a lake or pond is located closer to the incident. **Water availability**, then, relates to the quantity, flow, pressure, and accessibility of a water supply.
- *Supply reliability.* The reliability of a source should also be considered when selecting a water supply. The **supply reliability** is the extent to which the supply will consistently provide water. Another way of looking at it is the extent to which the supply fluctuates in flow, pressure, and quantity. For example, a river or pond may not be a reliable supply if the water level changes frequently or is frozen, and tidal water supplies may change dramatically within a short period of time.

NOTE

Preplanning will make the task of securing a water supply a little smoother and less stressful.

- *Supply layout.* Finally, the **supply layout** hose should also be considered when selecting a water supply. Supply layout is the required supply hose configuration necessary to secure the water supply efficiently and effectively. The layout of supply hose may be affected by the type of source, the size of hose, the hose appliances available, and the number and size of intakes on the pump. Therefore, a variety of supply hose configurations can often be used. Keep in mind, though, that the more elaborate the supply configuration, the longer it will take to set it up. For example, connecting a supply hose to a pump located near the hydrant will not take as long to secure as a relay or **tanker shuttle** operation. In general, the supply configuration should be sufficient to provide the flow needed for an operation.

Although pump operators should be prepared to establish each kind of water supply system, preplanning will make the task a little smoother and less stressful. This is especially true when considering the compressed time in which decisions must be made during an emergency. Pump operators should be familiar with the different types of water supply systems, considerations for their use, and considerations for securing them.

ON-BOARD WATER SUPPLIES

The **on-board supply** is simply the water carried in a tank on the apparatus (**Figure 7-2**). The on-board water supply is used for several reasons. First, it is used when no other water supply is available. In locations where this is a common occurrence, apparatus

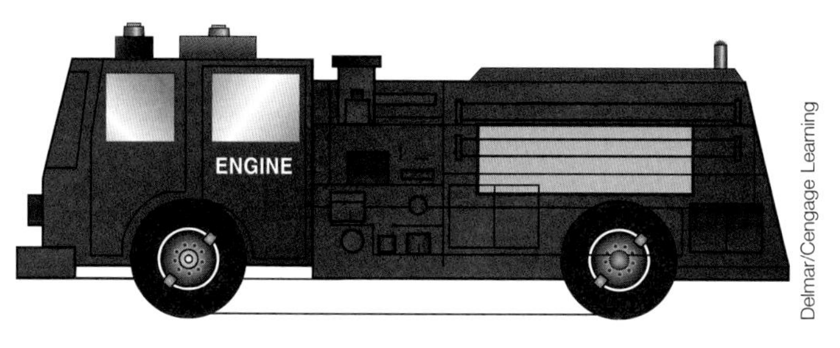

Delmar/Cengage Learning

FIGURE 7-2 Most pumping apparatus carry at least a minimum on-board water supply.

typically have larger on-board tanks. Second, it is used when an incident requires only a small quantity of water, such as car fires and small brush fires. This saves time and energy when a supply source is not readily available. Third, it is used when an immediate water supply is deemed more critical than the time it would take to secure a supply with greater flows. In this situation an alternate supply is secured while utilizing the on-board source. Fourth, it may be used to circulate water through the pump at times when water is not used for firefighting, such as when lines are relocated and during overhaul. Finally, on-board water supplies can be used as a backup or emergency water supply in the event the other water supply is interrupted.

Water Availability

An immediate and readily available supply of water is the main advantage of the on-board source. Although readily available, the on-board supply is the most limited in terms of quantity of water. Most pumping apparatus have at least a small tank of water on board. NFPA 1901 specifies minimum tank capacity for fire apparatus as follows:

Initial attack apparatus	200 gallons (800 litres)
Pumper fire apparatus	300 gallons (1,200 litres)
Mobile water apparatus/tanker	1,000 gallons (4,000 litres)
Quint fire apparatus	300 gallons (1,200 litres)

Tank capacities of 500 to 1,000 gallons (2,000 to 4,000 litres) are common on many pumper fire apparatus, and many mobile water apparatus have tank capacities over 2,000 gallons (8,000 litres). NFPA 1901 also requires that the piping between the tank and the pump must be capable of flowing at least 250 gpm (1,000 Lpm) for tank capacities less than 500 gallons (2,000 litres) and must be able to flow 500 gpm (2,000 Lpm) for tank capacities of 500 gallons (2,000 litres) or larger. Greater flows from the tank can be achieved by specifying a larger-diameter pipe between the tank and the pump or multiple tank-to-pump connections.

The length of time water can be supplied when utilizing the on-board water depends on the size of the tank and the flow of discharge lines. For example, a 1¾-inch (45-mm) preconnect flowing 125 gpm (500 Lpm) can be sustained for 8 minutes with a 1,000-gallon (4,000-litre) tank, 6 minutes with a 750-gallon tank (3,000-litre), and 4 minutes with a 500-gallon (2,000-litre) tank. In comparison a 2½-inch (65-mm) line flowing 250 gpm (1,000 Lpm) can be maintained for 4, 3, and 2 minutes with 1,000-, 750-, and 500-gallon (4,000-, 3,000-, 2,000-litre) tanks, respectively.

In general, the quantity of water in the on-board water tank is insufficient to fight all but the smallest fires. It should, however, be large enough to start an initial attack and supply water while an additional water source is being established.

Supply Reliability

The on-board water supply is generally considered to be a reliable source. A potential concern, however, is not having water in the tank when arriving on-scene. This can occur when the tank is accidentally left empty or when the tank's level indicating device is malfunctioning, showing the tank to be full when it is not. To guard against this problem, the apparatus inspections described in **Chapter 2** recommended checking the tank level via the tank stack.

Another potential concern is that air pockets may impede pumping of the tank water. Air pockets may occur for several reasons: when the tank and piping develop leaks and holes, when air is trapped in the piping and pump, or when the pump is left dry to guard from freezing. Priming the pump to eliminate air pockets is usually all that is needed to begin pumping the tank water.

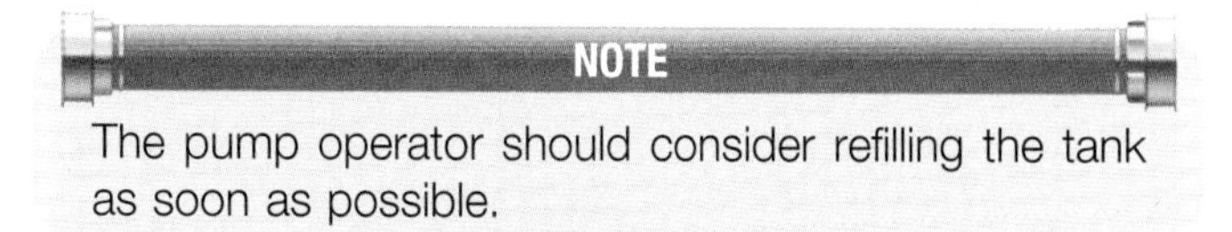

The pump operator should consider refilling the tank as soon as possible.

Supply Layout

The on-board water supply is by far the fastest supply to secure in that the supply line is permanently attached. All that is required to secure the supply is to open the tank-to-pump control valve. The tank-to-pump control valve opens and closes a passage from the tank to the pump. The tank is usually mounted higher than the pump to allow gravity to move water to the intake side of the pump. A control valve is also typically installed in the piping from the pump to the tank and has several names: pump-to-tank, recirculating valve, and tank fill. This valve opens and closes a passage from the pump to the tank, allowing the tank to be refilled from the pump. The two valves working together circulate water between the tank and the pump. In addition, a tank level indicator is usually provided to monitor water levels in the tank. The tank-to-pump and pump-to-tank control valves as well as the tank level indicator are commonly mounted on the pump panel (**Figure 7-3**).

In most cases, only small discharge lines are used in conjunction with the on-board tank. Typically, the discharge lines are preconnected and are either 1-inch (25-mm) booster lines, or 1½-inch or 1¾-inch (38-mm or 45-mm) attack lines. When using the on-board tank, a good habit is to plan ahead for an alternate water supply should additional water be required. When an alternate supply is secured or when the incident is over, the pump operator should refill the tank as soon as possible.

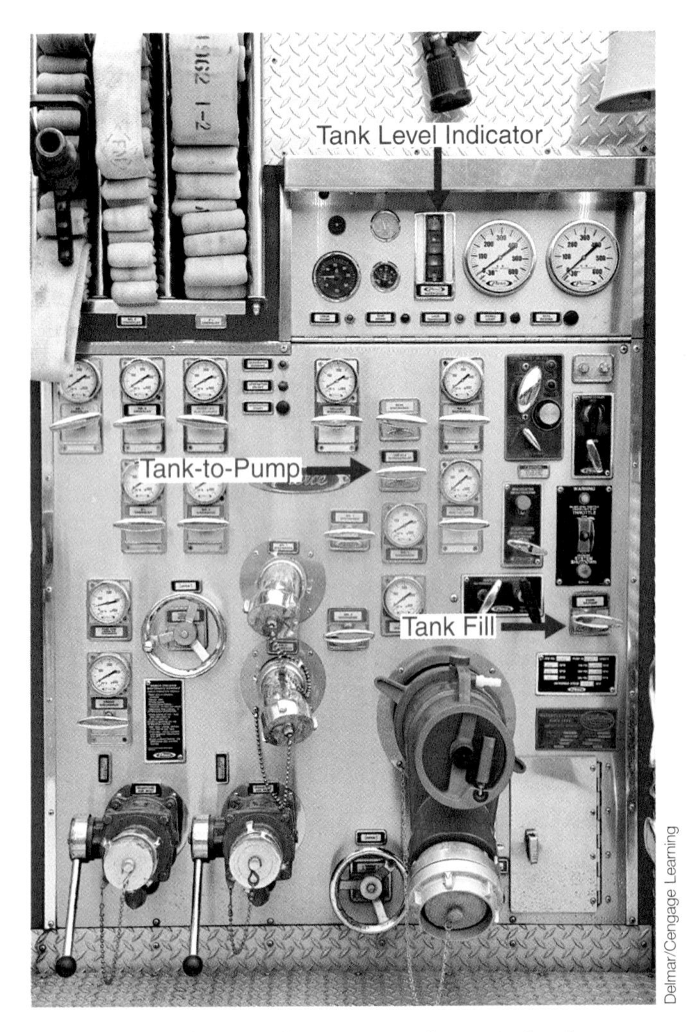

FIGURE 7-3 On-board water supply control valves are typically located on the pump panel.

Frequently, the on-board supply is used to provide immediate water to the incident. A difficult decision is how quickly water is needed at the scene. The quickest delivery of on-board water is simply to drive the apparatus to the scene and start pumping. The major concern with this approach is the ability to secure another source before the on-board supply runs out. The second quickest way is to lay a line from the hydrant on the way to the scene. Although this approach takes a little longer to start flowing the on-board water, an alternate water supply can usually be secured within a reasonable time frame. This timing is critical when the total quantity of water needed exceeds tank capacity. Whether the first- or second-arriving engine connects to the hydrant is often determined by how quickly after the first engine the second engine is expected to arrive.

MUNICIPAL SYSTEMS

Most densely populated areas in the North America utilize a municipal water supply system. In many cases, a city, county, or special water district utilizes a public works or water department to design, maintain, and test the municipal water system. Often, the difficult part is to not only maintain existing systems but to expand systems in population growth areas as well as to upgrade systems as they wear out or become obsolete.

Municipal supplies deliver water to the intake side of the pump under pressure. They are used when water demands exceed the on-board water supply or when the distance from the source to the incident is great. Hydrants are connected via a water-main distribution system. Both hydrants and their water mains are components of a municipal water supply system. Pump operators should be familiar with the municipal water system, its distribution system, and the types of hydrants used in their service area.

Municipal Water Systems

Municipal water supplies commonly provide water for two purposes. Primarily, they provide water for normal consumption such as household and industrial uses. Second, they provide water for emergency use to hydrants and fixed fire protection systems. In some locations, two completely separate systems are used, one to provide water for normal consumption and one to provide water for emergency use. More commonly, the same system provides water for both domestic consumption and emergency use. When this is the case, fluctuating hydrant pressures and flows can be expected with the changes in domestic consumption. The basic components of municipal water systems include a water supply source, processing facilities, and a distribution system that includes hydrants.

Municipal Water Supply Sources

The water supply source for a municipal system is obtained from either surface water such as reservoirs (also called impounded water supplies), lakes and ponds, or groundwater such as wells or springs. In some cases, a municipal system will use water from both surface and groundwater sources. The water can be moved from the source to the distribution system by means of a pumping system, gravity system, or a combination of the two. Pumping systems utilize pumps of sufficient size and number to meet the consumption demands of the service area. Typically, backup electrical power in the form of diesel generators and redundant pumps is maintained in case the primary electrical power or pumps fail. Gravity systems use elevation, typically of at least a couple hundred feet, as a means to move water from the supply to the distribution system. A mountain lake providing water to a city in the valley is one example of a gravity system. Another example of a gravity system is an elevated storage tank (**Figure 7-4**). Elevated storage tanks are common in many areas and can either be located on a hill or mountain, or the tank itself may be elevated. Sometimes a combination pumping system and gravity system is used. In combination systems, the elevated storage tanks are used to assist pumping systems with meeting peak demand needs or as an emergency or backup water supply. In many cases the elevated storage tanks are filled at night when demands are lower and pressures are higher. During the day, elevated storage tanks use gravity to assist pumps in meeting increased water demands.

Delmar/Cengage Learning

FIGURE 7-4 Elevated storage tanks are sometimes located on higher elevations to increase gravity flow further.

Processing Facilities

Potable (drinking) water passes through a processing facility. The processing facility treats water by a number of different techniques, depending on its facilities and the quality of the source water. Types of treatments include:

- Coagulation, where chemicals, bacteria, or other organisms are added to the water, which causes particles suspended in the water to collect into large particles that are easier to remove from the water.
- Sedimentation, where slow-moving water is given time for particles to settle out of the water.
- Filtration, where water moves through a porous material such as sand to remove particles in the water.
- Adding chemicals for public safety, including disinfectants such as chlorine and fluoride for dental health.

Typically these facilities are designed to process water with sufficient capacity to handle peak domestic consumption and fire suppression needs; however, areas of rapid growth place a heavy burden on processing facilities to the extent that they may no longer be able to provide the required flow to the community and/or for fire suppression efforts. Natural disasters, equipment malfunctions, and terrorist activities may also impact the ability of the processing facility to provide adequate water flows. Preplanning is therefore important to determine alternate water supplies. See **Figure 7-5** for an illustration of the three types of systems.

Distribution System

Municipal water distribution systems consist of a series of pipes that decrease in size, called mains, and control valves that move water from the source or treatment facility to individual hydrants and commercial and residential occupancies. The distribution system typically starts with large-sized pipes, 16 in. (400 mm) or larger, called feeder mains, which

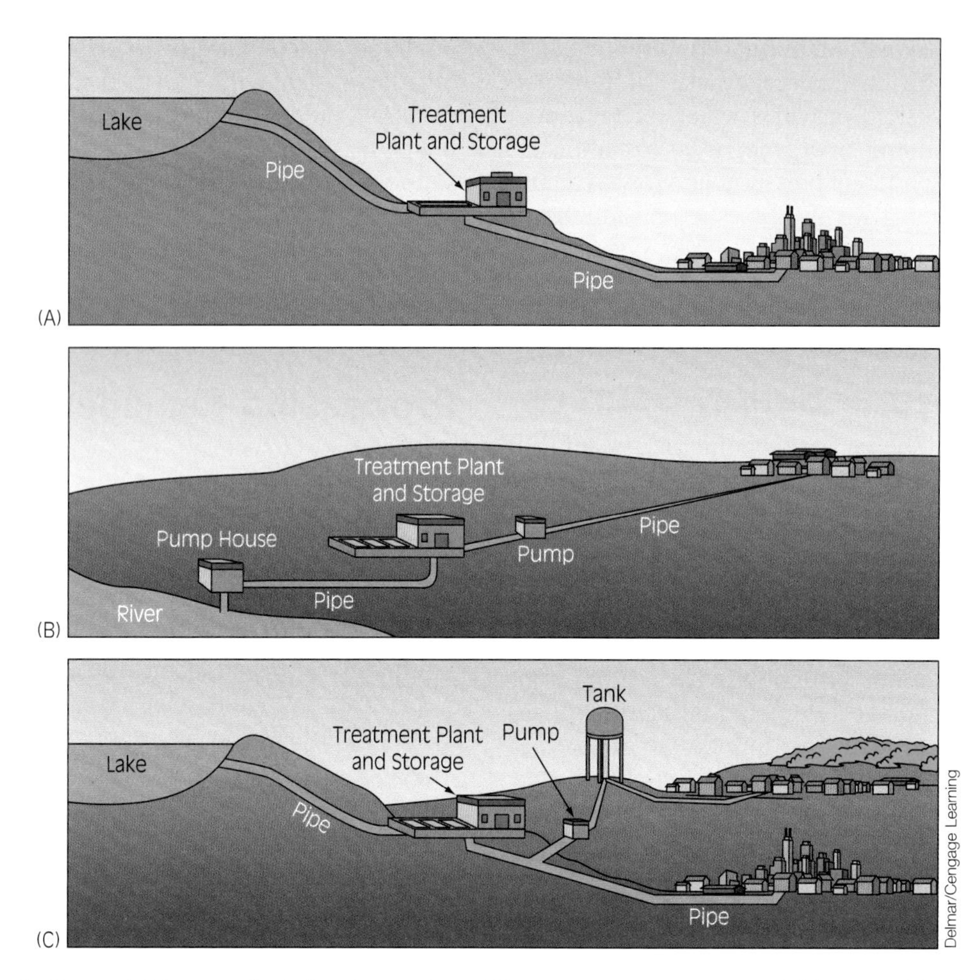

FIGURE 7-5 Illustrations of (A) a gravity-fed water system, (B) a pumping system, and (C) a combination system.

carry water from the source or treatment facility to various locations within the distribution system. Larger-sized pipes are used to reduce the loss of pressure caused by friction and can carry water long distances. Because of this, feeder mains are typically spaced farther apart. Secondary feeder mains of intermediate size, 12 to 14 inches (300 to 355 mm) in diameter, carry water to a great number of locations within the distribution system. In addition, 12-inch (300-mm) mains are also used where long mains are needed and on primary streets. Finally, distributors, the smallest size of pipe in the system, complete the system, with 8-inch (200-mm) pipes providing water to businesses and industries and 6-inch (150-mm) pipes connecting directly to hydrants and residential customers (**Figure 7-6**). Actual pipe sizes for a distribution system can differ considerably between one system and another and are in part based on the age of the system. For example, old systems, built to a prior standard, may still have 4-inch (100-mm) pipes supplying water to individual homes and hydrants. Hydrants in an area containing only 2½-inch (65-mm) discharges and no steamer connections are often a sign of smaller water mains in the area. The major factor for pipe size in all systems is the ability to provide the required flow. Simply put, larger demands require larger-sized pipes within the distribution system.

Often, feeder mains and distributors are interconnected, allowing water to be delivered to the same location through alternate routes. This interconnected system, sometimes called a loop or grid or circulating feed system, has several important features. First, when the system supplies water from two or more directions, it helps reduce friction loss. Second, damage or maintenance on one part of the system can be isolated using control valves so that the distribution system remains in operation. Third, a high demand in one area will not adversely affect another area as much as it would if a grid system were not in place. Basically, water can flow from multiple locations to compensate for a high-consumption area. The ability to provide water from multiple locations can also be used to help maintain adequate water flow and pressure to high-risk areas. Where these interconnections cannot be made at least every 600 feet (180 m), larger-diameter pipes are often used.

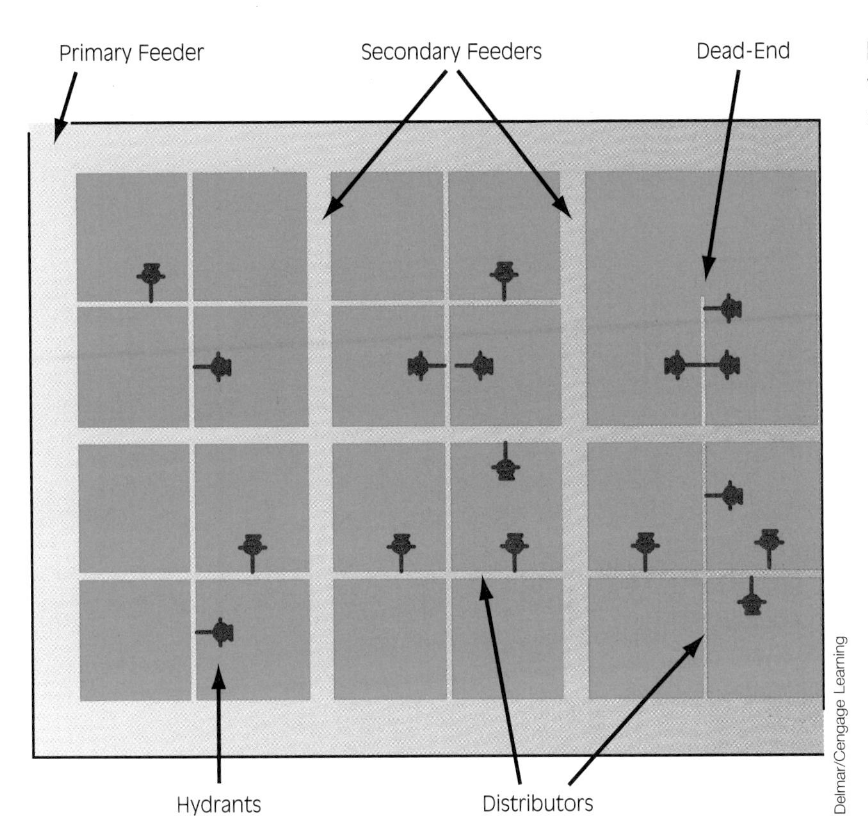

FIGURE 7-6 A grid distribution system provides better water supplies because several mains can feed one hydrant.

A hydrant supplied from only one direction is on a dead-end main, and hydrants located on this type of main are known as dead-end hydrants (see **Figure 7-6**). Should the dead-end main become damaged, all hydrants from that point to the end of the main become inoperable. Dead-end hydrants are more affected by flow obstructions such as tuberculation, which is the corrosion or rust in cast iron pipes, and encrustation, which is caused by tuberculation and sedimentation (**Figure 7-7**). Also, if a pumper begins pumping large quantities of water from one of the first hydrants on a dead-end main, all subsequent hydrants will realize a significant reduction in water pressure and flow. The more water that is pumped from the hydrant, the less water is available for the remaining hydrants on the dead-end main.

FIGURE 7-7 The roughness in the interior of this pipe (which is called encrustation) is due to a combination of tuberculation and sedimentation. It greatly increases friction loss.

Control valves are used to direct water within a distribution system. These valves can be opened and closed to redirect water within the system to isolate sections of the grid or to enhance flow and pressure to sections of the grid. Control valves therefore should be placed in strategic locations within the grid to provide maximum flexibility. In some locations only the public works or water department can operate control valves within the distribution system. Other locations allow fire department personnel to operate control valves. Regardless, preplanning is essential, and pump operators should be familiar with all aspects of the distribution system, including control valve locations.

With few exceptions, such as an area under construction, all valves in a water system should be fully open. To ensure valves are fully open, personnel normally operating valves should have maps showing valve locations, size, and number of turns between closed and open. Valves that are not fully open may not appear to have any effect on a water system during normal usage but may create a significant flow obstruction at the time of higher flows used for firefighting. Ensuring that such restrictions do not exist is one of the reasons for flow testing hydrants. Flow testing also removes a lot of sedimentation from water mains, which both improves flow rates and reduces the amount of sedimentation that goes through the pump, thus reducing pump wear.

Indicating and nonindicating control valves are the two broad types of control valves used in distribution systems. As the name implies, indicating valves provide a visual indication on the status of valves—whether open, closed, or in between—and are more commonly found in private distribution systems. The common types of indicating valves include the outside screw and yoke valve, commonly referred to as an OS&Y valve (**Figures 7-8** and **7-9**); the post-indicating valve, referred to as a PIV (**Figure 7-10**); and the wall-indicating valve, referred to as a WIV (**Figure 7-11**). The OS&Y valve, typically used in sprinkler systems, is made up of a threaded stem connected to a gate. The location of the stem within the yoke indicates the location of the gate. The PIV consists of a stem within a post that is attached to the valve. When the valve is fully open or closed, the words "open" or "shut" appear in the PIV. PIVs are commonly used as sectional and isolation valves. A PIV, WIV, or OS&Y valve may be used as a riser control valve. Nonindicating control valves are more commonly found in municipal distribution systems and are usually either buried or installed within manholes. Gated valves and butterfly valves are the two most common types used in distribution systems.

FIGURE 7-8 The two OS&Y valves on either side of the silver back-flow preventer are in the open position. The threaded stem rising above the black hand wheel is the indicator that the valve is in the open position. The red electrical box just below each hand wheel is a tamper switch. The tamper switch allows a remote notification of any attempt to close this valve.

FIGURE 7-9 This OS&Y valve is in the closed position. The threaded stem being even with the black hand wheel is the indicator that the valve is in the closed position.

FIGURE 7-10 The PIV is one of the two basic types of indicating valves. The lock, which is designed to prevent vandalism, not to provide security, is precut (see blue arrows). This allows the lock to be easily broken with a flathead axe or Halligan bar should the fire department need to operate the valve.

FIGURE 7-11 These wall-indicating valves (WIVs) are in the open position. The text in the window is the indicator of the valve's position. The text will read either "OPEN" or "SHUT." Since the words have no letters in common, even if only one letter can be read due to a foggy window, the position of the valve is known. The red electrical box just above each valve is a tamper switch. The tamper switch allows a remote notification of any attempt to close the valve.

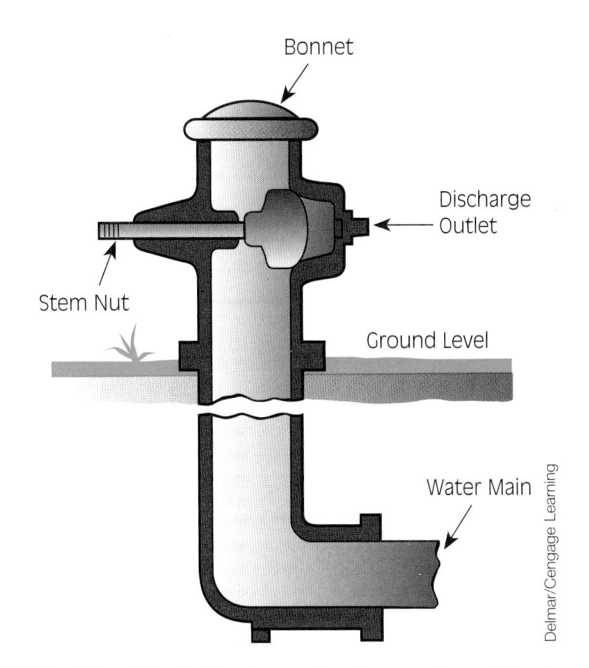

FIGURE 7-12 Typical schematic of a wet barrel hydrant.

Hydrants

The two basic types of hydrants used by the fire service are wet and dry barrel hydrants, the latter being the most common. Both are typically made of cast iron except for important working parts that are made of bronze. **Wet barrel hydrants** are only used where freezing is not a concern. This type of hydrant is usually operated by individual control valves for each outlet (**Figure 7-12**). As the name implies, water is typically maintained in the hydrant at all times. Because each outlet is individually operated, supply hose can be connected and charged independently.

NOTE

The two basic types of hydrant used by the fire service are wet and dry barrel hydrants.

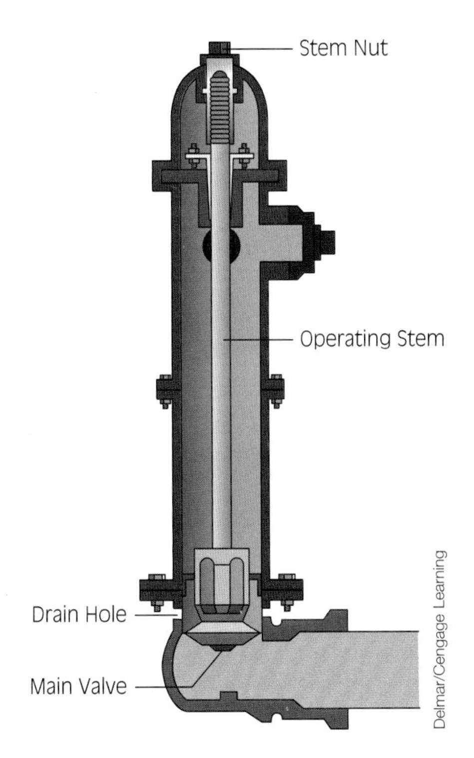

FIGURE 7-13 Typical schematic of a dry barrel hydrant.

Dry barrel hydrants must be used in areas where freezing is a concern. This type of hydrant is operated by turning the stem nut located on the bonnet of the hydrant, which opens the main valve at the base (**Figure 7-13**). The main valve is located at the bottom of the hydrant below the frost line. When the main valve opens, water enters the barrel for use through all the outlets. Because of this, supply hose should be attached to outlets prior to opening the main valve. In addition, gated valves should be attached to unused outlets (**Figure 7-14**). In doing so, the outlets will be available for use after the main valve is opened. When the main valve closes, the drain, also known as the weep valve, opens. The open drain allows water to escape and returns the barrel to its normal dry state. One potential concern is that if the hydrant is not fully closed or opened when in use, the drain valve may remain slightly open, allowing water to flow from the hydrant and causing erosion and potential damage.

Both wet and dry barrel hydrants have a variety of outlet configurations and can be found in a variety of

FIGURE 7-14 Additional lines can be attached to a dry barrel hydrant while in use if gated valves are attached to unused outlets before the outlet is charged.

FIGURE 7-15 Some hydrants have only two 2½-inch (65-mm) outlets.

positions. The normal configuration is one large outlet (4 inches [100 mm] or greater) called a steamer and two 2½-inch (65-mm) outlets (see **Figure 7-14**); however, hydrants with only two 2½-inch (65-mm) outlets are also found (**Figure 7-15**). Most of these outlets use American National Fire Hose connection screw threads (NH) couplings, although some locations still have special threads. Typically, hydrants are installed with the large outlet facing the street and the 2½-inch (65-mm) outlets parallel with the street. In addition, the outlets should be installed so they are not obstructed. They are not, however, always properly installed. Outlets can face almost any direction and can be obstructed in any number of ways. Some of the more common obstructions occur when hydrants are improperly installed (**Figure 7-16**). Even properly installed hydrants may become inoperable (**Figure 7-17**). Hydrants therefore should be serviced, inspected, and operated twice a year.

Hydrants can be located almost anywhere and are often hidden from view (**Figure 7-18**); therefore, fire departments utilize a variety of methods to assist in the quick identification of hydrant locations. One common method is to use reflective paint when painting the hydrant bonnets. Another common method is to place three reflective bands on street poles, signs, or other tall objects located near a hydrant. The reflective strips can also match the color code of the hydrants (see **Table 7-1**). Another

FIGURE 7-16 Improper installation can reduce the usability of a hydrant. Hydrants should allow for a hydrant wrench to turn fully without removing it.

Delmar/Cengage Learning

FIGURE 7-17 Hydrants should be serviced, inspected, and operated semiannually. These procedures will catch unreported accidents and allow scheduling of any needed repairs.

Delmar/Cengage Learning

FIGURE 7-18 This hydrant (opposite the blue marker in the street) is totally blocked from view (see insert) by the mailbox on the right until you have just about passed it.

method gaining popularity is to place traffic reflective markers in the middle of the road adjacent to a hydrant (**Figure 7-19A**). To identify on which side of the road the hydrant is located, the reflector can be offset on the side of the hydrant. Another method for identifying on which side of the road a hydrant is located is to use two different colors on either side of the reflector, one color to indicate the left side and another color to indicate the right side. Road reflectors don't work well in areas where snow plows routinely work the streets in the winter. In these areas, a tall reflective flag or other indicating device is used. The device must be tall enough to be seen over snow drifts and snow banks (**Figure 7-19B**).

TABLE 7-1 NFPA 291 Suggested Classification and Color Coding of Hydrants[HS20]

Class	Rated Capacity	Color of Bonnets and Nozzle Caps
AA	1,500 gpm (6,000 Lpm) or greater	Light blue
A	1,000-1,499 gpm (4,000-5,999 Lpm)	Green
B	500-999 gpm (2,000-3,999 Lpm)	Orange
C	Less than 500 gpm (2,000 Lpm)	Red

Reproduced with permission from NFPA 291-2010, Recommended Practice for Fire Flow Testing and Marking of Hydrants, *Copyright © 2009, National Fire Protection Association, Quincy, MA. This is not the complete and official position of the NFPA on the referenced subject, which is represented only by the document in its entirety.*

The most important consideration a pump operator has about a particular hydrant is whether it can deliver the amount of water that may be needed. Cases exist where firefighting activities have been hampered by lack of available water only to find out in a post-incident analysis that a high-capacity hydrant was nearby. To help operators select the most appropriate hydrant quickly, NFPA 291 recommends a color scheme (see **Table 7-1**) providing operators with information on how much water the hydrant can deliver before they even connect to it.

The 291 standard paints bonnets and caps different colors based on the amount of water they will flow with a 20-pound (140-kPa) residual pressure.

STREETSMART

As a memory aid to remembering hydrant classifications and colors, remember that each class goes up by 500 gpm (2,000 Lpm), about the amount of water required to flow the three required preconnect lines. (Assuming 250 gpm [1,000 Lpm] for the 2½-in. [65-mm] preconnect and 125 gpm [500 Lpm] for each of two 1¾-in. [45-mm] preconnects.) Next, think of grading (for classes) and a traffic light (for colors). A hydrant that provides less than 500 gpm (2,000 Lpm) is unable to provide enough water for all preconnects, is a barely passing C grade, and is painted red as a reminder to STOP and think if a better hydrant is nearby. A hydrant providing 500 to 999 gpm (2,000 to 3,999 Lpm) would get an acceptable grade of B and be painted orange for you to use CAUTION and think if a better hydrant is nearby. A hydrant providing 1,000 to 1,499 gpm (4,000 to 5,999 Lpm) and able to supply a master stream device would get an excellent grade of A and be painted green to remind you to go for it. As many pumps got larger and were able to move 1,500 gpm (6,000 Lpm) or more from a hydrant, there was a need to add a fourth superior class without changing the existing class definitions. So, like food grades, a class AA was invented; class AA hydrants are painted blue to remind you of all the water in the hydrant or, alternatively, that the sky is the limit in what to expect from the hydrant.

In all cases, operators should be familiar with the hydrant color scheme used in their district. Some examples of alternative schemes include painting private and public hydrants differently, color codes that indicate the hydrants' normal operating pressure, or the size of the main it is connected to, or whether it is connected to a high-pressure main, and so forth. Operators should also be familiar with which parts of their districts have strong and weak flows. Common reasons for differences include elevation differences, age of the water mains (mains installed to newer standards are more robust than those put in using prior standards), growth in the town that has the system attempting to provide more water than its original design, and main conditions such as tuberculation, encrustation, and sedimentation.

Private Water Systems

Industrial, commercial, and large complexes or facilities often maintain their own private water supply system. In most cases these private water systems are similar to municipal water systems. They can either receive their water from a municipal water system or use surface or groundwater sources. In addition, private water supply systems can use a pumping system, gravity, or a combination of the two to move water from the source to the hydrant, standpipe system, or sprinkler system. Typically, private water supply systems provide water for industrial/manufacturing use, employee use, and fire protection. Private fire protection systems are generally maintained separately from other water systems. As with municipal systems, private fire protection systems consist of a series of decreasing-sized pipes, called mains, and control valves that move water from the source to individual hydrants and/or sprinkler and standpipe systems. The primary disadvantage of private water systems is the limited quantity of available water from some smaller systems that do not receive water from a large municipal system and adequate maintenance of

(A)

(B)

Delmar/Cengage Learning

FIGURE 7-19 (A) Hydrant traffic reflectors can help pump operators locate hydrants from a distance, especially at night. Traffic reflectors are usually located in the road adjacent to the hydrant. (B) Poles can help locate hydrants that may be buried with snow.

some systems. In particular, systems protecting vacant and abandoned facilities may not be adequately maintained.

Water Availability

Municipal supplies, as a water source, have three main advantages. First, hydrants provide a readily available supply of water over an expanded geographical area. How available this water supply is depends on the extent to which hydrants are distributed within an area. Second, hydrant flow capacities can be determined in advance through flow tests. NFPA 291, *Recommended Practices for Fire Flow Testing and Marking of Hydrants*, identifies testing procedures and suggests that hydrants be classified and color coded based on their rated capacity (**Table 7-1**). Finally, municipal systems can provide a sustainable supply for several hours. It is important to realize that municipal water supplies are typically designed to provide required fire flows for a specific time frame, usually between 2 and 10 hours. When pumping less than the required flow, hydrant operations can be sustained for longer periods. If pumps are used in the water system to move water, several types of pumps are installed. Some of the pumps run constantly, others turn on when additional pressure and volume is required, and still others are maintained as emergency backup pumps complete with emergency electrical power sources.

Hydrant Flow Test

Hydrant flow testing provides the means to determine the available pressure and flow within municipal and private distribution systems. The information collected during flow tests is critical for identifying and selecting hydrants that can provide the estimated fire flows calculated during pre-fire planning activities. In some locations the public works or water department is responsible for conducting hydrant flow tests. In others the pump operator conducts or assists with hydrant flow testing. Regardless, pump operators should be familiar with hydrant flow test procedures.

Essentially, the flow test consists of measuring static and residual pressures from one hydrant and measuring flow from one or more other hydrants. The data collected during tests provide the means to calculate available hydrant flow at specific residual pressure. NFPA 291, *Recommended Practices for Fire Flow Testing and Marking of Hydrants*, provides excellent information on all aspects of conducting flow tests, as discussed next.

Preparation (Prior to Conducting the Flow Test)

Several considerations and actions to complete before conducting the flow test include the following:

- Hydrant flow testing during periods of ordinary water demand will yield more accurate results based on realistic conditions.
- Consideration of test hydrant location may help to ensure minimum impact on traffic and surface areas.
- Use of hydrant diffusers can reduce the impact of erosion.
- Select a test hydrant, sometimes referred to as the residual hydrant, and one or more flow hydrants.
- Place a cap with a pressure gauge onto the 2½-inch (65-mm) test hydrant outlet.
- To increase the accuracy of the reading, ensure individuals at flow hydrants have received training and have practice taking pitot gauge readings.

General Procedure

The general procedures for conducting a hydrant flow test are as follows:

1. The test begins with a static pressure reading at the test hydrant. The reading is taken after fully opening the hydrant valve and removing the air.
2. Open flow hydrant(s) one at a time until a 25% drop in residual pressure is achieved.
3. After a sufficient drop is noted, continue flowing to clear debris and foreign substances.
4. Take all readings at the same time: a residual reading at the test hydrant and flow readings using the pitot gauges at each of the flow hydrants.
5. Record the exact interior size, in inches (millimetres), of each outlet flowed.
6. After recording all readings, slowly shut down the hydrants one at a time to reduce the likelihood of a water hammer or surges within the system.

Equipment

The following equipment is often required during hydrant flow testing:

- Pressure gauge mounted on an outlet cap, with the pressure gauge having been calibrated within the past 12 months
- Pitot gauge for each hydrant (note: NFPA 291 requires gauges calibrated within the past 12 months)

- A hydrant diffuser, which can be used to reduce damage caused by large-volume flows from hydrants
- Hydrant wrenches
- Portable radios

Calculating Results

First, calculate the discharge from hydrants used for flow during the test. The flow from hydrant pitot readings can be determined by either a chart, see Table 4.10.1(a) within NFPA 291, or by the following formula:

$$Q = 29.84 \times c \times d^2 \times \sqrt{p} \qquad (7\text{-}1)$$

$$Q = 0.067 \times c \times d^2 \times \sqrt{p} \qquad (7\text{-}1m)$$

where

c = coefficient of discharge

0.90 for smooth and rounded outlets (**Figure 7-20A**)

0.80 for square and sharp outlets (**Figure 7-20B**)

0.70 for square outlets projecting into the barrel (**Figure 7-20C**)

d = diameter of the outlet in inches (mm)

p = pitot pressure in psi (kPa)

When multiple hydrants are used in the flow test, calculate (or look up) each flow and then add them together.

Next, calculate the discharge at the specified residual pressure and/or desired pressure drop using the formula:

$$Q_R = Q_F \times \left(\frac{h_r^{0.54}}{h_f^{0.54}}\right) \qquad (7\text{-}2)$$

where

Q_R = predicted flow at specified residual pressure

Q_F = total flow from hydrants

h_r = pressure drop to desired residual pressure (initial static pressure reading – desired residual pressure)

h_f = pressure drop measured during test (initial static pressure reading – final residual pressure reading)

The formula can be computed using a calculator capable of logarithms or by looking up the values of pressure readings to the 0.54 power in a table. NFPA 291 suggests both a form to use to document flow test data and a form to graph results. Various computer software programs are available to assist with hydrant flow calculations and reporting.

Reliability

Municipal systems generally provide a reliable supply of water; however, several factors may reduce the reliability of this supply source. First, the flow from hydrants may decrease over time based on gradual increases in municipal consumption or through the normal deterioration of piping and components. The color coding of hydrants, as suggested by NFPA 291, provides flow rates for a single hydrant. When multiple hydrants are used, individual hydrant flows may change dramatically. In addition, the color code indicates flows during normal municipal consumptions. During peak use hours, hydrant flow rates may again change dramatically. Finally, leaks, preventive maintenance, scheduled outages, and damaged or broken components may all lead to reduced flow capacity of hydrants from time to time.

Hydrant Capability

When operating from a hydrant, the operator needs to be able to estimate how much additional water can be supplied should a request be made to supply additional hoselines. One does not want to find out that a water supply problem exists after taking the time and effort to put an additional line in place and charge it.

To determine how much additional water a hydrant can supply, the operator needs to know three things: the static pressure before fire flows were started, the current residual pressure, and how much water is flowing. The static pressure can be read on the pump's intake gauge after connecting to the hydrant and opening the intake valve and before starting to flow water to pump discharges. It does not matter if the pump is engaged, only that discharges are not flowing. If the pump is not engaged, both the intake and discharge gauges will show the static pressure and should show the same reading. If the pump is engaged, the intake gauge will show the static pressure and the discharge gauge will show a higher pressure, or more specifically, intake pressure plus net pump discharge pressure, which is called pump discharge pressure.

Once discharges have been opened and water starts flowing, the intake pressure will decrease to the residual pressure, which is always lower than

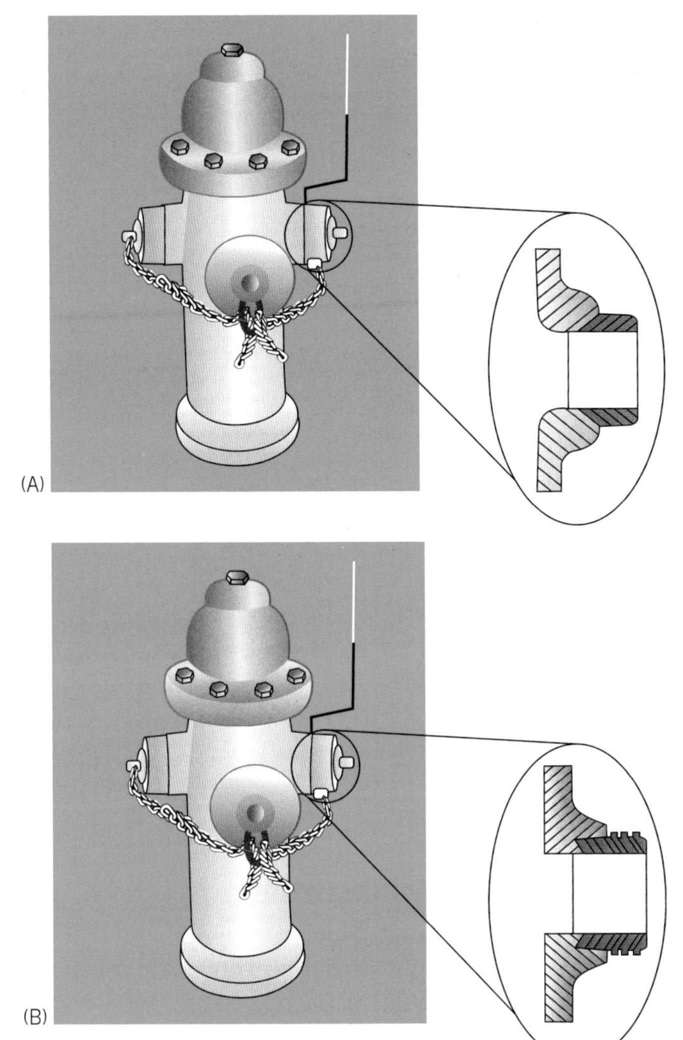

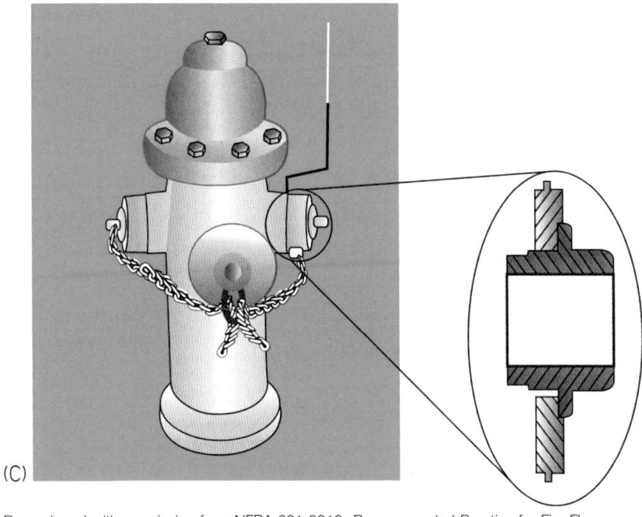

FIGURE 7-20 These figures show three different hydrant outlet types. Since each one has a different amount of friction, they each allow a different amount of water to flow. Style A with its rounded corners allows the greatest flow. Style B with its square corners allows an in-between flow, and style C with its intrusion into the barrel of the hydrant allows for the least flow.

the static pressure. The available hydrant water is the amount of water that is flowing when the decreasing residual pressure has dropped to 20 psi (140 kPa), the minimum acceptable intake pressure to avoid possible cavitation, collapse of intake hoses, and problems within the municipal water system. So, the biggest factor affecting hydrant capacity is how much the residual pressure drops when water flow begins. The larger the residual pressure drop, the weaker the hydrant. The second factor affecting hydrant capacity is its initial static pressure. The higher the static pressure, the more the residual pressure can drop before reaching the hydrant's capacity.

As an example, think of a hydrant that was fed by a high-pressure garden hose instead of a water main. While nothing was flowing, the garden hose would have zero friction loss and the hydrant would have a great static pressure. Once an attempt to flow water from the hydrant started, however, the garden hose would not be able to supply the hydrant adequately and the garden hose would have a very high friction loss, resulting in a very low residual pressure at the hydrant. Exactly the same thing happens when small water mains in the street supply a dead-end hydrant.

As a second example, think of a hydrant fed by a 16-inch (400-mm) water main and right next to a large ground-level water storage tank. Because of limited head pressure (the height of the water in the storage tank above the hydrant), the hydrant would have a low static pressure. Opening the hydrant, however, will have very little effect on the residual pressure since the size of the hydrant outlet is so small relative to the size of the water main feeding it and its supply.

Understanding these concepts combined with some practice allows pump operators to make judgments about the adequacy of their water supply routinely without performing any math. In some situations, the answer is not obvious. In these situations, the operator needs to use a more formal mathematical equation. Three different formulas known as the percentage method, first-digit method, and squaring-the-lines method will be presented. All three do a good job of predicting hydrant capacity. They differ in whether the formula is easier to calculate with a pocket calculator or in your head. After learning all three, many operators find a favorite and use a second to confirm their result, or, in some cases, department policy may specify a particular method to be used consistently within a jurisdiction. All three formulas use the key elements of static pressure, residual pressure, and quantity of water flowing. Due to differences in the formulas, slight differences in the predictions are expected.

First-digit Method

This method, which does not have a metric equivalent, is considered the easiest to do in your head by some since it does not require squaring of numbers or division. To use this method:

1. Find the psi drop, which is the difference between the static and residual pressure.
2. If the psi drop is equal to or less than the first digit of the static pressure multiplied by:
   - 1, then three times the current flow is possible
   - 2, then two times the current flow is possible
   - 3, then one times the current flow is possible

Static pressure is 65 psi; 250 gpm is flowing; and the residual pressure is 58.

1. psi drop = static pressure – residual pressure

   psi drop = 65 – 58

   psi drop = 7

2. First digit of static pressure times 1

   $$6 \times 1 = 6$$

   7 (the psi drop) is not less than 6 but is less than 12 (6 × 2), so two more lines at 250 gpm can be added.

Percentage Method

This method uses a simple formula that is easy to do on a simple calculator or even a cell phone.

1. Determine the percentage drop using the formula:

   $$(\text{static pressure} - \text{residual pressure}) \times \left(\frac{100}{\text{static pressure}}\right) = \text{percentage drop}$$

2. Determine the amount of additional hydrant capability using the percentage drop and the following table:

Percentage Decrease in Pressure	Additional Water Available
0–10%	3 times
11–15%	2 times
16–25%	1 time
Over 25%	Less than original flow

Static pressure is 60 psi (420 kPa). One line is flowing, and the residual pressure is 48 psi (336 kPa). Can one additional line be added?

1. $$(\text{static pressure} - \text{residual pressure}) \times \left(\frac{100}{\text{staticpressure}}\right) = \text{percentage drop}$$

   $$(60 - 48) \times \left(\frac{100}{60}\right) = 20\%$$

   $$(420\,\text{kPa} - 336\,\text{kPa}) \times \left(\frac{100}{420\,\text{kPa}}\right) = 20\%$$

One additional line can be added.

Squaring-the-lines Method

This method uses the same information you will learn about what happens to friction loss as flow increases to determine hydrant capacity. Some people like it best because it saves learning another type of formula. For this method the new pressure drop is the number of new lines squared, which means the number of lines multiplied by the number of lines or itself, times or multiplied by the original pressure drop.

Static pressure is 60 psi (420 kPa); one 250-gpm (1,000-Lpm) line is flowing; and the residual pressure is 52 psi (364 kPa). What would the residual pressure be if two 250-gpm (1,000-Lpm) lines are flowing?

1. The static pressure – the residual pressure = the original pressure drop

   $$60 - 52 = 8$$

   $$420\,\text{kPa} - 364\,\text{kPa} = 56\,\text{kPa}$$

2. $(\text{The new number of lines})^2 \times \text{original pressure drop} = \text{new pressure drop}$

   $$2^2 \times 8 = \text{new pressure drop}$$

   $$56\,\text{kPa} = \text{new pressure drop}$$

   $$4 \times 8 = 32$$

   $$4 \times 56\,\text{kPa} = 224\,\text{kPa}$$

3. The static pressure – the new pressure drop

= the new residual pressure

60 – 32= 28

420 kPa – 224 kPa = 196 kPa

Since the new residual pressure is higher than the minimum acceptable residual pressure of 20, the additional hoseline can be supported.

Supply Layout

Hydrant supply hose configurations can vary depending on the flow and pressure of the hydrant, the position of the apparatus, and the number and size of intakes available on the apparatus. The more complex the configuration, the longer it will take to set up the supply operation. The first step, though, is to select a hydrant. In some cases, there is no choice in that only one hydrant is available. In general, hydrants closer to the incident should be selected to reduce the time to set up the operation as well as to reduce the loss of pressure due to friction; however, in some cases a stronger hydrant (with more volume and pressure) that is farther away may be picked over weaker hydrants closer to the incident. In addition, the number and size of hydrant outlets may influence which hydrant to select. Finally, the pump capacity and the size of supply lines available will influence the selection of a hydrant.

Once the hydrant is selected, the next step is to determine supply line configurations. In general, larger supply lines should be used when possible. Supply lines should be laid to minimize bends and kinks in order to reduce losses in pressure from friction.

The pump can be positioned at the hydrant for several reasons. One reason is that the hydrant is relatively close to the scene and attack lines can be advanced from that position. Another reason is when a **reverse lay** is used. A reverse lay (see **Figure 7-21**) is when the apparatus stops at the scene; drops off attack lines, equipment, and personnel; and then advances to the hydrant. Another reason for the pump being located at the hydrant is when it is the first pump in a **relay operation**. Finally, the pump may be located at a hydrant when increasing hydrant pressure to another pump by use of a four-way hydrant valve (see **Figure 7-22**).

Because hydrant outlets and pump intakes vary, a wide range of configurations is possible when the pump is located at a hydrant. The quickest way is to use a supply hose to connect directly into one of the main intakes. Another way is to use a 50-foot (15 metre) section of LDH. When smaller-diameter hose is used, two or more lines can be connected from the hydrant. This can be accomplished using wyes on the hydrant and Siamese on the pump intake or using multiple hydrant outlets and pump intakes. Because hydrants are located at different distances from the curb and outlets can face just about any direction, pump operators must realize that positioning of the apparatus at hydrants will vary; nevertheless, the driver should be able to position the apparatus such that the hydrant can be reached with a single hoseline without repositioning the apparatus or being so close that kinks are unavoidable. When possible, the configuration should provide enough water to pump capacity or deliver the required flow.

If the pump is not positioned at the hydrant, either a **forward lay** has been conducted or the pump is being supplied as part of a relay operation.

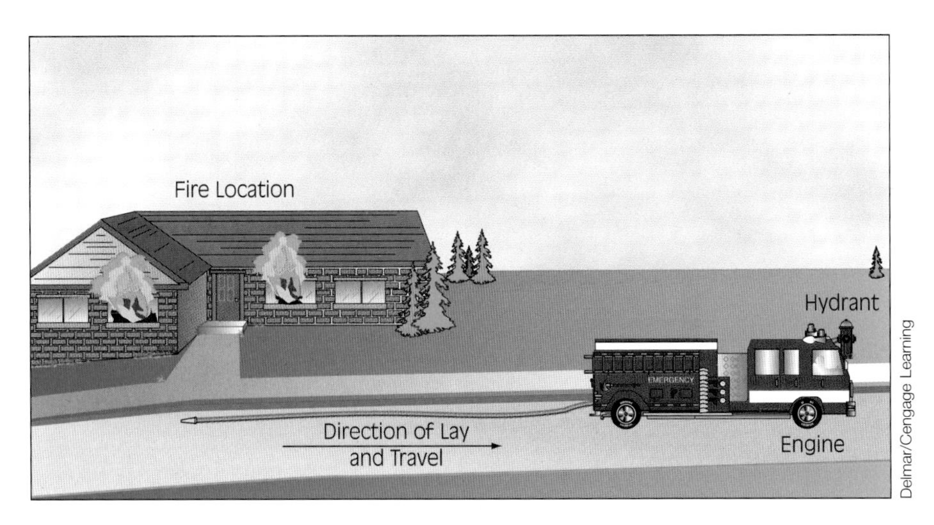

FIGURE 7-21 A reverse hose lay.

FIGURE 7-22 The four-way hydrant valve allows the attack engine to connect to the hydrant and to establish a water supply. Subsequently another engine can connect to the four-way valve and boost flow and pressure without ever interrupting the flow to the attack engine.

A forward lay (see **Figure 7-23**), sometimes referred to as a straight lay, is when the apparatus stops at the hydrant and lays a supply line to the fire. If the apparatus proceeds directly to the scene, a second apparatus can lay a line to or from a hydrant. When a supply line is laid to the pumper, the line is connected to one of the pump's intakes.

A **split lay** includes components of both a forward lay and a reverse lay. For example, a forward lay could be deployed by an attack pumper going up a long driveway from the street or down a dead-end street from the intersection. A second apparatus would deploy a reverse lay from the forward lay coupling to the hydrant (**Figure 7-24**). The forward lay and reverse lay couplings are connected and a completed lay from the water source to the attack pumper exists.

STATIC SOURCES

Static sources are those supplies, such as ponds, lakes, rivers, and swimming pools, that generally require drafting operations (**Figure 7-25**). Even a shallow, fast-moving stream may be dammed with a ladder and savage cover and used for a drafting operation. **Drafting** is the process of moving or drawing water away from a source by a pump. Static sources are used for several reasons. First, they are used when hydrants are not available. Many communities simply cannot afford or don't need a municipal water system. Second, static sources are used when available hydrants cannot provide the needed flows. Finally, they are used when natural disasters or mechanical failures cause hydrant systems to shut down.

Drafting operations require the use of a special supply hose called suction hose (**Figure 7-26**; see also "Classification of Hose" in **Chapter 6**). Suction hose is used because supply line pressures will be at or below atmospheric pressure within the hose. It is important to match the size of suction with the rated capacity of the pump (**Table 7-2**). Suction hose too big or too small may make it difficult or impossible to prime the pump. In addition, the ability to pump at capacity may be significantly reduced. Pumpers that comply with NFPA 1901 will have the appropriate size of suction hose for the rated capacity of the pump.

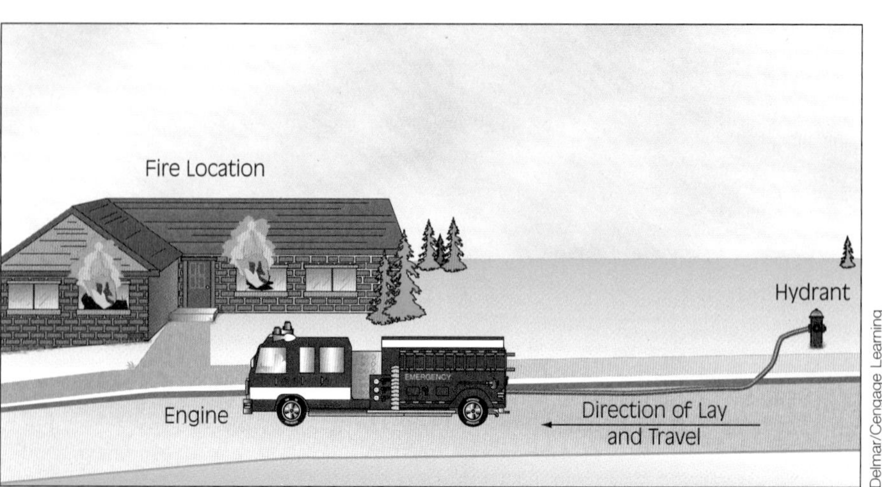

FIGURE 7-23 A forward or straight hose lay.

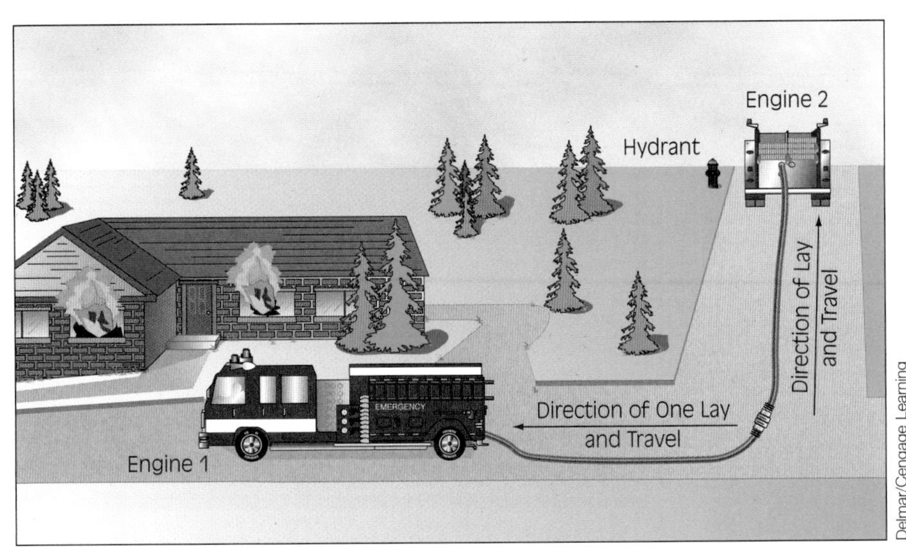

FIGURE 7-24 The split hose lay.

FIGURE 7-25 Example of a static water supply.

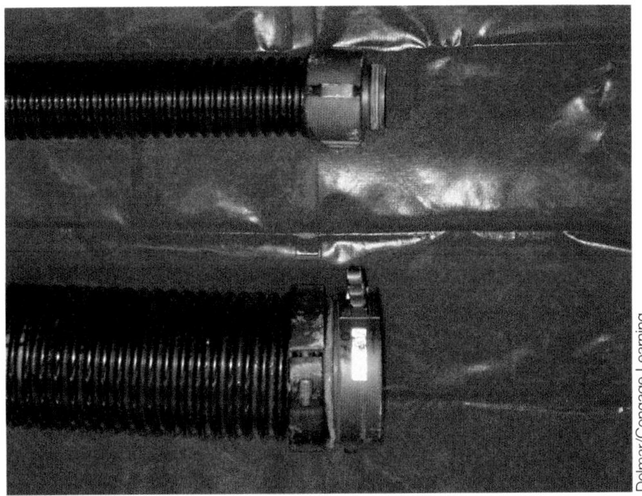

FIGURE 7-26 Drafting operations require hard suction hose. Pictured are the newer, more maneuverable sections of suction hose. This style of vinyl hose is, however, not usable with pressurized water.

To keep debris from the static source from entering the pump, strainers are connected to the end of the suction hose. Note the strainer attached to the bottom section of suction hose in **Figure 7-27**. Care must be taken to keep the strainer off the bottom of the water source to prevent clogging. In addition, if the strainer is too close to the surface, whirlpools may develop, allowing air to enter the pump (**Figure 7-28**). The result may be either inefficient pumping or loss of prime. Drafting operations typically require apparatus to be positioned fairly close to the source, as shown in **Figure 7-28**, because the height to which water can be drafted is limited. Recall that water is not sucked or pulled into the pump. Rather, when the priming system reduces the pressure in the pump (below 14.7 psi or 101 kPa) at sea level, water is forced into the pump by atmospheric pressure (14.7 psi or 101 kPa). Water will rise approximately 2.3 feet for each 1 psi of pressure, or 1 metre for every –10 kPa negative pressure. As a rule of thumb, water will rise about 1 foot for each inch of vacuum.

TABLE 7-2 Rated Capacity of Hard Suction Hose

Rated Capacity	Suction Size
750 gpm (3,000 Lpm)	4½ inch (110 mm)
1,000 gpm (4,000 Lpm)	5 inch (125 mm)
1,250 gpm (5,000 Lpm) and above	6 inch (150 mm)

Delmar/Cengage Learning

FIGURE 7-27 This strainer also keeps the suction hose off the bottom to prevent clogging.

If the priming device reduces the pressure inside the pump from 14.7 psi to 9.7 psi (a 5-psi reduction), the atmospheric pressure, now 5 psi greater than that inside the pump, will force water to a height of 11.5 feet ($2.3 \times 5 = 11.5$ feet). If the priming device reduces the pressure inside the pump from 101 kPa to 81 kPa (a 20-kPa reduction), the atmospheric pressure, now 20 kPa greater than that inside the pump, will force water to a height of 2 metres (.1 metre/kPa $\times$ 20 kPa $=$ 2 metres).

The height water rises (lift) is measured from the water's surface.

Delmar/Cengage Learning

FIGURE 7-28 The strainer being used in this drafting operation floats on the surface. It is designed to skim debris-free surface water while controlling whirlpool production.

If the priming system reduced the pressure to 0 psi, a perfect vacuum, how high would atmospheric pressure force the water? This height is called theoretical lift and can be calculated by multiplying 14.7 psi (101 kPa) (atmospheric pressure) by 2.3 feet per psi (0.1 m/kPa), approximately 33.81 feet (10 metres). Obviously, priming systems on apparatus come nowhere near creating a perfect theoretical lift.

STREETSMART

The general rule of thumb is not to pump more than 22.5 feet (6.6 m), or two-thirds of the theoretical lift, with a centrifugal pump. The two-thirds rule takes into account intake hose friction loss, even lower pressures that occur near the impeller's eye, cavitation, and effects on the pump's capacity caused by the amount of lift. While theoretical lift cannot be exceeded, some equipment, such as the vacuum tankers discussed later in this chapter, can operate at a greater percentage of theoretical lift.

Capacity tests for pumps are conducted at draft. In general, pumps are required to flow at capacity through 20 feet (6 metres) of suction with a maximum height of 10 feet (3 metres). Greater heights and longer supply lines reduce the ability to flow at capacity.

Availability

Regional differences determine the availability of static sources. Some areas have many sources, while others may have only a few. Where static sources are

FIGURE 7-29 The banks of this static water supply may not support the weight of a pumper, particularly when the ground is wet.

available, access to the supply may be limited. For example, the banks of rivers may be too steep or too soft to support the weight of the apparatus (**Figure 7-29**). The available flow from different static sources also varies as well. The available water from a small pond will not be as great as that from a large river. Pump operators should be familiar with drafting locations and available flows for their service area.

Access and flow for static sources can be enhanced in several ways. One way is to provide an adequate road to the drafting site (**Figure 7-30**). In doing so, access can be increased during seasonal and weather changes. Another way is to provide a stable area close to the source where drafting operations can take place safely. Dredging the immediate area surrounding the drafting location is another way to enhance the static source by helping to provide unrestricted flows. Finally, **static source hydrants**, sometimes called **dry hydrants**, can be installed (**Figure 7-31**). These hydrants are simply pre-piped lines that extend into the static source (**Figure 7-32**). Static hydrants can be used for natural static sources (ponds, lakes, and rivers) as well as artificial sources such as underground water tanks. These hydrants can be beneficial in that they reduce setup times, can increase flows by using large lines, and can increase the number of access points to a static source. A disadvantage is sediment that may form in the piping, requiring a back-flush prior to priming the pump.

Reliability

Static water supply reliability is affected by three factors. The first factor is the supply itself. Static

FIGURE 7-30 Providing an adequate road will enhance the availability of a static source, especially in areas where seasonal conditions impair access.

FIGURE 7-31 Example of a static source dry hydrant.

supplies can change based on environmental and seasonal conditions. For example, droughts tend to diminish water supplies, while excessive rains may flood normal drafting locations or soften the ground, thus restricting access. In addition, excessive silt and debris may render the source unusable. In the winter the water may freeze, hindering access to the supply. Tidal water supplies can change dramatically within a short period of time. The second factor is the pump and equipment used to draft the static source. Drafting is demanding for both the pump and the pump operator. Pumps and equipment must be maintained in good working order to efficiently and effectively use static sources as a supply. Finally, the pump operator is a factor. Pump operators must be thoroughly

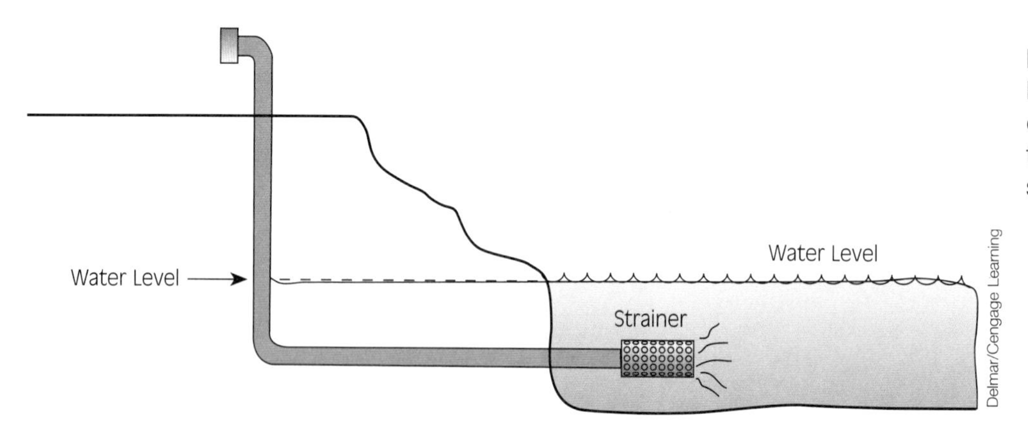

FIGURE 7-32 Static source hydrants, also called dry hydrants, are pre-piped lines that extend into the static source.

FIGURE 7-33 Allow at least 2 feet (0.6 metres) in every direction from the barrel strainer to ensure unrestricted flow.

familiar with pump operations, priming operations, and drafting operations to use static sources reliably.

The suction hose should not be placed over an object that is higher than the pump intake because air pockets are likely to form that may hinder priming and pumping from the static source.

Supply Layout

Setting up a drafting operation takes time, skill, and at least one person to assist the pump operator. The first step is to position the apparatus as close to the source as possible. Next, the suction hose and the strainer should be coupled, making sure all connections are tight. Then the suction hose is connected to the pump intake and the strainer is placed into the water. The suction hose should not be placed over an object that is higher than the pump intake. When this occurs, air pockets are likely to form that may hinder priming and pumping from the static source. In addition, the strainer should be placed to ensure unrestricted flow. For a typical barrel strainer, 2 feet (60 cm) should exist in every direction from the strainer to ensure unrestricted flow (**Figure 7-33**). The last step is to ensure that the pump is airtight. This means that every possible location where air can enter the discharge or intake piping should be checked. Inlet and outlet caps should be tightened and all control valves should be checked to make sure they are completely closed.

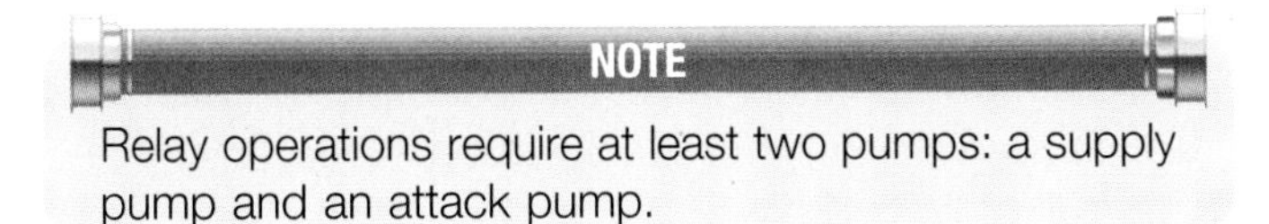

Relay operations require at least two pumps: a supply pump and an attack pump.

HIGH-CAPACITY RURAL WATER SUPPLIES

Moving large quantities of water from over a distance from a source to a fire requires large amounts of equipment, personnel, and practice. The amount of practice to perform these operations efficiently is so high that it is the one item that ISO requires a department to demonstrate during a rating process if the department wishes to get credit for its water-moving abilities. As a result, when high-capacity rural water supplies are put in place, water supply officers and separate radio channels are normally assigned.

High-capacity operations use either a relay or water haul method, with different jurisdictions tending to focus on one of these methods and obtaining the equipment best suited for their chosen method. A third method known as **nurse feeding** is often used for smaller-capacity requirements or during the time it takes to set up a high-capacity operation. The remaining topics in this chapter will provide information on each of these operations.

RELAY OPERATIONS

Relay operations are those operations in which two or more pumpers are connected in-line to move water from a source to a discharge point. Relay operations are used for several reasons. First, they are used when a water source is distant from the incident and cannot provide appropriate flows and pressures to the scene. Second, they are used when a water source is relatively close to the incident but lacks pressure. Third, relay operations are used to overcome loss of pressure from elevation gains. For example, skyscrapers use building fire pumps to relay water from street-level pumps to upper stories of the building.

The original water supply can be either a static source or a hydrant in a municipal source. Therefore, the considerations for static sources and hydrants are the same for relay operations as they are for drafting operations. This section discusses the unique considerations and setup requirements of relay operations.

Relay operations require at least two pumps: a supply pump or source pump and an attack pump. The supply pump, obviously, is positioned at the supply, while the attack pump is the last pump in the relay. Additional pumps in the relay are called in-line or relay pumps. Relay operations tend to take longer to set up and are usually more complicated to operate (**Figure 7-34**).

One of the major difficulties in relay operations is how to control changes in pressures and flows.

A relay operation is similar to the operation of a two-stage pump operating in the series mode. Recall from **Chapter 4** that in the series mode, water is discharged from one impeller to the intake of the second impeller. In a relay operation, water is discharged from one pump to the intake of a second pump. However, water flows through supply hose between the two pumps rather than through fixed piping within the pump. In the series mode, the flow available to the second impeller is limited by the flow produced by the first impeller. The same

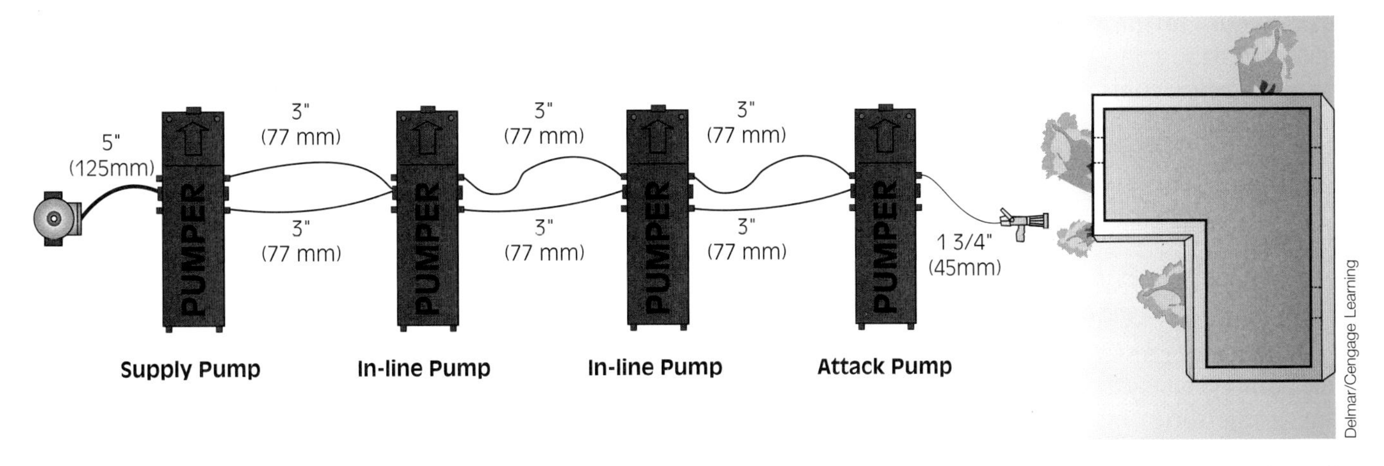

FIGURE 7-34 Relay operations require at least two pumps, usually take longer to set up, and are more complicated to operate.

holds true for relays in that the flow available to the second pump is limited by the flow generated by the first pump. Furthermore, if the first pump is operating from a draft, it needs to use additional power to lift water into the pump, thus producing a greater NPDP to create the same PDP as other pumpers in the relay. For this reason, when drafting, the largest pumper should be placed at the source when possible. When using a pressurized source, such as a hydrant, with a residual pressure of greater than 20 psi (140 kPa), the smallest pumper is used at the source since it will have the least NPDP, and therefore need the least capacity. In series mode, pressure is doubled as the second impeller takes the pressure generated by the first impeller and adds the pressure it generates. The same concept holds true for relay operations, in that the second pump can take advantage of the pressure provided by the first pump. In a relay, however, the pressure generated by the first pump will be reduced by friction in hoselines and by increases in elevation before it enters the intake of the second pump.

The major difference between a relay operation and a pump operating in the series mode is the ability to control changes in pressure and flow independently. In a series mode, each impeller rotates at the same speed; therefore, a change in flow in the first impeller will automatically be pumped by the second. Further, a change in pressure by the first impeller will automatically be doubled as it passes through the second impeller. In each case, the change is predictable and automatic. In a relay operation, the impellers in one pump do not rotate at the same speed as other pumps and are therefore able to generate different flows and pressures. When the pumps are connected and flowing, changes in pressures and flows in one pump affect the entire system.

One of the major difficulties in relay operations is how to control changes in pressures and flows. When additional water is needed for the incident, flow must increase from the supply pump and all other pumps in the relay. If an attack line is shut down, pressure can increase throughout the system. The manner of controlling pressures and flows in a relay depends, in part, on the type of relay used. In general, relay operations can be either open or closed.

Closed Relay Operations

In a **closed relay** system, water enters the relay at the supply and progresses through the system to the attack pump. This system is similar to the pump operating in the series mode in that all of the water and pressure is delivered from one pump to the next pump in the relay. Because all of the water is contained, controlling pressure and flow requires that each pump be changed. For example, if an attack line is shut down, the attack pump would compensate for reduced flows, as would each of the pumps in the relay. Controlling pressure and flow with this system is difficult, at best, in that changes occur quickly and affect the entire relay operation. Pump operators would find it difficult to keep up with needed changes at each pump as well as to coordinate efforts with other pumps in the relay.

Open Relay Operations

In an **open relay** system, flow and pressure are not contained within the total system. In other words, all of the flow and pressure from one pump is not always delivered to the next pump. This design allows a continuous flow through the relay, which reduces the need to make constant changes and reduces the overall effect of pressure changes within the system. Open systems can be set up according to one of three methods.

One method simply dedicates and maintains an open discharge line from the attack pump to flow excess water. In this case, once flow is initiated, only the attack pumper is required to make changes. If one or more attack lines are shut down, the pump operator simply increases flow through the dedicated line. This also has the benefit of quicker responses to change as well as having additional water readily available. Care must be taken concerning where the excess water is flowing. This system tends to pump more water than is actually needed.

Another method is to have the relay deliver its water directly into a **portable dump tank** rather than directly into the intake of the attack pump. The attack pump would simply draft from the portable tank. In this case, the relay operates continuously without needing constant changes. If attack lines are shut down, the attack pump reduces flow and the portable tank simply overfills. This system also provides ready access to additional water if needed.

Alternatively, the relay can be set up to take advantage of new automatic intake and discharge relief valve requirements. In this case, each pump sets its relief valves(s) to control pressure increases within the relay. When pressures rise above the setting, relief valves automatically open to dump excess pressure. This system has the added benefit of being able to increase flows automatically by simply increasing the relief valve setting. Relief valve requirements are discussed in both NFPA 1901 and NFPA 1962.

Finally, the relay can be set up using apparatus equipped with quick-acting electronic governors that will adjust engine RPM to maintain the same

PDP. Relay relief valves or intake relief valves help by dumping water before water with too high a pressure enters a pump.

Engines are typically used for relay operations; however, any apparatus with a pump can be used—so, for example, a quint or quad could be used. While it is convenient to have all in-line pumps be of the same size, actual pump sizes can vary.

SAFETY

As noted, a relay is like a multistage pump, with each pump adding to the prior pump's' discharge pressure. A dangerous problem can occur when lines on the attack pumper are shut down. When this happens, friction losses in each of the hoselines decreases, resulting in a higher intake pressure at the next pump as less of the discharge pressure of the prior pump is lost to friction. As each pump continues to add pressure, the pressure can quickly reach dangerous levels, causing sudden, catastrophic equipment failures. To prevent such failures, every pressure relief valve must be set and all pressure governors must be operating in pressure mode. Constant monitoring of pressure gauges and making adjustments as necessary to ensure that safe discharge pressures are never exceeded are primary responsibilities of each pump operator. Intake pressure at each relay pumper should not be allowed to drop below 20 psi (140 kPa) or exceed 50 psi (350 kPa).

Designing Relays

Step One

The first step in relay design is to evaluate each of the following factors:

- Amount of water to flow
- Available water at the supply
- The size and number of pumps available
- The size and length of supply hose available
- The distance from the source to the incident

The weakest of these factors will be the limiting factor of the relay design. For example, a relay operation capable of flowing 1,000 gpm (4,000 Lpm) is limited when the supply can only provide 500 gpm (2,000 Lpm). In addition, a 1,000-gpm (4,000-Lpm) pump equipped with 500 feet of 2½-inch (150 m of 65-mm) hose is not able to pump the quantity or distance as the same pump equipped with 500 feet of 4-inch (150 m of 100-mm) hose.

Two of the more important factors in the design of a relay are the amount of water the relay is expected to flow and the water available at the supply.

When designing a relay, it is important to keep in mind the relationship between flow and pressure.

Two of the more important factors to consider in the design of a relay are the amount of water the relay is expected to flow and the water available at the supply. The basic design of the entire system will be affected by these factors. Another important consideration is the number of pumps available and their rated capacity. Obviously, pumps must be of sufficient capacity to provide the required flow in the relay. Pumps must also be able to generate sufficient pressures to move this flow over a distance. When designing a relay, it is important to keep in mind the relationship between flow and pressure. When pressure increases, flows decrease. Recall, as mentioned earlier in this chapter, that pumps at draft are expected to flow 100% capacity at 150 psi (1,050 kPa), 70% capacity at 200 psi (1,400 kPa), and 50% capacity at 250 psi (1,750 kPa). In essence, increased pressures reduce the quantity of water that can be pumped, which is often the case during relay operations.

The size of hose available for use in the relay is another important consideration from two perspectives. First, it is important to keep in mind the relationship of hose size to the amount of water the hose can flow and the friction loss it develops (see classification of hoses in **Chapter 6**). In general, larger-diameter hose will flow more water and have less friction loss than smaller-diameter hose. Second, the highest operating working pressure of the hose must be considered. For medium- and small-diameter hose attack lines, the highest operating pressure is 250 psi (1,750 kPa), while the highest operating pressure for LDH supply lines is 185 psi (1,295 kPa), according to NFPA 1961 and 1962. Because of the nature of relay operations, supplying water over distances, pressures can easily exceed the 185-psi (1,295-kPa) operating pressure of LDH supply hose. To utilize LDH safely, NFPA 1962 requires a discharge relief device with a maximum setting no higher than the service test pressure of the hose in use.

There are two additional considerations related to pressure in relay operations. The first relates to the gain or loss in pressure resulting from changes in elevation within the system. In general terms, pumping uphill is harder than pumping downhill; more pressure is needed to pump water uphill and less pressure is needed to pump water downhill. For every 1 foot (metre) of elevation gain, discharge pressure will increase 0.434 psi (10 kPa). If the average height of a single story is 10 feet (3.5 m), then the pressure gain will be 4.34 psi (10 kPa) (0.434 psi/ft × 10 ft = 4.34 psi) (10 kPa/metre × 3.5 m = 35 kPa/story minus 1). As a rule of thumb, add or subtract 5 psi (35 kPa) for each story (or each 10 feet) (or 10 kPa for each metre) in elevation gain or loss. (Additional information on pressure gain and loss is presented in **Section IV** of this book.) Another consideration in the design of a relay is the pressure required at the intake of each pump within the relay. When one pump flows water to the next, friction loss in the hose will reduce the discharge pressure. The goal is to provide enough pressure from the first pump to cover the loss in pressure from friction so that at least 20 psi (140 kPa) remains when water enters the second pump. The purpose of maintaining a minimum of 20 psi (140 kPa) is to ensure that the pump will not cavitate. Recall that cavitation can damage pumps, cause a loss of prime, and reduce pumping efficiency. Cavitation is discussed in greater detail in **Chapter 9**.

Finally, the distance between the source and the incident is important. In general, the farther the distance, the more resource-intensive the relay will be. When large flows are required, either large pumps with LDH will be spaced quite some distance apart or smaller pumps with medium-diameter hose will be spaced rather close together in the relay (**Figure 7-35**). In reality, a variety of combinations can occur, each depending on the weakest component in the system.

Step Two

When all of these factors have been considered, the next step is to determine the distance between

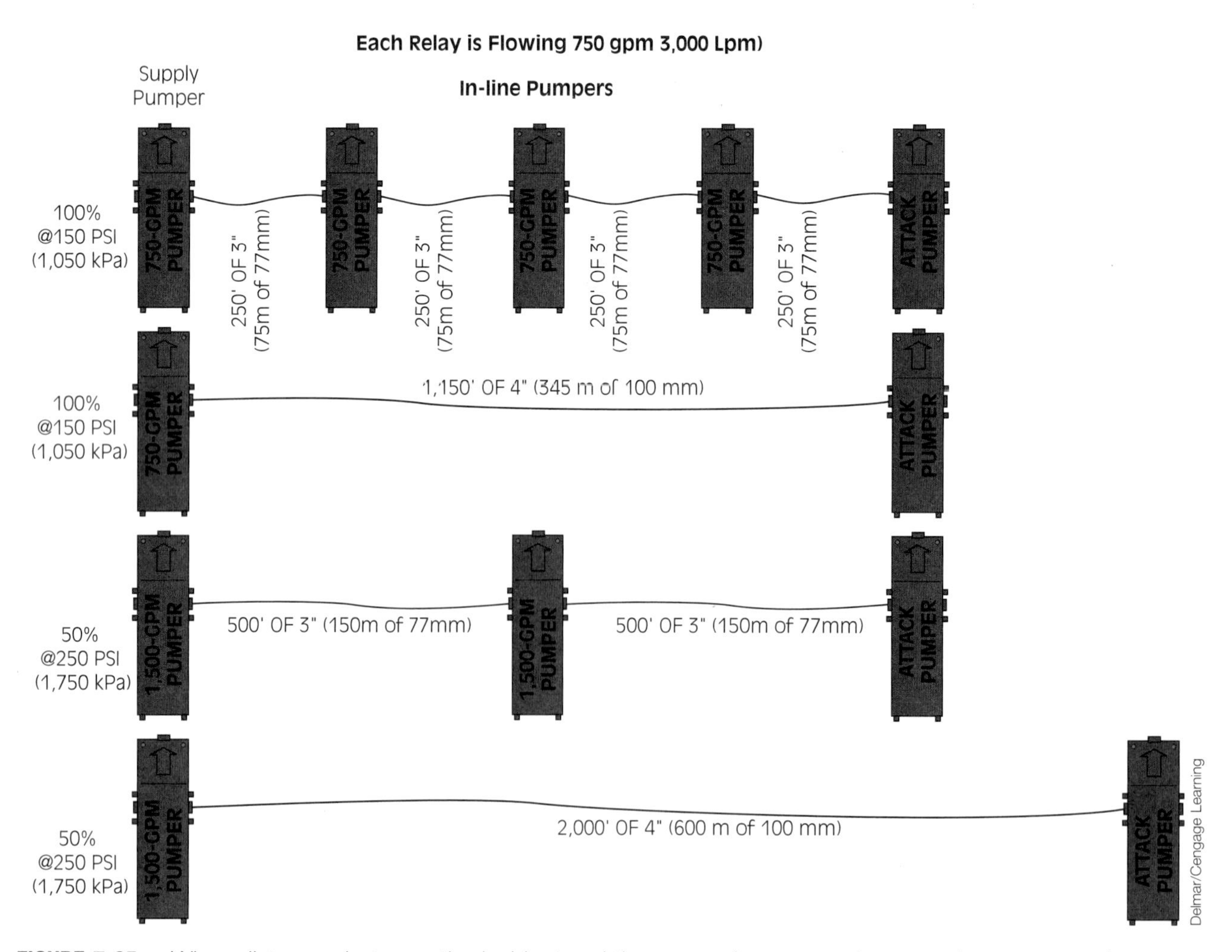

FIGURE 7-35 When distances between the incident and the source increase, relay operation resources increase as well.

pumpers. Unless each pump has the same capacity and available hose, the distance between one pump and the next will be different. The goal is to maximize the distance between pumps while providing the required flow. This is accomplished by first determining the pump discharge pressure that can provide the required flow in the relay. For example, a relay operation requiring 500 gpm (2,000 Lpm) can be provided by a 750-gpm (3,000-Lpm) pump operating at 200 psi (1,400 kPa) (70% capacity) or a 1,000-gpm (4,000-Lpm) pump at 250 psi (1,750 kPa) (50% capacity).

The next step is to determine the distance between pumps. When the water leaves the discharge side of the pump, the pressure will be reduced by friction as water travels to the next pump. The question, then, is how far can the water travel until the pressure is reduced to 20 psi (140 kPa) (the minimum intake pressure for pumps in a relay)? The formula in equation (7-3) can be used to determine this distance.

$$\text{Distance} = (PDP - 20) \times \left(\frac{100}{FL}\right) \tag{7-3}$$

$$\text{Distance} = (PDP - 140\text{kPa}) \times \left(\frac{30\text{m}}{FL}\right) \tag{7-3m}$$

where

PDP = pump discharge pressure

20 = reserved intake pressure (psi) (kPa) at the next pump

100 = length (feet) (metres) of one section of hose (the most common length of hose is 50 feet [15 m]; however, friction loss calculations use 100-foot [30-m] increments/sections)

FL = friction loss (psi) (kPa) per 100-foot (30-m) section of hose

Take, for example, a 750-gpm (3,000-Lpm) pump flowing 500 gpm (2,000 Lpm) through 4-inch (100-mm) hose at 200 psi (1,400 kPa) pump discharge pressure. The friction loss in 4-inch (1,000-mm) hose flowing 500 gpm (2,000 Lpm) is 5 psi (35 kPa) per 100 feet (30 m) (see chart on friction loss in **Appendix G**). In this example, the pressure in the hose will be 20 psi (140 kPa) after water travels a distance of 3,300 feet (990 metres):

$$(185 - 20) \times \left(\frac{100}{5}\right) = 3,300 \text{ feet}$$

$$(1,295 - 140) \times \left(\frac{30}{35}\right) = 990 \text{ metres}$$

Note that the pump discharge pressure was reduced to 185 psi (1,295 kPa) to comply with supply hose maximum operating pressure. What distance will water travel if the 500-gpm (2,000-Lpm) relay flow is provided by a 1,000-gpm (4,000-Lpm) pump with a 3-inch (77-mm) supply line? First, the 1,000-gpm (4,000-Lpm) pump can provide 500 gpm (2,000 Lpm) at a pump discharge pressure of 250 psi (1,750 kPa). Next, the friction loss in 3-inch (77-mm) hose flowing 500 gpm (2,000 Lpm) is 20 psi per 100 feet (140 kPa per 30 m). In this example, water will travel 1,150 feet (345 metres) before it reaches 20 psi (140 kPa):

$$(250 - 20) \times \left(\frac{100}{20}\right) = 1,150 \text{ feet}$$

$$(1750 - 140) \times \left(\frac{30}{140}\right) = 345 \text{ metres}$$

In comparison, if the 3-inch (77-mm) hose is replaced with 2½-inch (65-mm) hose in the last example, water will travel 460 feet (140 metres) when the pressure is reduced to 20 psi (140 kPa):

$$(250 - 20) \times \left(\frac{100}{50}\right) = 460 \text{ feet}$$

$$(1750 - 140) \times \left(\frac{30}{350}\right) = 138 \text{ or } 140 \text{ metres}$$

Thus the factor limiting the maximum distance water can be pumped in a supply line is based on the necessary residual pressure, typically 20 psi (140 kPa), at the discharge end of that line. Two items affect the residual pressure: total pressure loss in the line and the discharge pressure of the upstream engine that supplies the other end of the line.

Three factors may be changed to increase the maximum distance water can be pumped:

1. The discharge pressure of the upstream pumper can be increased. These increases are limited by the maximum safe operating pressure and by the fact that increased pressure reduces the upstream pumper's capacity. Pressure cannot be increased beyond the point where the pump's reduced capacity would exceed the desired flow rate.

2. The friction loss may be reduced either by using a larger-diameter hoseline or by using multiple hoselines.
3. The flow rate (gpm or Lpm) may be reduced since the largest factor affecting friction losses for a given hose lay is the flow rate. *Note*: For a given fire a required fire flow exists that must be achieved for effective extinguishment. The change in flow rate as used here applies when using items such as relay valves. With relay valves an initial flow is established to slow the fire spread, and that flow rate is increased as additional relay pumpers arrive. As the additional relay pumpers enter they provide shorter relay distances between pressure boosts. The shorter distance allows for an increased friction loss per foot (metre) from a higher flow rate while still maintaining the required residual pressure. Later, as flow requirements decrease, some relay pumpers may be released since with reduced flows a longer distance is possible.

Alternate Step Two

The preceding step two calculates the maximum distance between pumpers, which is excellent for minimizing the number of pumpers. It is useful in areas that use hose tenders, a type of apparatus carrying primarily hose and usually more than a mile of hose. These hose tenders lay hose and perhaps relay valves and manifolds. When laying hose, they need to allow for separately arriving relay pumpers to connect and pump the water with the maximum possible distance between relay pumpers. Sometimes, however, the relay may contain more than the minimum number of pumpers. This may happen either because a relay pumper is placed after laying all of its available hose even if not yet needed or because the flow rate has been reduced on an in-service relay. In these cases, the pump operator needs to determine the pump discharge pressure that will leave a 20 psi (140 kPa) residual at the next pumper using the formula:

$$PDP = FL \times Length + 20$$

$$PDP = FL \times Length + 140\,\text{kPa}$$

where

PDP is the pump discharge pressure
FL is the hose friction loss per 100 feet (30 metres)
$Length$ is the length of hose in hundreds of feet (metres)

Take, for example, a 1,000-gpm (4,000-Lpm) flow pumped by a 1,250-gpm (5,000-Lpm) pump through 2,000 feet of 5-in. (600 m of 125-mm) hose. At 1,000 gpm (4,000 Lpm), 5-in. (125-mm) hose has a friction loss of 8 pounds per 100 feet (65 kPa per 30 metres). So:

$$\begin{aligned} PDP &= FL \times Length + 20 \\ &= 8 \times 20 + 20 \\ &= 180 \end{aligned}$$

$$\begin{aligned} PDP &= FL \times Length + 140\,\text{kPa} \\ &= 56 \times 20 + 140 \\ &= 1,260\,\text{kPa} \end{aligned}$$

Since the 180-psi (1,260-kPa) PDP is less than the 185-psi (1,295-kPa) maximum pressure for 5-in. (125-mm) LDH hose, this is an acceptable pump pressure.

Step Three

The last step is to lay the supply lines and position the pumps. The largest pump should be positioned at the water source whenever possible. The next pump in the relay simply lays a line equal to the predetermined distance. When all of the lines are in place, the relay operation can begin. The steps for initiating relay flows are discussed in Chapter 8.

TANKER SHUTTLE OPERATIONS

Tanker shuttles are those operations where apparatus equipped with large tanks transport water from a source to the scene (**Figure 7-36**). Tanker shuttle

FIGURE 7-36 Apparatus equipped with large tanks are used in shuttle operations.

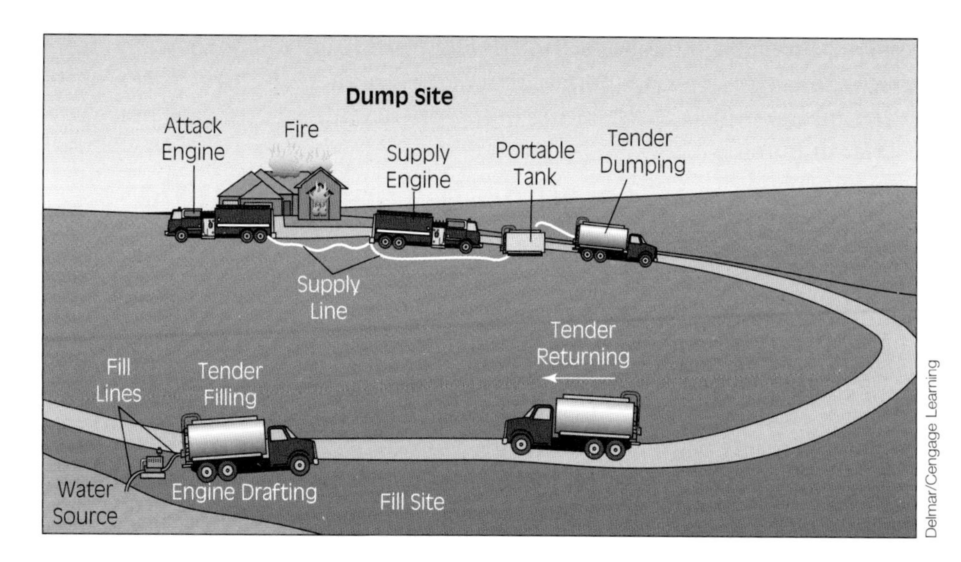

FIGURE 7-37 Major components of a tanker shuttle operation.

operations are used for two general reasons. First, they are used when the required pressures or flow cannot be achieved by the supply source or pump due to the limitations of hose size and/or distance to move water from the supply to the incident. Second, they are used when obstacles such as elevation, winding roads, intersections, and railroad crossings limit the use of other supply methods.

The components of a tanker shuttle include multiple tankers, pumps, a **fill site** where tankers receive their water, and a **dump site** where tankers unload their water (see **Figure 7-37**). As with relay operations, the fill site supply can be either a static source or a hydrant. This section, then, discusses the unique considerations and setup requirements for tanker shuttle operations.

Shuttle Equipment

The equipment used in shuttle operations includes pumpers, tankers, portable dump tanks, **jet siphons**, and gravity dumps. At least one pumper is required at the scene and uses the water delivered by the shuttle to supply attack and exposure lines. Typically, the pumper will either draft from a portable tank or will be supplied under pressure by a nurse tanker. Other pumps can be located at the fill site to assist with rapid filling of tankers or at the dump site to assist in the movement of water to the scene. For example, limited access to the scene may require that tankers deliver their water some distance from the scene. In this case, an additional pumper is required to move water from the dump site to the attack pumper.

Tankers used in shuttle operations should be sufficient in capacity to provide required flows. According to NFPA 1901, tankers should have a minimum capacity of 1,000 gallons (4,000 litres). In addition, the tank must be able to both fill and unload at a minimum rate of 1,000 gpm (4,000 Lpm), with typical tankers ranging from 1,500 to 3,000 gallons (6,000 to 12,000 litres) and a few over 5,000 gallons (20,000 litres). Adequate ventilation is also required to ensure that the tank is not damaged during filling and unloading operations. The weight of the tanker is an important consideration for maintaining control as well as for weight limits of roads and bridges. Consider the weight of water in a 1,000-gallon tanker. Since the weight of 1 gallon of water is 8.35 pounds, the weight of 1,000 gallons is 8,350 pounds (8.35 lbs per gal × 1,000 gal = 8,350 lbs), or 4.175 tons (8,350 lbs (2,000 lbs per ton = 4.175 tons). Since the weight of 1 litre of water is 1 kilogram, the weight of 4,000 litres is 4,000 kilograms (1 kg per litre × 4,000 litres = 4,000 kg), or 4 tonnes (1,000 kg/tonne). In general, smaller tankers are better for short-distance shuttles because they are able to load and unload water at a faster rate than larger tankers, which are better suited for longer distances. Another tanker consideration is the vacuum tanker. This type of tanker is relatively new to the fire service and provides a viable alternative to the traditional style of tanker. The apparatus carries water in a completely enclosed tank that is pressurized. The speed at which it fills and offloads is faster than that of a traditional tanker, and it is capable of lifting water up to 28 ft (8.5 m), which is beyond most current nonpressurized tanker capabilities. This faster fill/offload ability makes it a consideration in sustaining higher gpm rates during shuttle operations. In addition, the lack of overflow openings makes transport safer due to lack of water spillage on the roadway.

Portable dump tanks and jet siphons are valuable pieces of equipment for shuttle operations. Portable tanks provide the most efficient means to unload tanker water for use by a pump (**Figure 7-38**). The tank provides the space (water storage) and means (it is faster to dump water out of a tanker than to pump it out) to reduce the amount of time the tanker spends at the dump site. These tanks can be set up in almost any location where the ground is relatively level. Most portable dump tanks have a capacity of 1,000 to 3,000 gallons (4,000 to 12,000 litres). Some are equipped with special devices to ease filling and transferring from one tank to another. Jet siphons (**Figure 7-39**) are devices that help move water quickly without generating a lot of pressure (**Figure 7-40**). They are used to move water from one portable tank to another (**Figure 7-41**) or to assist in the quick off-loading of tanker water.

Delmar/Cengage Learning

FIGURE 7-38 Portable tanks (which are also called dump tanks) are an essential piece of equipment in most shuttle operations.

Delmar/Cengage Learning

FIGURE 7-39 This low-level strainer incorporates a jet siphon. It is used in a dump tank to remove as much water as possible. When the 1½-inch (38 mm) line connected in the rear is charged, some water comes out of the small hole in the center of the LDH connection. That water brings much more water with it and transfers it via the connected hard hoseline.

Designing a Tanker Shuttle

Although the specific design of a tanker shuttle will vary from one operation to another, three basic components exist in all shuttle operations: the fill site, the dump site, and the **shuttle flow capacity**. Each of these components must be carefully coordinated in order to supply water efficiently and effectively. A shuttle with all tankers of approximately the same capacity generally increases the efficiency of the shuttle. When some tankers are significantly larger than others, using different fill sites for tankers of different sizes helps to reduce bottlenecks.

Courtesy of Donald Fisher

FIGURE 7-40 Jet siphons use a small amount of water and pressure to transfer a larger amount of water. The type shown in the photo is used to transfer water between portable water tanks. Jet siphons are often the most efficient method of transferring water between portable water tanks.

Delmar/Cengage Learning

FIGURE 7-41 Jet siphons move water from one drop tank to another.

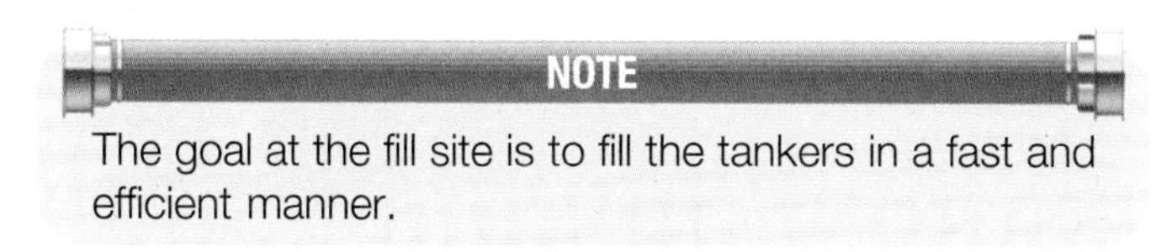

NOTE

The goal at the fill site is to fill the tankers in a fast and efficient manner.

Fill Site

Refilling tankers occurs at the fill site. Obviously, the goal is to fill the tankers in a fast and efficient manner. Minimizing hose movement reduces make-and-break time, and placing a traffic cone at the exact place apparatus should stop helps minimizes apparatus positioning time. Another consideration for a fast and efficient fill operation is the source used to supply the tankers. Both hydrants and static sources can be used to fill tankers. When hydrants are used, using an engine to boost pressure on all but the strongest hydrants will decrease tanker fill time. The most efficient connection on a tanker is the connection that leads directly into the tank. These are typically located on the rear of a tanker and labeled "Direct Fill." When a strong hydrant is used, it should be connected to the intakes leading directly to the tank (**Figure 7-42**). When hydrant flows are low, a portable dump tank can be used to allow a pumper to draft with greater flows.

A second consideration for a quick and efficient fill operation is to provide adequate access to the fill site. In addition, the fill station should be set up to connect and disconnect fill lines quickly. If possible, the fill site should have two complete sets of fill lines available. While one tank is being filled, the second tanker can be connected and standing by. Filling two tankers at the same time may actually lengthen fill times.

Finally, safety should be considered at fill sites. One potential safety concern is that apparatus will be moving in close proximity to personnel and equipment. If a fill site is going to involve multiple apparatus, a safety officer should be considered. This person is responsible for directing staging and incoming and outgoing apparatus to help minimize traffic concerns. Another concern is adequate venting when a tank is being filled. Improper venting can be a significant hazard to both equipment and personnel since it can result in a ruptured tank.

Delmar/Cengage Learning

FIGURE 7-42 The fill site is the location where tankers receive water.

SAFETY

Tanker operations should be improved by decreasing fill and dump timings, *not* by questionable driving techniques. Both tanker weight and the fact that the weight (water) moves make safe tanker driving difficult. About a quarter of firefighter deaths occur while in transit, with tankers being the apparatus type most often involved with fatalities.

NFPA has mandated changes that help reduce such incidents, including better brakes than commercial vehicles, lower top speeds, additional tank baffles, and seatbelt warning systems. Nothing, however, is more important than how the driver operates the apparatus.

Dump Site

The dump site is the location where tankers deliver their water. The goal, again, is to unload the water quickly and head back to the fill site. As with the fill site, issues of access and safety must be considered. For example, use of traffic cones at the dump site both helps drivers quickly and accurately position their apparatus and can be used to keep people out of the area in which vehicles are moving. For positioning, a cone just outside the driver's door allows the driver to stop at the same spot each time.

Several options are available when delivering the tanker's water. Tankers can deliver their water directly to the attack pump, to another tanker (called a nurse tanker or nurse feeding), or to a portable dump tank (**Figure 7-43**). Each tanker should carry a portable dump tank so it can quickly unload its water and head to the fill site. Portable tanks should be large enough to allow uninterrupted dumping of the largest tanker. Thus the tank capacity must be at least as large as the largest tanker or a little larger since it will hold less than its rated capacity if the ground it is used on is not perfectly level. In some cases, the dump site will have multiple portable dump tanks, allowing several tankers to unload their water at the same time. The dump site should be located where adequate room is available for tankers to maneuver.

FIGURE 7-43 Tanker dumping water directly into a portable dump tank.

FIGURE 7-44 Multiple tanks using a jet siphon may be set up together.

When multiple dump tanks are used, a means must be provided to move water between dump tanks. Common methods for transferring water between dump tanks include: connecting dump tank drains together (if more than two tanks are going to be used in this manner, tanks must have multiple drains), U-tubes (typically PVC pipe in a U shape to allow siphoning between tanks), and jet siphons, which provide the most efficient water movement between tanks.

Jet siphons typically use hard suction hose between tanks and may be connected to some low-level strainers (**Figure 7-44**). They use a small amount of water from the pump to transfer a lot of water between tanks. Advantages of jet siphons, particularly when used with a low-level strainer, include faster transfer rates than other methods; the ability to get the water levels lower in the tank being siphoned than a draft pumper can achieve (which ends up allowing more water to be drafted instead of left in a tank); and allowing tanks to be at different levels (lowering the level in one tank makes more room for the dump operation so it does not need to wait as water is used and increases the level in the tank the engine is drafting from, allowing more water to be used for firefighting while waiting for the next tanker).

Ideally, tankers delivering water to a dump tank can dump water out of either side or the rear. When tankers do not have this ability, or the dump rates vary significantly based on which dump is used, additional accommodations in how the dump site is set up must be made to allow the tankers to dump. One advantage to side dumps is that they can prevent accident-prone backing. Alternatively, some tankers use a rear camera to allow drivers to see where they are backing.

Tankers may directly pump their water into a dump tank, although this is typically the slowest method and limited by the tank-to-pump flow rate, which NFPA only requires to be 500 gpm (2,000 Lpm) versus the NFPA required dump rate of 1,000 gpm (4,000 Lpm); they also may jet dump it or gravity dump it. Jet dumps were very common because they used a small amount of pressurized water from the pump water to drive a lot of water out of the tank and had an advantage of removing the vast majority of the tank water at a high rate. In the past, the pump valves were set and the driver would stay in the cab, engage the pump, and depress the fuel pedal to a predetermined RPM.

Newer (non-pump-and-roll) apparatus disable the fuel pedal when the pump is engaged. The current need for a jet dump operator to get out of the cab and go to the pump panel to increase the throttle has led to the replacement of jet dumps with gravity dumps in many newer apparatus. Depending on department-specified options, gravity dumps can be completely controlled from the cab by the driver. Gravity dumps use a larger opening to allow water to empty from the apparatus tank quickly. Since their flow rate is dependent on head pressure, or the height of the water in the booster tank, they start with a fast dump rate that helps when dump tank levels are low. The disadvantage of gravity dumps is that their dump rates get progressively slower as the tank empties. Because of the reducing dump rates, it does not make sense to wait an extended period of time for the last drop of water to come out of a gravity dump system. Once the dump rate has slowed significantly, the driver should stop dumping and return to the fill site for another load of water.

Shuttle Flow Capacity

Shuttle flow capacity is the volume of water that can be pumped without running out of water. The flow capacity of a shuttle is limited by the volume of water being delivered and the time it takes to complete a shuttle cycle. The volume of water being delivered depends on the size and number of tankers. The **shuttle cycle time** is the total time it takes to dump water and return with another load. The cycle time includes the time it takes to fill the tanker, the time it takes to dump its water, and the time it takes to travel between the fill and dump stations. In addition, the time a tanker must wait, if any, while other tanks are filling or dumping is included in the cycle time. The use of appropriate portable dump tanks tends to increase shuttle flow capacity by allowing the tanker to dump at its fastest possible rate and to reduce wait time at the dump site, allowing a quicker release of the tanker to get the next load of water.

Improper venting while filling a tank can be a significant hazard to both equipment and personnel.

Individual shuttle tanker flow capabilities can be determined by dividing tank volume by the time it takes to complete a cycle. For example, a 1,500-gallon (6,000-litre) tanker that can complete a cycle in 10 minutes will have a shuttle flow rating of 150 gpm (600 Lpm) (1,500 ÷ 10 = 150 gpm) (6,000 ÷ 10 = 600 Lpm). Cycle times may vary because different-size tanks can fill and dump their water at different speeds. In addition, several fill sites may be used that will have different travel times. A 1,000-gallon (4,000-litre) tanker with an 8-minute cycle time will have a shuttle flow capacity of 125 gpm (1,000 Lpm). When the two tankers are utilized in the same shuttle, the combined shuttle flow will be 250 gpm (1,000 Lpm).

Combined shuttle flows can be calculated by adding individual tank flows (as in the previous example) or by adding the volumes of all tankers and dividing the average cycle time for all tankers. In this case the total volume would be 2,500 gpm (10,000 litres) and the average cycle would be 9 minutes, for a total combined flow of 277 gpm (1,111 Lpm) (2,500 ÷ 9 = 278 gpm) (10,000 ÷ 9 = 1,111 Lpm). The slight difference between the two methods is the result of averaging the cycle times.

NURSE FEEDING

Nurse feeding is the process of providing water under pressure to the attack engine from another engine. It is commonly used in areas without hydrants when limited amounts of water are needed. Its advantages include quick setup time and simplification of the tasks the attack pump operator needs to perform. With nurse feeding, the attack engine receives pressurized water from the booster tank of a pumper-tanker so the attack engine does not need to draft. In cases where more than one tanker's water is required, use of a Siamese allows transfer from one supply tanker to another without interrupting the availability of pressurized water to the attack engine (**Figure 7-45**). When used in this manner, the two tankers supplying the Siamese should use different pump discharge pressures. The goal is to use all of the water from the first tanker before using any water from the second tanker. With the first tanker using a pump discharge pressure that is 20 to 30 pounds (140 to 210 kPa) higher than the second tanker, the clapper on the Siamese to the second tanker remains closed until the first tanker runs out of water. At that point, the clapper automatically opens, allowing water from the second tanker to flow. At this point, the second tanker increases its discharge pressure to the level the first tanker was using (commonly 80 psi [560 kPa]) and the first tanker shuts down, disconnects, and goes to the fill site. A third tanker can take the place of the first tanker and use the lower PDP until the second tanker runs out of water. At that point the process is repeated.

Delmar/Cengage Learning

FIGURE 7-45 The use of a Siamese and multiple supply tankers allows the continued replacement of tankers when they become empty with full tankers without interrupting the availability of pressurized water to the attack pumper.

LESSONS LEARNED

The first step in fire pump operations is securing a water supply. The basic water supplies available to pump operators are on-board tanks, static sources, and municipal systems. These sources can be used in combination with relay operations and tanker shuttle operations. In reality, any and all of the water supplies can be used in a variety of configurations. The on-board water source is by far the fastest, yet most limiting. Municipal water supplies are by far the most common. Various hose layouts are available depending on required flow, pump size, hose size, and appliances available. Drafting operations from static sources tend to take longer to set up and are demanding on both the apparatus and the pump operator. Relay operations and tanker shuttle operations should be designed rather than just pieced together.

The most important concept in securing a water supply is knowing the expected flow. Simply selecting the first available source may lead to inefficient utilization of the source and pump. It is also important to understand the considerations for each of the water sources. Preplanning will make the hectic first minutes of an incident a little less stressful for the pump operator.

KEY TERMS

Closed relay Relay operation in which water is contained within the hose and pump from the time it enters the relay until it leaves the relay at the discharge point; excessive pressure and flow is controlled at each pump within the system.

Drafting The process of moving or drawing water away from a static source by a pump.

Dry barrel hydrant A hydrant operated by a single control valve in which the barrel does not normally contain water; typically used in areas where freezing is a concern.

Dry hydrant A piping system for drafting from a static water source with a fire department connection at one end and a strainer at the water end.

Dump site Location where tankers operating in a shuttle unload their water.

Dump tank See **Portable dump tank**.

Fill site Location where tankers operating in a shuttle receive their water.

Forward lay Supply hoseline configuration in which the apparatus stops at the hydrant and a supply line is laid to the fire.

Jet siphon Device that helps move water quickly without generating a lot of pressure and that is used to move water from one portable tank to another or to assist with the quick off-loading of tanker water.

Municipal supply A water supply distribution system provided by a local government and consisting of mains and hydrants.

Nurse feeding A method of water supply where one tanker pumps the water in its booster tank into another tanker or engine.

On-board supply The water carried in a tank on the apparatus.

Open relay Relay operation in which water is not contained within the entire relay system, excessive pressure is controlled by intake relief valves and pressure regulators, and dedicated discharge lines are used to allow water to exit the relay at various points in the system.

Portable dump tank A temporary reservoir used in tanker shuttle operations that provides the means to unload water from a tanker for use by a pump.

Relay operation Water supply operations where two or more pumpers are connected in-line to move water from a source to a discharge point.

Relay valves Work like a four-way hydrant valve in that water can initially flow through them and subsequently an additional pumper can connect to boost pressure without interrupting flow; the only difference from a four-way hydrant valve is that the side that would connect to a hydrant on a four-way hydrant valve instead connects to a supply line.

Required flow The estimated flow of water needed for a specific incident.

Reverse lay Supply hoseline configuration where the apparatus stops at the scene; drops attack lines, equipment, and personnel; and then advances to the hydrant laying a supply line.

Shuttle cycle time The total time it takes for a tanker in a shuttle operation to dump water and return with another load; includes the time it takes to fill the tanker, to dump the water, and the travel time between the fill and dump stations.

Shuttle flow capacity The volume of water a tanker shuttle operation can provide without running out of water.

Split lay A hose lay that includes both a forward lay and a reverse lay component.

Static source Water supply that generally requires drafting operations, such as ponds, lakes, swimming pools, and rivers.

Static source hydrants or dry hydrants Pre-piped lines that extend into a static source.

Supply layout The required supply hose configuration necessary to secure the water supply efficiently and effectively.

Supply reliability The extent to which the supply will consistently provide water.

Tanker shuttle Water supply operations in which the apparatus are equipped with large tanks to transport water from a source to the scene.

Water availability The quantity, flow, pressure, and accessibility of a water supply.

Wet barrel hydrant A hydrant operated by individual control valves; it contain water within the barrel at all times; only used where freezing is not a concern.

CHAPTER FORMULAS

7-1 Determine the flow from hydrant pitot readings.

$$Q = 29.84 \times c \times d^2 \times \sqrt{p}$$

$$Q = 0.067 \times c \times d^2 \times \sqrt{p} \qquad \text{7-1m}$$

7-2 Calculate the discharge at the specified residual pressure and/or desired pressure drop.

$$Q_R = Q_F \times \left(\frac{h_r^{0.54}}{h_f^{0.54}}\right)$$

7-3 Determine the distance between pumps.

$$\text{Distance} = (PDP - 20) \times \left(\frac{100}{FL}\right)$$

$$\text{Distance} = (PDP - 140\,\text{kPa}) \times \left(\frac{30\,\text{m}}{FL}\right) \qquad \text{7-3m}$$

REVIEW QUESTIONS

Multiple Choice

Select the most appropriate answer.

1. The estimated flow of water needed for a specific incident is called
   a. available flow. **c.** critical flow.
   b. required flow. **d.** incident flow.
2. Each of the following factors contributes to the availability of water, *except*
   a. quantity. **c.** accessibility.
   b. flow. **d.** pump capacity.
3. Which of the following water supplies is usually the easiest to secure yet often the most limited?
   a. municipal systems **c.** rivers
   b. ponds **d.** on-board tank
4. How long can a 1,000-gallon (4,000-litre) on-board water tank sustain two 150-foot, 1¾-inch (45-m, 45-mm) hoselines each flowing 125 gpm (500 Lpm)?
   a. 2 minutes **c.** 6 minutes
   b. 4 minutes **d.** 8 minutes
5. The type of hydrant usually operated by individual control valves for each outlet is called a(n)
   a. individual outlet hydrant (IOH).
   b. dry barrel hydrant.
   c. wet barrel hydrant.
   d. None of the above is correct.
6. Which of the following NFPA standards suggests that hydrants be classified and color coded based on their rated capacity?
   a. 291 **c.** 1002
   b. 1500 **d.** 1901
7. A hydrant with a red bonnet would most likely flow __________ according to NFPA's hydrant classification system.
   a. less than 500 gpm (2,000 Lpm)
   b. between 500 and 750 gpm (2,000 to 3,000 Lpm)
   c. between 500 and 1,000 gpm (2,000 to 4,000 Lpm)
   d. greater than 1,000 gpm (4,000 Lpm)

8. If the pump is located at the hydrant, which of the following has *not* occurred?
   a. reverse lay
   b. forward lay
   c. first pump in a relay
   d. boosting pressure using a four-way hydrant valve
9. If the priming system reduced the pressure to 0 psi (kPa), a perfect vacuum, water would be forced to a height of ______, also known as theoretical lift.
   a. 7.35 ft (2.2 m)
   b. 14.7 ft (4.4 m)
   c. 22.5 ft (6.8 m)
   d. 33.81 ft (10 m)
10. Because priming systems come nowhere near creating a perfect theoretical lift, the general rule of thumb is to attempt to pump no more than two-thirds of the theoretical lift, which is
   a. 7.35 feet (22 m).
   b. 14.7 feet (4.4 m).
   c. 22.5 feet (6.8 m).
   d. 33.81 feet (10 m).
11. Which of the following best describes a static hydrant?
   a. pre-piped line that extends into a water source
   b. wet or dry barrel hydrant when no water is flowing
   c. hydrant that is identified as being out of service
   d. There is no such hydrant.
12. All of the following are factors that affect static water supply reliability, *except*
   a. environmental and seasonal conditions.
   b. condition of hydrants and water mains.
   c. pump and equipment used to draft.
   d. pump operator's drafting knowledge and skill.
13. One of the major difficulties in relay operations is
   a. walking from one pumper to the next.
   b. having enough pumpers to use in the relay.
   c. ensuring that an adequate water supply is chosen.
   d. how to control changes in pressures and flows within the system.
14. A relay where water enters at the supply and progresses through the system to the attack pump is called a(n)
   a. relay system.
   b. open relay system.
   c. traditional relay.
   d. closed relay system.
15. According to NFPA 1961, supply hose operating pressure should not exceed
   a. 150 psi (1,050 kPa).
   b. 185 psi (1,295 kPa).
   c. 200 psi (1,400 kPa).
   d. 250 psi (1,750 kPa).
16. The second pumper, all in-line pumpers, and the attack pumper should maintain at least ________ intake pressure.
   a. 5 psi (35 kPa)
   b. 20 psi (140 kPa)
   c. 35 psi (245 kPa)
   d. 50 psi (350 kPa)
17. The weight of water in a 1,500-gallon (6,000-litre) tanker weighs a little over
   a. 2 tons (2 tonnes).
   b. 4 tons (4 tonnes).
   c. 6 tons (6 tonnes).
   d. 8 tons (8 tonnes).
18. A hydrant with an orange bonnet can deliver which of the following flow ranges?
   a. 200 to 500 gpm (800 to 2,000 Lpm)
   b. 500 to 999 gpm (2,000 to 3,996 Lpm)
   c. 1,000 to 1,500 gpm (800 to 2,000 Lpm)
   d. 1,500 gpm or greater (6,000 Lpm)

Short Answer

On a separate sheet of paper, answer/explain the following questions.

1. Explain the importance of preplanning water supplies.
2. When the required flow is unknown, what should govern the selection of a water source?
3. What effect do higher pump pressures have on the ability of a pump to flow capacity?
4. What effect does the size of hose have on the ability to move water?
5. What factors contribute to "water availability"?
6. What affects supply reliability for the water sources?
7. What affects the supply hose layout?
8. Why is it important to open or close dry barrel hydrants fully?

9. List four factors that may reduce the reliability of municipal water supply systems.

10. If the pressure inside a centrifugal pump is reduced to 8.7 psi (61 kPa), how high will atmospheric pressure raise water?

11. Explain the limitations of using LDH as a supply between pumps in a relay.

12. List two reasons for using a tanker shuttle operation.

ACTIVITIES

1. Identify all usable static sources within your response district.
2. Determine two potential relay operations within your response district. If hydrants are the predominate water supply, assume several hydrants are out of service.
3. Develop specific procedures and forms to conduct flow testing, using NFPA 291 as a guide.

PRACTICE PROBLEMS

1. Determine how long a 500-gallon (2,000-litre) tank and a 1,000-gallon (4,000-litre) tank can sustain two 1½-inch (38-mm) lines flowing 100 gpm (400 Lpm) each.
2. Design a relay using the following information and provide the information requested:
   - Required relay flow of 750 gpm (3,000 Lpm)
   - Distance between pump and incident is 1,500 feet (450 m)
   - Water supply is a hydrant with an orange bonnet
   - Pumpers and hose available:

   P1 = 750-gpm (3,000-Lpm) pump with 500 feet of 2½-inch (150 m of 65-mm) hose

   P2 = 750-gpm (3,000-Lpm) with 1,000 feet of 4-inch (300 m of 100-mm) hose

   P3 = 1,500-gpm (6,000-Lpm) with 600 feet of 3-inch (180 m of 77-mm) hose

   a. Where will you place each of the pumpers?

   b. What will be the distance between each pump?
3. Determine the individual and combined shuttle flows using the following information, where fill times and dump times include connecting and disconnecting hose:

T1:	Tank size	1,000 gallons (4,000 litres)
	Fill time	3 minutes
	Dump time	2 minutes
	Travel time	5 minutes round-trip
T2:	Tank size	1,500 gallons (6,000 litres)
	Fill time	3 minutes
	Dump time	4 minutes
	Travel time	5 minutes round-trip
T3:	Tank size	2,000 gallons (8,000 litres)
	Fill time	4 minutes
	Dump time	4 minutes
	Travel time	6 minutes round-trip

ADDITIONAL RESOURCES

ISO. *Fire Suppression Rating Schedule*. New York: Insurance Service Office, 1980.

(Water supply section provides information on how water supplies affect rating.)

NFPA 291, *Recommended Practice for Fire Flow Testing and Marking of Hydrants*. National Fire Protection Association, Quincy, MA, 2010.

NFPA 1961, *Standard on Fire Hose*. National Fire Protection Association, Quincy, MA, 2007.

NFPA 1962, *Standard for the Inspection, Care, and Use of Fire Hose, Couplings, and Nozzles and the Service Testing of Fire Hose*. National Fire Protection Association, Quincy, MA, 2008.

Website http://www.firehydrant.org

8 Pump Operations

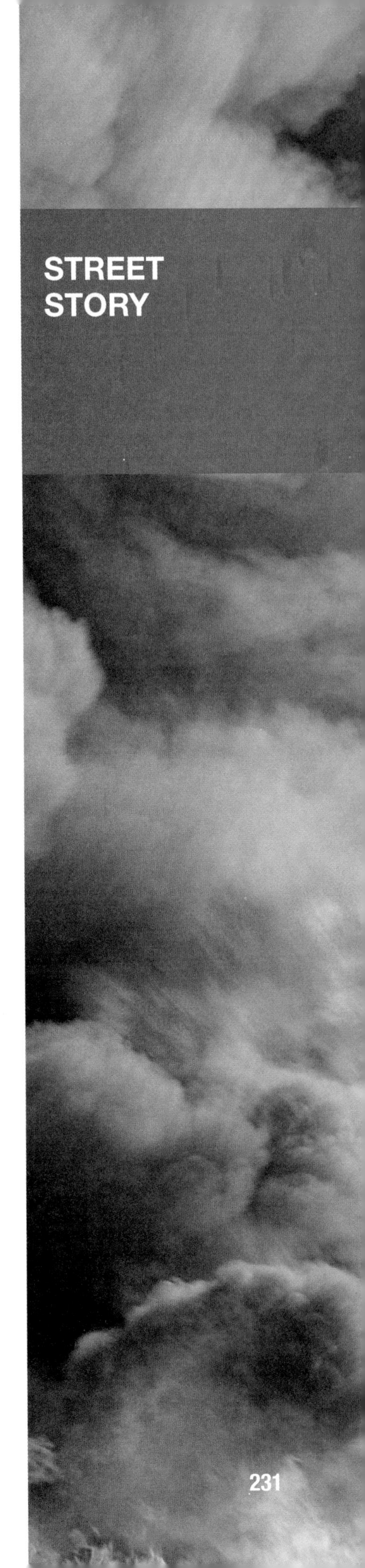

We were called to a two-story colonial-style single-family dwelling with fire through the roof at 0400 hours. While en route, we received information indicating that there had been an explosion, and that nobody had made it out of the house. Upon arrival as the first engine, we laid a 4-inch (100 mm) supply line. The pre-arrival information was confirmed, and we observed an advanced fire rapidly consuming the second floor.

We immediately began trying to begin a rescue from the only remaining survivable location possible. This included advancing a 1¾-inch (45 mm) attack line flowing 150 gpm (1,050 kPa) to the second floor; while this line was being positioned, we were engaging in a blitz attack from the exterior from a portable 500-gpm (2,000 Lpm) master stream. This had no affect on the fire. Once the attack line made it to the second floor, the crew engaged the rapidly spreading fire, again without any effect. (It was later determined that this was a gasoline-accelerated fire with six points of origin, which explains why the blitz had no effect.) Still, during this effort, the rescue company made it down the hall to search the far bedroom.

On the exterior of the house, as our tank water started to diminish, I began to pull the supply line off to make the connection to my intake to bring in a water supply. In the midst of this step I heard the operator of the second-arriving engine announce, "Here comes your water."

Our policy is that the second-due engine establishes a water supply and sends the water once the arriving-engine operator calls for the water. In this case, the premature arrival of the water created a situation in which I was wrestling with the 4-inch (100 mm) supply line. This prevented me from returning to the pump panel or communicating with my crew.

Fortunately, one of the attack crew members came out of the house to get more hose to further the advance under rapidly deteriorating conditions, and noticed what had happened. The crew member called command to report the problem, and to have them shut down the supply. Then the crew member notified the attack and rescue crews—who had just run out of water and were forced to evacuate the house. Once they exited the structure the entire second floor ignited.

The nature and complexity of the incident required everything to be perfectly executed, including proper apparatus positioning, proper attack line selection and advancement, and proper tactics. But the lack of coordination—largely due to an inexperienced water supply engine driver—substantially changed the outcome of the incident. It could have been even worse. You never want your guys inside to run out of water. The water is the only protection they have. The lesson I would like to share from this experience is to connect your supply line to the intake or clamp it as soon as possible. There is a lot going on for the first-arriving engine driver and it is necessary that you maximize every step, as this is when the fire is at its most dangerous. The result, if you don't, may be firefighter injuries or not being able to perform a rescue.

—Street Story by Mike Feaster, Apparatus Technician,
Fairfax County Fire & Rescue, Fairfax, Virginia

LEARNING OBJECTIVES

After completing this chapter, the reader should be able to:

8-1. List and explain the factors used to determine where to place an apparatus prior to beginning pumping operations.

8-2. Properly position a pumping apparatus at a hydrant so that all supply connections can be made.

8-3. Explain the safety considerations that must be addressed when placing a pumping apparatus into operation.

8-4. Discuss the steps to place the following typical fire pumps into operation: midship, front-mount, PTO-driven, auxiliary-engine driven.

8-5. List and discuss the three primary methods used to transfer power from the engine to the pump.

8-6. Demonstrate the steps in each of the following basic pump procedures: engaging the pump, priming the pump, operating the throttle control, setting the pressure regulating device, operating the transfer valve, operating the auxiliary cooling system.

8-7. Explain pump procedures to establish fire flows when utilizing: the internal tank water supply, a hydrant or pressurized water supply, and static water supply.

8-8. Explain the differences in pump procedures when utilizing the on-board water tank, hydrants, and drafting operations.

8-9. Transition from an internal to an external water source while supplying an effective fire stream.

8-10. Transfer from an internal tank to an external supply source while supplying an effective hand or master stream, so that all safety considerations are addressed, the pressure control device is set, the rated nozzle flow is achieved and maintained, and the apparatus is monitored for potential problems.

8-11. Demonstrate the ability to draft from a static water source using a fire apparatus pump.

8-12. Properly position a pumping apparatus at a static water source to enable a drafting operation.

8-13. Perform the following tests: dry vacuum; pressure regulating devices—pressure governor (if applicable), pressure relief valve (if applicable), intake relief valve (if applicable); transfer valve (if applicable).

*The driver/operator requirements, as defined by the NFPA 1002 Standard, are identified in black; additional information is identified in blue.

INTRODUCTION

After a water supply has been selected and secured, the next activity in fire pump operations is to operate the pump. This activity moves water from the intake to the discharge side of the pump. Although various pump sizes and configurations exists, the same basic steps must be taken to move water from the supply to the discharge point.

Basic Steps to Move Water from the Supply to the Discharge Point

1. Position the apparatus, set the parking brake, and let the engine return to idle.
2. Engage the pump.
3. Provide water to the pump (prime if necessary).
4. Set transfer valve (if so equipped).
5. Open discharge line(s).
6. Throttle to desired pressure.
7. Set the pressure regulating device.
8. Maintain appropriate flows and pressures.

In general, these steps are presented in the order in which they are carried out during pump operations. The order, however, may vary slightly depending on department policy, manufacturer recommendations, and type of water supply being used. Regardless of the information provided, pump operators should always follow department standard operating procedures (SOPs) and especially manufacturers' recommendations. Typically, pump manufacturers will provide operating and maintenance manuals as well as other pump training material such as PowerPoint presentations and videos. Pump operators should be intimately familiar with fire department SOPs, and department training manuals can provide excellent information on operating procedures.

This chapter presents the second task of fire pump operations: operating the pump. Basic pump procedure is discussed first, and then specific pump operations are presented for each of the five water supplies described in **Chapter 7**.

BASIC PUMP PROCEDURES

The difficulty in discussing pump procedures is that procedures can vary from department to department and from manufacturer to manufacturer and by the date of manufacture due to standards and technology changes. Terms such as "always" and "never" must be used with extreme care. Several caveats that come close to being universal among departments and manufacturers (or at least should be) for centrifugal pump operations include the following:

- Never operate the pump without water.
- Always keep water moving when operating the pump.
- Avoid cavitating the pump.
- Never open, close, or turn controls abruptly.
- Always maintain awareness of instrumentation during pumping operations. This includes understanding what acceptable readings are and what to do to avoid unacceptable readings.
- Never leave the pump unattended.
- Always maintain constant vigilance to safety.

Beyond these, there are few ironclad rules for carrying out pump procedures. Basic pump procedures, then, are those general procedures commonly utilized during most pumping operations. Understanding these procedures will increase the ease and reduce the complexity of demanding pump operations.

A few conditions exist that also lead to excessive pump wear. When unavoidable, corrective actions are required. Specifically,

- *Avoid corrosive water.* Corrosive water is water that is acidic, base, or contains salts. Sometimes an alternative water source can be used. If corrosive water must be used, the pump should be thoroughly back-flushed following its use and refilled with clean water. Anodes, which are also called zincs, should also be installed to help counteract the ill effects of electrolysis if hard water is present in your area.
- *Avoid abrasive water.* Abrasive water is water containing suspended solids such as sand and silt. Examples include the first very discolored water that comes out of a hydrant and the normally clear pond water that turns a muddy color right after a heavy rain.

In essence, the suspended particles hitting the pump, which is running at a very high speed, sandblast it. Since the critical parts of the pump have very tight tolerances and are made of bronze, a soft metal, it does not take a lot of this abuse before the metal removed increases tolerances and the pump is no longer able to pass its annual service test.

This is why hydrants should be flushed until clear water comes out before connecting to an engine. The flushing that should be part of routine hydrant maintenance helps remove sediments for water mains, thus increasing flow rates and reducing pump wear. It also means that the day after it rains would not be an ideal time to do a day-long training session on drafting at a local pond using water that looks particularly muddy.

Positioning Pumping Apparatus

The final task of the driver/operator's responsibility of getting the apparatus and crew safely to the scene is to position the apparatus correctly at the scene. Exact positioning at the scene depends on a number of items, including order of the unit's arrival, other en-route units, characteristics of the property to which you are dispatched, existing preplans for the property, and departmental SOPs/SOGs and specific orders from someone at a higher position within the incident command system.

Attack Pumper

Typical structure fire positional recommendations include leaving the best position for expected aerial apparatus since they are most limited in terms of the distances from which they can effectively work. Apparatus should not park beneath overhead utility lines, particularly power lines connected to an involved structure. The first arriving apparatus should provide as complete of an initial size-up as possible. One way of doing this is to drive past the structure so that three sides of the structure can be observed. The engine should stop on the same side of the street as the involved structure. The wheels should be turned towards the curb at a 45-degeree angle so that if the apparatus rolls, it will hit the curb and not move into another traffic lane. Whether a stopped vehicle faces or opposes normal traffic flow does not matter. This leaves the other lane open for other emergency vehicles even after attack lines have been stretched from the engine to the structure.

The driver also needs to consider tactical priorities and water supply. For example, an engine being supplied by a water shuttle will need more space than one supplied by a nearby hydrant. If a hydrant is available on the same side of the road as the pumper, its use increases the maneuvering flexibility for other apparatus.

The distance between the engine and the seat of the fire is important: too far and none of the preconnected lines will reach, resulting in delays as other lines are deployed; too close and the apparatus may be damaged by radiant heat from the fire. While considering radiant heat, consider fire extension and what the fire may grow to, not just the current fire size. Cases do exist where apparatus have been moved, but remember, needing to move an apparatus while the attack of a growing fire is under way is time consuming, manpower intensive (shutting off, moving, and recharging hoselines), and hampers operations. Conditions are even worse if multiple apparatus need repositioning because they are blocking each other.

Repositioning as the scene winds down to restore normal traffic flow is not bad. To paraphrase a slogan, getting everything perfect the first time is priceless.

Another important consideration of the surroundings is the potential for collapse. When possible, the apparatus should be placed outside the collapse zone. Building corners are a structure's strongest area; combined with the advantages of being able to view two sides, corners are often a desirable apparatus location. Preplanning can help determine the appropriate collapse zone distance. At a minimum, the collapse zone should be at least one and a half times the height of the structure. Other considerations include:

- *Power lines.* Don't park under them.
- *Escape route.* Consider developing one before one is needed.
- *Wind direction.* Try to position the vehicle upwind.
- *Terra firma.* The harder the ground, the better; and the higher the apparatus, the easier it is to pump and the less likely it is that hazardous runoff will flow under the apparatus.

Supply Pumper

When a second engine is expected to arrive shortly after the first, it often has the responsibility of establishing a water supply. This may include connecting to a hydrant and stretching a supply line from the hydrant to the attack engine. Often the second engine will stage at the hydrant until the first engine determines if a water supply is necessary and calls for the second engine. When stretching a line from a hydrant, the line should be stretched on the side of the street containing the hydrant, regardless of whether this has the engine traveling with or against normal traffic flow. If the line is stretched on the opposite side of the street, the street is effectively closed off at the hydrant due to the hose crossing the street. Stretching on the side of the street containing the hydrant allows additional emergency vehicles to approach the fire scene.

Whether the second engine stays at the hydrant position and boosts water pressure or stretches hose from the hydrant to the attack pumper is typically part of a department's operating procedures. Considerations often evaluated when setting those procedures include:

- Amount of additional time before the expected arrival of the supply pumper
- Hydrant pressure
- Expected length of supply lines, which is based on hydrant spacing and the distance between the roadway and structures

(For step-by-step photos of Positioning an Apparatus at a Hydrant Using MDH, please refer to pages 260–261.)

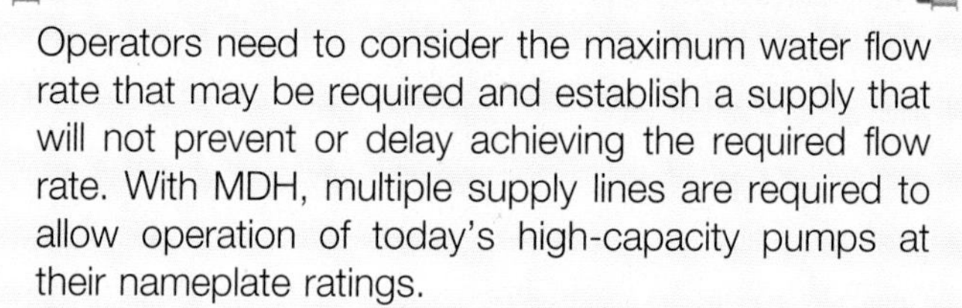

Operators need to consider the maximum water flow rate that may be required and establish a supply that will not prevent or delay achieving the required flow rate. With MDH, multiple supply lines are required to allow operation of today's high-capacity pumps at their nameplate ratings.

(For step-by-step photos of Positioning an Apparatus at a Hydrant using LDH, please refer to pages 262–263.)

In areas with good hydrant pressures, short supply lines, and quickly arriving supply pumpers, the supply pumper often connects to the hydrant and lays the supply line. In areas with weaker hydrants, longer supply lines, and/or delays in the arrival of the supply pumper, four-way hydrant valves are often used. These valves (**Figure 8-1**) allow the attack engine to connect to the hydrant and establish a water supply. Once an additional pumper arrives, it stops at the hydrant and is able to connect to the hydrant valve and increase the pressure and flow to the attack engine.

(For step-by-step photos of Using a Four-way Hydrant Valve, please refer to page 264.)

FIGURE 8-1 The four-way hydrant valve allows the attack engine to connect to the hydrant and to establish a water supply.

FIGURE 8-2 Pavement markings aid in positioning the pumper at a dry hydrant.

FIGURE 8-3 Range marks can aid in correct apparatus positioning. Ideally, both side and forward ranges exist.

SAFETY

With safety as the driver's primary responsibility, before leaving any crew at the hydrant, the driver should be able to see crew members and should have received a radio or hand signal from them indicating that it is safe to proceed. Not doing so can result in an injured firefighter, either from being hit with a hose coupling or tripped by a dragged hose as the apparatus moves away.

Dry Hydrant Pumper

When positioning at a dry hydrant for drafting, the operator should be able to position the apparatus such that once parked, repositioning is not necessary. Ideally, aids such as pavement markings (**Figure 8-2**) or range markers (**Figure 8-3**) exist to help the operator position the vehicle; however, typically it is not possible to make such markings on public roadways.

Generally a 10-foot (3-m) section of hard suction hose is used to connect between the dry hydrant and the apparatus. Shorter sections typically increase the connection time since the exact apparatus position becomes more critical and it takes longer to make connections due to the stiffness of the hard suction hose. Multiple sections of hard suction may also be necessary based on how close the apparatus can get to the hydrant. Positioning the apparatus so that a slight bend exists in the suction hose instead of the hose running directly from the apparatus to the hydrant eases connections and allows the apparatus position to be less precise.

(For step-by-step photos of Positioning an Apparatus at a Dry Hydrant Using Hard Suction Hose, please refer to pages 265–266.)

Static Source Pumper

Static sources, including oceans, rivers, lakes, streams, ponds, swimming pools, and reservoirs, all may be drafted from providing they have adequate depth and quantity and a close enough approach. The less hard suction that can be used the better, both from friction loss and setup time perspectives; however, the length of hose is far less important than the amount of lift and the ability of the ground to support the apparatus even if the ground is wet. In general, any draft site should be first tried under nonemergency conditions to establish and practice the most efficient setup procedures applicable to the particular site. For example, at boat trailer launching ramps, procedures should address having adequate redundancy in place to make sure the apparatus does not inadvertently roll down the ramp and into the water.

(For step-by-step photos of Positioning an Apparatus at a Static Water Source Using Hard Suction Hose, please refer to pages 267–268.)

Other positioning considerations exist when positioning at a water source. These considerations will be discussed later with the specific operations for each type of water source.

NFPA 1901 does not allow a manufacturer to place any discharge larger than 2½ in. (65 mm) at a place where the pump operator is stationed, and NFPA 1911 also recommends that pressurized hoselines not be connected to the pump where the operator is stationed. Both of these recommendations are intended to protect the operator from possible injury should a hoseline, particularly a large-diameter line, burst. For the same reasons, when possible, operators should connect pressurized lines to other connections and not circumvent NFPA's recommendations on LDH placement through the use of adapters.

Pump Engagement

One step required in most every pump operation is **pump engagement**. Typically this means providing power to the pump from either a separate engine or from the engine that drives the apparatus (often

FIGURE 8-4 The pump may be directly connected to a separate engine, sometimes called an auxiliary engine.

referred to as the drive engine). When powered by a separate engine, sometimes called an auxiliary engine, the pump is typically connected directly to it (**Figure 8-4**). In this case, the pump can be operated independently of the drive engine. Auxiliary engine sizes range from small gasoline engines, such as used in wildland vehicles, to large diesel engines used in airport rescue and firefighting (ARFF) apparatus. Several benefits of this system include:

- Increased versatility in placing/locating the pump on the apparatus
- Allows pump-and roll-capability—important for wildland and ARFF apparatus
- Pump pressure is not controlled by the speed of the apparatus

Disadvantages of direct drive auxiliary engines include:

- Most expensive (due to the cost of the additional engine)
- Heavier

When powered by the drive engine, the pump is typically connected indirectly to the engine. Most structural firefighting pumps receive their power in this manner. The procedure for engaging the pump will depend on the location of the pump and how power is transferred from the drive engine. The power to pump can be transferred from the drive engine by one of three general methods: power takeoff (PTO), front crankshaft, or split-shaft.

Power Takeoff (PTO)

The **PTO method** transfers power to the pump through a PTO on the transmission (**Figure 8-5**). The PTO method provides the ability to operate the pump either in a stationary position or while the vehicle is in motion, a capability known as pump-and-roll.

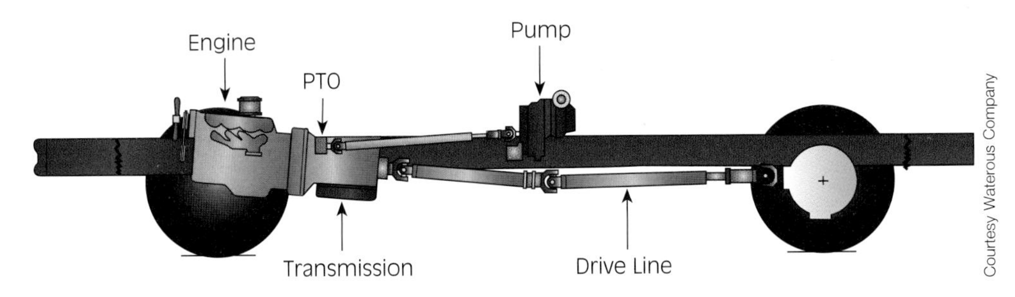

Courtesy Waterous Company

FIGURE 8-5 The PTO method transfers power to the pump through a PTO on the transmission. It allows the pump to be operated in a stationary position or while the vehicle is in motion (pump-and-roll).

Wildland, mobile water supply, and, increasingly, initial attack apparatus use the PTO method to transfer power from the drive engine to the pump. Older vehicles that use the PTO method tend to power smaller pumps up to 500 gpm (2,000 Lpm). Today, pumps up to 1,250 gpm (5,000 Lpm) can be powered using the PTO method. When the pump is engaged and the vehicle is in motion, the speed of both the pump and the apparatus is controlled by the speed of the engine. When the apparatus speeds up, therefore, the pump will speed up as well.

Advantages of a PTO-driven pump include:

- Able to handle pump-and-roll applications such as wildland fires
- Less expensive than other approaches

The primary disadvantage of a PTO-driven pump is the restriction on maximum pump size.

Basic Steps to Engage a Pump Utilizing a PTO

1. Bring the apparatus to a complete stop and let the engine return to idle speed.
2. Disengage the clutch (push in the clutch pedal).
3. Place the transmission in neutral.
4. Operate the PTO lever.
5. **a.** For pump-and-roll operations, place the transmission in the proper gear.
   b. For stationary pumping, place the transmission in neutral and be sure to set the parking brake.
6. Engage the clutch slowly.

In some cases, the pump-and-roll capability has been eliminated on PTO-driven pumps for safety reasons. For example, on a 60,000-plus-pound (27,000-plus-kilogram) tanker with a small 500-gpm (2,000-Lpm) pump designed for stationary use, eliminating pump-and-roll means a runaway truck will not occur from an in-gear engine overpowering the parking brake, nor will transmission damage occur from accidentally running in gear for an extended time without moving. It also eliminates potential liability claims against the manufacturer, the department, and the driver who might accidentally leave the transmission in gear.

To disengage the pump, simply return the engine to idle, disengage the clutch, and operate the PTO lever.

Front Crankshaft

Another way to transfer power to the pump is the **front crankshaft method**. In this method, the pump is connected directly to the crankshaft located at the front of an engine. This method is used when the pump is mounted on the front of the apparatus (**Figure 8-6**). As with the PTO method, the pump can be operated while in motion or in a stationary position. Pump capacities for older vehicles are typically limited to pumps up to 750 gpm (3,000 Lpm), while newer pumps up to 1,250 gpm (5,000 Lpm) can be powered in this manner.

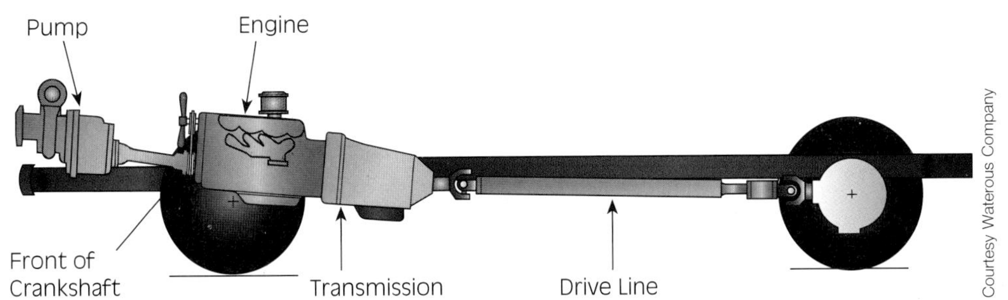

Courtesy Waterous Company

FIGURE 8-6 Front-mounted pumps receive power from the front of the engine crankshaft.

Advantages of this approach include:

- Least-expensive approach
- Better scene visibility than side-operated midship pumps

Disadvantages of this approach include:

- The restriction on maximum pump size
- More susceptible to freeze damage because it is more exposed
- More expensive repairs from minor front-end collisions

Basic Steps to Engage a Pump Connected to the Front Crankshaft

1. Bring the apparatus to a complete stop, apply the parking brake, and let the engine return to idle speed.
2. Put the transmission in neutral.
3. Operate the pump control (may be in the cab or on the front of the pump). If the pump control is located on the pump, a "pump engaged" warning light is usually provided in the cab.
4. For pump-and-roll pumping, place the transmission in the proper gear.

To disengage the pump, return the engine to idle and operate the pump control.

Split-Shaft

Finally, the **split-shaft method** can be used to transfer power to the pump. The split-shaft method is used for both midship and rear-mounted pumps. Midship pumps are mounted just behind the cab. Pump controls are either on the driver's side, called a side-mount pump, or on the top, called a top-mount pump. Top-mount pumps may have access to their controls enclosed within the cab. Such enclosures are most common in areas known for extreme weather conditions.

The split-shaft method is used for large pumps mounted in the middle or toward the back of the apparatus. It allows power from the apparatus transmission to be delivered to either the drive axle(s) or to the pump through a sliding clutch gear (**Figure 8-7**) in the pump transmission. Based on the position of the sliding clutch gear, the apparatus is referred to as either being in **road gear** or **pump gear**. The major concern with transferring power in this manner is to ensure that the drive axle(s) are properly disconnected when operating the pump. If the drive axle(s) are not properly disconnected, the apparatus may attempt to move when the pump throttle is increased. Often, indicating lights are provided to let the pump operator know when the drive axle(s) are properly disconnected and when the pump is properly engaged. (New apparatus contain a number of interlocks that prevent operating the pump while in road gear.)

Advantages of this approach include:

- Support of larger-size pumps
- Easier to add protection of the pump from extreme cold weather conditions
- When used with a top-mount pump:
  - Operator protection from traffic
  - Better operator visibility of the fire scene

Disadvantages of this approach include:

- More expensive implementation
- Long apparatus length, particularly with the top-mount configuration

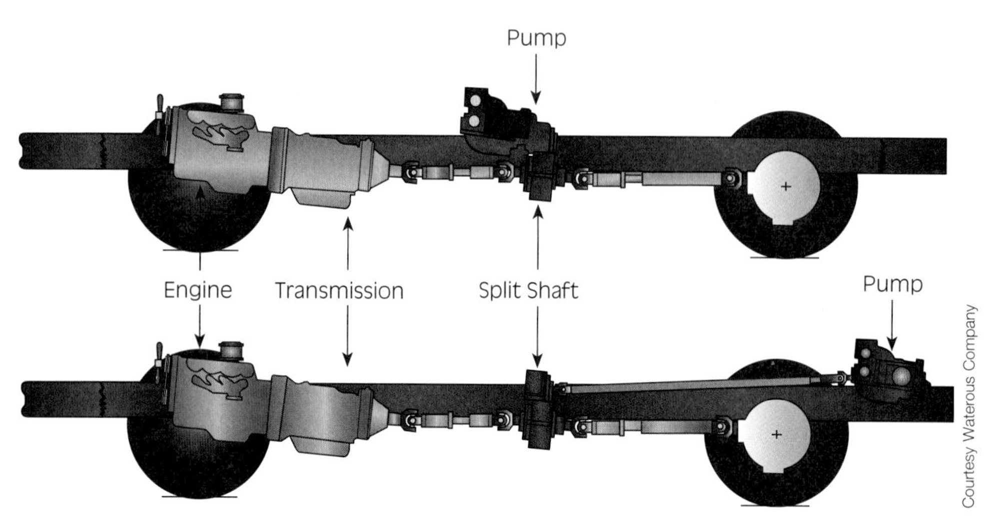

FIGURE 8-7 The most common and versatile manner to power the main pump is by using a split-shaft transmission.

Basic Steps to Engage a Pump Utilizing a Split-shaft Arrangement

1. Bring the apparatus to a complete stop.
2. Engage the parking brake. With the exception of a few older vehicles, a red parking brake light should come on.
3. Let the engine return to idle speed.
4. Put the transmission in neutral.
5. Apply the parking brake.
6. Move the pump shift control from the "ROAD" position to the "PUMP" position (see **Figure 8-8**).
   Note: The shift-indicating light should come on, often a green "Pump Engaged" light, which indicates the shift is complete (if so equipped).
7. Shift transmission into pumping gear. (This is usually "drive" or fourth on transmissions with a digital display. It is the gear where one revolution of the engine shaft results in one revolution of the drive shaft. Higher gears like fifth or overdrive result in more than one revolution of the drive shaft for each revolution of the engine and are not used for pumping.)
   Note: An indicating light should come on, often a green "OK to Pump" light (indicates the road transmission is properly disengaged). Also, the speedometer should increase (5-15 MPH or 8-25 KPH) and the truck should not attempt to move if the engine speed is increased.

To Disengage the Pump

1. Return the engine to idle.
2. Put the transmission in neutral.
   Note: You may need to wait 5 to 10 seconds to allow the drive shaft to stop turning.
3. Move the pump shift lever from "PUMP" to "ROAD" position.
   Note: The shift warning light should turn off.

(For step-by-step photos of Engaging a Pump, please refer to pages 269–272.)

FIGURE 8-8 Indicating lights on pump shift control devices let the pump operator know when the pump is operating.

Whenever the pump operator leaves the driver's seat, he or she should be sure the apparatus is chocked.

Safety Considerations

When the apparatus comes to a stop at the desired location, the parking brake should be set. Prior to exiting the apparatus, a check should be made to assure that no vehicles are moving in close proximity. As mentioned in **Chapter 3**, a good safety habit to practice is to ensure the apparatus is chocked whenever the pump operator leaves the driver's seat. This is especially important during pumping operations, in that improper pump engagement may not disengage the drive axle(s). In addition, setting of the parking brake may be accidentally overlooked or the parking brakes may fail, allowing the apparatus to move. Depending on the location, traffic cones can be used to define a safe area around the apparatus or to keep traffic from coming near the apparatus. As the operator charges hoselines, he or she should take care to stand to one side or the other of the discharges. This safety measure should be considered due to possible hose and/or coupling failure during charging.

Priming the Pump

Recall that priming is simply the act of getting air out of and water into a pump. When water is routinely carried in a pump, priming may not be necessary. When using the on-board tank or a hydrant as a water source, priming can occur as the supply pressure forces air out.

Steps for Priming the Pump Using the Supply

1. Open the water supply to the intake side of the pump.
   a. On-board source: open the tank-to-pump valve.
   b. Hydrant: open the intake after bleeding air from the supply line.

2. Open the discharge gate (to allow supply pressure to force air out).
3. Engage the pump, if not already engaged.
4. Slightly increase the engine speed.
   Note: A corresponding increase in the master discharge gauge means the pump is primed.

NOTE

Operating the pump dry for even short periods of time may quickly damage it.

If these simple steps fail to get all of the air out of the pump, the positive displacement priming pump must be used. This is usually the case if the pump is maintained without water. If the pump is being primed in conjunction with a drafting operation, the priming system is usually the fastest and most effective means to prime the pump. Some pump manufacturers suggest that priming occur before the pump is engaged to prevent damage caused by overheating. When the pump is engaged during the priming process, some experienced pump operators suggest slightly increasing engine RPMs to decrease the priming time. The engine speed, however, should not be increased over 1,200 RPMs until primed. Keep in mind that operating the pump dry even for short periods of time may quickly damage it. According to NFPA 1901, *Automotive Fire Apparatus*, a dry pump must be able to achieve a prime within 30 seconds for pumps rated at less than 1,500 gpm (6,000 Lpm), and within 45 seconds for pumps rated at 1,500 gpm (6,000 Lpm) or larger. An additional 15 seconds is also allowed when a front or rear large-diameter suction exists. Longer priming times are required when hard suction hoses longer than those specified by NFPA are needed. Typically, priming should be achieved within approximately 15 seconds.

Steps to Prime the Main Pump Using the Priming Pump

1. Secure a water source (connect supply lines).
2. Engage the pump.
   Note: Some pump manufacturers recommend priming the pump before it is engaged.
3. Close all discharge control valves.
4. Open an intake for the water supply (tank-to-pump, intake valve).
5. Engage the primer.
   Note: The primer should not be operated more than is necessary.
6. Ensure the pump is primed. If the pump is not engaged, you can tell that it's primed when the discharge from the priming pump changes from an air/oil mixture to a water/oil mixture. If the pump is engaged, when the master discharge gauge reads a positive pressure, sometimes a change in engine sound can be heard as it is placed under load when water enters the pump, indicating that it's primed.
7. Engage the pump if not already engaged.

STREETSMART

Priming from draft has a reputation for being one of the more difficult pump operations. In fact, when everything goes right, it is a simple operation accomplished in a matter of seconds. Its bad reputation comes both from the fact that many operators do not draft frequently enough to stay proficient and because a number of problems can occur. The secret to consistent priming success is to:

- Practice priming frequently.
- Understand all the things that can go wrong.
- Understand how to isolate quickly what has gone wrong and correct it.

This section will list possible problems, and explain how to narrow down the list so you can quickly focus on the correct ones. Air leaks are the most common problem that prevents priming. Any leak that occurs below water does not create a priming problem. Air may enter the pump from any of the following:

- Pump drains that are not fully closed (a common problem for departments that keep their pumps dry and leave the pump drain open)
- Leaking or open valves (solved by capping and closing drains and bleeders)
- Leaking booster tank valves, including tank fill, jet dump, and tank-to-pump (These are only a problem if the tank water level is below where the water enters the pump; hence, if priming with a full tank, these do not need to be considered.)
- Leaking suction hose fitting, cap, or gasket
- Leaking intake drain
- Leaking intake relief valve (These are often labeled "Do not cap"; however; it is acceptable to cap these if their leaking is preventing priming. In this case, after drafting, they should be uncapped and repaired.)

- Pump cooler not closed (Again this depends on where the pump cooler is connected. Many connect to the tank fill line, so if the tank is filled to the point this line is below water, it does not matter.)
- Leaking pump seal or packing
- Missing test gauge plugs and caps (This usually only occurs right after service testing if someone forgot to replace these after removing the test gauges.)

There are also a number of items that will not cause priming problems. Understanding which items can be ignored saves time. Pump components that will not affect priming include:

- Engine cooler valve (because it is a closed loop connecting the pump discharge to the pump intake)
- Discharge caps, drains, and bleeders, as long as the valve is closed and not leaking

Many other items can be easily eliminated. For example, if your pump is kept wet and you don't find water under it (which should be the norm), assume you do not have a problem with leaking packing. Some procedures can also help. First, before priming, make sure all valves are closed. Redundancy in the pump panel will help avoid some failures, for example, capped discharges with closed valves, drains, and bleeders need at least two problems before they would affect priming. As a result, these items do not typically cause a leak. To eliminate all of the apparatus-based problems quickly, after closing all valves and before removing the cap for the hard suction hose or connecting it, pull the primer for just a few seconds. The master gauge needles should be seen dropping below zero and they should stay below zero. If this does not happen, an apparatus side leak exists, and, unless that leak is from a booster tank that is leaking water into the pump, you will have problems priming. As previously mentioned, most discharges will not create a problem, but an exception and common problem is the deck gun. Since the deck gun typically has a slow-close valve to prevent water hammer, closing it is often a little harder and the valve may be cracked open. Furthermore, instead of being capped, it often has a nozzle on it allowing air to enter the pump through a leaking valve. If you believe the deck gun is leaking and cannot close the valve, the problem can be circumvented by removing the deck gun nozzle and either capping it or placing a closed ball valve on it.

Once you know that no apparatus leaks exist, you are ready to connect the hard suction hose. Each section should contain a good gasket that is neither hard nor cracked. A missing gasket will not allow a leak-free connection. Couplings should be tight but do not need pounding to the point of pushing the gasket into the waterway. Once the strainer is placed in the water and the primer is pulled, water should immediately start filling the suction hose. Water not entering the suction hose, or draining out of the suction hose after it has entered, indicates either a leak or, if the master gauges on the apparatus drop below zero and water does not enter the hose, a closed valve to the intake hose.

Open discharge lines with connected soft hoselines should not be a problem, as long as they are not leaking at the pump coupling. Air cannot be pulled through a 50-ft (15-m) section of soft hose. As the hose collapses, it prevents further air from going through it.

A common suction hose used today is made from vinyl. Among the advantages of this type of hose is the ability to see the water inside and see any possible leaks. Bubbles traveling upstream in the hose indicate a leak downstream (away from the pump). Once you get to the correct coupling, you will be able to see bubbles above the coupling but not below it. Watching the water level drop, but not seeing bubbles, indicates a leak upstream (toward the pump).

Some departments also use Storz connections on either their hard suction hose or adapters connected to it. When Storz couplings are used, the type of Storz gasket used is important. Both pressure and suction Storz gaskets are available, and suction gaskets are required for nonpressurized Storz connections (**Figure 8-9**). Another effective way to

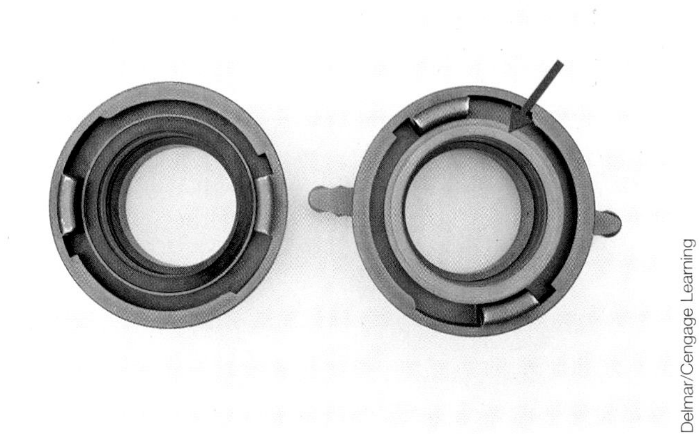

FIGURE 8-9 This photo shows both a pressure and a suction Storz gasket. The one on the left is a pressure gasket and the one on the right is a suction gasket. The difference is the added hump around the outside edge of the suction gasket (see red arrow). Suction gaskets should be used on discharge caps and at least one side of any Storz connection whose pressure could drop below 0.

prevent leaks in Storz couplings is to run a strip of vinyl electric tape completely around the seam between two couplings.

Transfer Valve

As mentioned in **Chapter 4**, two-stage centrifugal pumps typically provide the ability to operate in either a pressure or volume mode. The transfer valve is used to switch between the two modes. Recall that when pumping in the volume mode, each impeller is providing half of the flow while generating the same pressure. When pumping in the pressure mode, one impeller pumps to the next impeller, where the same flow is pumped but the pressure is doubled.

In essence, when a large volume of water is expected to be pumped, the pump should be placed in the volume mode. When higher pressures are needed, the pump should be placed in the pressure mode. The general rule of thumb is to pump in volume when expected flows will be greater than 50% of the pump's rated capacity and when pressures will be less than 200 psi (1,400 kPa); otherwise, the pump should be operated in pressure mode. For example, a 1,000-gpm (4,000-Lpm) pumper expecting to flow 750 gpm at 140 psi (3,000 Lpm at 980 kPa) should have the transfer valve in the volume position. If the same pumper expects to flow 400 gpm at 220 psi (1,600 Lpm at 1,540 kPa), the transfer valve should be in the pressure mode position.

The transfer valve should be operated before pump pressure is increased. If the transfer valve is manually operated, higher pressures will make switching modes nearly impossible. When power-assisted transfer valves are provided, switching modes at higher pressures is possible. However, switching from the volume mode to the pressure mode may significantly increase pressures. In addition, switching modes may cause a loss in prime during drafting operations. As a general rule, slow the pump speed to idle and then change the mode.

Abrupt operation of control valves can cause extensive damage and excessive wear and tear.

Control Valve Operations

Recall from **Chapter 6** that control valves start, stop, or direct water and should always be operated slowly. They are designed to be operated by either pulling or pushing (**Figure 8-10**), or by rotating up and down (**Figure 8-11**) or sideways (**Figure 8-12**).

FIGURE 8-10 This type of valve is operated by either pulling or pushing on the T-handle. When pulled, as shown in the photograph, the valve is open. When pushed, the valve is closed. When in between, the valve is neither fully open nor fully closed and is referred to as "gated" or "feathered." The valve handle may also be rotated to lock or unlock the handle. The handle should only be pushed or pulled in the unlocked position and should be locked whenever water is flowing through it.

Some pump panel designs incorporate electronic control valve operation. The benefits of these electronic control valve actuators include slow activation to prevent water hammer, resistance to valve movement after setting, and ease of operation with higher pressures. Electrically controlled valves will include at least three lights (**Figure 8-13**): a red indicating the valve is completely closed, a green indicating the valve is completely open, and one or more amber lights indicating the valve is throttled or partially opened. Control valves are provided with a locking mechanism to ensure that they do not move after being set. Control valves can typically be locked by twisting the handle. Some control valves automatically lock, as is the case with most electronic actuators. Care should be taken to unlock the handle prior to changing the valve position. Most important, though, control valves should be operated in a slow and smooth manner. Abrupt operation of control valves can cause extensive damage and excessive wear and tear.

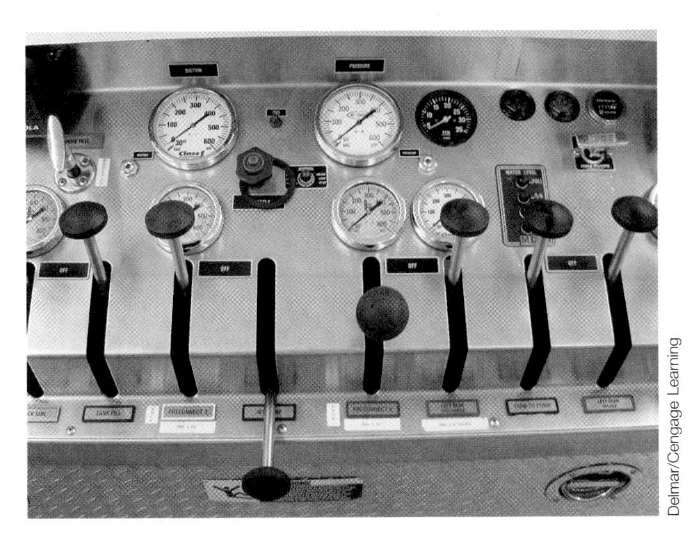

Delmar/Cengage Learning

FIGURE 8-11 This type of control valve is operated by rotating it to lock or unlock it and moving it up and down to open and close it. Prior to moving it, make sure it is unlocked and make sure to lock it once it is in position.

Delmar/Cengage Learning

FIGURE 8-12 This type of control valve is operated by moving it sideways. By design, it will stay wherever it is left, so an additional locking device is not required.

Throttle Control

The purpose of the **throttle control** is to control engine speed, which in turn controls the speed of the pump. When the throttle is increased, the engine RPMs increase and subsequently the speed of the pump impellers increases. The end result is an increase in water discharge pressure. The total discharge pressure of a pump is often referred to as engine pressure (EP), engine discharge pressure, or pump discharge pressure (PDP). The term *engine discharge pressure* describes the relationship

Delmar/Cengage Learning

FIGURE 8-13 Electrically controlled valves such as this will have at least three indicator lights: green for open, yellow for throttled, and red for closed. To prevent water hammer and still be responsive, they will not move from fully opened to fully closed or fully closed to fully opened in less than 3 seconds or more than 10 seconds.

Delmar/Cengage Learning

FIGURE 8-14 The throttle is located on the pump panel and controls engine speed, thereby controlling the speed of the pump.

between the engine RPMs and the discharge pressure. In general, an increase or decrease in the throttle produces a corresponding increase or decrease in the discharge pressure. This increase or decrease in engine pressure will be indicated on the master discharge gauge located on the pump panel.

The throttle is typically found on the pump panel (**Figure 8-14**). To increase pressure, the throttle is usually turned clockwise, and to decrease pressure it is usually turned counterclockwise. The throttle should be increased and decreased slowly. A red emergency button is often found on the throttle. Depressing the button allows the throttle to be moved in or out without turning. The intent of the button is to allow the pump operator to reduce engine pressure quickly in cases of emergency. Normal pump operations do not require the use of this button. When the button is used to increase pressure, high-pressure surges and damage to equipment may occur. When the pump is dry or when no water is flowing, the throttle should not be increased to high speed or damage may result. The most common throttle device is depicted in **Figure 8-14**; however, electronic throttle devices are increasingly being installed on newer pump panels (**Figure 8-15**).

When throttling up, pump operators should listen for increased engine RPMs. In addition, the master intake and discharge gauges should be constantly scanned (**Figure 8-16**). The master discharge gauge should show a corresponding increase as the throttle is increased. If a corresponding increase is not observed when first initiating flows, most likely the pump is not in gear or is not primed. Check the pump and initiate pump engagement procedures if it is not properly engaged. If the pump is properly engaged, it must be primed. If a corresponding increase in pressure is not observed during an operation, the pump is flowing all of the water from the supply. In this case the throttle should be slowly decreased until the pressure begins to drop. If the throttle is increased past the point of a corresponding increase in discharge pressure, the pumping operation may fail and result in:

- Pump cavitation
- Loss of prime
- Intake line collapse
- Damage to municipal water mains

When pumping from a hydrant, the master intake gauge will show a positive pressure. As the throttle is increased and more water is being pumped, the pressure reading will drop. When the pressure reading nears zero, the pump is again flowing all of the water that is available from the supply. When using hydrants as a supply source, the intake pressure

Delmar/Cengage Learning

FIGURE 8-16 When increasing the throttle, pump operators should listen to engine RPMs and scan discharge gauges.

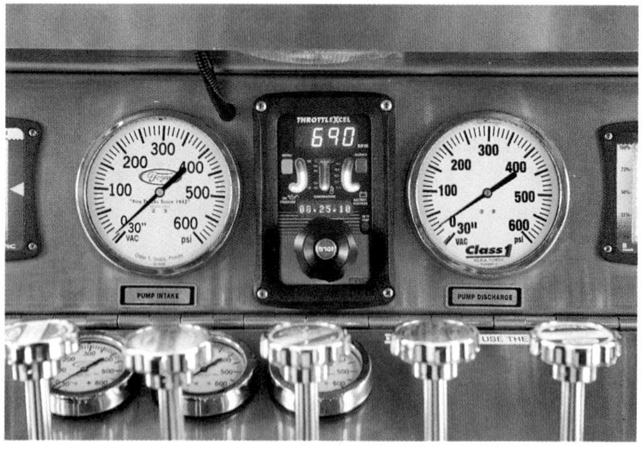

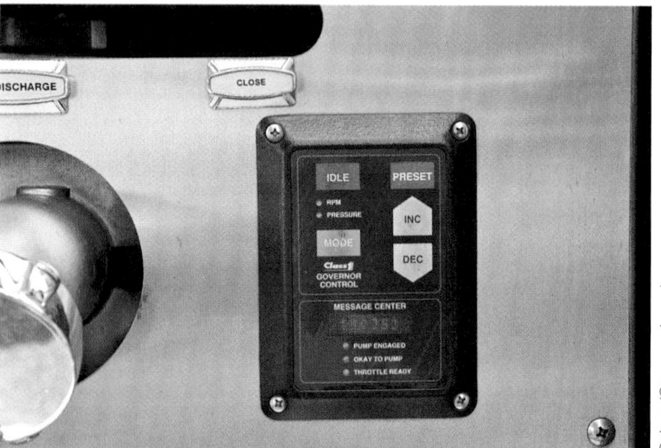

Delmar/Cengage Learning

FIGURE 8-15 Electronic throttles are common on newer pump panels.

should not be reduced below 20 psi (140 kPa), according to NFPA 291, *Recommended Practice for Fire Flow Testing and Marking of Hydrants.* Maintaining 20-psi (140-kPa) intake pressure will help guard against the pump cavitating or collapsing the intake lines. Finally, if the engine speed begins to increase without a corresponding increase in pressure, the pump may be cavitating. If this occurs, the engine speed should be reduced until the discharge pressure begins to drop. Cavitation is discussed in detail in **Chapter 9**.

When the throttle is increased and a corresponding increase in pressure is not observed, or when the master intake gauge is reduced to 20 psi (140 kPa), the pump is flowing all of the water available from the supply. Pump operators can compensate for this occurrence by increasing the supply flow. Increasing the supply flow can be accomplished by adding another supply line, increasing the size of the supply line, or increasing the discharge pressure of an upstream pumper. In addition, one or more discharge lines can be either closed or replaced by smaller lines. Finally, nozzle flows can be reduced by using smaller-flow nozzles, changing the setting on nozzles to lower flows, or simply decreasing pressure for automatic nozzles, which will adjust for reduced flows.

If the engine speed begins to increase without a corresponding increase in pressure, the pump may be cavitating.

Pressure Regulator Operation

Pressure regulating devices provide a degree of safety to both personnel and equipment. Recall from **Chapter 5** that two types of devices are used on fire department pumps: pressure governors and pressure relief valves. Pressure governors regulate pressure by controlling engine speed. Pressure relief valves regulate pressure by opening a valve to release excessive pressure from the discharge side of the pump to the intake side of the pump or to the atmosphere. The general steps for their use are provided next.

Pressure Governor

1. Turn on the pressure governor.
2. Put the governor control into the RPM mode.
3. Set the discharge pressure.
4. Put the governor control into the pressure mode.

Main Discharge Pressure Relief Valve

1. Increase the throttle to the desired operating pressure.
2. Turn on the pressure relief valve.
3. Turn the relief valve handle counterclockwise slowly until the valve opens and the light comes on. The discharge pressure will decrease and you may hear a change in engine sound.
4. Turn the relief valve handle clockwise until the light goes off and the discharge pressure returns to the desired operating pressure. This should not exceed one revolution.
5. The relief valve is now set slightly above normal operating pressure.
   Note: When the relief valve directs excessive pressure to the intake side of the pump, the ability to operate properly may be reduced when excessive intake pressure occurs.

Intake Pressure Relief Valve

NFPA 1901 now requires the use of **intake pressure relief valves** (**Figure 8-17**). The intent is to dump excessive pressure before it reaches the pump. According to NFPA 1901, intake relief valves must be adjustable. Some relief valves are, however, field adjustable during an operation while others are not. Field-adjustable intake relief valves (**Figure 8-18**) can be controlled by turning a handle (or knob) to one of several preset pressures closest to the maximum intake pressure desired. Non-field-adjustable intake relief valves must be set using an external source and gauge. These intake relief valves are set

FIGURE 8-17 Intake relief valves control excessive pressures before entering the intake side of the pump. Based on a recent change to NFPA 1901, intake pressure relief valves are now required.

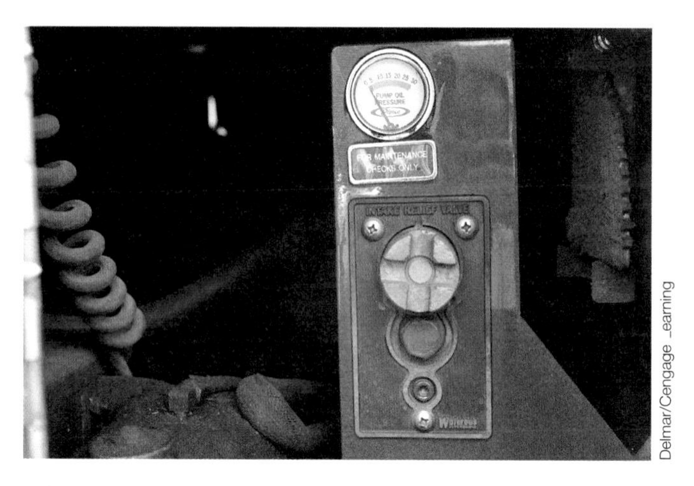

FIGURE 8-18 A field-adjustable intake relief valve.

FIGURE 8-19 The pump engine should contain a secondary cooling system. It contains a liquid-to-liquid heat exchanger and is normally located near the radiator.

in the same basic manner as the main discharge relief device.

General Steps for Setting a Non-field-adjustable Intake Relief Valve

1. Connect a line from the discharge of the assisting pump to the intake containing the relief valve.
2. Ensure that all discharge and intake valves are closed.
3. Increase the discharge pressure of the assisting pump to the desired intake relief valve setting.
4. Set the intake relief valve.
   a. If the intake relief valve is already open, turn the control handle clockwise until the valve closes.
   b. If the intake relief valve is not open, turn the control handle counterclockwise until the valve opens.

Attack crews must be notified in advance of the tank water running out.

Auxiliary Cooler

The pump's engine should contain a secondary cooling system, usually labeled "Auxiliary Cooling," "Aux Cooler," or "Engine Cooler." This system contains a liquid-to-liquid heat exchanger, or one set of pipes inside of another pipe (**Figure 8-19**). The heat exchanger is often located near the radiator. Engine coolant passes through the larger pipe, where it transfers some of its heat to the other pipe and cooler pump water. The smaller pipes are connected to hoses leading to the pump's discharge and intake manifolds. Higher-pressure water from the discharge manifold passes through the heat exchanger, picking up heat from the pipes heated by the engine coolant, and returns to the lower-pressure pump intake. A valve at the pump panel allows the operator to either allow or disallow the flow of pump water. When the engine is at or below its normal operating temperature, this valve is normally left closed. If the engine temperature starts to rise above its normal operating temperature, the valve should be opened. Essentially, the engine cooler provides additional cooling to the engine. It substitutes for the reduced air that goes through the radiator when the apparatus is stationary instead of going down the road, where its speed forces additional cooling air through the radiator.

ON-BOARD WATER PROCEDURES

Operating the pump with the on-board tank as a water supply is perhaps the easiest of all the pumping procedures. Pump operators should know in advance how long the supply will last. Care must be taken to ensure that the pump does not operate without water, or extensive damage will likely occur. In addition, attack crews must be notified in advance of tank water running out.

General Steps for Pumping the On-board Tank Water

1. Position the apparatus, set the parking brake, and let the engine return to idle.
2. Engage the pump.

3. Upon exiting the vehicle, chock the vehicle, and ensure that the pump panel "Throttle Enabled" light is on.
4. Set the transfer valve (if so equipped). In most cases, this should be in series mode based on the limited supply of water in the tank.
5. Open the tank-to-pump control valve.
6. Check to see that the master discharge gauge shows pressure, and if it does not, prime the pump (it may have air leaks). You will have to prime the pump if it was stored dry.
7. Connect the discharge lines.
8. Open the discharge control valves.
9. Increase the throttle.
10. Set the pressure regulating device.
11. Plan for more water!
    With an additional water supply secured, the following steps should be taken:
12. Ensure that the pressure regulator is properly set.
13. Open the supply intake valve.
    Note: Intake pressure will increase, causing the governor to reduce engine speed or the relief valve to open, depending on which one is used.
14. If equipped with a relief valve, reduce engine speed to the lowest RPM that will deliver the desired discharge pressure.
15. Open the pump-to-tank valve slightly to refill the on-board tank, closing it once the tank is full.

HYDRANT PROCEDURES

Hydrant operations range from simple to complex. The most time-consuming aspect is connecting the supply lines. When the apparatus is positioned next to the hydrant, large-diameter hose (LDH) is often used. Care should be taken to position the vehicle close enough to the hydrant, yet not too close. Positioning too close to the hydrant may cause kinking in the supply hose and cramped working conditions for the pump operator. When connecting non-Storz supply hose to the side intake, the vehicle should be positioned just before or after the hydrant in relation to the intake to help reduce kinks in the supply hose. Placing a couple of counterclockwise twists in the hose will help prevent kinks from forming while opening the hydrant, and the twists have the effect of causing the hose to tighten on the intake. Clockwise rotations prior to charging can cause the hose to loosen on the intake and potentially come off, causing serious injury.

No rotations should be made on hoses using Storz couplings.

When connecting to the front intake, the vehicle must be positioned a few feet (about a metre) before the hydrant to reduce kinks in the hose. Positioning the vehicle with the wheels turned at an angle allows minor distance adjustments to be made if necessary. After supply lines are connected, use the following procedure.

Basic Steps for Operating the Pump with a Hydrant as a Supply Source

1. Position the apparatus, set the parking brake, and let the engine return to idle.
2. Engage the pump.
3. Upon exiting the cab, chock the wheels and ensure that the pump panel "OK to Pump" light is on.
4. Remove the hydrant cap and the flow hydrant quickly to remove debris.
5. Connect the hose from the hydrant to the intake side of the pump.
6. Slowly open the hydrant.
7. Bleed off air in the supply line.
8. Slowly open the intake valve.
9. Set the transfer valve to volume or pressure (if so equipped).
10. Slowly open the discharge(s).
11. Throttle up to the desire pressure.
    Note: Ensure that the master intake gauge does not fall below 20 psi (140 kPa).
12. Set the pressure regulating device.

TRANSITION FROM TANK TO HYDRANT OR OTHER PRESSURIZED WATER SOURCE

Often, pump operators must transition from on-board tank water to an external supply of water such as a hydrant. The pump operator must be able to make the transition to an external water supply without significant increase or decrease in pressure. Inadvertent flow interruptions could be extremely dangerous to internal attack crews. External water supplies should be connected as soon as possible to ensure tank water is not exhausted.

Basic Steps for Transitioning from Tank Water to Hydrant Water

1. Connect the supply hose to the intake as soon as possible after initiating flow from the tank.
2. Charge the hydrant slowly.
3. Bleed off the air in the supply hose.
4. Make sure the pressure regulating device is set.
5. Open the intake valve slowly while constantly monitoring the discharge pressure. On apparatus equipped with a relief valve, the throttle will need to be adjusted to keep the correct pressure and the relief valve closed. On governor-equipped apparatus in pressure mode, the governor should automatically adjust the throttle.
6. Crack the pump-to-tank valve (sometimes labeled "Tank Fill"), or tank-to-pump valve if no check valve is installed, to refill the tank.

Some pumps may not have a check valve or may have a leaking check valve in the tank-to-pump line. For these pumps, the tank-to-pump valve needs to be closed as the intake valve is opened. Most pumps have a check valve; on these pumps, leaving the tank-to-pump valve open provides a safety backup water supply. As long as the pressurized water is available, the check valve remains closed, and hydrant water is used; should a problem occur, such as a supply line coupling failure, the check valve will automatically open and continue to supply water while the operator takes corrective actions to reestablish a water supply. In the case of a governor pressure control device, the engine will even speed up to maintain the set discharge pressure.

TRANSITION FROM PRESSURIZED WATER TO TANK WATER

Operators may also change from a pressurized source to tank water. This can occur with hydrants during cleanup phases as attempts are made to open roadways or if additional hoses need to be added to valveless hydrant outlets and with nurse feed operations as transitions are made between empty and full tankers. These transitions should also be made without significant changes in discharge pressures. Pressure governors can make these transitions almost effortless by handling the engine RPM as the operator slowly closes the intake valve while the tank-to-pump valve is open.

On relief-valve-equipped apparatus, the operator will need to monitor the discharge pressure carefully and increase the throttle to keep the pressure constant as the intake valve is closed.

Intake Water

Fundamentally, only two types of intake water exist. It is either pressurized or it needs to be drafted. Pressurized water is able to use stored potential energy, which exists in the water in the form of pressure, to push the water into the pump. Pressurized water displaces any air in the pump by pushing it out through any opening the pump operator provides. Typically, when priming the pump from a pressurized source, the opening provided is one of the pump discharges, which includes tank-fill. Examples of pressurized water include: hydrant water, water from relay pumpers, nurse feed water, water from portable pumps, and water from industrial fire pumps.

Drafted water needs to be pushed into the pump by atmospheric pressure. For atmospheric pressure to push water into the pump, the pressure inside the pump needs to be lower than the pressure outside the pump. Unlike the case with a pressurized source where opening a discharge will release any air left in the pump, allowing it to prime, opening a discharge when attempting to prime with drafted water will let air into the pump, allowing water to seek the lowest level and preventing a prime. The number of places air can enter the pump to prevent successful priming combined with the fact that it is harder to detect unintended air entering the pump through an opening than unintended water leaking out of an opening (as would occur with a pressurized water source), results in many operators feeling that drafting is too hard. In fact, when everything is done correctly and works, drafting is not difficult.

To reduce the pressure inside the pump successfully, all unintended ways air could enter the pump must be sealed. A primer is then typically used to remove air from inside the pump. An open intake with a hard suction hose allows the removed air to be replaced with water pushed through the hose by atmospheric pressure outside the pump. Examples of drafted water sources include lakes, ponds, rivers, dry hydrants, cisterns, swimming pools, and dump tanks.

Booster tank water is often considered a third category. In reality it starts as a pressurized source, albeit at a very low static pressure that is generally below 2 psi (14 kPa). Since it has a positive pressure a primer pump is not required to start a water flow from a booster tank; however, some departments that keep their pump dry do use their primer pump as a method to remove air from the pump quickly. Booster tank water also differs from other supplies because as flows are increased, resulting in friction losses, the pressure at the impeller switches from a positive pressure to a partial vacuum. The limiting factor of the tank-to-pump flow rate is in fact caused by this friction loss. Departments requiring large tank-to-pump flow rates may either increase the diameter of the tank-to-pump plumbing or provide multiple tank-to-pump connections.

DRAFTING PROCEDURES

Drafting operations tend to be complex and time consuming.

Procedure for Pumping from Draft

1. Position the apparatus (as close to the static water source as safety permits) and let the engine return to idle.
2. Set the parking brake.
3. Engage the pump.
   Note: Some manufacturers recommend that the pump be engaged after priming.
4. Connect hard suction from the intake to the static water source.
   Note: Some departments require the hard suction to be connected before final positioning of the apparatus. After the hard suction is connected, the apparatus slowly moves into position. A rope can be tied to the end of the hard suction hose to help maneuver it into the water and help keep the strainer from resting on the bottom. Another method to keep the strainer off the bottom is to use a float or ground ladder.
5. Ensure that all discharges, caps, and drains are closed.
6. Prime the pump (engage priming device).
   Note: If the pump is engaged, do not increase engine speed over 1,200 RPMs.
7. The intake master gauge should read a negative pressure.
   Note: The primer should not be operated for more than 30 seconds for pumps rated at less than 1,500 gpm (6,000 Lpm), 45 seconds for 1,500-gpm (6,000-Lpm) or larger pumps. If the intake does not read negative pressure or if water is not discharging on the ground within the these time limits, disengage the priming system and look for leaks.
8. Verify that the pump is primed.
   a. If the pump is engaged, the discharge pressure gauge will increase.
   b. If the pump is not engaged, you may hear or see water discharging on the ground.
9. Engage the pump if not already engaged.
10. Set the transfer control valve.
11. Gradually open the discharge outlet.
12. Slowly increase pump speed.
    Note: Monitor hard suction to ensure that drafting does not cause a whirlpool and that debris does not block the screen.
13. Set pressure regulating device.

RELAY OPERATIONS

Relay operations are required when more than one pumper is needed to move water from the supply to the incident. Once the relay is designed and the lines have been laid, flow can be initiated. Attempts should be made to keep apparatus and hoses on one side of the road so the other side of the road may be used by other apparatus if needed. When initiating flow, during the operation, and when shutting down the operation, communications between each of the pump operators within the relay is vital. Pressure can quickly rise within the system, requiring constant evaluation of gauges and communications between the pump operators.

Basic Steps to Initiate Flow in a Relay Operation

1. The supply pump secures the water source (can be either hydrant or static source).
2. Each pump in the relay, except the supply pump, opens one discharge outlet and sets an intake relief valve to 50 psi (350 kPa) (if adjustable).
3. The supply pump initiates flow by opening a discharge.
4. The supply pump increases the discharge pressure to the desired setting (keep in mind that the next pump in the relay should have a minimum of 20 psi (140 kPa) intake pressure).
5. When water reaches the second pump, close the discharge gate.
6. Engage the pump if not already engaged.

7. The second pump increases its discharge pressure to the desired setting.
   Note: The second pump must ensure that the intake pressure does not drop below 20 psi (140 kPa).
8. Set the pressure regulating device.
9. Repeat steps 5 and 8 for each pump in the relay.

The system should now be flowing water. The use of pressure regulating devices will help reduce constant changes in relay pressures; however, all pump operators must continuously monitor instrumentation.

DUAL AND TANDEM PUMPING OPERATIONS

Two special pumping configurations are dual and tandem pumping. **Dual pumping** operations are similar to the volume mode in a multistage pump, whereas **tandem pumping** is similar to the pressure mode or a type of relay with just a source and attack pumper. While many firefighters confuse these terms, a thorough understanding of the principles and knowledge of when to use each concept will make a difference in the operator's success in a number of demanding situations.

Tandem pumping is commonly used and may be the department's standard procedure, particularly in areas with low hydrant pressures, although it may be called something else, such as relay pumping or supply and attack engines. (In areas with a high residual hydrant pressure, the water department is essentially performing the job of the first engine.) It also may be performed using a four-way hydrant valve. If not used when required, the attack engine will not be able to:

- Maintain the necessary 20+ psi (140+ kPa) intake pressure since friction loss in the intake (supply) lines will cause the hydrant's residual pressure to drop even more before arriving at the attack engine's intake;
- Maintain the necessary discharge pressure since with a lower intake pressure resulting from not having the first pumper, the attack pumper will need to develop a higher NPDP, requiring more work from its pump to achieve the same required PDP; and
- Maintain the necessary discharge volume (gpm or Lpm) since the volume a given pumper can achieve is affected by its intake pressure, with a lower intake pressure resulting in a lower possible volume.

Furthermore, because these effects become more pronounced as more water is discharged; it may initially appear that operations are successful, only to find out about the problems as additional fire streams are put in service.

On the other hand, few firefighters have ever used dual pumping on an actual fire. A number of reasons have led to this. Today's pumps are much larger than those of a generation ago, with a typical current pump rated at 1,500 gpm (6,000 Lpm) and capable of a larger flow when connected to a strong hydrant. So a truly exceptional hydrant is needed to support two such engines at the same time, and the hydrant will start experiencing other restrictions such as flow restrictions created by the steamer connection itself. Typically, on such a large fire, other pumpers will be operating from nearby hydrants, further reducing the available water at a normally strong hydrant. Dual pumping does drive home important characteristics about pumper design and may help with large fires needing high water flows in an area with strong hydrants but a limited number of them. If implementing a dual pumping environment, be sure to understand the items described in the following safety notes to prevent creating other problems.

Dual pumping is when one hydrant supplies two pumps. This process is used when the hydrant is strong in terms of pressure and volume and when large quantities of water are required. In the basic setup, one pumper connects to a hydrant and then connects another section of hose from its unused intake to the intake of the second pumper (**Figure 8-20**). In this process the first pumper is not pumping water to the second pumper. Rather, the excess water not used by the first pumper is diverted to the second pumper. Note the similarity to the volume mode: both pumps receive water and then discharge the water independently of each other. Add the total discharge flow from each pumper to determine the total flow from the hydrant.

FIGURE 8-20 Example of a dual pumping operation. Excess water from the first (leftmost) engine is transferred to second (rightmost) engine.

DUAL PUMPING SAFETY CONSIDERATIONS

A number of steps were followed to ensure firefighter safety in the configuration shown in Figure 8-20. Many pumps do not have valves on their steamer connection. Traditional instructions told the user in this case to partially close the hydrant, allowing the residual pressure gauge on the first pumper to drop to zero while it continued to pump water. It was also traditionally done with a type of hard hose that was rated for both suction and pressure. Today, the most common type of hard hose is not rated for pressurized applications and will fail if used under pressure. Furthermore, most authorities currently recommend against connecting any hard hose, even pressure-rated hose, directly to a hydrant. If the pressure on a soft hose is reduced to zero, instead of the normally recommended minimum of 20 psi (140 kPa), the risk of hose collapse affecting the water supply to firefighters on the first pumper is high.

Once a near-zero residual pressure is achieved, the valveless cap on the other side of the first pumper can be removed without water coming out since all water that enters the first pumper would go through the pump and out a discharge. The instructions also acknowledge that a slight residual pressure, say 5 psi (35 kPa), might exist. While each of these statements is true, they have several potential problems. They assume that one can get to zero and maintain residual pressure. Due to items such as gauge error (remember we would not be in the center ⅓ where gauges are most accurate) and the ability to read the gauge exactly, one might not get to exactly 0 psi (kPa). Even at just 5 psi (35 kPa), the pressure on a typical 6-in. (150-mm) cap, which is affected by its surface area, is about 180 pounds (80 kg), a significant force for a firefighter to handle, particularly with a task like threading a LDH coupling with a hose having the extra weight of water in it. Even worse would be a slight negative pressure, causing air to enter the pump through the uncapped discharge and affecting the firefighters using water supplied by the first pump. Additional problems are created by changing hydrant pressure. Particularly as other pumpers start operating in the area, hydrant pressure changes may be much greater than assumed for successful evolution with a valveless and uncapped intake. And during the entire connection operation, pressurized water coming out of the drain on a dry barrel hydrant is undermining its support. All of the preceding problems are resolved if operations permit making all connections prior to opening the hydrant and starting water flow or if the first pumper at least has a valve on the steamer connection used to connect the first and second pumpers. In the case of the engines shown in **Figure 8-20**, all three steamer connections used had valves.

In following the spirit of a NFPA 1901 manufacturer's requirement not to place LDH discharges where the operator will stand, thus protecting the operator from injury from a possible hose break, the first pumper used in the scenario in **Figure 8-20** was a top-mount pump so that the operator is standing on the catwalk and not near the visible LDH hoselines. On the second pumper with a side-mount pump, the hose was connected on the officer side with the operator on the opposite side of the apparatus safely protected.

Basic Procedures for Setting Up a Dual Pumping Operation

1. Connect supply hose from the hydrant to the first pumper (this first pumper can begin pumping).
2. Position the second pump close to the first pumper and connect the supply hose from the unused intake of the first pumper to the intake of the second pumper.
   Note: If the unused intake on the first pumper has a gated valve, the valve should be opened after making the hose connections.
3. The second pumper can then begin pumping water as described earlier in this chapter.

Tandem pumping is similar to dual pumping in that one hydrant typically supplies two pumps. The difference is that in tandem pumping the first pumper pumps all its water to the second pumper as in a relay operation (see **Figure 8-21**).

Delmar/Cengage Learning

FIGURE 8-21 Example of a tandem pumping operation. The discharge pressure includes pressure added by both pumps.

Note the similarity with operating in the pressure mode of a multistage pump. The pressure provided by the first pumper is added to the pressure generated by the second pump. Tandem pumping is used when higher pressures are required than can be provided by a single pump. Because of this, caution must be exercised to ensure hoseline pressures do not exceed their designed operating pressures. Higher pressures are sometimes required when supplying high-rise sprinkler systems, standpipes, or long hose lays.

Basic Procedures for Setting Up a Tandem Pump Operation

1. Connect supply hose from the hydrant to the first pumper.
2. Connect supply hose from the discharge gate(s) of the first pumper to the intake of the second pumper.
3. Initiate flow from the first pumper and bleed air from the supply hose at the second pumper.
4. The second pumper can then begin pumping water as described earlier in this chapter.

Tandem pumping, like relay pumping, runs the risk of exceeding the pressures that hose, appliances, and even the pumps themselves have been tested to withstand. The pump operator needs to make sure these pressures are not exceeded to prevent injuries, equipment damage, and other safety problems. In general, medium-diameter hose and pump pressures should not exceed 250 psi (1,750 kPa) and large-diameter hose and appliances should not exceed 185 psi (1,295 kPa).

PUMP TEST PROCEDURES

Pump tests are conducted to determine the performance of the pump and related components. In general there are five types of tests performed on pumps: manufacturer's tests, certification tests, delivery or acceptance tests, annual service tests or performance tests, and weekly/monthly tests. The manufacturer's, certification, and acceptance tests are typically conducted on new apparatus. The basic intent of these tests is to verify apparatus and pump performance prior to acceptance by a fire department. NFPA 1901, *Standard for Automotive Fire Apparatus,* specifies the requirement for each of the tests. The annual service test and weekly/monthly tests focus on all in-service pumps. The basic intent of these tests is to ensure the pump continues to perform at an acceptable level. NFPA 1911, *Standard for the Inspection, Maintenance, Testing, and Retirement of In-Service Fire Apparatus*, contains the requirements for annual service tests, while manufacturer's recommendations typically determine monthly/weekly tests.

Manufacturer, Certification, and Acceptance Tests

The requirements of these tests are clearly delineated in NFPA 1901. Manufacturer's tests are typically conducted by the manufacturer at the manufacturer's facilities and, except for the addition of road tests and hydrostatic tests, are almost identical to certifications tests. Manufacturer's tests include:

- Vehicle stability, demonstrated either by a tilt table, where the entire apparatus is placed on a table and tilted 26.5 degrees to both sides; a calculated or measured center of gravity that is not higher then 80% of the rear axle width; or by equipping the apparatus with an enhanced roll stability system.
- A braking test, where the loaded apparatus is able to make a complete stop within 35 ft (10.5 m) on level, dry pavement.
- A parking brake test, where the parking brake will hold the apparatus on a 20% grade.
- An acceleration test, where the apparatus must be able to achieve 35 MPH (55 KPH) within 25 seconds on flat pavement. The apparatus must also be able to reach at least 50 MPH (80 KPH) but not exceed 68 MPH (110 KPH). Apparatus with a gross vehicle weight rating (GVWR) over 50,000 pounds (22,680 kg) and a tank capacity over 1,250 gallons (5,000 Litres) cannot exceed 60 MPH (100 KPH), according to the new NPFA 1901 standard.
- Booster tank capacity.
- An electrical reserve capacity test that shows the engine will start after the batteries have supplied power to the electrical loads of the apparatus for 10 minutes without a charging system.
- An alternator performance test showing both its idle and full-load performance.
- A low-voltage alarm test.
- Tests of foam proportioning, CAFS, generators, and breathing air systems for those apparatus equipped with these systems.
- A 3-minute 250-psi (1,000-kPa) hydrostatic test of the fire pump pumping.

Some certification tests must be conducted by an independent testing organization, usually Underwriters Laboratory (UL). Independent certification tests includes a pump test that is very much like the annual service test described below except for much longer durations. The certification test for pumps rated between 750 gpm (3,000 Lpm) and 3,000 gpm (12,000 Lpm) runs for 2 hours at 100% of the pump's rated capacity with a 150-psi (1,050-kPa) NPDP, followed by a half hour at 70% of the pump's rated capacity with a 200-psi (1,400-kPa) NPDP, followed by a half hour at 50% of the pump's rated capacity with a 250-psi (1,750-kPa) NPDP and is also called the 3-hour test. A 10-minute test at 100% of the pump's rated capacity and a 165-psi (1,155-kPa) NPDP is also run between the 100% and the 70% test.

The acceptance test is normally conducted upon delivery of a new apparatus to a fire department to verify and document stated performance levels of the apparatus, pump, and related components. In addition, the test provides a benchmark for comparison of future pump tests. Finally, the test allows pump operators to become familiar with the new apparatus. In essence, the acceptance test is simply a repeat of some of the manufacturer and certification tests. An apparatus that cannot pass its acceptance test should be rejected by the purchaser.

Annual Service or Performance Test

SAFETY

As with any activity carried out by pump operators, safety should be considered when conducting tests. One way to help ensure safety is not to rush through the tests. Hurrying to finish can increase the risk of an accident, increase the chance of inaccurate results, and increase the chance that a safety problem is overlooked. Another important safety consideration is to ensure that the work area is free from hazards. This can be accomplished by walking around the apparatus, looking under and above for slippery surfaces and loose equipment. Finally, increased safety can be achieved by always keeping safety in mind. Several common test safety considerations include:

- Wear protective gloves, helmet, and hearing protection.
- Slowly open and close all control valves to prevent dangerous water surges.
- Allow only essential personnel within the testing area.
- Do not straddle hoselines that are under pressure.
- Ensure hoselines and nozzles are properly secured.
- When possible, make hoseline connections to discharge away from the pump control.

According to NFPA 1911, Chapter 18, all pumps rated at 250 gpm (1,000 Lpm) or higher must be tested annually and when any major repair or modification has taken place to ensure the pump maintains appropriate performance levels. Several different methods are commonly used to accomplish this by different departments. **Table 8-1** highlights advantages of each of the common methods.

CAUTION

While service testing ensures that the engine is still able to perform at the appropriate level, it is also the most stressful operation most engines go through all year. Extreme vigilance in monitoring apparatus gauges is required during the test to make sure that the apparatus is not run outside of acceptable operating ranges, which would result in excessive wear and/or damage requiring repairs.

Site Requirements

NFPA 1911 Section 18.3 suggests that the test be conducted at draft to help determine the true performance of the pump. When conducting the test at draft, the water supply should be clear, at least 4 feet (1.2 m) deep, not more than 10 feet (3 m) below the pump, and the suction hose must be at least 2 feet (0.6 m) below the surface of the water. Because environmental conditions affect pumping ability, NFPA 1911 Section 18.4 specifies acceptable ranges for service testing. These include:

- Air temperature must be between 0°F and 110°F (–18°C and 43°C).
- Water temperature must be between 35°F and 90°F (2°C and 32°C). (Water at higher temperatures is likely to cause cavitation during the pump test, causing a pump that should pass to fail.)
- Barometric pressure must be a minimum of 29 in. Hg (737 mm) corrected to sea level. (Lower pressures reduce both the pump's capacity and the engine's horsepower.)

Environmental data should be recorded both before and after the test and should not go out of acceptable ranges any time during the test.

Discharge hose must be provided that will flow the rated capacity through the nozzle(s) or other

TABLE 8-1 Service Testing Methods

Service Testing Method	Method Advantages
Service test is performed by the engine's assigned crew	■ Adds to the crew's proficiency. Since the service test operates the engine at its capacity, operational errors that could go uncorrected at a typical scene where the pump is operated at a fraction of its capability must be corrected to be successful. ■ Makes sure the crew members are aware of the pump's true capability, which is helpful when they assess whether they have adequate capability to handle a particular situation.
Service test is performed by a qualified mechanic with assistance from the engine's assigned crew	■ Should something need repair, the mechanic sees the problem first-hand and can make the necessary repair. ■ If a failure is reported that the mechanic did not witness, he or she needs to both re-create the problem and perform a service test before returning the apparatus for service to confirm that the repair was successful. That could be less efficient and more time consuming than taking care of everything at once. ■ The crew's involvement provides its members with the same advantages as if they ran the test themselves. ■ Since the mechanic is typically involved in more service tests per year than a typical apparatus crew, the mechanic is often more familiar with some of the test procedures, which helps ensure everything is done properly.
Service test is performed by a qualified fire mechanic	■ Often helps with maintaining district coverage. Service tests are often done at a single designated test site within an area. Having the engine's crew at this area means backup coverage needs to be arranged for the crew's normal response district.

flow measuring device without exceeding 35 ft/sec (10.5 m/sec) flow velocity. This requires one 2½-in. (65-mm) hoseline for each 500 gpm (2,000 Lpm) of flow. The flow measuring device can be either a pitot tube or flow meter with an accuracy of ±5%. When a pitot tube is used, it should be a fixed UL-certified pitot (see **Figure 8-22**), not the handheld type used for hydrant testing since that type does not provide the necessary accuracy. When a pitot tube is used, smooth-bore nozzles connected to a monitor, or otherwise properly secured, must be used, and the specific nozzle size and hose configuration must be calculated or looked up in a chart to ensure the specific capacity ratings are achieved during the test. In most cases, this may mean shutting down the pump to change the discharge nozzle and hose configuration. When flow meters are used, the device directly measures flow and no calculations are required. In most cases this means the test is performed more quickly in that the same discharge configuration is used for each capacity test and shutting down the pump is not necessary. All test gauges must be calibrated within 60 days of the annual pump test.

The following tests are included in the annual service test with the order of some of the tests specified by NFPA. In general, the order is such that if one test's success depends on another item working correctly, that item is tested first. For example, the vacuum test is done before the priming test. Problems that would not allow the vacuum test to succeed

FIGURE 8-22 If using a pitot tube, it should be a fixed, UL-certified rather than a handheld one. The fixed tube provides much greater accuracy than the handheld devices.

could also cause the priming test to fail. By doing tests in the specified order, if a repair is necessary, fewer items that could have caused the problem need to be looked at. After the priming test, the pumping test is performed.

The pumping test order must be 100% test (also call the capacity test), overload test (if needed), 70% test, and, finally, the 50% test. Furthermore, the engine must not be stopped or throttled down except as necessary to change tips during the series of tests. One reason for this requirement is to make sure the engine is able to handle the workload without overheating.

Engine Speed Check

The no-load governed engine speed shall be checked and compared to the results taken when the apparatus was new. Any variance not within ± 50 RPM must be investigated and corrected prior to starting any test.

Prior to performing the engine speed test, the engine should be run long enough to reach its normal operating temperature. With the transmission in neutral and the parking brake set, the operator slowly presses the accelerator while watching a tachometer. If for some reason the engine should exceed the manufacturer's specified no-load RPM plus 50, the test should be discontinued. At this point, the test has failed and a mechanic needs to fix the problem that is allowing the engine RPMs to go this high. Continuing to press the throttle to see just how high the RPM might go can lead to engine failure and very expensive repairs. If the engine is more than 50 RPM below the manufacturer's specified RPM, the test has failed. Additional testing should not be attempted until this problem is fixed since the engine may not be developing the required power to complete the additional tests.

Pump Engine Control Interlock Test

This test confirms that certain interlocks that inhibit the pump panel throttle are working for apparatus equipped with the interlocks. The required interlocks were introduced in the 1991 edition of NPFA 1901 for electric and electronic engine controls. Safe completion of this test requires one person in the apparatus to depress the brake pedal while another person confirms that the pump panel throttle is disabled.

- With the pump shift selector in road gear, the transmission in neutral, and the parking brake released (be sure to press the brake pedal and have wheel chocks in use so the apparatus does not move), advance the pump panel throttle. The apparatus passes this test if the engine RPMs do *not* increase.
- With the pump shift selector in road gear, the transmission in pump gear (generally fourth), and the parking brake applied, advance the pump panel throttle. The apparatus passes this test if the engine RPMs do *not* increase.

Pump Shift Indicator Test

This test confirms that the "Pump Engaged," "OK to Pump," and "Throttle Enabled" indicator lights work correctly. Operation of these indicators can be observed as the pump is engaged for other tests.

Vacuum Test

This test ensures the interior of the pump can maintain a vacuum. The test begins by operating the primer, in accordance with manufacturer's recommendations, to obtain a minimum vacuum of 22 in. Hg (75 kPa). The vacuum must be maintained for 5 minutes with no more than a 10-in. Hg (34 kPa) drop in vacuum. After the test begins, the primer cannot be operated.

SAFETY

- This vacuum test is performed with the pump transmission in neutral. Remember that engaging the pump without water flowing through it to cool and lubricate it should be avoided due to the overheating and excessive wear it creates.
- If the apparatus has a way of obtaining a high idle, and most do, increasing the idle to 1,200 RPM helps with this test. For electric-driven primer motors, 1,200 RPM increases alternator output, which helps priming and reduces the chance of burning out either the primer motor or the alternator. For air-driven primers, 1,200 RPM increases air compressor output, which helps the air primer.

Basic Steps for Conducting the Vacuum or Dry Vacuum Test

1. Position the apparatus on a level surface, leave the vehicle running, and chock the wheels.
2. Drain all water from the pump.
3. Inspect and connect 20 feet (6 metres) of suction hose to the intake and place a cap on the open end.
4. Ensure that the intake valve is open, all connections are tight, all discharge control valves are closed, and discharge control valve outlet caps are removed.
5. Ensure that the priming pump oil reservoir is full.
6. Connect the vacuum test gauge to the test gauge connection.
7. Engage the priming device until the test gauge reads 22 in. Hg. (558 mm). Note the time and

compare the test gauge reading with the master intake gauge.

8. Turn off the engine and listen for air leaks.
9. After 5 minutes, note the test gauge reading.

Priming Device Test

This test is conducted to ensure the priming device is able to develop a sufficient vacuum to draft. This is a timed test and consists of priming the pump and discharging water. The timed test begins when the primer is started and ends when water is discharging. The time must not exceed 30 seconds for pumps rated at 1,250 gpm (5,000 Lpm) or less and not exceed 45 seconds for pumps rated at 1,500 gpm (6,000 Lpm) or greater. For pump systems with 4-inch (100-mm) or larger intake pipes, an additional 15 seconds is allowed.

Steps for Conducting the Priming Test

1. Set up the intake hose for a drafting operation and at least one discharge line.
2. Open the intake valve, close all discharge valves, and remove discharge caps.
3. Engage the priming device as previously discussed in this chapter and note the time.
4. Open a discharge line to increase the pressure and flow.
5. Note the time when the discharge pressure/flow is achieved.

Gauge and Flow Meter Test

This test is conducted to ensure all pressure gauges and flow meters provide an appropriate level of accuracy. Any pressure gauge reading off by more than 10 psi (70 kPa) and flow meter reading off by more than 10% should be recalibrated, repaired, or replaced.

Steps for Testing Pressure Gauges

1. Cap all discharge outlets and open the discharge control valves.
2. Increase the pressure until the calibrated test gauge reads 150 psi (1,050 kPa).
3. Verify that each discharge pressure reading is within ±10 psi (70 kPa).
4. Repeat step 3 for a pressure of 200 psi (1,400 kPa) and again at 250 psi (1,750 kPa).

Steps for Testing Flow Meters

1. Connect a hose and smooth-bore nozzle to each discharge outlet; it is not necessary to test all flow meters at the same time.
2. Establish flow at each discharge and verify via pitot tube reading at the nozzle. NFPA 1911 suggests specific test flows for various pipe sizes. Several examples include 1-inch (19-mm) pipes at 128 gpm (512 Lpm), 2½-inch (65-mm) pipes at 300 gpm (1,200 lpm), and 3-inch (77-mm) pipes at 700 gpm (2,800 Lpm).

Pumping Test

The annual pumping test helps ensure the pump is capable of flowing its rated capacity. The overall operating condition of the pump is established and can be compared to previous years' tests to determine performance trends. The annual pump test is a 40-minute test consisting of the following:

- 20 minutes: 100% rated capacity at 150 psi (1,050 kPa)
- 10 minutes: 70% rated capacity at 200 psi (1,400 kPa)
- 10 minutes: 50% rated capacity at 250 psi (1,750 kPa)

Pumps with a rated capacity of 750 gpm (3,000 Lpm) must also undergo an overload test or excess power test consisting of flowing the rated capacity for 5 minutes at 165 psi (1,155 kPa). This test is conducted immediately following the 20-minute 100% rated capacity test. For two-stage pumps, the transfer valve shall be in volume mode for the 100% rated capacity test, either volume or pressure for the 70% rated capacity test, and in the pressure mode for the 50% rated capacity test. During the 20-minute test, a complete set of readings shall be taken and recorded a minimum of five times (or every 5 minutes, including one at the beginning and one at the end). During the 10-minute tests, readings shall be taken and recorded a minimum of three times. Any 5% or greater variance in pressure or flow readings shall be determined and corrected and the test continued or repeated. The only time engine pressure should be reduced is to change hose and nozzle configuration between capacity tests. During the pump test, the operator should scan all instrumentation to ensure the engine and pump are within normal operating ranges.

The 150-, 165-, 200-, and 250-psi pressures quoted above are all NPDPs. Since most service tests are conducted at a draft, or with a negative intake pressure, the actual reading on the discharge gauge

PUMPING TEST TIMINGS AND TIPS

While the pumping tests are called a 20-minute (100%), two 10-minute (70% and 50%), and 5-minute (overload) test, it in fact takes more than 45 minutes to run these tests. The actual timed testing time does not begin until all measured values are correct. For a new pump operator, it often takes a while to make all of the necessary adjustments so that all gauges read correctly; this lengthens the time it takes to complete a pump test. Some of this additional time is because of differences between adjustments for service testing and adjustments for firefighting. When firefighting, the operator only needs to adjust the throttle and valve positions to obtain the desired discharge pressure; when service testing, the operator needs to obtain both the desired net pump discharge pressure and the correct flow rate. To do this, the operator needs to look at additional gauge(s)–either flow meter(s) or pitot gauges. Fortunately several techniques are available to simplify and speed the process of getting the correct pump settings. A recommended procedure is as follows:

1. Determine the number of hoselines that will be used (based on one 2½-in. [65-mm] line for each 500 gpm [2,000 Lpm]). Open the discharge valves for these lines halfway.
2. Increase the RPM to the RPM used in the prior year's service test or the manufacturer's test if the prior year's data are unavailable.
3. Adjust individual discharge valves according to **Table 8-2** to get the correct pressure. When adjusting valves, alternate between different valves, keeping each valve gated back at about the same percentage.

is less than the NPDP due to the negative intake pressure that is caused by friction losses in the intake hose and lift. When lift remains constant for a given test (and sometimes it changes due to water spillage), intake pressure still drops by about 1 pound (.45 kg) for the 70% test and 2 pounds (.9 kg) for the 50% test due to the lower flow rates.

At the completion of the pumping test, the engine and transmission should continue to be operated at an idle for 5 to 10 minutes. This gives them a chance to cool down while lubricating oil and engine coolant continue to circulate, removing heat from hot spots.

When pumping, the operator should allow lock valves in their desired position. Doing so prevents the valves from moving to an incorrect position. During the pumping test this is particularly important since many valves are likely to be gated back and have high flows going through them, exactly the conditions under which the valve positions are most likely to move. If these do move, they most often slam shut, causing a significant water hammer that will actually rock the apparatus and can cause water hammer damage to multiple components. (At a fire, it also leaves the nozzle crew with no protection from the fire they are fighting.)

The pump test can begin immediately following the priming device test previously discussed.

Basic Steps for Pumping Test

1. Verify hose and nozzle configure that will enable 100% rated capacity flow (hose and nozzle configures can be calculated as discussed in **Section IV** or looked up in charts and tables).
2. Open the intake valve and gradually increase pump speed while slowly opening discharge valves. Continue until the discharge valve is fully open and 150-psi (1,050-kPa) pump discharge pressure is obtained. (For two-stage pumps the transfer valve should be in volume mode.)
3. Verify that 100% capacity is flowing by either using a pitot tube gauge to verify pre-established pressure is obtained or a flow meter to verify discharge rate. Adjust engine pressure speed and discharge control valves to obtain 100% rated capacity flow at 150-psi (1,050-kPa) discharge pressure.
4. The test begins upon verification of pressure and flow.
5. During the test, record a complete set of readings about every 5 minutes.

TABLE 8-2 Flow Test Adjustments

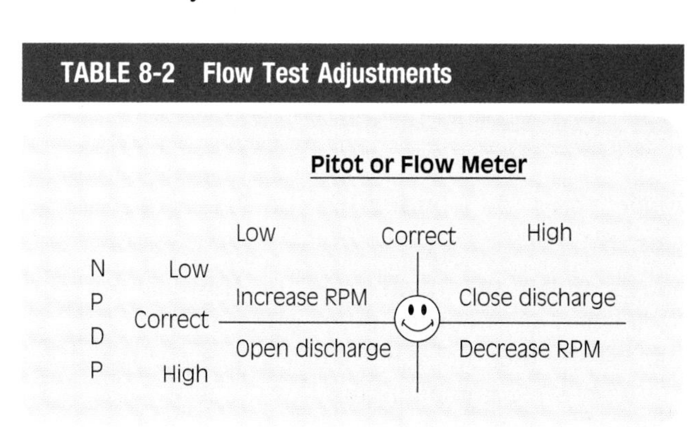

6. After 20 minutes, increase the discharge pressure to 165 psi (1,155 kPa) to conduct the overload test for 5 minutes (only required for pump rated capacities of 750 gpm [3,000 Lpm] and higher).
7. Increase pump pressure to 200 psi (1,400 kPa) and verify a flow of 70% rated capacity. For two-stage pumps, either pressure or volume mode can be used. In some cases, hose and nozzle configures may need to be changed. Record a complete set of readings at least three times during the 10-minute test.
8. Increase the pump pressure 250 psi (1,400 kPa) and verify a flow of 50% rated capacity. As in step 7, the hose and nozzle configuration may need to be changed and a minimum of three readings should be recorded. Two-stage pumps should be in the pressure mode.
9. After 10 minutes at 250 psi (1,750 kPa), the test is complete and the engine pressure should be decreased and all discharge and intake valves closed.

Pressure Control Test

This test is conducted to ensure the pressure control device is capable of adequately maintaining safe discharge pressures. Regardless of whether the pressure control device is a mechanical pressure governor, electronic pressure governor, or relief valve, NFPA requires it to pass the same pressure control test. The pressure control device is tested at the following pressures:

- 150 psi (1,050 kPa) flowing 100% of rated capacity
- 90 psi (630 kPa) (this is achieved by flowing rated capacity with the discharge pressure reduced from 150 psi to 90 psi [1,050 kPa to 630 kPa] by using the throttle only)
- 250 psi (1,750 kPa) flowing 50% of rated capacity

While operating at each of these pressures, the pressure control device is set according to the manufacturer's recommendations. Next, the discharge valves are closed no faster than in 3 seconds and no more slowly than in 10 seconds. To pass the test, the discharge pressure shall increase no more than 30 psi (210 kPa).

Basic Steps for Conducting the Pressure Control Test

1. Establish the rated capacity flow at 150 psi (1,050 kPa) net pump pressure.
2. Set the pressure control device per the manufacturer's recommendations.
3. Slowly close all discharge control valves and note the rise in discharge pressure; the rise should not exceed 30 psi (210 kPa).
4. Open all discharge control valves and re-establish 100% capacity flow at 150 psi (1,050 kPa).
5. Using the engine throttle only, reduce the pump discharge pressure to 90 psi (630 kPa) (do not change the discharge valve setting, hose, or nozzles).
6. Set the pressure control device per the manufacturer's recommendations.
7. Slowly close all discharge control valves and note the rise in discharge pressure; the rise should not exceed 30 psi (210 kPa).
8. Open all discharge control valves and establish 50% rated capacity flow at 250 psi (1,750 kPa).
9. Set the pressure control device per the manufacturer's recommendations.
10. Slowly close all discharge control valves and note the rise in discharge pressure; the rise should not exceed 30 psi (210 kPa).
11. The test is complete; slowly open the discharge control valves and reduce the engine speed to idle.

Tank-to-Pump Piping Flow Test

This required test is conducted to ensure the piping between the on-board water supply and the pump is capable of flowing the manufacturer's designed flow rate.

Steps for Conducting the Tank-to-Pump Piping Flow Test

1. Fill water tank until it overflows.
2. Close all intakes, the fill line, and the bypass cooling line.
3. Connect discharge lines (hose and smooth-bore nozzle) capable of flowing the specific flow rating of the tank-to-pump piping.
4. Open the discharge control valves and increase the engine speed until the maximum consistent pressure reading is obtained.
5. Without changing the engine speed, close the discharge valves and refill the water tank. If necessary, the bypass valve can be temporarily operated to maintain acceptable water temperature range.
6. Completely open the discharge control valves and take pitot tube or flow meter readings. The engine

speed can be adjusted to maintain specified flow rates.

7. After the flow rate is recorded, the test is finished.

Other Tests

Depending on the type of operation, environmental conditions, or specific needs, other tests can be included in the annual service tests. For example, if the need for foam operations is significant, foam system tests can be included in the annual service test. In addition, intake relief devices, when equipped, should be tested to ensure they meet the manufacturer's operations and specifications.

Weekly/Monthly Tests

Weekly, monthly, and periodic tests are conducted for two primary reasons. First, they are conducted to ensure components are in proper working order. Second, they are conducted in order to keep them working properly. For example, testing the priming pump ensures that it is operating properly while also lubricating its close-fitting parts. These tests are also called floor tests. Several examples of weekly and monthly tests typically required by manufacturers include the following:

- Testing pressure regulating device
- Operating priming system
- Conducting a dry vacuum test
- Operating the transfer valve

SKILLS NOTE

Many minor variations exist based on the specific configuration and options of a particular apparatus and the edition of the NFPA specification to which it was built. The information provided in the following skills is representative of many recent models. Manufacturer-provided owner manuals should be the authoritative source for information about any specific apparatus.

SKILL 8-1

Positioning an Apparatus at a Hydrant Using MDH

Delmar/Cengage Learning

A The operator should be able to position the apparatus such that repositioning is not necessary. It should be both close enough to and far enough from the hydrant that the provided hose will reach the hydrant without kinking. The steering wheels should be set at a 45-degree angle to the curb so that if the apparatus rolls it will roll into the curb, not oncoming traffic.

Delmar/Cengage Learning

B The driver should set the parking brake, put the transmission in neutral (newer models will do this automatically), and set wheel chocks upon exiting the apparatus.

Delmar/Cengage Learning

C Replace cap(s) from the hydrant with valve(s) and open the hydrant to flush any sediment and debris. (Note: Using multiple valves allows changing which lines are connected to the hydrant without interrupting flow to other connected lines.) The amount of water discharged from the hydrant should be strong enough to remove sediment and discharge heavier items that may be in the hydrant, including small stones and trash. It should also provide some level of confidence that a problem does not exist in the hydrant's water supply that would prevent it from adequately supplying an engine.

SAFETY

When flushing the hydrant, the operator should be aware of where the flushed water will be discharged, making sure to not create another hazard, such as discharging into traffic or creating road ice during freezing temperatures.

Delmar/Cengage Learning

D Close the valve on the hydrant. The valve allows the hydrant to be closed faster. It should, however, not be closed so fast as to create a water hammer. Taking at least 3 seconds will prevent a water hammer.

Delmar/Cengage Learning

E Connect a hose from the hydrant to the intake side of the pump. Placing a counterclockwise twist in the hose helps prevent kinks and tightens the hose connections.

SKILL 8-2

Positioning an Apparatus at a Hydrant Using LDH

Delmar/Cengage Learning

A The operator should be able to position the apparatus such that repositioning is not necessary. It should be both close enough to and far enough from the hydrant that a provided hose will reach the hydrant without kinking.

Delmar/Cengage Learning

B The driver should set the parking brake, put the transmission in neutral (newer models will do this automatically), and set wheel chocks upon exiting the apparatus.

Delmar/Cengage Learning

C Remove the steamer cap from the hydrant and open the hydrant to flush any sediment and debris. The amount of water discharged from the hydrant should be strong enough to remove sediment and discharge heavier items that may be in the hydrant, including small stones and trash. It should also provide some level of confidence that a problem does not exist in the hydrant's water supply that would make it unusable for supplying an engine.

SAFETY

When flushing the hydrant, the operator should be aware of where the flushed water will be discharged, making sure to not create another hazard, such as discharging into traffic or creating road ice during freezing temperatures.

Delmar/Cengage Learning

D Close the hydrant valve.

Delmar/Cengage Learning

E Connect a hose along with any necessary adapters from the hydrant to the intake side of the pump. Placing a counterclockwise twist in the hose helps prevent kinks and tightens the hose connections.

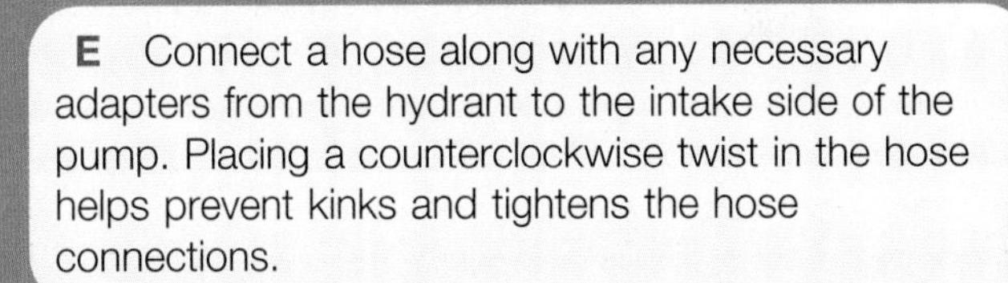

SAFETY

Do *not* twist hoses using Storz fittings.

Delmar/Cengage Learning

F Once everything is in place, the hydrant should be opened and air bled out of the line at the engine prior to opening the engine's intake valve.

SKILL 8-3

Using a Four-way Hydrant Valve

A The first-in engine should stop just past the hydrant on the same side of the street as the hydrant and allow a firefighter to disembark with the engine's hydrant tool bag, four-way hydrant bag, and hose. Once the firefighter has wrapped the hose around the hydrant, the firefighter should signal the driver to proceed to the fire while deploying a forward lay from the hose bed on the same side of the street as the hydrant.

B Once the engine has laid at least 100 ft (30 m) of hose or stopped at the fire scene, the firefighter should flush the hydrant, connect the four-way hydrant valve to the hydrant, connect the supply hose for the first engine to the four-way hydrant valve discharge, and confirm that the four-way hydrant valve handle is in the correct position. The hydrant should be opened when the first engine indicates that it is ready for water.

C When the second engine arrives at the hydrant, a hose should be connected between its intake and the port on the four-way hydrant valve designed for its intake. Another line should be connected between the second pump's discharge and the four-way hydrant valve's pump discharge connection. When the second engine's pump operator is ready, the valve handle on the four-way hydrant valve should be turned to the position that will charge the line to the second engine without interrupting the flow to the first engine.

D Once the operator of the second engine has bled air from the intake line and charged the discharge line with a pressure above the hydrant's pressure, the check valve in the four-way hydrant valve will automatically switch to flow to the first engine from the hydrant to the second engine's discharge.

SKILL 8-4

Positioning an Apparatus at a Dry Hydrant Using Hard Suction Hose

Delmar/Cengage Learning

A The operator should be able to position the apparatus such that repositioning is not necessary. It should be both close enough to and far enough from the hydrant that the provided hose will reach.

Delmar/Cengage Learning

B The driver should set the parking brake, put the transmission in neutral (newer models will do this automatically) and set wheel chocks upon exiting the apparatus. On public roads, traffic cones should also be set.

Delmar/Cengage Learning

C Remove the hydrant cap, confirm that the strainer and hose gasket are in place, and attach the hard suction hose.

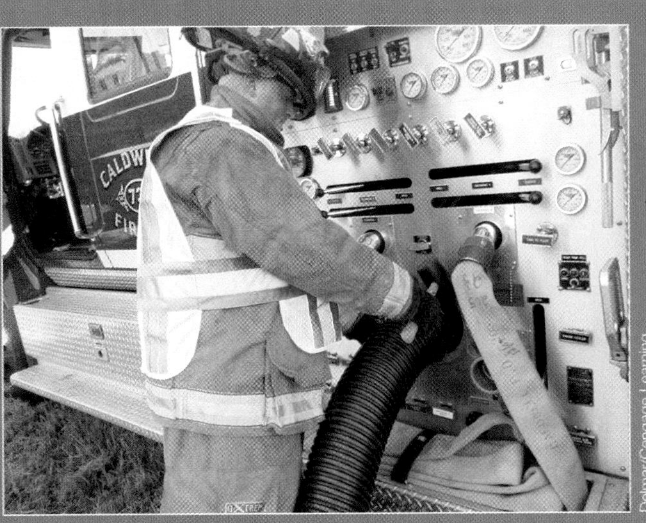

Delmar/Cengage Learning

D Remove any intake caps or appliances and connect any necessary adapters and the hard suction hose. Make sure all connections are tight. This is often done by tapping long-handle hose lugs with a rubber mallet.

(Continues)

SKILL 8-4

Positioning an Apparatus at a Dry Hydrant Using Hard Suction Hose (*Continued*)

STREETSMART

Most appliances, unless designed explicitly for full-flow drafting operations, will significantly reduce the amount of water a given pump can flow; removal of these appliances increases the available pump capacity. It also eliminates possible leak points.

Delmar/Cengage Learning

E Unless the dry hydrant is known to be debris free from recent usage, it is a good idea to back-flush it to reduce the possibility of items entering the pump. Back-flush by fully opening the tank-to-pump valve and any valves to the attached intake. After allowing half of the tank's water to drain, close the tank-to-pump valve, engage the pump, and prime it.

SKILL 8-5

Positioning an Apparatus at a Static Water Source Using Hard Suction Hose

Delmar/Cengage Learning

A The placement of a hard suction and strainer directly into a water source requires engine placement near the water source at an angle that allows the hose to run easily between the engine intake and water source.

Delmar/Cengage Learning

B Connect the hard suction to the engine intake.

Delmar/Cengage Learning

C Attach a strainer and guide rope to the other end of the hose.

Delmar/Cengage Learning

D The crew places the hard suction and strainer into the water, making sure that sand, mud, and other debris do not enter the strainer so that they will not be drawn into the pump. (This photo shows the use of a floating strainer.)

(Continues)

SKILL 8-5

Positioning an Apparatus at a Static Water Source Using Hard Suction Hose (*Continued*)

Delmar/Cengage Learning

E If a non-floating strainer is used, the rope is used to keep the strainer from resting on the bottom. Traditional barrel strainers should have 24 inches (60 cm) of water around them in all directions to prevent both air and debris from entering the pump.

NOTE

A ladder may also be used to support the hard suction hose and keep the strainer off the bottom.

SKILL 8-6

Engaging a Pump

NOTE

The actual time to engage a pump should be quicker than the time it takes to read these instructions. Details are provided here to allow corrections of problems that may occasionally occur. Safety interlocks on new apparatus will actually prevent some actions from occurring if the preceding steps were not successful. Effective pump operators recognize these problems immediately and take corrective action quickly.

Delmar/Cengage Learning

A Apply the parking brake. On newer air-brake-equipped vehicles, this is accomplished by pulling a yellow diamond-shaped knob. On older vehicles, this may be accomplished by moving a lever.

Delmar/Cengage Learning

Delmar/Cengage Learning

B You should observe a red parking brake applied light (this could differ on some older vehicles) and hear the air being released from the spring brakes on air-brake-equipped vehicles. Failure to apply the parking brake will prevent transmission lock-up on newer apparatus.

C Make sure the transmission is in neutral. Note: Continuing without being in neutral will not work and may cause physical damage. If pressing the "N" selector button does not cause an "N" to display on an electronic transmission, you may try stopping the truck engine and turning the master power switch off for 5 seconds. When you restart the engine, the transmission should be in neutral. Be sure to report this problem so repairs can me made.

(Continues)

SKILL 8-6

Engaging a Pump (*Continued*)

Delmar/Cengage Learning

D Move the pneumatic pump power shift switch from "Road Gear" (top position) to "neutral" (middle position) and, after all air exhausts, to the "Pump" (bottom position) by pulling the yellow collar with your first two fingers toward the black handle end. Your thumb should be on the black handle.The "Pump Engaged" light should come on.

Delmar/Cengage Learning

E Push "D" on the Allison Pushbutton Selector. The "Digital Display" on the Allison Pushbutton Selector should show a "4." The "OK to Pump" light should come on, and the speedometer should read at least 10 MPH (16 KPH).A "5" showing on the Allison digital display would normally mean a preceding step was not successfully completed prior to shifting. In addition to correcting the preceding step, the transmission will need to be returned to neutral for several seconds. Not staying in neutral long enough for internal transmission parts to stop spinning may cause repeated failures when returning to drive even if no other problems exist.
Note: At this point the throttle pedal and fast idle switch in newer vehicles should be disabled.

Delmar/Cengage Learning

F Exit the cab and set the wheel chocks.
Note: Even if the apparatus is facing uphill, a wheel chock should be placed in front of a wheel. Since the vehicle is in drive, this helps prevent forward movement in the case of a problem with the power transfer. The one behind the wheel prevents reverse movement in the case of a brake problem.

Delmar/Cengage Learning

G At the pump panel the "Throttle Engaged" light should be on. (In bright sunlight this may be hard to see. If everything else appears correct, assume this is on.) If other problems are occurring, make an effort to determine if this light is on.

Delmar/Cengage Learning

H Open the tank-to-pump valve(s) fully. Open the tank fill slightly (slightly can be defined as until a small pressure drop is noted on the discharge gauge).

Delmar/Cengage Learning

I Advance the throttle to the desired pressure. (Per NFPA 1901, most pumps have only been hydrostatically tested to 250 psi [1,750 kPa]. Operators should never exceed the pressure to which their pump has been hydrostatically tested. Exceeding this pressure can result in blown seals and expensive, time-consuming repairs and have the truck out of service for an extended time period.) If advancing the throttle does not result in an increased RPM and the "Throttle Engaged" light is on, confirm that the throttle was screwed all the way in before you started. If not, screw it all the way in (low idle) and then start advancing it. The engine speed should now increase if the "Throttle Enabled" light is on.

(Continues)

SKILL 8-6

Engaging a Pump (*Continued*)

Delmar/Cengage Learning

J Set the pressure relief valve or governor. Note: This should be set even if only a single line is in use.

LESSONS LEARNED

Operating the pump, the second activity of fire pump operations, moves water from the intake side of the pump to the discharge side of the pump. In general, the same basic steps must be taken to move water from the intake to the discharge regardless of the size and configuration of the pump. However, pump procedures may vary based on different pump manufacturers and different fire departments.

KEY TERMS

Dual pumping A hydrant that directly supplies two pumps through intakes.

Front crankshaft method (pump engagement) Method of driving a pump in which power is transferred directly from the crankshaft located at the front of an engine to the pump; this method of power transfer is used when the pump is mounted on the front of the apparatus and allows for either stationary or mobile operation.

Intake pressure relief valve Pressure regulating system that protects against excessive pressure buildup on the intake side of the pump.

PTO method (pump engagement) Method of driving a pump in which power is transferred from just before the transmission to the pump through a PTO; a method of power transfer that allows either stationary or mobile operation of the pump.

Pump engagement The process or method of providing power to the pump.

Pump gear Indicates that power from the transmission will be transferred to the pump on a split-shaft pump.

Road gear Indicates that power from the transmission will be transferred to the drive axle(s) on a split-shaft pump.

Split-shaft method (pump engagement) Method of driving a pump in which a sliding clutch gear transfers power to either the road transmission or to the pump transmission; this method of power transfer is used for stationary pumping only.

Tandem pumping A hydrant that directly supplies one pumper and then discharges to the second pumper's intake.

Throttle control Device used to control the engine speed, which in turn controls the speed of the pump, when engaged, from the pump panel.

Transfer valve Control valve used to switch between the pressure and volume modes on two-stage centrifugal pumps.

REVIEW QUESTIONS

Multiple Choice

Select the most appropriate answer.

1. The power to main pumps can be transferred from the drive engine by each of the following methods, *except*

- **a.** split-shaft.
- **b.** power takeoff (PTO).
- **c.** front crankshaft.
- **d.** midship transfer (MST).

2. When the pump is powered by the drive engine through a PTO, the transmission is placed in ________ for stationary pumping.

- **a.** low gear
- **b.** high gear
- **c.** neutral
- **d.** either fourth or fifth gear

3. A pump connected to the front crankshaft of the drive engine is usually mounted

- **a.** on the front of the apparatus.
- **b.** toward the back of the apparatus.
- **c.** in the middle of the apparatus.
- **d.** on top of the drive engine.

4. A pump connected to a split-shaft transmission is usually mounted
   a. on the front of the apparatus.
   b. toward the back of the apparatus.
   c. in the middle of the apparatus.
   d. either toward the back or the middle of the apparatus.

5. Which NFPA standard focuses on the annual service test of pumps?
   a. 1901
   b. 1911
   c. 1921
   d. 1500

Short Answer

On a separate sheet of paper, answer/explain the following questions.

1. List the four ways a pump can receive its power.
2. Can a pump be primed without using the priming system? If not, why? If so, how?
3. How can you tell if the pump is primed?
4. How do you determine in which mode (pressure or volume) to operate a two-stage centrifugal pump?
5. Why is it important to set the transfer valve before pressure is increased?
6. How can you tell if the pump is flowing all the water it is receiving from the supply?
7. Explain how to set a pressure relief valve.
8. When using a hydrant as a supply source, the intake should not be reduced below what pressure?
9. When conducting a drafting operation, should the pump be engaged before or after the pump is primed?
10. Before increasing the throttle on the pump panel, what important safety item should the pump operator look for?
11. How long can the priming pump be safely activated?
12. How can you tell if the pump is properly engaged when power is transferred using the split-shaft method?
13. What is the purpose of the throttle control on a pump panel?
14. What should occur when the throttle is increased?
15. If the throttle is increased past the point of a corresponding increase in discharge pressure, what might be occurring?
16. List three tests required by NFPA 1901.
17. List the various tests associated with the annual service testing of pumps.

ACTIVITY

1. Obtain a pump manufacturer's operating manual and develop a step-by-step procedure for pumping the on-board tank, a hydrant, and a drafting operation.

PRACTICE PROBLEMS

1. Which position should the transfer valve be in for the following pumping operations (each pumper is drafting from a static water supply):
   a. 1,000-gpm (4,000-Lpm) pumper flowing the following lines:
      - Two 1¾-inch (45-mm) preconnect lines; each line is 150 feet, flowing 120 gpm @ 133 psi (45 m, flowing 480 Lpm @ 931 kPa)
      - One 2½-inch (65-mm) line, 400 feet (120 m), flowing 250 gpm (1,000 Lpm) @ 150 psi (1,050 kPa)
      - One 3-inch (77-mm) line, 500 feet (150 m), flowing 300 gpm (1,200 Lpm) @ 136 psi (952 kPa)

b. 750-gpm (3,000-Lpm) pumper flowing the following lines:

- Two 1¾-inch (45-mm) preconnect lines; each line is 200 feet (60 m), flowing 125 gpm (500 Lpm) @ 148 psi (1,036 kPa)
- One 2½-inch (65-mm) line, 1,200 feet (360 m), flowing 275 gpm (1,100 Lpm) @ 182 psi (728 kPa).

ADDITIONAL RESOURCES

NFPA 1901, *Standard for Automotive Fire Apparatus*. National Fire Protection Association, Quincy, MA, 2009.

NFPA 1911, *Standard for the Inspection, Maintenance, Testing, and Retirement of In-Service Automotive Fire Apparatus*. National Fire Protection Association, Quincy, MA, 2007.

9 Discharge Maintenance and Troubleshooting

- Learning Objectives
- Introduction
- Initiating and Changing Discharge Flows
- Dealing with High-pressure Supplies
- Drafting
- Maintaining Pump Operations
- Shutting Down and Post Operations
- Pump Operation Considerations
- Supporting Sprinkler and Standpipe Systems
- Troubleshooting
- Lessons Learned
- Key Terms
- Review Questions
- Activities
- Practice Problems
- Additional Resources

STREET STORY

Having been a fireman for a busy engine company, I knew how important it was to have a good engineer: someone who will drive you to the scene quickly and safely, someone who will be able to obtain a positive source of water quickly, someone who will be able to engage the pump and begin flowing the proper amount of water quickly so that the entire operation is seamless. I had worked with several good and several bad engineers. On the days I was working with a bad engineer I would sometimes pray we didn't get a fire.

In 2007, I was promoted to engineer myself. And I knew I wanted my crews to see me as a good engineer.

When we get promoted in Chicago, we get assigned to a particular district as a relief engineer—so I go to a different firehouse within that district every work day. It took me 6 to 12 months to get comfortable with my new position. I had to become comfortable driving every day. I had to learn my way around. I had to learn what each officer and crew expected of me, and how they liked to operate. Most important, I had to become proficient in my new skills. It was very stressful at first. It's also an adjustment being on the other side of the hoseline.

As that new engineer, all you can do is call on your past experiences and training to guide you. I can recall one of my first fires as an engineer. We had responded on a 2-11 alarm for a fire that was burning in a common basement of a row of stores. We were ordered to the front of the building to feed a tower ladder. After positioning my engine, another, more experienced engineer came over to assist me with my connections. I was so excited I was tripping and fumbling over everything. After some words of encouragement from him, I calmed down. We completed our task, and I was finally flowing water. He then told me to look at the guys in the basket of the tower ladder putting the water on the fire. "That fire is going out because of you," he said. He was right!

Now that I have experienced a few more fires and have become more comfortable with my position, I can watch from the outside and see the fire going out, or watch the smoke changing colors, and know that all of that is happening because of me. Nothing is more rewarding than having your guys come out of a building and give you a smile and a thumbs up. You know then that you did your job: The fire is out and everyone is safe. None of that would have been possible without you.

—Street Story by Scott Shawaluk, Engineer, Chicago Fire Department, 4th District Relief, Chicago, Illinois

LEARNING OBJECTIVES

After completing this chapter, the reader should be able to:

9-1. Produce an effective hand stream from the internal tank of a pumping apparatus, so that all safety considerations are addressed, the pressure control device is set, the rated nozzle flow is achieved and maintained, and the apparatus is monitored for potential problems.

9-2. Produce an effective hand or master stream from a pressurized source, so that all safety considerations are addressed, the pressure control device is set, the rated nozzle flow is achieved and maintained, and the apparatus is monitored for potential problems.

9-3. Produce an effective hand stream from a static water supply, so that all safety considerations are addressed, the pressure control device is set, the rated nozzle flow is achieved and maintained, and the apparatus is monitored for potential problems.

9-4. List and explain the procedures for maintaining pump operations.

9-5. Explain the basic process of troubleshooting pumping problems.

9-6. Define *water hammer* and explain how to prevent its occurrence.

9-7. Define *cavitation* and explain how it can be detected and controlled.

9-8. Explain the unique concerns with hot and cold weather operations.

9-9. List the basic components of a sprinkler system.

9-10. Explain the operating principles of a sprinkler system.

9-11. Describe the various types of sprinkler systems and explain the differences in the ways they operate.

9-12. Calculate the pump discharge pressure for a given sprinkler or standpipe system evolution.

9-13. Describe the considerations for hose layouts when connecting to a fire department connection.

9-14. Discuss considerations when supplying water to sprinkler systems.

9-15. Given system information, supply water to a fire sprinkler at the correct volume and pressure.

9-16. Explain the alternative supply procedures if a fire department connection is unusable.

9-17. List the basic components of a standpipe system.

9-18. Describe how a standpipe operates.

9-19. Discuss considerations when supplying water to standpipe systems.

9-20. Given system information, supply water to a standpipe at the correct volume and pressure.

***The driver/operator requirements, as defined by the NFPA 1002 Standard, are identified in black; additional information is identified in blue.**

INTRODUCTION

When the water supply is secured and pump procedures are initiated, the next task of fire pump operations is **discharge maintenance**. In essence, discharge maintenance is simply ensuring that pressures and flows on the discharge side of the pump are properly initiated and maintained. To accomplish this task, pump operators must understand how to determine appropriate discharge settings, set and adjust these settings, temporarily pause operations, and maintain the pump engine within safe operating parameters. In addition, pump operators must understand, recognize, and minimize conditions such as cavitation and environmental factors that affect pump operations. Finally, pump operators must be able to troubleshoot pump problems quickly and efficiently. Effectively carrying out the task of discharge maintenance requires pump operators to continuously monitor instrumentation and to plan ahead for potential changes.

Monitoring pump instrumentation is important for ensuring that discharge pressure and flow settings are maintained. Gradual changes in pressures and flows can occur on both the intake and discharge sides of the pump. Abrupt changes to pressures and flows can also occur on both the intake or discharge sides. For example, supply or discharge lines can break, seriously affecting pumping operations. Constant monitoring of instrumentation is essential for detecting and compensating for these gradual and abrupt changes. Finally, constant evaluation of engine instrumentation may detect potential problems. For example, a gradual increase in engine temperature could cause serious problems if not detected early enough.

Pump operators should plan ahead for potential changes in pump operation. Planning ahead will ensure that changes to pump operations are completed in an efficient and effective manner. Common tasks that may require planning include:

- Pausing operations
- Extending pump operations
- Securing an alternate supply source should the current supply fail
- Determining what to do should a supply line or discharge line break
- Judging the ability to add lines and the best manner to supply them

This chapter presents the task of discharge maintenance by first discussing procedures for maintaining flows and pressures, then pump operation considerations, sprinkler and standpipe operations, and, finally, pump troubleshooting. The process for determining discharge settings is presented in **Section IV**, **Chapter 12**.

FIGURE 9-1 Discharge lines of different pressure and volume require feathering of discharge control lines.

Delmar/Cengage Learning

INITIATING AND CHANGING DISCHARGE FLOWS

Discharge flows are controlled by pump speed and discharge control valves. Pump speed is increased and decreased to control total flow from the pump. As discussed in **Chapter 5** of this text, the primary means to control pump speed is the throttle or governor located on the pump panel. Discharge control valves are opened and closed to provide individual control over flows from discharge outlets. Used in combination, a variety of flows and pressure settings can be achieved. Recall that the master discharge gauge indicates total discharge pressure from the pump, while individual discharge gauges indicate pressure readings for their specific discharge.

Multiple lines of different pressures and flows require a technique called **gating** of discharge control valves. Gating is simply the process of partially opening or closing control valves to regulate pressure and flow for individual lines (**Figure 9-1**). The terms *gating back, throttling* and **feathering** are also used to describe this process.

Gating, or feathering, a valve is most often used when initiating, changing, or shutting down a discharge line while other discharge lines are flowing.

NOTE

The following sections are repeated twice. Once for pumps using relief valves and once for pumps using governors. Traditionally most pumps used relief valves; today, however, most pumps are being equipped with electronic governors, with many of both styles of pumps in active service. With either type of pressure control device, the same pump discharge pressures need to be maintained to accomplish the same operation and flow the same handlines. The exact pump panel operations to accomplish these results are different, with each being covered in the corresponding section.

Initiating Discharge Flows on Pumps with Relief Valves

Initiating flow on the discharge side of the pump entails opening the discharge control valve, increasing engine speed (RPM) to the desired pressure, and setting the relief valve. When a discharge line is first charged, water enters the hose and pressure increases to the discharge pressure of the pump. Pump operators should check the hose for sharp bends and kinks that restrict flow. Kinks and sharp bends restrict flow and cause higher pressure readings. When the kinks and bends are removed, less restriction will increase flow, resulting in a reduced pump panel pressure reading. In addition, when nozzles are momentarily opened and closed, pressure readings on the pump

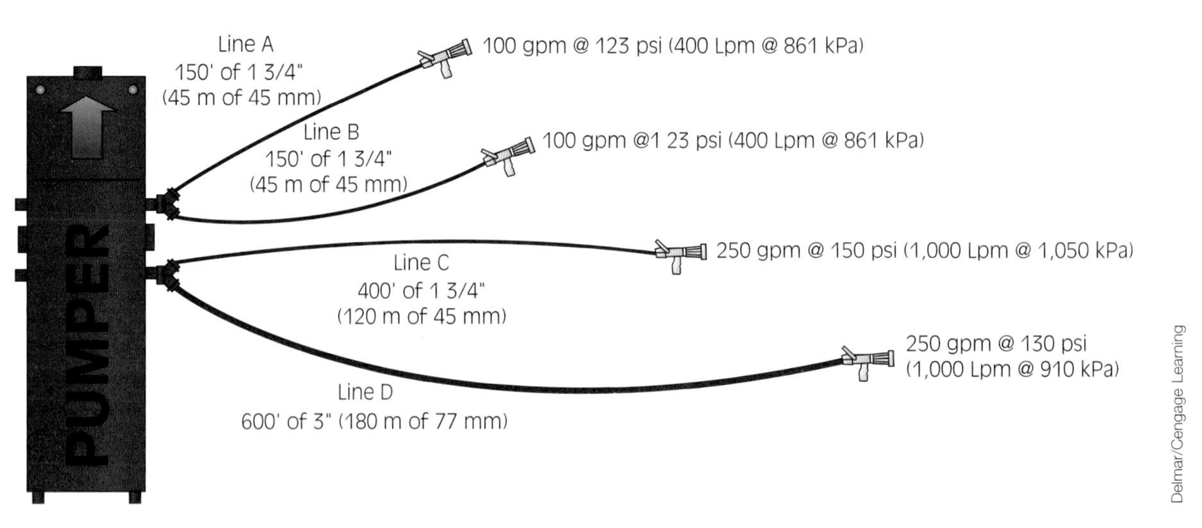

FIGURE 9-2 In this example, the procedure for placing any one of the lines in service will vary based on lines already in operation.

panel will rise and fall, creating difficulty in setting appropriate pressures. The best and easiest time to set pressures occurs when nozzles are open and water is flowing. This is especially true when multiple lines of varying pressures are used.

Pump operators must recognize and compensate for these difficulties in setting discharge lines. A good pump operator will not have a noticeable increase or decrease in pressure (no more than 5 to 10 psi or 35 to 70 kPa) when adding or removing lines. The discharge lines in **Figure 9-2** will be used to explain procedures for initiating and changing discharge flows. Pressure settings for these lines were determined using the friction loss formula cq^2L, which is discussed in the section on friction loss calculations in **Chapter 11**.

Initiating One Line

When one line is placed in service, the pump operator first fully opens the appropriate discharge control valve. Remember, in order to ensure the safety of personnel and equipment, all discharge control valves should be operated in a slow and smooth manner. Next, the throttle is slowly increased until the desired pressure setting is reached. For example, to initiate line A in **Figure 9-2**, the pump operator would simply open the discharge control valve for line A and then increase the throttle to 123 psi (861 kPa).

(For step-by-step photos of Producing an Effective Hand Stream from the Internal Tank, please refer to page 298.)

Initiating Two Lines

Multiple lines of the same configuration (same hose size, length, and nozzle setting) require the same pressure and flow setting; therefore, the lines can be opened and the discharge pressure set at the same time. For example, to place both lines A and B into service at the same time, the pump operator would simply open the control valves for lines A and B and increase the throttle until the pressure is increased to 123 psi (861 kPa).

Multiple lines of different configurations may require different pressure and flow settings. In this case, the line with the lower pressure must be feathered. For example, to place lines B and C into service at the same time, the pump operator first opens both control valves for lines B and C. Then, the pressure is increased to 123 psi (861 kPa) in both lines. The next step is to continue increasing the pressure to 150 psi (1,050 kPa) for line C while maintaining 123 psi (861 kPa) for line B. Maintaining 123 psi (861 kPa) on line B is accomplished by slowly closing (feathering) the discharge control valve for line B.

Changing Discharge Flows on Pumps with Relief Valves

More often than not, discharge lines are added and removed during pumping operations. When this occurs, discharge flow settings must be changed at the pump panel. The procedure for changing discharge flow settings depends on the pressure and flow of the line being added or removed.

Adding a Line of Less Pressure

When adding a line of less pressure than those currently in operation, the pump operator simply feathers the control valve open for the new line until the proper setting is reached. Because the total gallons-per-minute (litres-per-minute) flow has increased, the pump speed may need to be increased to maintain the discharge pressure of existing line(s). This occurs because pressure is a function of restriction. When an additional line is opened, less restriction occurs on the discharge side of the pump, resulting in a reduction in pressure.

Take for example, placing line D in **Figure 9-2** in service when lines A and C are currently in operation. To accomplish this task, the control valve for line D is slowly feathered open to 130 psi (910 kPa) while increasing the throttle to maintain 123 psi (861 kPa) on line A and 150 psi (1,050 kPa) on line C.

Adding a Line of Equal Pressure

When adding a line of equal pressure, the new line is simply feathered open until the desired pressure setting is achieved, while the throttle is increased to maintain the original pressure setting. For example, to initiate flow to line B when line A is currently in operation, the pump operator simply feathers open line B to 123 psi (861 kPa) while maintaining 123 psi (861 kPa) on line A by increasing the throttle.

Adding a Line of Greater Pressure

When adding a line of greater pressure than those currently in operation, the pump operator must feather existing lines. For example, to place line C in service while line D is currently in operation, the pump operator slowly feathers open the control valve for line C while maintaining 130 psi (910 kPa) on line D by simultaneously increasing the throttle. When line C reaches 130 psi (910 kPa), line D must be feathered closed as the pressure in line C is increased to 150 psi (1,050 kPa). Adding a line of higher pressure requires the pressure regulating system to be reset.

Removing a Line

Removing lines from service requires the pump operator to slowly close the control valve of the line being removed. If the line being removed is the highest of those flowing, the throttle should be reduced to the new highest operating line. If the line being removed is a lower-pressure line, no change in engine pressure is required. When the line is removed, however, the pressure may increase as a result of reduced flow. If the rise in pressure is great enough to cause the pressure regulator to activate, the throttle should be slowly reduced to the appropriate pressure setting. In doing so, the pressure regulator should reset itself. No adjustment to the pressure regulator itself should be necessary.

Changing the Relief Valve Setting

The relief valve should be set as soon as hoselines are placed into operation. Prior to engaging the relief valve, it should be set to a higher pressure than the PDP will be set to. Once engaged, the relief valve pressure setting is slowly lowered until the relief valve just starts to open. As the relief valve opens, three things should happen: the pump discharge pressure drops slightly, a relief valve indicator light comes on, and the sound of the pump changes, with the pump discharge pressure being the most significant item and the one used for setting the relief valve. At this point, the relief valve setting should be raised about a half turn so that the PDP, light, and sound returned to normal. Relief valves are normally set slightly above the desired PDP to prevent excessive operation of the relief valve. It is important to adjust the relief valve slowly to avoid overcompensating and dropping the PDP too much. Relief valves are intentionally dampered, or slow to react, to prevent them from overreacting to transient pressure changes. The same dampering results in overreaction if one attempts to adjust it too fast.

The function of the relief valve is often described as protecting line crews from excessive pressure, which could occur when other hoselines are shut down. It is still important to set the relief valve even if only one line is operating, although some of the reasons are less obvious. The first line is often put in operation from the booster tank and then a connection is made to a water supply. When making the connection to the water supply, the relief valve offers some protection from an increase in PDP, although the operator should be carefully controlling the throttle as the intake valve is opened. In addition, the relief valve offers protection from changes in the intake pressure that may occur due to changes that are occurring with the water supply system.

Relief valves should not be used to change the discharge pressure setting. Rather, the control valves and the pump panel throttle should be used. Using the relief valve to control discharge pressures can result in excessive pump speeds, which increase pump wear and tear, as well as increasing safety concerns should the regulating device fail or be accidentally turned off.

RELIEF VALVE CONTROL ADJUSTMENT

Operators need to avoid the temptation to "overadjust" pump panel controls. Controls should be adjusted when strategic changes are made; however, changing for each short-term interim operational change leads to an operator who is always late in providing necessary support.

For example, if a line that was gated back because it is operating at a lower pressure than other lines is temporarily shut down to be repositioned, several things will occur at the pump panel. The pressure on the line gauge for the line that was shut down will increase, and the relief may partially open due to the reduced flow. With a correctly set relief valve doing its job, the PDP should stay about the same and all other lines should remain at the correct pressure and flow. The relief valve setting should certainly not be adjusted since the desired PDP has not changed. If the engine throttle is reduced, the relief valve will close again; however, once the line is reopened, the throttle will be too low, PDP will drop, and everyone will have insufficient pressure until the throttle is increased to its original setting. Any attempts to adjust the gating on the closed line will be ineffective. Gated lines can only be adjusted while they are flowing. If a gated line is closed, the line gauge will read the same as the PDP. If it was not adjusted, its reading will return to exactly where it was once the line is reopened. If it was adjusted, it will need a second adjustment once reopened and will not be working at the correct pressure until adjusted for this second time.

If the pump discharge pressure is increased, the pressure regulating system must first be reset above the new discharge pressure setting. The pressure regulating system should be turned off before pressure is increased. Turning the pressure regulator off when it is activated will cause an increase in pressure on the discharge side of the pump, which may cause a water hammer intense enough to damage equipment and injure personnel. Water hammer is further discussed later in this chapter.

After the discharge pressure is increased, the pressure regulating system is then set to the new pressure setting. When the pump discharge pressure is decreased, the pressure regulating setting must be reduced as well. The need to change pressure regulating settings can arise from:

- Adding a line of higher discharge pressure
- Increasing the pressure of an existing line
- Removing the highest-pressure discharge line
- Reducing the pressure of an existing line

Initiating Discharge Flows on Pumps with Pressure Governors

Initiating flow on the discharge side of the pump entails opening the discharge control valve, increasing engine speed (RPM) to the desired pressure, and setting the pressure governor. When a discharge line is first charged, water enters the hose and pressure increases to the discharge pressure of the pump. Pump operators should check the hose for sharp bends and kinks that may restrict flow. Kinks and sharp bends that restrict flow will cause higher pressure readings. When the kinks and bends are removed, less restriction will reduce pressure. Two common methods exist for the initial setting of an electronic pressure governor. Many have a preset button. With these governors, once the initial discharge valve is open, press the preset button if it has been preset to the desired pressure. Normally the pressure is set for flowing the engine's customary first attack line. Alternatively, and when a different pressure is required, the operator presses an increase throttle button until the desired PDP is achieved. Then a "set" button is pressed and the governor will adjust the engine throttle as required to maintain the desired PDP.

A good pump operator will not have a noticeable increase or decrease in pressure (no more than 5 to 10 psi or 35 to 70 kPa) when adding or removing lines. The discharge lines in **Figure 9-2** will be used to explain procedures for initiating and changing discharge flows. Pressure settings for these lines were determined using the friction loss formula cq^2L, which is discussed in the section on friction loss calculations in **Chapter 11**.

Initiating One Line

When one line is placed in service, the pump operator first fully opens the appropriate discharge control valve. Remember, in order to ensure the safety of personnel and equipment, all discharge control valves should be operated in a slow and smooth manner. Next, either the throttle is increased until the desired pressure setting is reached (123 psi or 861 kPa for line A in **Figure 9-2**) and the set button is pressed on the governor or if the preset pressure is correct, press the preset button on the governor.

(For step-by-step photos of Producing a Hand or Master Stream from a Pressurized Source, please refer to page 299.)

Initiating Two Lines

Multiple lines of the same configuration (same hose size, length, and nozzle setting) require the same pressure and flow setting. Therefore, the lines can be opened and the discharge pressure set at the same time. For example, to place both lines A and B into service at the same time, the pump operator would simply open the control valves for lines A and B. Next, either the throttle is increased until the desired pressure setting is reached (123 psi or 861 kPa for lines A and B in **Figure 9-2**) and the set button is pressed on the governor or if the preset pressure is correct, press the preset button on the governor.

Multiple lines of different configurations may require different pressure and flow settings. In this case, the line with the lower pressure must be feathered. For example, to place lines B and C into service at the same time, the pump operator first opens both control valves for lines B and C. Then, the pressure is increased to 123 psi (861 kPa) in both lines. The next step is to continue increasing the pressure to 150 psi (1,050 kPa) for line C while maintaining 123 psi (861 kPa) for line B. Maintaining 123 psi (861 kPa) on line B is accomplished by slowly closing (feathering) the discharge control valve for line B.

Changing Discharge Flows on Pumps with Pressure Governors

More often than not, discharge lines are added and removed during pumping operations. When this occurs, discharge flow settings must be changed at the pump panel. The procedure for changing discharge flow settings depends on the pressure and flow of the line being added or removed.

Adding a Line of Less Pressure

When adding a line of less pressure than those currently in operation, the pump operator simply feathers the control valve open for the new line until the proper setting is reached. Because the total gallons-per-minute (litres-per-minute) flow has increased, the pump speed may need to be increased to maintain the discharge pressure of existing line(s). This is a perfect scenario for a pressure governor. Since the pressure governor automatically adjusts pump speed as necessary to maintain PDP, the operator only needs to open the valve and the governor will make sure the pressure remains constant.

Take, for example, placing line D in **Figure 9-2** in service when lines A and C are currently in operation. To accomplish this task, the control valve for line D is slowly feathered open to 130 psi (910 kPa) while the pressure governor adjusts the throttle to maintain a PDP of 150 psi (1,050 kPa). As long as the PDP does not change, individual line pressures will remain constant, so line A will have 123 psi (861 kPa) and line C will have 150 psi (1,050 kPa).

Adding a Line of Equal Pressure

When adding a line of equal pressure, the new line is simply opened slowly while the pressure governor adjusts the throttle to maintain the original PDP pressure setting. For example, to initiate flow to line B when line A is currently in operation, the pump operator simply opens line B to 123 psi (861 kPa) while the pressure governor adjusts the throttle to maintain the PDP, which keeps 123 psi (861 kPa) on line A.

Adding a Line of Greater Pressure

When adding a line of greater pressure than those currently in operation, the pump operator must feather existing lines. For example, to place line C in service while line D is currently in operation, the pump operator slowly feathers open the control valve for line C while the pressure governor automatically maintains a PDP of 130 psi (910 kPa) by simultaneously increasing the throttle. When line C reaches 130 psi (910 kPa), line D must be feathered closed as the pressure in line C is increased to 150 psi (1,050 kPa). This requires adjusting the pressure governor manually to the higher pressure. When complete, make sure the pressure governor is in pressure mode, not RPM mode.

Removing a Line

Removing lines from service requires the pump operator to close the control valve of the line being removed slowly. If the line being removed is the highest of those flowing, the pressure governor should be reduced to the new highest operating line while concurrently opening the valves on the remaining line so they maintain their same line pressures. If the line being removed is a lower-pressure line, no change in engine pressure is required and the pressure governor will automatically adjust pump speed as required for the reduced flow. No adjustment to the pressure governor itself should be necessary.

PRESSURE GOVERNOR CONTROL ADJUSTMENT

Operators need to avoid the temptation to "overadjust" pump panel controls. Controls should be adjusted when strategic changes are made; however, changing for each short-term, interim operational change leads to an operator who is always late in providing necessary support.

For example, if a line that was gated back because it is operating at a lower pressure than other lines is temporarily shut down to be repositioned, several things will occur at the pump panel. The pressure on the line gauge for the line that was shut down will increase and the speed may reduce due to the reduced flow. With a correctly set pressure governor doing its job, the PDP should stay about the same and all other lines should remain at correct pressure and flow. The pressure governor setting should certainly not be adjusted since the desired PDP has not changed. Any attempts to adjust the gating on the closed line will be ineffective. Gated lines can only be adjusted while they are flowing. If a gated line is closed, the line gauge will read the same as the PDP. If it was not adjusted, its reading will return to exactly where it was once the line is reopened. If it was adjusted, it will need a second adjustment once reopened and will not be working at the correct pressure until adjusted for this second time.

Changing Pressure Governor Settings

Unlike a relief valve, a pressure governor is used to change the discharge pressure setting. Remember that for the protection of line crews, the pressure governor needs to be in pressure mode. In RPM mode, no pressure protection is provided.

If the pump discharge pressure is increased, the pressure regulating system must be reset to the new discharge pressure setting. When the pump discharge pressure is decreased, the pressure regulating setting must be reduced as well. The need to change pressure regulating settings can arise from:

- Adding a line of higher discharge pressure
- Increasing the pressure of an existing line
- Removing the highest-pressure discharge line
- Reducing the pressure of an existing line

DEALING WITH HIGH-PRESSURE SUPPLIES

Many jurisdictions are blessed with at least some hydrants that have great pressures. Using these hydrants, however, may require some changes to how pump panel valves are set. For example, if connected to a hydrant with a 100-psi (700-kPa) residual pressure and a desire to flow line A in **Figure 9-2** with a discharge pressure of 123 psi (861 kPa), the pump will create a PDP of greater than 123 psi (861 kPa). Some operators have mistakenly handled this by gating back the intake; however, intakes should *not* be gated back. Gating back runs the risk of cavitating the pump should the flow increase or a drop in hydrant pressure occur. This drop could occur either due to water consumption or other engines operating in the area. Instead, the discharge line should be gated back and the pressure control device set for the actual PDP. In essence, it's like flowing multiple lines of different pressures with the actually line(s) being the lower pressure line(s) and an imagery high pressure discharge line.

DRAFTING

Drafting is the process of pumping water from a static source such as a lake or dump tank where the water's surface is lower than the pump's intake. It requires the use of atmospheric pressure to push water into the pump. In order to get the water pushed into the pump, the pressure in the pump needs to be lowered by removing the air in the pump, thus allowing that air to be replaced with water. The process of replacing air in the pump with water is called priming.

(For step-by-step photos of Producing an Effective Hand Stream from a Static Source, please refer to page 300.)

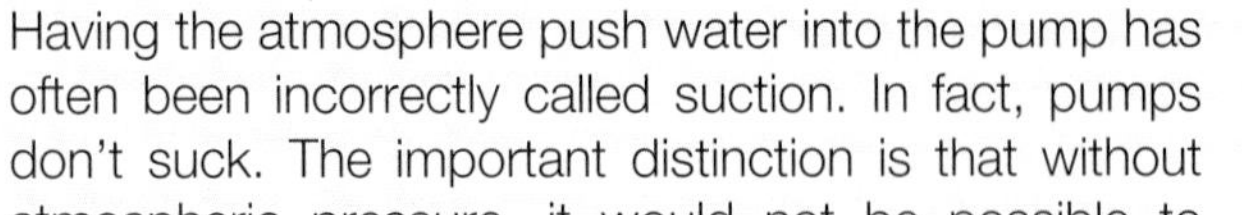

Having the atmosphere push water into the pump has often been incorrectly called suction. In fact, pumps don't suck. The important distinction is that without atmospheric pressure, it would not be possible to get water to rise above its surface and enter a pump. It also means that how high water can be lifted is effected by atmospheric pressure, with maximum lift being more restrictive in the mountains due to the reduced atmospheric pressure at higher elevations.

MAINTAINING PUMP OPERATIONS

The pump operator does not relax after pump operations are initiated. Maintaining pump operations is an important activity that requires constant attention and planning. Common activities for maintaining pump operations include temporarily pausing operation, planning for extended operations, and maintaining the engine and pump within operating parameters.

Pausing Operations

If discharge lines will not be flowing water for a short period of time, the pumping operation should be paused to reduce wear and tear on the pump and to avoid damage from overheating and cavitation. When pausing pump operations, slowly reduce engine speed and slowly close discharge control valves. If drafting from a static source, a slight discharge pressure should be maintained. This pressure can be accomplished by opening an unused discharge outlet with a hose directing water back into the static source. This procedure is also a good practice with reduced flows, such as during overhaul. If the operation will be paused for an extended period, the pump should be disengaged.

Extended Operations

Special consideration should be given to pump operations that are expected to last for 4 hours or more. Long pumping operations tend to reduce attention to instrumentation, especially when the operation does not require constant changes in flow. Special attention should be given to the engine and pump to ensure that overheating does not occur, and attention should be given to oil pressure and fuel levels. Modern electronic instrumentation often provides visual and audible warning signals as systems and components move outside normal operating parameters. Arrangements should be made well in advance if refueling is required.

Operators must constantly monitor all instrumentation to allow timely, proactive actions preventing apparatus damage and safety problems for the crew.

Maintaining Engine and Pump Operating Parameters

Even during short pump operations, engine and pump parameters can change quickly. Pump operators must be diligent in monitoring both pump and engine instrumentation. Of particular concern during pump operations is the detrimental effect of excessive heat.

Engine Cooling Systems

As discussed in **Chapter 5**, engine cooling systems are designed to maintain normal operating temperatures while the vehicle is in motion or while the vehicle is stationary at normal idle speed. During stationary operations under load, as is the case during pumping operations, the engine cooling system is inadequate to remove excessive heat buildup, so pumping apparatus are equipped with an auxiliary cooling or engine cooling system (**Figure 9-3**). Water from the pump is diverted through piping to the engine, absorbs heat, and is then returned to the pump. This system should be activated when the engine temperature first begins to rise. In most situations, turning on the auxiliary cooling system is all that is required to maintain proper engine temperature. When the cooling system is not effective in maintaining engine temperatures within normal operating parameters, other measures must be taken. Such measures could include bringing in another engine to handle some of the load. Since such measures can take a while to implement, it is important that the pump operator constantly monitor the gauges so that actions can be taken before a problem becomes a crisis. In the past, one option was to open the engine

FIGURE 9-3 Supplemental cooling systems are required to make sure pumping apparatus can be kept within normal operating temperatures. The "engine cooler," which is sometimes labeled "auxiliary cooling," uses pump water to supplement what the radiator does to remove heat for the engine coolant. The "pump cooler" allows a small amount of water to leave from the pump discharge header so that water is always circulating through the pump, keeping it cool.

compartment to increase ventilation; however, this is not a recommended practice with current engines. Today's engines run at hotter temperatures, closer tolerances, and a more carefully designed air flow than prior engines. On today's engines, opening covers may lead to damage from "hot spots." Fortunately, with today's improved designs, overheating is less likely than it was in the past. If measures taken to reduce engine temperatures to safe levels fail, the operation should be shut down to protect the engine from extensive damage. Conversely, care should also be taken to avoid cooling the engine below its normal operating range because excessive cooling can reduce efficiency, cause excessive wear and tear, and potentially damage the engine.

Pump Cooling Systems

When the pump is engaged and water is flowing, the temperature inside the pump will normally be maintained within the normal temperature operating range. When discharge lines are closed or flows are reduced to a minimum, however, the temperature of the water in the pump can quickly rise. Increased water temperature inside the pump can cause serious injury to personnel through burns and can cause significant damage to the pump. To compensate, simply maintain water movement within the pump. Either flow water from an unused discharge or open the tank-to-pump and the pump-to-tank valves to recirculate water.

Recirculate Lines

One way of ensuring enough water is always flowing through the pump to keep it cool is the use of **recirculate lines**, which are small handlines or booster lines that discharge water in desirable places for operations. When drafting, recirculate lines normally discharge water at the water source, such as a dump tank, where the water is reused (recirculated). When used with hydrant-supplied water, recirculate lines are normally placed in a storm drain.

Cracking the tank fill is a great way to circulate water when water is coming from the booster tank; however, when water is coming from an external source, including hydrants and dump tanks, cracking the tank fill will eventually lead to overflowing the booster tank. While this works acceptably in warm climates, in freezing weather it creates unnecessary ice problems on the road from the overflowing water. With rural water supplies, it also is an unacceptable waste of water. Recirculate lines solve both of these problems in addition to keeping the pump cool.

SHUTTING DOWN AND POST OPERATIONS

Pumping operations are not complete until the pump is shut down and properly readied for the next operation. The desire to return to the station can tempt pump operators to wrap up the operation quickly. Shutting down pumping operations requires the same care and diligence to safety that is required when initiating them.

Shutting Down Operations

The steps to shutting down operations are similar to those used to initiate the operation but in reverse order.

Basic Steps to Shut Down a Pumping Operation

1. Slow the engine to idle speed.
2. Close all discharge control valves.
3. Check the water tank level (refill if necessary).
4. Disengage the pump drive.
5. Bleed off any remaining pressure from the discharge line.
6. Turn off pressure regulating devices and engine cooling systems (return them to their normal position).
7. Close the intake supply.
8. Disconnect intake and discharge hoses.
9. Remove wheel chocks before moving the apparatus.

Although the sequence of these steps may change depending on department policy and the pump manufacturer's recommendations, care should be taken regardless of the sequence. Carelessness in the shutdown sequence can cause damage and even injury to personnel resulting from:

- Excessively high pressure
- Generation of steam or near-boiling water temperatures
- Operating the pump dry
- Water hammer
- Cavitation

Post-operation Activities

Post-operation activities ensure that the pump is returned to a ready status. Pump operators should follow the pump manufacturer's recommendations for post-operation activities. Post-operation activities are discussed in the following subsections.

Flushing the Pump

The pump should be thoroughly flushed if foam, saltwater, dirty water, or contaminated water was used. When flushing the pump, use a clean water supply. **Back-flushing** may be required and is simply reversing the flow so that water enters the discharge side and leaves the intake side of the pump. Back-flushing is helpful in removing debris such as grass and leaves that entered the pump but did not go through the pump. While in the pump, grass and leaves can significantly reduce the pump's performance.

Inspect and Fill Liquid Levels

All liquid levels should be inspected and replenished as necessary. For example, the tank water level and the priming oil tank reservoir should be checked and filled if necessary. In addition, engine coolant, oil, and fuel levels should be checked and filled if necessary.

Intake and Regulator Screens

If water used during pump operations is suspected of containing dirt, sand, or other debris, intake screens, as well as regulator screens, should be inspected and cleaned if necessary. Back-flushing may also be required.

Control Valves

Pump operators should ensure that all discharge and intake control valves, drain valves, transfer valves, and other components are returned to their normal positions. If the primer was used during pump operations, it is a good idea to operate the priming device briefly. In doing so, the close-fitting parts of the priming device are lubricated, helping to ensure the system remains in good operating order.

STREETSMART

When operating oil-less primers, it is important to allow water to flow through the primer pump. If an oil-less primer is operated dry, the vanes create dust inside the primer pump. If the dust is allowed to accumulate in the primer, it will result in sticking vanes and primer pump failures. Water flowing through the primer pump flushes the dust from the pump and alleviates the problem.

PUMP OPERATION CONSIDERATIONS

Several conditions may be encountered that affect the ability to conduct pump operations safely and efficiently. Pump operators must be familiar with these conditions in order to detect them and compensate for them. The most common conditions include **water hammer**, **cavitation**, and the effects of environmental factors.

Water Hammer

Water hammer is a surge in pressure created by the sudden increase or decrease of water. This surge is created by the velocity of the water as well as its weight (quantity) within a hose. The potential for damage from water hammer is significant given the velocity and weight of water during pumping operations. The velocity of water on the discharge side of the pump is often high, with normal pump pressure in excess of 100 psi (700 kPa). In addition, the weight of water in the hose during normal pumping operations is also high. For example, a 400-foot (120-m) section of 3-inch (77-mm) hose filled with water weighs 1,253 lbs (568 kg). Imagine the effect of a little more than half a ton of water at 100 psi (700 kPa) coming to an immediate stop!

Obviously, this surge in pressure can be high enough to damage the pump and hose as well as cause personnel to lose control of the hose. Pump operators can reduce the likelihood and extent of water hammer in several ways. The best protection, perhaps, is simply knowing that it can occur. All personnel, not just the pump operator, should understand the effects and consequences of water hammer because hydrants, nozzles, and appliances are not always operated by just the pump operator. The best way to reduce water hammer is to start and stop water flows slowly by opening and closing control valves slowly for pump intakes and discharges, as well as hydrants, nozzles, wyes, hose clamps, and other similar appliances. The use of pressure regulating devices can also help reduce the effects of water hammer.

As a result of past water hammer damage, numerous standards have changed to reduce the likelihood of a serious water hammer occurring. Today, valves that are 3 inches (77 mm) or larger in diameter must not be able to be opened or closed in less than 3 seconds and must be able to be operated in 10 seconds or less. Large-diameter appliances and pumps must also have relief valves that will discharge water to the ground and help to abate the higher pressures that accompany water hammers.

In spite of manufacturer-implemented safety devices, water hammer can still occur, particularly when filling long hoselines. Be sure to open a hoseline valve only partially until you have seen pressure building in the hoseline. Remember, if a water hammer does occur, it not only occurs at the end of the hoseline but then moves back through the hoseline to the pump, supply lines, hydrant, and water mains. Water hammer has sometimes even broken the water main in the ground.

Cavitation

The damage caused by cavitation can be significant, resulting in reduced pump efficiency and, when severe enough, extensive pump repairs. Often referred to as "the pump running away from the water supply," cavitation occurs when a pump attempts to flow more water than the supply can provide. Most often, cavitation occurs when using nonpressurized intakes but can just as easily occur when using a hydrant as a supply source. The primary cause of cavitation is the effect of pressure zones in the pump as they act on water.

Pressure Zones

When more water is being pumped than the supply can deliver, two pressure zones are produced within the pump (**Figure 9-4**). One is a low-pressure zone developed at the eye (center) of the impeller. This low-pressure zone occurs when the supply is unable to deliver the amount of water being discharged. The other is a high-pressure zone that is the result of increased velocity (pressure) as water moves from the center of the impeller to the outer edge of the impeller. The outer edge of the impeller has greater velocity than the center of the impeller, thus creating more velocity (pressure) toward the outer edge.

Effects of Pressure on Water

At normal atmospheric pressures (14.7 psi, [760 mm, or 101.325 kPa] at sea level), water boils at 212 degrees Fahrenheit (100 degrees Celsius). As shown in **Table 9-1**, the boiling point of water is affected by pressure. When pressure increases, the boiling point of water increases, meaning that water boils at higher temperatures. Conversely, when pressure decreases, the boiling point of water is reduced. When this occurs water boils at lower temperatures. The rate at which water evaporates is closely associated with

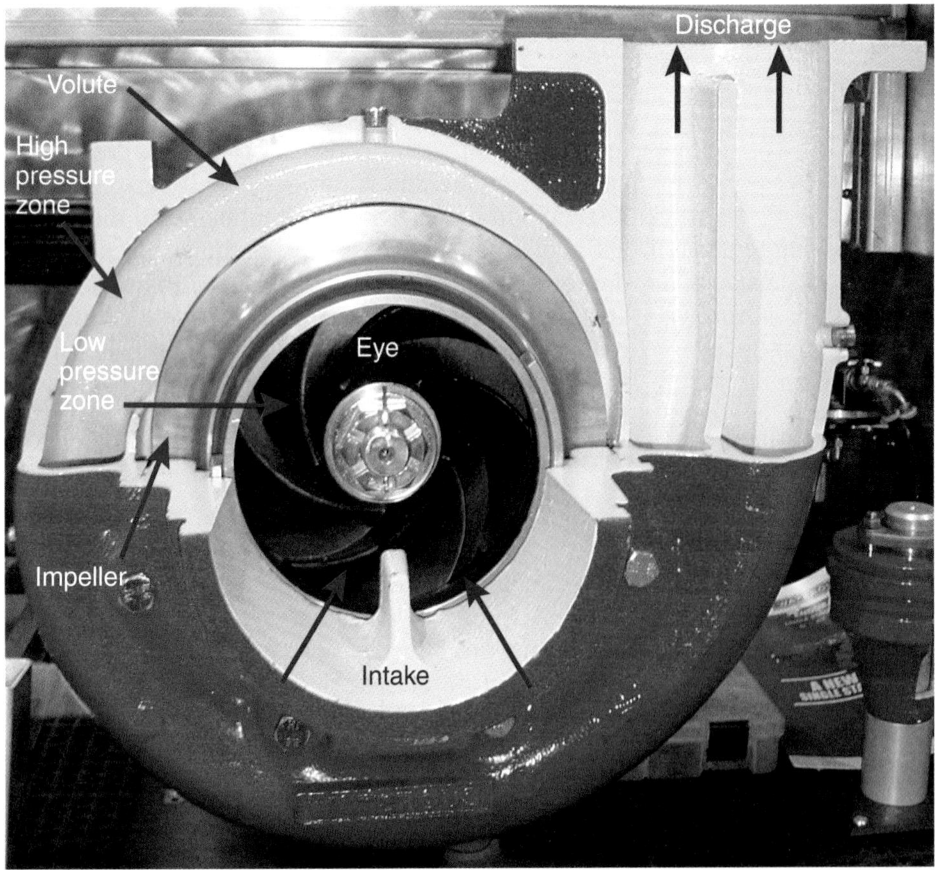

FIGURE 9-4 Two pressure zones are created within the pump by the spinning impeller. When the supply cannot provide enough water to the discharge, air bubbles form in the low-pressure zone near the eye of the impeller. These air bubbles implode or collapse in the high-pressure zone near the outer vanes of the impeller. The creation and imploding of air bubbles is known as cavitation.

Courtesy of Waterous Company

TABLE 9-1 Boiling Points of Water

Vacuum/Pressure		Degrees F	Degrees C
29 in. Hg	736 mm	79.03°F	26.1°C
25 in. Hg	635 mm	133.76°F	56.5°C
20 in. Hg	508 mm	161.49°F	71.9°C
15 in. Hg	381 mm	179.14°F	81.7°C
10 in. Hg	254 mm	192.37°F	89.0°C
5 in. Hg	127 mm	203.06°F	95.4°C
0 psi	0 kPa	212.00°F	100.0°C
5 psi	35 kPa	227.96°F	108.9°C
10 psi	70 kPa	240.07°F	115.6°C
15 psi	105 kPa	250.33°F	121.3°C
20 psi	140 kPa	259.28°F	126.3°C
25 psi	175 kPa	267.25°F	130.7°C
30 psi	210 kPa	274.44°F	134.7°C

its boiling point at a specific temperature. The closer water comes to boiling, the greater the rate of evaporation. Increased pressure reduces the ability of a liquid to vaporize, while decreased pressure increases the ability to vaporize. In other words, the more pressure there is pushing down on water, the less it can evaporate and, consequently, higher temperatures are required for it to boil. When less pressure is pushing down, it is easier for water to evaporate and, consequently, lower temperatures are required for it to boil.

Effects of Cavitation

When water passes through the low-pressure zone, its boiling point is reduced and its ability to vaporize increases. The low-pressure zone allows water to vaporize readily and form vapor pockets. When these vapor pockets pass through the high-pressure zone, the boiling point of water increases and the ability of water to vaporize decreases. The high-pressure zone forces the vapor pockets back into the original liquid state. Often the return of vapor pockets to a liquid is forceful and is an implosion.

STREETSMART

Operators need to avoid cavitation where gaseous water vapor forcefully returns to a liquid in an implosion. This causes permanent pump wear, eventually requiring expensive avoidable repairs.

Thousands of small vapor pockets commonly form in the low-pressure zone typically described as a region. Consequently, thousands of implosions occur in the high-pressure region of the pump. These small implosions can cause pitting on the smooth surface of the impeller, resulting in increased friction loss, and can cause the impeller to become unbalanced. In either case, the efficiency of the pump is diminished.

Signs of Cavitation

In some cases the only warning sign of cavitation is the knowledge that pumping more water than the supply can handle will cause cavitation. That is to say, cavitation may occur with no outward warning signs. In other cases the signs that cavitation is occurring are obvious. Signs include:

- No corresponding increase in pressure as engine speed is increased
- Engine speed automatically increases during a pumping operation
- Fluctuating discharge pressure readings
- Rattling sounds like sand or gravel going through the pump
- Excessive pump vibrations
- Sudden pressure or capacity loss
- Too high a vacuum reading on the master intake gauge (intake vacuum should not exceed 20–25 in. Hg or 508–636 mm)

STREETSMART

Most signs of cavitation become more severe as the cavitation becomes more severe. So, for example, just because you do not hear rattling sounds like gravel going through the pump does not mean cavitation is not occurring; the noise may just not be audible over the sound of the engine. The two most reliable signs of cavitation are no increase in pressure as engine speed is increased and too high a vacuum reading. The other signs may be thought of as "late" signs or signs of severe cavitation.

Stopping Cavitation

The quickest way to stop cavitation is simply to reduce the pump speed. This results in the immediate reduction or elimination of the two pressure zones. Cavitation can also be stopped by increasing the flow of the supply to meet the demands of discharge flows adequately. This is usually done by adding an additional supply hoseline. Finally, cavitation can be

EFFECTS OF CAVITATION

Think of the water implosions, or the filling of the air bubbles created by cavitation, as little hammers hitting the impeller. As cavitation grows from mild to severe, the size of the air bubbles increases, just as the size of air bubbles increases in a pot of boiling water on the stove when it moves from a slow boil to a rapid boil. The bigger bubbles of severe cavitation are like big hammers hitting the impeller and become so noisy that they can even be heard over the noise of the apparatus engine.

Now think of the effects a hammer hitting the body of your car would have. Each place the hammer hit the car would leave a dent, and each dent would remain on the car unless a repair action was completed. The same is true with a pump impeller. While cavitation should be minimized, any pump can take occasional mild cavitation, just like a car will still work after getting hit in a parking lot by a shopping cart. With proper care a pump's impeller should last the 25-year life of an apparatus. With too much cavitation, however, the pump's impeller will need to be replaced due to the cumulative damage from each time the pump was cavitated.

So remember, just because you do not notice a problem immediately after a pump has been cavitated, that does not mean that damage has not been done to the pump. In fact, the same cumulative effects apply to other wear items. For example, pump performance decreases as wear ring clearances increase due to sediment that goes through the pump and effectively sandblasts the soft internal bronze parts of the pump.

stopped by reducing the amount of water being discharged. Examples for reducing discharge flows include:

- Removing one or more discharge lines
- Reducing the size of discharge lines
- Reducing nozzle gallons-per-minute (litres-per-minute) settings for one or more lines
- Reducing pump discharge pressure

Cavitation may also be stopped by reducing lift while drafting or when this is not possible using a lift pump.

Environmental Considerations

Normal changes in weather conditions typically do not adversely affect pumping operations. Cold weather can actually help engine performance. Hot weather effects can usually be controlled by the auxiliary cooler and by recirculating pump water. The extremes, however, can be menacing to pump operators.

Cold Weather Operations

Extremely cold weather can cause major problems with pumps. When the pump is susceptible to freezing, all water must be removed from the pump and associated piping, such as discharge lines, auxiliary cooling system, pressure regulating line, and primer lines. During cold weather operations, intake and discharge lines should be laid out so that water can flow as soon as the pump is primed. The movement of water in the pump should not be stopped until the operation is complete and the pump and piping can be drained. Whenever possible, leaks should be stopped to reduce the accumulation of ice.

Steps That Should Be Taken to Drain the Pump

1. Open all intake and discharge control valves and their caps.
2. Open the main pump drain valve (some pumps may require discharge drain valves to be opened as well).
3. For multistage pumps, operate the transfer valve several times.
4. Drain the pressure regulating devices (if required by the manufacturer).
5. After the pump is drained, close all discharge and intake control valves as well as the drain valves.

In many jurisdictions with frequent severe cold weather, the department may also include a number of options, such as insulated and perhaps heated pump compartments instead of compartments whose bottoms are open to the environment. The operator's compartment may also be incorporated into the cab instead of being outside.

Hot Weather Operations

Extremely hot weather can also cause problems during pump operations. As discussed earlier in this chapter, both the engine and the pump must be maintained within normal operating temperatures. When operating in hot weather, pump operators must take into consideration the temperature of the supply water, especially if the supply is a static source. The major concern is that warmer water is more susceptible to cavitation because less energy is required to bring the water to its boiling point.

SUPPORTING SPRINKLER AND STANDPIPE SYSTEMS

Pump operators may be called upon to support sprinkler and standpipe systems. Sprinkler and standpipe systems are fixed systems within buildings and structures that are used to help protect lives and property by providing immediate detection and/or extinguishing capabilities. To support these fixed systems properly, the pump operator must be familiar with the basic types of systems, their operations, and support considerations. NFPA 13E, *Recommended Practice for Fire Department Operations in Properties Protected by Sprinkler and Standpipe Systems*, suggests that inspecting and preplanning sprinkler and standpipe systems is critical to the development of appropriate support procedures. Selecting a water source, locating where to connect to the system, and determining what pump pressure to use should be accomplished during preplanning and should be incorporated into standard operating procedures.

Sprinkler Systems

According to NFPA 13E, when sprinkler systems are properly designed, installed, and maintained, they can provide water to the fire in a more effective manner than using manual fire suppression methods. Sprinkler systems are designed to move water from a source, through piping, to one or more discharge points (sprinkler heads). The water supply for most sprinkler systems is designed to supply only a small number of sprinkler heads. This reduces both water requirements and water damage. Even a small strip mall may have several hundred heads that, if all were flowing, would require many thousands of gallons (litres), exceeding the capabilities of most departmental pumps. For larger fires, when multiple sprinkler heads are activated, the pump operator must provide additional water to the system. Sprinkler systems can be installed for both interior and exterior use. Interior systems are designed to keep fire growth contained or to extinguish the fire. Exterior sprinkler systems are designed primarily to protect exposed properties from the spread or extension of fire.

The typical sprinkler head has a ½-inch (13-mm) orifice and a flow rate of 25 to 40 gpm (100 to 160 Lpm) depending on pressure. These heads come in several styles, including: pendant, which hang down and are often used with drop ceilings; upright, which are placed above a branch line sprinkler pipe and are the only type used with dry-pipe systems; and sidewall, which are mounted on walls. Any of these types are available with either fusible link or glass bulb (sometimes called an ampule) devices. These devices detect the temperature present and activate the sprinkler at the correct temperature. Sprinkler heads set for different temperatures are available to meet the needs of the different locations where they will be installed.

There are several common parts or components to most sprinkler systems (**Figure 9-5**). One common component is a water supply source. Sprinkler systems are connected to a main water supply, typically either a municipal or private water supply source. In many cases, particularly with larger systems, they are connected to multiple water supply sources. Sprinkler systems may get their water from a number of different sources, including public or municipal water supplies, elevated or gravity tanks, ground-level water tanks, and ponds, lakes, rivers, and streams. When multiple water supplies exist, the water on the property that is dedicated to fire protection is referred to as the **primary water supply**. The municipal water system is referred to as the **secondary water supply**. When both supplies exist, the secondary supply is the preferred water supply. This prevents depleting the primary water supply. If the only water supply for a sprinkler is a public or municipal water supply, it is considered the primary water supply. In addition, sprinkler systems have a fire department connection (FDC) that provides both a means to increase the pressure of the water within the system and an alternative way to supply water. NFPA 13, *Standard for the Installation of Sprinkler Systems*, requires that FDCs have caps that can easily be removed by fire department personnel. Most FDCs are Siamese with two, 2½-inch (65-mm) female connections, although a number of areas allow or require LDH Storz fittings on new FDCs. One-way check valves are most commonly installed on a sprinkler system, especially if the system is connected to a municipal water system. These valves, which are also called backflow preventers, prevent water from within the sprinkler system from entering and possibly contaminating the connected water supply. Each type of sprinkler system has a valve that is used to initiate the flow of water from the connected water supply to the system. The basic types corresponding to the types of

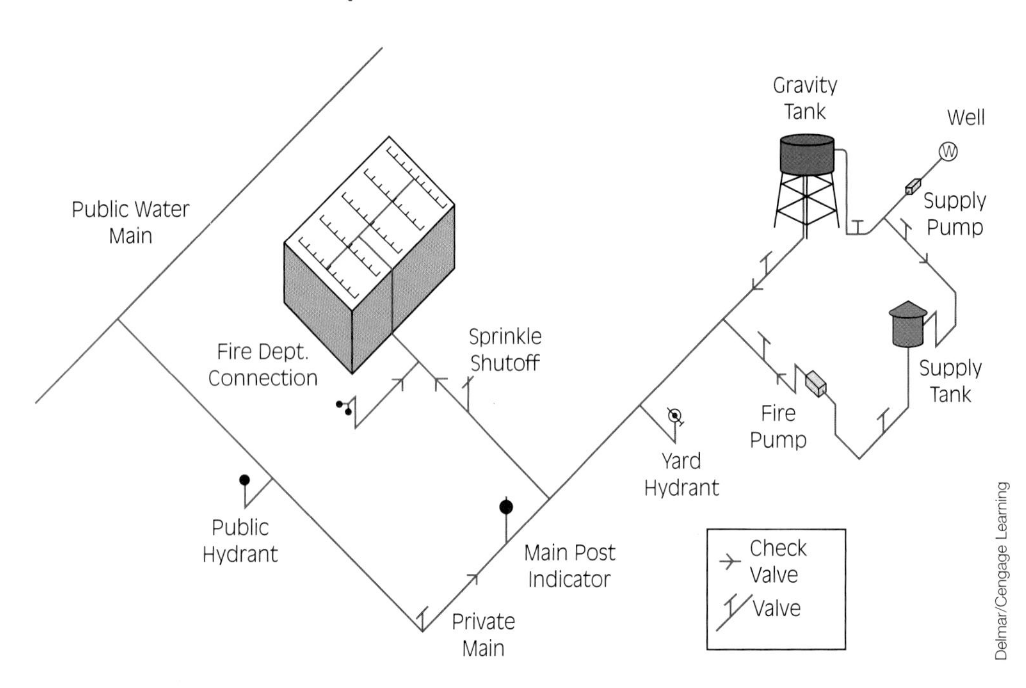

FIGURE 9-5 The diagram shows the common parts of a sprinkler yard system.

systems include: main operating valve or wet-pipe system, dry-pipe valve, preaction valve, and deluge valve.

Piping or tubing moves the water from the supply to one or more sprinkler heads. First, the riser moves water from the supply source to the feeder mains, then to the cross mains, and finally to the branch mains onto which sprinkler heads are connected. The water supply control valve is usually located on the riser. This valve is most often an outside stem and yoke (OS&Y) indicating valve and allows for quick visual determination of the valve position (open or shut). A main drain control valve allows the system to be drained for maintenance or repair. Finally, an inspector's test valve, located at the farthest end of the system, is used to simulate the single head flow to test system response (**Figure 9-6**).

Sprinkler systems are also equipped with water flow alarms that activate when the system is in operation. According to NFPA 13, water flow alarms must activate when the flow of water is equal to or greater than that from a single sprinkler head with the smallest orifice. The alarm must initiate within 5 minutes after flow begins and continue until the flow stops. A retard chamber is usually provided to help prevent the system alarm activation from sudden water supply pressure surges. This is accomplished by collecting water within the chamber prior to activating the alarm. Water supply surges will normally not provide the sustained flow into the chamber that is required to activate the system. A small opening at the base of the chamber allows the water to drain from the retard chamber.

Finally, all sprinkler systems have sprinkler heads. Most sprinkler heads, except those installed on deluge systems, are closed heads equipped with some sort of heat-sensitive device or operating mechanism. This sensing device is usually a fusible

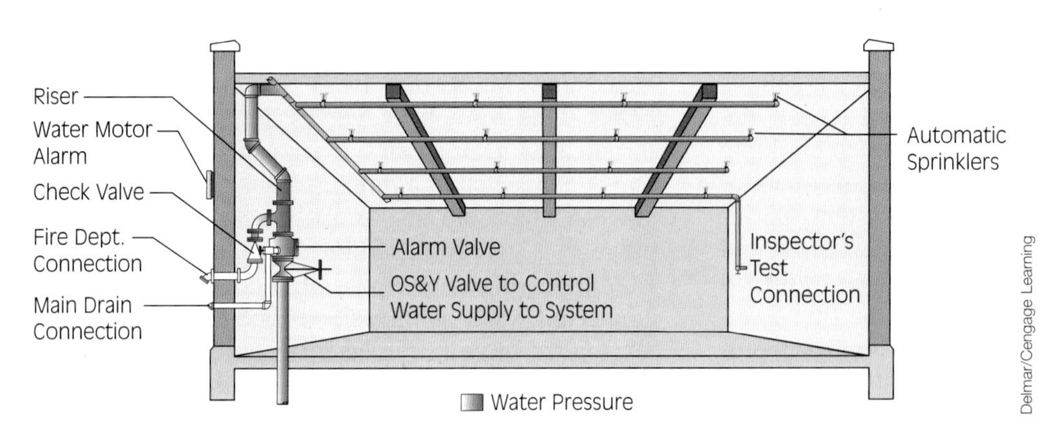

FIGURE 9-6 The diagram shows the common parts of a sprinkler system.

TABLE 9-2 NFPA 13 Required Sprinkler Head Temperature Rating, Classification, and Color Coding

Maximum Ceiling Temperature		Temperature Rating		Temperature Classification	Color Code	Glass Bulb Colors
°F	°C	°F	°C			
100	38	135–170	57–77	Ordinary	Uncolored or black	Orange or red
150	66	175–225	79–107	Intermediate	White	Yellow or green
225	107	250–300	121–149	High	Blue	Blue
300	149	325–375	163–191	Extra high	Red	Purple
375	191	400–475	204–246	Very extra high	Green	Black
475	246	500–575	260–302	Ultra	Orange	Black
625	329	650	343	Ultra high	Orange	Black

link or a chemical pellet that melts at a fixed temperature, or a liquid-filled tube that bursts at a fixed temperature. Sprinkler heads come in a variety of temperatures ratings. The NFPA 13 required sprinkler head temperature ratings and classifications are provided in **Table 9-2**. The three basic sprinkler head designs are upright, sidewall, and pendant (see **Figures 9-7**, **9-8**, and **9-9**). Note that each style of sprinkler head has a different deflector design based on how the sprinkler head will be mounted. Finally, sprinkler heads operate at a fixed discharge rate and pattern.

FIGURE 9-7 Upright sprinkler heads are mounted on top of sprinkler piping and are the only type of head used on a dry-pipe sprinkler system.

FIGURE 9-8 Sidewall sprinkler heads are mounted through the wall of a building.

FIGURE 9-9 Pendant sprinkler heads drop below the sprinkler pipe. They must be used where a sprinkler pipe is above a ceiling and only the head will show through. In some cases a decorative cover hides the sprinkler head. Pendants are the most common type of sprinkler head.

There are four basic types of sprinkler systems: wet-pipe systems, dry-pipe systems, deluge systems, and preaction systems.

STREETSMART

Since sprinkler systems have a check valve called a backflow preventer that prevents water in the sprinkler system from flowing back into the domestic water system, higher pressures within the sprinkler system do not drop along with drops in domestic water pressure. So the sprinkler system pressure remains constant while the domestic pressure varies along with demand changes in the domestic system. If the domestic system pressure ever exceeds the sprinkler system pressure, the check valve opens, allowing water to flow into the sprinkler system, thus causing the pressure in the sprinkler system to match the highest pressure in the domestic system. Maximum domestic pressure generally occurs in the early morning, just before people wake up and start using water.

Wet-pipe Systems

As the term implies, wet-pipe systems contain water within the piping from the water source to the sprinkler head. Because of this, these systems are mainly used within heated structures or where freezing is not a concern. The water within the system is under pressure and the sprinkler heads are closed. Before activation, the pressure within the system is higher than the pressure from the connected water supply. Pressure gauges are usually located on either side of the main check valve, providing direct reading of both pressures.

Heat from the fire will melt the fusible element on the sprinkler head, which opens a discharge orifice. When a sprinkler head is activated, water immediately discharges from the system. As the sprinkler system pressure drops, several actions occur within the system. First, the main check valve opens, allowing water from the supply source to flow into the system. The check valve prevents backflow from the system to the water supply when the system operates and in some cases also operates a flow alarm. Second, water flows into the retard chamber and activates the alarm system. The alarm system can be configured for a local alarm, such as a water gong, and/or can send an electronic signal to a monitored or supervised system or to notify a fire department. Water continues to flow through the fused sprinkler heads until the water supply control valve is closed, sprinkler wedges are inserted into flowing sprinklers, or sprinkler heads are replaced. Many departments prefer the use of sprinkler wedges since it allows the sprinkler system to continue to protect the remainder of the facility and the fire department does not assume the liability for repairing the system.

Wet-pipe systems are less expensive to install than other types of sprinklers, are reliable when properly maintained, and are by far the most common type of sprinkler system.

Dry-pipe Systems

Dry-pipe systems contain compressed air as opposed to water within the system. These systems are used where freezing temperatures occur. The compressed air in the system is of sufficient pressure to cause the water supply control valve to remain closed. This is usually accomplished by a differential valve, which allows a smaller amount of air pressure within the system to hold back the higher pressure of the connected water supply. When a sprinkler head opens, air initially escapes from the head. The pressure in the system is reduced enough to cause the dry-pipe control valve (water supply valve) to open, allowing water to enter the sprinkler system. Because air is initially contained in the system, water does not immediately discharge from the sprinkler head as it does with wet-pipe systems. Some systems use an exhauster or accelerator to remove air from the system quickly in order to reduce the time it takes to start discharging water. According to NFPA 13, dry-pipe systems must be capable of discharging water to the most remote sprinkler in not more than 60 seconds.

Preaction Systems

A preaction system is similar to both a dry-pipe system and a wet-pipe system. First, as with a dry-pipe system, the piping is void of water and the sprinkler heads are closed. Air within the system may or may not be compressed. The main difference with this system is that a preaction valve is used to open the connected water supply. This preaction valve is activated via a connected detection system. When heat or smoke is detected, the preaction valve opens, allowing water to flow into the sprinkler system. At this point, the system operates similarly to a wet-pipe system. When a sprinkler head is fused, water immediately discharges. Preaction systems are normally used when an accidental discharge would cause an unacceptable loss, such as in a computer room. Since two separate items are required (fire detected and sprinkler head activated), no single failure will result in an unnecessary water discharge.

Deluge Systems

The difference between a preaction system and a deluge system is that all the sprinkler heads in a deluge system are open. A separate detection system is used to open the water supply valve. Water entering the system flows through all of the sprinkler heads at the same time. These systems are used when rapid fire spread is a concern, such as within an aircraft hanger. Because of the large water flow requirements of these systems, larger piping is typically installed and separate pumps may be required to ensure appropriate flows and pressures are provided to the system.

Support Considerations

As discussed earlier, sprinkler system support should be preplanned and incorporated into standard operating procedures. Several preplanning considerations include the following:

- Identify the primary and secondary water supplies for the system.
- Locate control valves. All control valves should be opened unless the valve is controlling flow to a part of the sprinkler system that is being repaired or renovated.
- Confirm that the secondary water supply will not degrade the system's primary water source. Open water supplies should not be used when the sprinkler system is directly connected to a municipal or potable water supply unless appropriate backflow devices are installed.
- Locate fire department connections and possible hose layout.
- Set appropriate discharge pressures and flow for system operations.

According to NFPA 13E, first-arrive companies should take prompt action to supply sprinkler systems. Specifically, at least one hoseline should be connected to the FDC and more lines added as required by fire conditions. Prior to connecting to the FDC, check to ensure no debris has accumulated within the Siamese (see **Figure 9-10**). The Siamese should have a clapper valve that allows one 2½-inch (65-mm) hoseline to be charged at a time. If the female swivel on the Siamese is frozen, a double male and double female from the apparatus may be added to it to provide a working swivel. If not, then connect both lines prior to charging the lines or place a gated valve on the second 2½-inch (65-mm) inlet. Unless otherwise indicated on the system, the supply line should be pumped at 150 psi (1,050 kPa). If hose streams are used, pump operators should secure a water supply that will not reduce water flow to the sprinkler system. One indication that water is actually flowing into a sprinkler system is a drop in intake pressure when the pump discharges are opened. No drop in intake pressure does not mean water is not flowing. It could just be too small of a flow to cause a drop in intake pressure such as when only one sprinkler is flowing.

FIGURE 9-10 The fire department connection should be inspected for debris prior to connection of the supply hose. In the photo, a discarded soft drink bottle has been placed in the sprinkler connection.

(For step-by-step photos of Connecting to a Sprinkler, please refer to page 301.)

Although it is possible to supply water to a sprinkler system that does not have an FDC by pumping water into a yard hydrant, the operator should understand the piping and that a backflow preventer is in place, preventing contamination of any connected public water supplies. This necessary information should be included in the site's preplan.

NFPA 13 suggests securing a water source from one of the following when hoselines will be used:

- Large mains that can support both hoseline operations and the sprinkler system
- Water mains not needed for sprinkler supply
- Drafting from static sources

Other considerations for sprinkler system support include the following:

- When fire and smoke are visible, start pumping to the sprinkler system immediately.
- If the sprinkler system has been activated but no fire or smoke is evident, wait until interior crews verify if a fire is present before pumping to the system. The system could have malfunctioned or extinguished the fire.
- Do not shut down a sprinkler system for improved visibility.
- When instructed, turn off the sprinkler system and stop flowing water to the FDC. Some FDC piping bypasses the sprinkler control valve.
- When using a two-stage pump, place the transfer valve into the volume position.
- Look to see if the discharge pressure is on the FDC.
- As a general rule of thumb, pump the FDC at 150 psi (1,050 kPa) unless otherwise noted.

The pump operator should only stop pumping water into a sprinkler system when instructed to do so by the incident commander. A number of cases exist where more damage than necessary has occurred when pumping was stopped as soon as crews thought the fire was under control, only to later learn that it was the sprinkler system that was keeping the fire from spreading further.

Standpipe Systems

Standpipe systems are installed to provide a pre-piped water supply to hoselines within large structures such as warehouses, high-rises, and industrial buildings. In much the same manner as sprinkler systems, standpipe systems move water from a source, through fixed piping, to hose connections. Standpipe systems can be supplied from one of two different water sources. First, the system can be permanently connected to a water supply source in the same fashion as automatic sprinkler systems. Second, the system can be connected to an FDC. In this case, all of the water required in the system must be provided by a pumper connected to the system.

(For step-by-step photos of Connecting to a Standpipe, please refer to page 302.)

With such systems, the fire department connection must be marked as "STANDPIPE" and the required pressure must be indicated. The system is comprised of control valves, check valves, risers, and cross mains, similar to sprinkler systems. The discharge points for standpipe systems are hose connections as opposed to sprinkler heads.

According to NFPA 14, *Standard for the Installation of Standpipe and Hose Systems,* there are three classes of standpipe systems that are based on their intended users. Class I standpipe systems provide 2½-inch (65-mm) hose connections and are intended to be used by firefighters or fire brigade members. Class II standpipe systems provide 1½-inch (38-mm) hose stations and are intended primarily for trained personnel during initial attack efforts. Some Class II systems may have 1-inch (25-mm) hose for areas of light-hazard occupancies. Class III standpipe systems provide 1½-inch (38-mm) and 2½-inch (65-mm) hose connections for use by firefighters and fire brigade members. Some Class III systems may also have 1-inch (25-mm) hose for light-hazard occupancies. Class I and III systems have a minimum flow rate of 500 gpm (2,000 Lpm), and each additional standpipe shall flow 250 gpm (1,000 Lpm), up to a total flow of 1,250 gpm (5,000 Lpm). Achieving such high flow rates may require multiple MDH supply lines or an LDH supply line connected through the pump's streamer intake. Class II standpipe systems have a minimum flow rate of 100 gpm (400 Lpm). Adequate water supply must be available to sustain the standpipe for a minimum of 30 minutes.

NFPA 14 also lists the five basic types of standpipe systems as follows:

1. *Automatic dry*. This dry-pipe system is filled with air under pressure. When a hose valve is opened,

the dry-pipe valve opens, allowing water to flow into the system. This water supply must be capable of supplying the required flow rate.

2. *Automatic wet.* An automatic wet system is filled with water. The water supply must be able to supply the required flow for the system automatically.
3. *Semi-automatic dry.* This system is a dry-pipe system with a deluge valve activated by a remote device at each hose connection. The water supply must also be capable of supplying the system demand.
4. *Manual dry.* This is a dry-pipe system without a permanently connected water supply. These systems require a fire department to supply the standpipe through an FDC.
5. *Manual wet.* This is a wet-pipe system without a permanently connected water supply that can fully provide the required fire flow. As with a manual dry-pipe system, manual wet-pipe systems require a fire department to supply the systems through an FDC.

Support Considerations

As with sprinkler system support, preplanning is critical to support standpipe systems properly. The same basic water supply considerations for sprinkler systems should be assessed when selecting a water supply source to support standpipe operations. Determining the correct pressure and flow for the standpipe system is of critical importance. NFPA 13E suggests that pump operators should consider the following when calculating pump discharge pressure:

- Friction loss in the hoseline connected to the fire department connection
- Friction loss in the standpipe itself
- Pressure loss due to elevation of nozzle or height of the standpipe when pressure reducing valves are used
- Number and size of attack lines operating from the standpipe
- Desired nozzle pressure

The best time to calculate pump discharge pressure is before an incident occurs. Friction loss calculations are discussed in detail in **Section IV** of this text.

TROUBLESHOOTING

Problems encountered during pumping operations can be reduced by becoming familiar and experienced with pumps and their systems, including knowledge of pump construction, operating principles, operating procedures, and proper preventive maintenance.

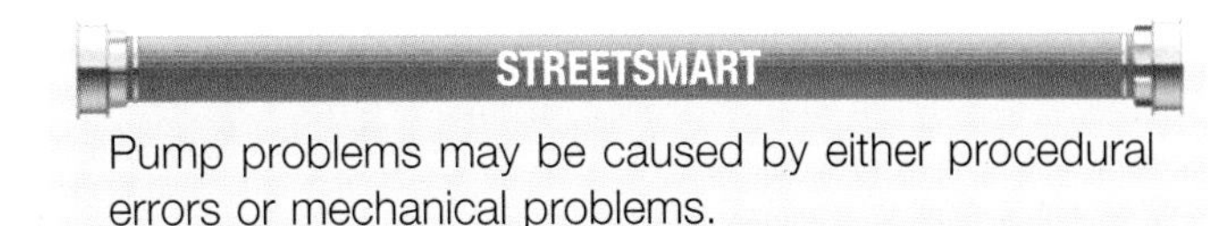

Pump problems may be caused by either procedural errors or mechanical problems.

In general, pump troubleshooting can be grouped into two categories: procedural problems and mechanical problems. If proper procedures were followed, it is more than likely that a mechanical problem exists. In either case, the best method for troubleshooting a pump problem is to follow the flow of water from the intake to the discharge while attempting to diagnose the problem. In most cases, problems are readily identifiable and correctable using this method. Pump manufacturers typically provide troubleshooting guides specific to their pumps (see **Appendix E**).

SKILL 9-1

Producing an Effective Hand Stream from the Internal Tank

SAFETY

When using the internal tank, both the maximum flow rate (gpm or Lpm) and supply (gallons or litres) are limited. For a single handline, the NFPA required 500-gpm (2,000-Lpm) flow rate, while less than the pump's capacity, will not be a limitation; however, the tank's capacity, often only several hundred gallons (about 1,000 litres), will prevent sustained handline use without an additional water supply. The pump operator should know how long the water supply will last and convert to an alternate water source without a change in discharge pressure prior to running out. When this is not possible due to the lack of an alternate source, the operator needs to make sure command has ample warning so that firefighters will not be subjected to any hazards due to a water supply restriction.

Delmar/Cengage Learning

A When directed to charge a deployed line that is completely clear of the hose bed and connected to an engine discharge with a pump in gear and the tank-to-pump valve open, the operator should slowly open that discharge, preventing any possible water hammer.

Delmar/Cengage Learning

B Increase the throttle or governor as required to build to the desired discharge pressure.

Delmar/Cengage Learning

C Set the pressure relief valve or governor for the desired pressure.

SKILL 9-2

Producing a Hand or Master Stream from a Pressurized Source

Delmar/Cengage Learning

Delmar/Cengage Learning

A When directed to charge a deployed line that is completely clear of the hose bed and connected to an engine discharge with a pump in gear and the pressurized intake valve open, the operator should slowly open that discharge, preventing any possible water hammer.

B Increase the throttle or governor as required to build the desired discharge pressure.

Delmar/Cengage Learning

C Set the pressure relief valve or governor for the desired pressure.

SKILL 9-3

Producing an Effective Hand Stream from a Static Source

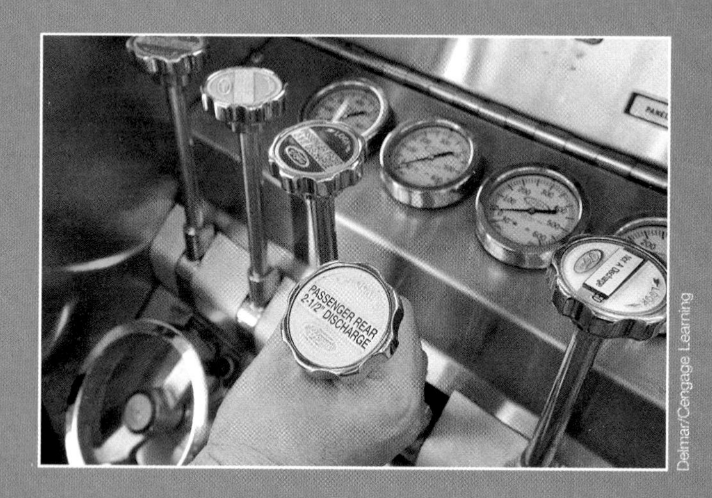

A When directed to charge a deployed line that is completely clear of the hose bed and connected to an engine discharge with a pump in gear and the pump primed, the operator should slowly open that discharge, preventing any possible water hammer.

B Increase the throttle or governor as required to build to the desired discharge pressure.

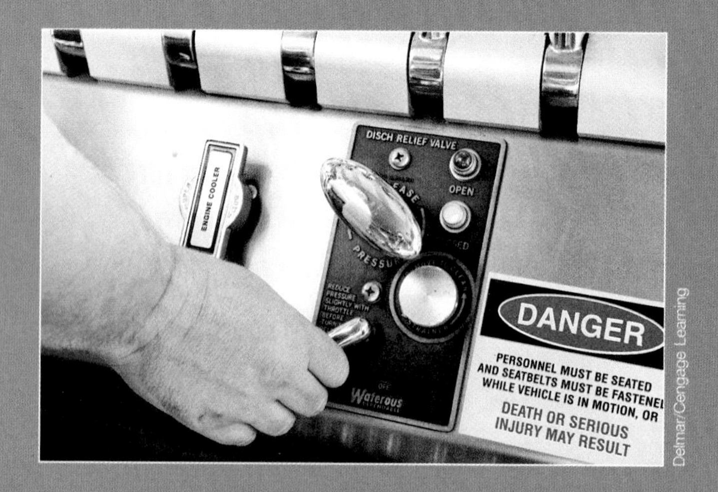

C Set the pressure relief valve or governor for the desired pressure.

SKILL 9-4

Connecting to a Sprinkler

Delmar/Cengage Learning

A After establishing a water supply, extend a hoseline with a compatible connection to the sprinkler. Note: Commercial sprinklers will use at least two medium-diameter connections or a large-diameter connection.

Delmar/Cengage Learning

B Remove the cap(s) from the sprinkler connection, remove any debris, confirm the clapper valve is not stuck in the closed position, and, if successful, connect the hoseline.

Delmar/Cengage Learning

C As soon as the hoseline is successfully connected, the pump operator should slowly charge the hoseline, avoiding a water hammer, build to the correct pressure (150 psi [1,050 kPa] unless otherwise noted), and set the relief valve.

Delmar/Cengage Learning

D When using MDH, a second hoseline should be connected and charged as soon as possible.

SKILL 9-5

Connecting to a Standpipe

A After establishing a water supply, extend a hoseline with a compatible connection to the standpipe. Note: Commercial standpipes will use at least two medium-diameter connections or a large-diameter connection.

B Remove the cap(s) from the standpipe connection, remove any debris, confirm the clapper valve is not stuck in the closed position, and, if successful, connect the hoseline.

C Connect the hose to the fire department connection (FDC).

D As soon as the hoseline is successfully connected, the pump operator should slowly charge the hoseline, avoiding a water hammer, build to the correct pressure, and set the relief valve.

LESSONS LEARNED

The task of discharge maintenance deals with those activities related to initiating and maintaining discharge lines. The procedures for initiating discharge lines vary depending on pressures and flows of both the new line(s) and those already in operation. Maintaining discharge lines focuses on those procedures that keep the engine and pump within normal operating parameters. Discharge maintenance also includes procedures for operating pumps over extended periods, pausing operations, shutting down operations, and post-operation activities. Two common problem conditions that can occur in almost any pump operation are water hammer and cavitation. Understanding why these conditions occur and how to prevent them is vital for the safety of both equipment and personnel. Preplanning is perhaps the most important activity required to ensure that sprinkler and standpipe systems are properly supported. Troubles encountered during pump operations are most likely the result of equipment malfunction or procedural error.

KEY TERMS

Back-flushing A process in which a pressurized water line is connected to a discharge while pump intakes are opened. The procedure allows removal of debris that has become trapped within the pump. The procedure is performed when the pump is not running.

Cavitation The formation and collapse of vapor pockets when certain conditions exist during pumping operations. Cavitation causes damage to the pump and should be avoided.

Discharge maintenance The process of ensuring that pressures and flows on the discharge side of the pump are properly initiated and maintained.

Drafting The process of pumping water from a static source such as a lake or dump tank where the water's surface is lower than the pump's intake. It requires the use of atmospheric pressure to push water into the pump.

Feathering or **gating** The process of partially opening or closing control valves to regulate pressure and flow for individual lines.

Gating or **feathering** The process of partially opening or closing control valves to regulate pressure and flow for individual lines.

Primary water supply A water supply used for a sprinkler system or standpipe system that exists on the establishment's property. If no water supply exists on the establishment's property, the public or only water supply is considered the primary water supply.

Recirculate line A line that carries a relatively small amount of water but enough water to make sure water is flowing through the pump, preventing it from overheating.

Secondary water supply A public or municipal water supply that backs up the primary water supply. Normal operating procedures should use the secondary water supply whenever possible to avoid depleting the primary water supply.

Water hammer A surge in pressure created by the sudden increase or decrease of water during pumping operations. It can damage hoses, appliances, pumps, water mains, and anything else in direct contact with the water.

REVIEW QUESTIONS

Multiple Choice

Select the most appropriate answer.

1. When the pump attempts to deliver more water than is provided by the supply, a _______ pressure zone is developed at the eye of the impeller, while a ______ pressure zone occurs at the outer edge of the impeller.

   a. high, low
   b. low, high
   c. vapor, cavitation
   d. cavitation, vapor

2. At normal atmospheric pressures (14.7 psi [101 kPa] at sea level) water boils at ____degrees Fahrenheit (____ degrees Celsius).
   - a. 100 (38)
   - b. 185 (85)
   - c. 212 (100)
   - d. 250 (121)
3. When pressure increases, the boiling point of water
   - a. increases.
   - b. decreases.
   - c. stays the same.
   - d. None of the above is correct.
4. Increased pressure will ______ the ability of a liquid to vaporize, while decreased pressure will ______ the ability of a liquid to vaporize.
   - a. increase, reduce
   - b. reduce, increase
   - c. initiate, stop
   - d. None of the above is correct.
5. Which of the following is *not* a method of stopping cavitation?
   - a. Reducing pump speed
   - b. Increasing the flow of the supply
   - c. Removing one or more discharge lines
   - d. All of the above are correct.

Short Answer

On a separate sheet of paper, answer/explain the following questions.

1. List five instrumentations that should be continuously monitored during pump operations.
2. Explain how discharge flows are controlled.
3. Discuss the general procedures for placing two discharge lines into service at the same time when one line requires a higher pressure setting than the other.
4. Explain how to place a discharge line into service when several lines are already in operation.
5. When should a pumping operation be paused? Explain the steps for doing so.
6. Explain how the drive engine can be maintained within the normal operating range.
7. Briefly explain how the pump water temperature can be prevented from rising.
8. What should be considered when the pumping operation is over?
9. What is water hammer? Explain how to prevent it from occurring.
10. Define *cavitation*, providing a brief explanation of the phenomenon.
11. List the warning signs of cavitation.
12. Explain how to control cavitation.
13. If problems occur during pumping operations, what are the most likely causes?
14. List the basic types of sprinkler and standpipe systems and discuss their components and operation.
15. Discuss basic support considerations for sprinkler and standpipe systems.

ACTIVITIES

1. Develop a lesson plan for teaching firefighters about water hammer. Include the following information in your lesson plan:
   - a. Definition of water hammer
   - b. Explanation of how water hammer occurs
   - c. Procedures for ensuring that water hammer does not occur
2. Develop a standard operating procedure for both a sprinkler system and a standpipe system installed in a local building or structure.

PRACTICE PROBLEMS

1. During a drafting operation, Engineer Smith could not secure a prime. He checked to ensure that no air leaks were present. In addition, the hard suction was checked for blockage and proper deployment. What really baffled Engineer Smith was that the priming pump had recently been replaced. In addition, the priming system had functioned properly during numerous training sessions within the last 2 weeks. What is Engineer Smith overlooking?
2. Develop a scenario similar to that presented in the preceding problem. Provide enough detail to narrow the problem down to one or two possible causes. Be sure to provide enough detail to determine the cause(s).

Refer to **Figure 9-11** for the remaining questions. Use a separate sheet of paper for your explanations.

3. When no lines are flowing, explain how to initiate flow for lines B and C at the same time.
4. When line D is flowing, explain how to initiate flow in line C.
5. Explain how to initiate flow in lines A and B when line C is already in operation.

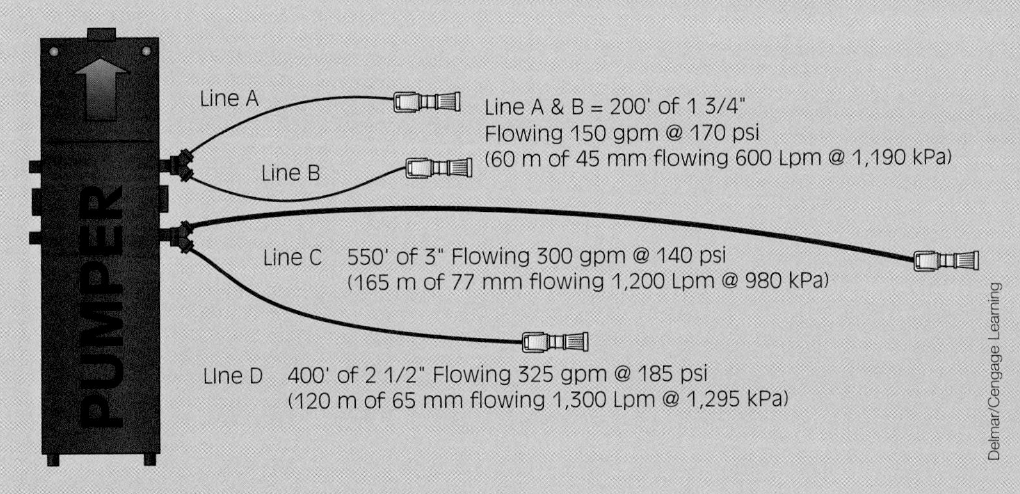

FIGURE 9-11 Typical hoseline configuration for pumping operation.

ADDITIONAL RESOURCES

FM Global. "Fighting Fire in Sprinklered Buildings." http://www.fmglobalcatalog.com

NFPA 13, *Standard for the Installation of Sprinkler Systems*. National Fire Protection Association, Quincy, MA, 2010.

NFPA 13E, *Recommended Practice for Fire Department Operations in Properties Protected by Sprinkler and Standpipe Systems*. National Fire Protection Association, Quincy, MA, 2005.

NFPA 14, *Standard for the Installation of Standpipes and Hose Systems*. National Fire Protection Association, Quincy, MA, 2007.

Water Flow Calculations

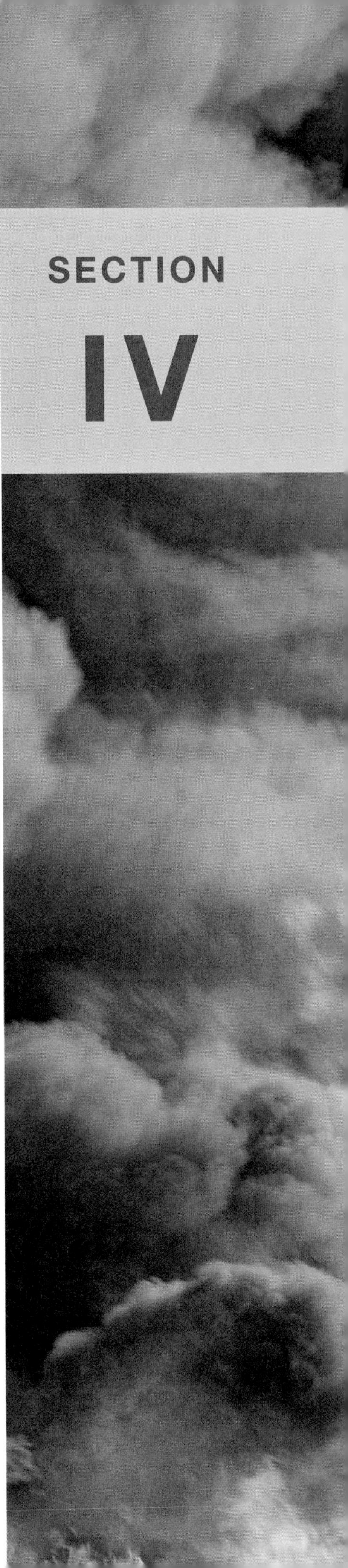

The first three sections of this book discuss pump operators and their three main duties: preventive maintenance, driving apparatus, and pump operations. In addition, how and why pumps work as well as the peripherals and equipment used during pump operations are discussed. Section IV now introduces a new element focusing on information related to water flow calculations. In general, concepts related to determining proper flows and pressures for fire pump operations are presented. **Chapter 10** focuses on definitions and concepts basic to the science of hydraulics; **Chapter 11** discusses methods for estimating the amount of water to deliver to the scene; and **Chapter 12** covers calculations specific to friction loss and engine pressure.

10 Introduction to Hydraulic Theory and Principles

- Learning Objectives
- Introduction
- Physical Characteristics of Water
- Density of Water
- Weight and Volume of Water
- Pressure
- Flow Concepts and Considerations
- Nozzle Reaction
- Lessons Learned
- Key Terms
- Chapter Formulas
- Review Questions
- Activities
- Practice Problems
- Additional Resources

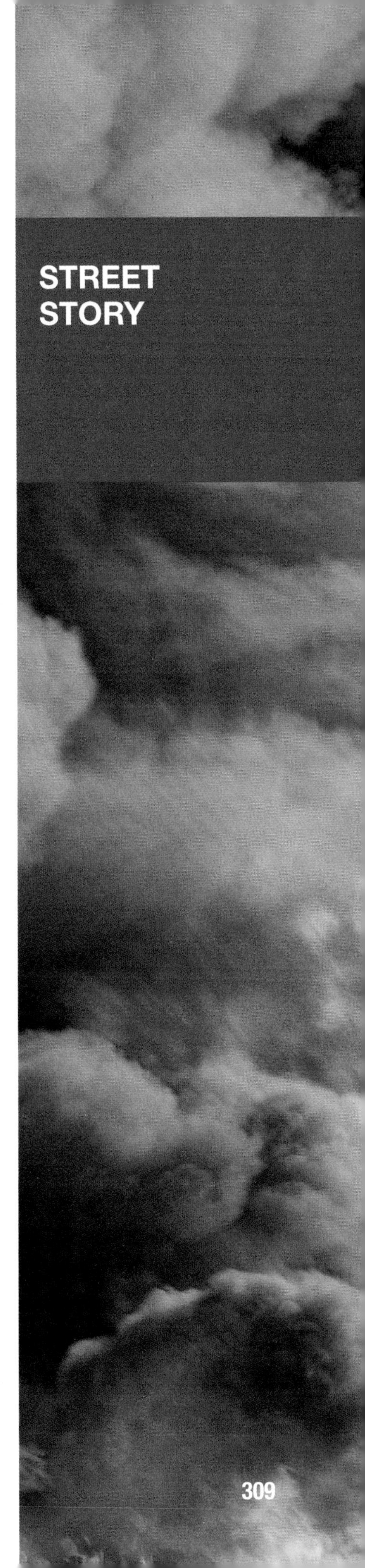

Three years ago, I was the relief pumper on a fire at a two-story house with a garage below. The alarm was dispatched as a confirmed structure fire. There was fire showing from the second-story windows upon arrival.

The crew had taken the first attack line into the house to extinguish the fire. A second line was pulled to protect the outside of the house. Often, I've found that when you pull the gate on a pump panel it doesn't happen very smoothly: It can be hard to pull, so then you pull too hard and risk pulling it out completely. So, generally, when I pump a fire, I pull the gate for the first line I'm charging with no pressure; then I put the pressure up so that it's not all of a sudden a lot on the nozzle. But it's harder to regulate on the second line. At this fire, I went to pull the other gate out and it started to stick; I tugged it too hard.

The problem was that the guy on the other shift had pulled up the truck in such a way that I couldn't see the fire from where I was. So I also couldn't see the guy I was giving pressure to, and I didn't know he had the nozzle open. I gave him all that pressure before he was ready, and he lost control of the hose. Another guy got hit in the head with the nozzle and was knocked out. A third guy rolled an ankle trying to avoid the hose. We're talking about a 2½-inch (65 mm) line—it took out three people before I even knew.

This fire validated the fear most apparatus engineers share, in that sometimes we don't always realize the sheer power of water. The velocity of the water these fire pumps discharge can sometimes pose just as serious a risk to firefighters as the fire itself. Especially on a bigger hose, you have to make sure the firefighter is prepared for it. It's pretty important when charging a 2½-inch (65 mm) line that the firefighter have the nozzle closed and be on ground with his or her weight on the hose. To make sure that's the case, I now try to park the truck in a way that I can see the fire *and* the firefighter.

—Street Story by Chris George, Apparatus Engineer, Quincy Fire Department, Engine Company, 3, Quincy, Illinois

LEARNING OBJECTIVES

After completing this chapter, the reader should be able to:

10-1. List and discuss the basic properties of water.

10-2. Define specific heat and latent heat of vaporization

10-3. Explain the difference between density, weight, and pressure.

10-4. Calculate volume and weight of water.

10-5. Calculate available water when given the dimensions of a static water source.

10-6. Discuss the five principles of pressure.

10-7. List and discuss the different types of pressures.

10-8. Discuss the four principles of friction loss.

10-9. Explain the concept of nozzle reaction.

*The driver/operator requirements, as defined by the NFPA 1002 Standard, are identified in black; additional information is identified in blue.

INTRODUCTION

Hydraulics is a branch of science dealing with the principles of fluids at rest or in motion. **Hydrodynamics** is the term given to the study of fluids in motion, while **hydrostatics** is the term given to the study of fluids at rest. The principles and laws associated with both hydrodynamics and hydrostatics are applicable to fire pump operations. A good grasp of hydraulic concepts will enable a better understanding of pump operations as well as of water flow calculations.

Hydraulic concepts and calculations are often perceived as being difficult and confusing. The reasons are many and varied. For some, the difficulty may stem from fear of math or lack of math skills. Others may try to learn rules of thumb without comprehending the basic principles that govern them. Admittedly, many of the equations are complicated and confusing to learn. Mastering this material requires thought, reflection, and dutiful study.

Three suggestions are offered to help reduce the difficulty and confusion of learning hydraulic concepts and calculations. First, thoroughly learn and understand the basic concepts, including their implications, followed by basic math skills. Understanding the concepts, including which one to apply, will allow checking the reasonableness of calculations. In many cases, it will alert you of an incorrect answer so it can be corrected before continuing.

Lack of basic math skills will make learning hydraulic calculations, as well as a whole array of other subjects, difficult. Consider taking at least a math course or using books such as those listed in this chapter's Additional Resources section. In addition, learn the basic concepts of hydraulics. Rules of thumb on the fireground are certainly important, but their usefulness will be increased if the basic concepts behind them are known and understood.

Second, keep the units straight. Hydraulic calculations involve units associated with the numeric value. For example, 1 gallon of water weighs 8.34 pounds (1 litre of water weighs 1 kilogram), which can be expressed as 8.34 lb/gal (pounds per gallon) (1 kilogram/litre [kilograms per litre]). Often, these units can be abbreviated in several ways. Consider the following examples:

Flow rate	gallon per minute = gpm or gal/min	(litres per minute = Lpm)
Pressure	pounds per square inch = psi or lb/in^2	(kilopascals per square centimetre = kPa)
Density	pounds per cubic foot = lb/cu ft or lb/ft^3	(kilograms per cubic metre = kg/m^3)

Learning to approach these calculations methodically and keeping track of units will ease a lot of the frustration and confusion. Consider the following equation used to determine the weight of 40 gallons (litres) of water:

$$40 \text{ gal} \times 8.34 \text{ lb/gal} = 333.6 \text{ lb}$$

$$40 \text{ litres} \times 1 \text{ kg/litre} = 40 \text{ kilograms (kg)}$$

Note that the volume units (gallons or litres) cancel each other and yield the described units of mass (lb [kg]). To avoid error and confusion, try to assign units consistently within all equations.

Third, practice...practice...practice. The more you practice, the easier it will become. This concept is certainly important while learning hydraulic calculations. It is equally important, if not more important, to maintain and increase the knowledge and skills developed.

This chapter first presents the basic physical characteristics of water, and then the closely related concepts of density, weight, and pressure are discussed. Finally, important friction loss concepts and nozzle reaction calculations are presented.

PHYSICAL CHARACTERISTICS OF WATER

Fire pump operations are all about water—moving it from a source to a discharge point. Pump operators should therefore have a solid grasp of water's physical characteristics. This is also essential for understanding hydraulic principles and concepts. Water is a chemical compound comprised of two parts hydrogen and one part oxygen (H_2O).

Pure water is a colorless, odorless, and tasteless liquid that readily dissolves many substances. The water used for fire suppression is rarely in a pure state, so values for fresh water (water prepared for domestic use) are typically used for fire protection hydraulic calculations. Water is a noncombustible liquid and is considered to be virtually incompressible.

Water exists in a solid state, called ice, at temperatures below 32 degrees Fahrenheit (°F) (0 degrees Celsius [°C]) and, unlike nearly all other compounds, expands when it freezes. Between 32°F and 212°F (0°C and 100°C), water exists predominately in a liquid state. At temperatures above 212°F (100°C), the **boiling point** of water, it exists almost exclusively in a vapor (gas) state called steam. Steam is a colorless gas, although the visible cloud that forms when cooling steam condenses back into water droplets is often incorrectly called steam. The temperatures at which these phase transitions occur are affected by **atmospheric pressure** and will vary relative to changes in **pressure**. Boiling points and freezing points are also affected by the purity of the liquid.

The physical change of state from a liquid to a vapor is known as evaporation. A very small amount of liquid water can produce a large amount of water vapor. This large increase in the **volume** occupied by water molecules can have the effect of increasing pressure when confined in a container. The closer the temperature of a liquid gets to its boiling point, the faster it tends to evaporate. It should be noted that **evaporation** takes place even with cold liquids. The pressure exerted on the atmosphere by molecules as they evaporate from the surface of the liquid (**Figure 10-1**) is known as **vapor pressure**. The higher the vapor pressure, the faster the rate of evaporation of the liquid. Conversely, the higher the atmospheric pressure, the slower the rate of evaporation.

Temperature affects the rate of evaporation and consequently vapor pressure. When heated, the vapor pressure of water increases, causing the water to evaporate at a faster rate. The temperature at which the vapor pressure of a liquid equals the surrounding pressure is known as its boiling point.

Atmospheric pressure can also affect the boiling point of water. Higher atmospheric pressure on a liquid reduces the rate of evaporation. For example, water in an open container boils at a lower temperature than in a closed container. Remember that boiling occurs when the vapor pressure of a liquid equals atmospheric pressure. Consider a pressure cooker. Water in the container boils at a temperature much higher than 212°F (100°C) because of the increase in pressure within the cooker. It stands to reason, then, that higher temperatures will be required to increase vapor pressure to equal that of the surrounding pressure (boiling point) within a closed container. It should also be pointed out that lowering the pressure will increase the rate of evaporation.

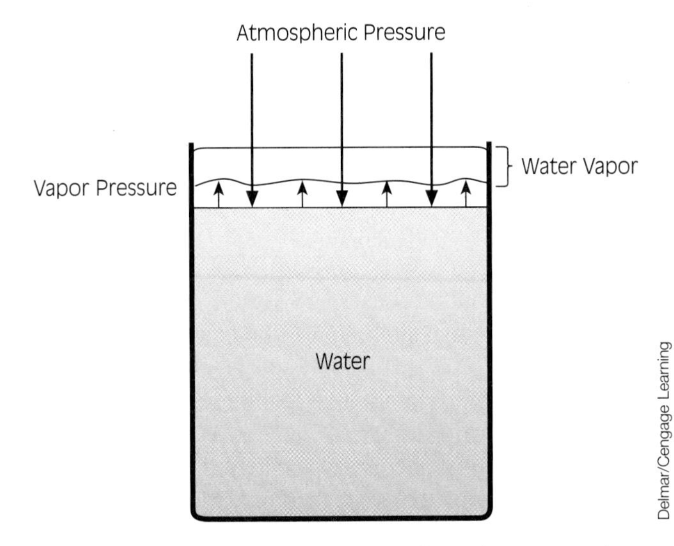

FIGURE 10-1 The pressure exerted on the atmosphere by molecules as they evaporate from the surface of a liquid is known as vapor pressure.

Recognition of how pressure and temperature affect vapor pressure and boiling point can be important considerations for pump operations, especially for understanding the phenomenon of cavitation. In addition, drafting operations can be affected by pressure and temperature. For example, when atmospheric pressure is reduced and the temperature of water increased, the height to which water can be drafted will decrease, and it is easier for cavitation to occur.

When water changes physical states (solid, liquid, and gas) it absorbs or releases heat. The amount of heat that is absorbed or released is called **latent heat of fusion** and is measured in **British thermal units (Btu)**. The Btu is defined as the amount of heat required to raise 1 pound of water 1 degree Fahrenheit (F). The equivalent metric measurement is the **calorie** and is defined as the amount of heat required to raise 1 gram of water 1 degree Celsius (C). One Btu is approximately equivalent to 252 calories or 1.055 kilojoules (KJ). The term **specific heat** refers to the amount of heat required to raise the temperature of a substance by 1°F (1°C). The specific heat of water is 1 Btu (1.055 KJ). For example, it takes 152 Btu (160 KJ) to raise 1 pound (0.45 kg) of water from 60°F to 212°F (15°C to 100°C) (ambient temperature to boiling point). If we use 8.34 pounds (3.78 kilograms) as the approximate **weight** of 1 gallon (3.78 litres) of water, then 1 gallon (3.78 kg) of water absorbs 1,268 Btu (1,338 KJ) (152 Btu/lb × 8.34 lb/gal = 1,267.65 Btu/gal [160 KJ/kg × 3.78kg/3.78 L = 1,338 KJ/3.78]). Another way of saying this is that it takes 1,268 Btu (1,338 KJ) to raise the temperature of 1 gallon (3.78 L) of water from 60°F to 212°F (15°C to 100°C). At sea level, 14.7 psi (101 kPa), water is ready to change from a liquid to a gas (or vapor). The term **latent heat of vaporization** refers to the amount of heat absorbed or released when changing from a liquid to a vapor state. When 1 pound (0.45 kg) of water changes from a liquid state to a vapor state it absorbs 970.3 Btu. Another way of saying this is that it takes 970.3 Btu (1,023 KJ) to convert 1 pound (0.45 kg) of water to vapor or, in fire suppression terms, steam. Again, if we use 8.34 pounds (3.78 kg) as the weight of 1 gallon (3.78 L) of water, then the amount of heat required to convert 1 gallon (3.78 L) of water from a liquid state to steam, sea level at 212°F (100°C), is 8,092.3 Btu (8,537 KJ) (970.3 Btu/lb × 8.34 lb/gal = 8,092.3 Btu/gal) (1,023 KJ/0.45kg × 3.78 kg/3.78 L = 8,537 KJ). Note that the amount of heat absorbed by 1 gallon (3.78 L) of water as it changes to a vapor state is more than six times as much as the amount of heat absorbed by the same amount of water as the temperature changes from 60°F to 212°F (15°C to 100°C). This high heat absorption coupled with the high expansion (approximately 1,700 to 1) makes water an excellent extinguishing agent.

Recognition of how pressure and temperature affect vapor pressure and boiling point can be an important consideration for pump operations.

DENSITY OF WATER

Density is the weight of a substance expressed in units of mass per volume. Typically, density is measured in pounds per cubic foot (kilograms per cubic metre = kg/m^3) and can be expressed as "lb/cu ft" or, more commonly, "lb/ft^3." Fresh water has a density of 62.4 lb/ft^3 when the water temperature is 50°F (10°C) and the atmospheric pressure is 14.7 psi (101 kPa sea level). Salt water has a density of 64 lb/ft^3 (1,000 kg/m^3). Water used in fire suppression is not always a constant 50°F (10°C) and is used at varying elevations. Although the density of water varies slightly with temperature and pressure, these differences are not significant for fire pump operation calculations. With this in mind, calculations in this text assume the density of fresh water to be 62.4 lb/ft^3 (1,000 kg/m^3).

WEIGHT AND VOLUME OF WATER

The most common U.S. unit of weight is the pound (lb) and other areas may use kilograms (kg), and it represents the downward **force** exerted on the object by the Earth's gravity. Consider a container measuring 1 foot wide by 1 foot long by 1 foot high (1 m × 1 m × 1 m) (**Figure 10-2**). When 1 cubic foot (ft^3) (cubic metre [m^3]) of water is added to this container, the weight (downward force) of water is 62.4 lb (1,000 kg). Now consider a second container of equal measure placed beside the first for a total of 2 ft^3 (2 m^3). The density of water is a constant and remains the same at 62.4 lb/ft^3 (1,000 kg/m^3). Therefore, the total weight of water increases to 124.8 lb for 2 ft^3 (2,000 kg for 2 m^3) (62.4 lb/ft^3 × 2 ft^3 = 124.8 lb) (1,000 kg/m^3 × 2 m^3 = 2,000 kg).

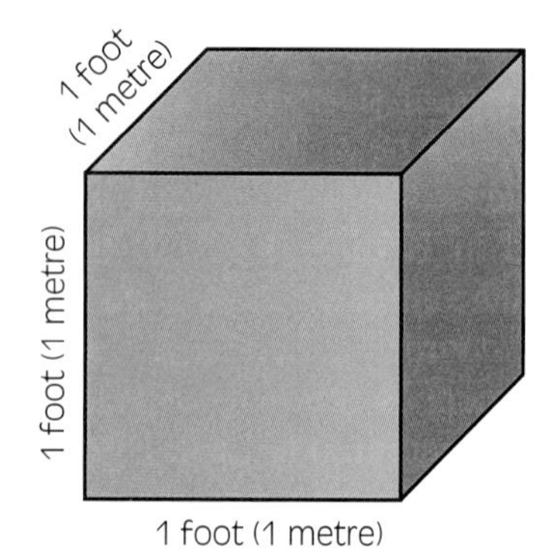

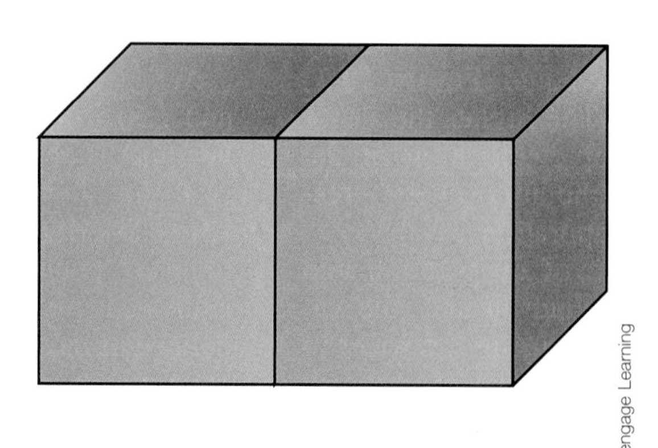

FIGURE 10-2 Weight is expressed in pounds (kilograms); density is expressed in pounds per unit of volume (kilograms per unit of volume).

Volume is the amount of space occupied by an object and is considered a three-dimensional measurement. The most common U.S. unit of volume used in the fire service is the gallon (gal), followed by cubic feet (ft^3) and cubic inches (in^3). For metric, the unit of volume is the litre, followed by cubic metres (m^3), cubic decimetres (dm^3), cubic centimetres (cm^3), or cubic millimetres (mm^3).

Calculating Weight and Volume

Determining the weight of small volumes of water can usually be done by simply placing the water on a balance. This is, however, impractical for larger quantities of water. To calculate the weight of water, the following formula can be used:

$$W = D \times V \qquad (10\text{-}1)$$

where

W = weight in pounds (lb) (kilograms [kg])

D = density (pounds [kilograms] per unit volume) (lb/ft^3) (kg/m^3)

V = volume. When the density of water is known, all that is needed to calculate the weight of water is the volume. It is important to make sure that the units of volume are the same as those given in the constant used for density.

NOTE

It is important to make sure that the units of volume are the same as those given in the constant used for density.

Weight Calculations Using Cubic Feet (Metres)

If the volume in cubic feet (metres) is known, the calculations are straightforward. For example, what is the weight of water in a vessel containing 50 ft^3 (5 m^3) of water? Recall that the density of water in each cubic foot (metre) of water is 62.4 (1,000 kg). The calculation, then, is as follows:

$$\begin{aligned} W &= D \times V \\ &= 62.4 \text{ lb/ft}^3 \times 50 \text{ ft}^3 \\ &= 3{,}120 \text{ lb} \end{aligned}$$

$$\begin{aligned} W &= D \times V \\ &= 1{,}000 \text{ kg/m}^3 \times 5 \text{ m}^3 \\ &= 5{,}000 \text{ kg} \end{aligned}$$

By using the appropriate value for the density of water, the cubic feet cancel and yield an answer with the units of pounds (lb) (kilograms [kg]), the desired units. The weight of 50 ft^3 (5 m^3) of water is 3,210 lbs (5,000 kg).

Volume Calculations Using Cubic Feet (Metres)

In many cases, the volume of a vessel is not known and will need to be calculated. A good example of this is a rectangular tank of unknown volume. The formula for calculating the volume of a square or rectangular vessel is as follows:

$$V = L \times W \times H \qquad (10\text{-}2)$$

To determine the volume of a square or rectangular tank in cubic feet, the units will be as follows:

V = volume in cubic feet (metres)

L = length in feet (metres)

W = width in feet (metres)

H = height in feet (metres)

To determine the weight of water in this tank, simply plug the volume into the formula used in the last example. What is the weight of water in an on-board tank measuring 4 feet (metres) long, 5 feet (metres) wide, and 3 feet (metres) high? The first step is to find the volume (number of cubic feet [metres] of water) of the tank.

$$\begin{aligned} V &= L(\text{ft}) \times W(\text{ft}) \times H(\text{ft}) \\ &= 4 \text{ ft} \times 5 \text{ ft} \times 3 \text{ ft} \\ &= 60 \text{ ft}^3 \end{aligned}$$

$$\begin{aligned} V &= L(\text{m}) \times W(\text{m}) \times H(\text{m}) \\ &= 4 \text{ m} \times 5 \text{ m} \times 3 \text{ m} \\ &= 60 \text{ m}^3 \end{aligned}$$

In this case the volume is 60 ft^3 (60 m^3). Notice that the correct answer will naturally include the described units (ft^3) (m^3). To find the weight of water, multiply the density of water (62.4 lb/ft^3) (1,000 kg/m^3) by the volume (60 ft^3) (60 m^3):

$$\begin{aligned} W &= D \times V \\ &= 62.4 \text{ lb/ft}^3 \times 60 \text{ ft}^3 \\ &= 3{,}744 \text{ lb} \end{aligned}$$

$$\begin{aligned} W &= D \times V \\ &= 1{,}000 \text{ kg/m}^3 \times 60 \text{ m}^3 \\ &= 60{,}000 \text{ kg} \end{aligned}$$

By using lb/ft^3 (kg/m^3) for the density of water and ft^3 (m^3) for the volume, the answer 3,744 lb (60,000 kg) includes the appropriate units of weight.

To find the volume in cubic feet (metres) for a cylinder, such as a municipal water tank or section of hose, use the following formula:

$$V = 0.7854 \times d^2 \times H \qquad (10\text{-}3)$$

where

V = volume in cubic feet (metres)

0.7854 = constant for calculating the area of a circle based on its diameter

d = diameter of the cylinder

H = height (or length) of the cylinder

What is the volume of water in a tank 50 feet (15 metres) wide by 75 feet (23 metres) tall?

$$\begin{aligned} V &= 0.7854 \times d^2 \times H \\ &= 0.7854 \times (50 \text{ ft})^2 \times 75 \text{ ft} \\ &= 0.7854 \times 2{,}500 \text{ ft}^2 \times 75 \text{ ft} \\ &= 0.7854 \times 187{,}500 \text{ ft}^3 \\ &= 147{,}262.5 \text{ ft}^3 \end{aligned}$$

$$\begin{aligned} V &= 0.7854 \times d^2 \times H \\ &= 0.7854 \times (15 \text{ m})^2 \times 23 \text{ m} \\ &= 0.7854 \times 225 \text{ m}^2 \times 23 \text{ m} \\ &= 0.7854 \times 5{,}175 \text{ m}^3 \\ &= 4{,}064 \text{ m}^3 \end{aligned}$$

Weight Calculations Using Gallons (Litres)

During fire pump operations, gallons (litres) rather than cubic feet (metres) are normally used to express volume. To calculate the weight of a volume expressed in gallons (litres), a different value for the density of water must be used. The same formula is used except the density is expressed in gallons (litres) as opposed to cubic feet (metres). In this case, D is expressed in pounds per gallon (kilograms per litre) rather than pounds per cubic foot (kilograms per cubic metre), and V is expressed in gallons (litres). The weight of a gallon (litre) of water can be calculated by dividing the weight of 1 cubic foot (metre) of water (62.4 lb/ft^3) (1,000 kg/m^3) by the number of gallons (litres) in a cubic foot (metre) (7.48 gal/ft^3) (1,000 L/m^3):

$$\begin{aligned} \text{Weight of 1 gallon of water} &= \frac{62.4 \text{ lb/ft}^3}{7.48 \text{ gal/ft}^3} \\ &= 8.34 \text{ lb/gal} \end{aligned}$$

$$1 \text{ kg} = 1 \text{ litre of water}$$

Note that the ft^3 cancel, yielding a new volume unit of lb/gal. One gallon (litre) of water, then, weighs 8.34 lb/gal (1 kg/L).

A variety of situations may occur when the volume, in gallons (litres), is known. When this is the case, simply multiply the density of water (lb/gal [kg/litre]) by the total number of gallons (litres). Typically, the number of gallons (litres) in an onboard tank or tanker will be known, so the weight of water can easily be calculated. Consider a 500-gallon (2,000-litre) on-board tank. Calculate the weight of water in this tank as follows:

$$\begin{aligned} W &= D \times V \\ &= 8.34 \text{ lb/gal} \times 500 \text{ gal} \\ &= 4{,}170 \text{ lb} \end{aligned}$$

$$\begin{aligned} W &= D \times V \\ &= 1 \text{ kg/litre} \times 2{,}000 \\ &= 2{,}000 \text{ kg} \end{aligned}$$

The units of gallon (litres) cancel, leaving the appropriate unit of weight in the answer as 4,170 lb (2,000 kg).

Another example in which the volume in gallons (litres) is typically known is that of fire streams. Consider, for example, a fire stream flowing 250 gpm (1,000 Lpm) into a structure.

If the stream is maintained for 5 minutes, the fire stream will deliver 1,250 gallons (5,000 L) of water into the structure (250 gpm × 5 minutes = 1,250 gallons) (5,000 Lpm × 5 minutes = 5,000 litres). Note that the units of minutes cancel, leaving the units of gallons (litres). The weight of water from a fire stream flowing 250 gpm (1,000 Lpm) for 5 minutes can now be calculated as:

$$\begin{aligned} W &= D \times V \\ &= 8.34 \text{ lb/gal} \times 1{,}250 \text{ gal} \\ &= 10{,}425 \text{ lb} \end{aligned}$$

$$\begin{aligned} W &= D \times V \\ &= 1 \text{ kg/L} \times 5{,}000 \text{ L} \\ &= 5{,}000 \text{ kg} \end{aligned}$$

Recall that 1 ton (tonne) is equal to 2,000 lb (1,000 kg). In this example, slightly more than 5 tons (tonnes) of water are flowing into the structure every 5 minutes.

Volume Calculations Using Gallons (Litres)

If the volume of a container is unknown, the first step is to calculate the volume in gallons (litres). For square or rectangular containers, simply find the volume in cubic feet (metres) and multiply by 7.48 (1,000) (7.48 [1,000] = number of gallons [litres] in 1 cubic foot [metre]). Next, calculate the weight ($W = D \times V$) using the appropriate units. For example, what is the weight of water in a container measuring 5 feet by 6 feet by 4.5 feet (1.5 metres by 1.8 metres by 1.4 metres)? The first step is to select the density and volume in appropriate units.

$$D = 8.34 \text{ lb/gal}$$

$$\begin{aligned} V &= (5 \text{ ft} \times 6 \text{ ft} \times 4.5 \text{ ft}) \times 7.48 \text{ gal/ft}^3 \\ &= 135 \text{ ft}^3 \times 7.48 \text{ gal/ft}^3 \\ &= 1{,}009.8 \text{ gal} \end{aligned}$$

$$D = 1 \text{ kg/litre}$$

$$\begin{aligned} V &= (1.5 \text{ m} \times 1.8 \text{ m} \times 1.4 \text{ m}) \times 1{,}000 \text{ L/m}^3 \\ &= 3.78 \text{ m}^3 \times 1{,}000 \text{ L/m}^3 \\ &= 3{,}780 \text{ litres} \end{aligned}$$

The next step is to calculate the weight using the familiar formula:

$$\begin{aligned} W &= D \times V \\ &= 8.34 \text{ lb/gal} \times 1{,}009.8 \text{ gal} \\ &= 8{,}421.7 \text{ lb} \end{aligned}$$

$$\begin{aligned} W &= D \times V \\ &= 1 \text{ kg/L} \times 3{,}780 \text{ L} \\ &= 3{,}780 \text{ kg} \end{aligned}$$

The weight of water in a container measuring 5 feet by 6 feet by 4.5 feet is 8,421.7 lb. (1.5 metres by 1.8 metres by 1.4 metres is 3,780 kg).

To find the number of gallons in a cylinder, such as a length of hose, use the formula:

$$V = 6 \times d^2 \times h \tag{10-4}$$

where

V = volume in gallons

6 = constant in gallons per cubic foot (gal/ft^3) derived by multiplying the constant 0.7854 (from the volume formula) by 7.48 (the number of gallons in 1 cubic foot) which equals 5.87 and is rounded up to 6 for ease of use on the emergency scene

d = diameter of the cylinder in feet (inches divided by 12 in/ft)

h = height or length of the cylinder in feet

$$V = 0.0007854 \times d^2 \times h \tag{10-4m}$$

where

V = volume in litres

0.0007854 = constant in litres per cubic metre (litres/metre^3) derived by multiplying the constant 0.7854 (from the volume formula) by 0.001 (the number of litres in 1 cubic metre times the number of metres in a millimetre)

d = diameter of the cylinder in millimetres

h = height or length of the cylinder in metres

Consider a 150-foot (45-m) length of 1½-inch (38-mm) hose filled with water. The calculation for the volume in gallons of water contained in this section of hose is given as follows:

$$\begin{aligned} V &= 6 \times d^2 \times h \\ &= 6 \text{ gal/ft}^3 \times \left(\frac{\text{in.}}{12 \text{ in./ft}}\right)^2 \times h \text{ (ft)} \\ &= 6 \text{ gal/ft}^3 \times \left(\frac{1.5 \text{ in.}}{12 \text{ in./ft}}\right)^2 \times 150 \text{ (ft)} \\ &= 6 \text{ gal/ft}^3 \times (0.125 \text{ ft})^2 \times 150 \text{ ft} \\ &= 6 \text{ gal/ft}^3 \times 0.0156 \text{ ft}^2 \times 150 \text{ ft} \\ &= 6 \text{ gal/ft}^3 \times 2.34 \text{ ft}^3 \\ &= 14.04 \text{ gal, or } 14 \text{ gal} \end{aligned}$$

$$\begin{aligned} V &= 0.0007854 \times d^2 \times h \\ &= 0.0007854 \times 38^2 \times h \\ &= 0.0007854 \times 38^2 \times 45 \\ &= 0.0007854 \times 1{,}444 \times 45 \\ &= 51.03 \\ &= 51 \text{ litres} \end{aligned}$$

Notice how the units cancel and yield an answer of 14 gallons (51 litres).

Calculating the weight of water in the hose can now be done by:

$$W = D \times V$$
$$= 8.34 \text{ lb/gal} \times 14 \text{ gal}$$
$$= 116.76 \text{ lb}$$

$$W = D \times V$$
$$= 1 \text{ kg/litre} \times 51$$
$$= 51 \text{ kg}$$

In this example, the 150-foot (45-m) length of 1½-inch (38-mm) hose contains 14 gallons (51 litres) of water, and the water weighs 116.76 lb. (51 kg).

PRESSURE

Fire pump operations develop pressure to move water from one location to another. The pressure developed by fire pumps to accomplish this task is expressed in pounds per unit area (kilopascals or kPa) (typically pounds [kilograms] per square inch [centimetre or millimetre], psi, or lb/in.^{2}). Often the terms *pressure* and *weight* are mistakenly used interchangeably. The concept of weight refers to the total force of attraction between an object and the Earth (gravity). Recall that the units of weight are pounds (kilograms) but do not involve a unit of surface pressure area. On the other hand, pressure combines the units of weight and area. Pressure, then, is the force exerted by a substance in units of weight per unit area, typically substance in units of pounds per square inch (psi) (kilopascals [kPa]). The formula for calculating pressure is as follows:

$$P = \frac{F}{A} \qquad (10\text{-}5)$$

where

P = pressure in force per unit of area (typically lb/in^2 or kPa)

F = force (weight, typically in pounds or kilograms)

A = surface area (typically in square inches, metres, decimetres, or centimetres)

Consider a vessel 1 foot by 1 foot by 1 foot containing 1 cubic foot (1 metre by 1 metre by 1 metre containing 1 cubic metre) of water (**Figure 10-3**). Note that the downward force (weight) of the water is spread out over a 1-square-foot (metre) area of surface area. Since there are 12 inches in a foot, the area in square inches can be found by multiplying 12 inches by 12 inches. It follows that the 1-square-foot (ft^2) area contains 144 square inches (or 144 in.^{2}/ft^2). Recall that 1 cubic foot of water weighs 62.4 lbs. Note that the weight is exerted over a square-foot area. The pressure exerted by this container can now be calculated:

$$P = \frac{F}{A}$$
$$= \frac{62.4 \text{ lb/ft}^2}{144 \text{ in.}^2/\text{ft}^2}$$
$$= 4.33 \text{ lb/in.}^2$$

$$P = \frac{F}{A}$$
$$= \frac{1{,}000 \text{ kg/m}^2}{1{,}000 \text{ dm}^2/\text{m}^2}$$
$$= 1 \text{ kg/dm}^2$$

In essence, weight is the total force and can be localized at one point, while pressure is the force averaged over a given surface area of the material. The pressure of water in a 1-ft^3 (dm^3) container is 0.434 psi (1 kPa). The formula for force is as follows:

$$F = P \times A \qquad (10\text{-}6)$$

where

F = force (weight, typically in pounds [kilograms])

P = pressure in force per unit of area (typically psi or kPa)

A = surface area (typically in square inches, metres, decimetres, or centimetres)

Using the previous example, what is the force of water in a 1-ft^3 (dm^3) container?

$$F = P \times A$$
$$= 0.434 \text{ lb/in.}^2 \times 144 \text{ in.}^2/\text{ft}^2$$
$$= 62.35 \text{ lb/ft}^2$$

(*Note:* rounded to 62.4 in this text.)

$$F = P \times A$$
$$= 1 \text{ kg/dm}^2 \times 1{,}000 \text{ dm}^2/\text{m}^2$$
$$= 1{,}000 \text{ kg/m}^2$$

NOTE

Weight is a total force and can be localized at one point, while pressure is the force averaged over a given surface of the material.

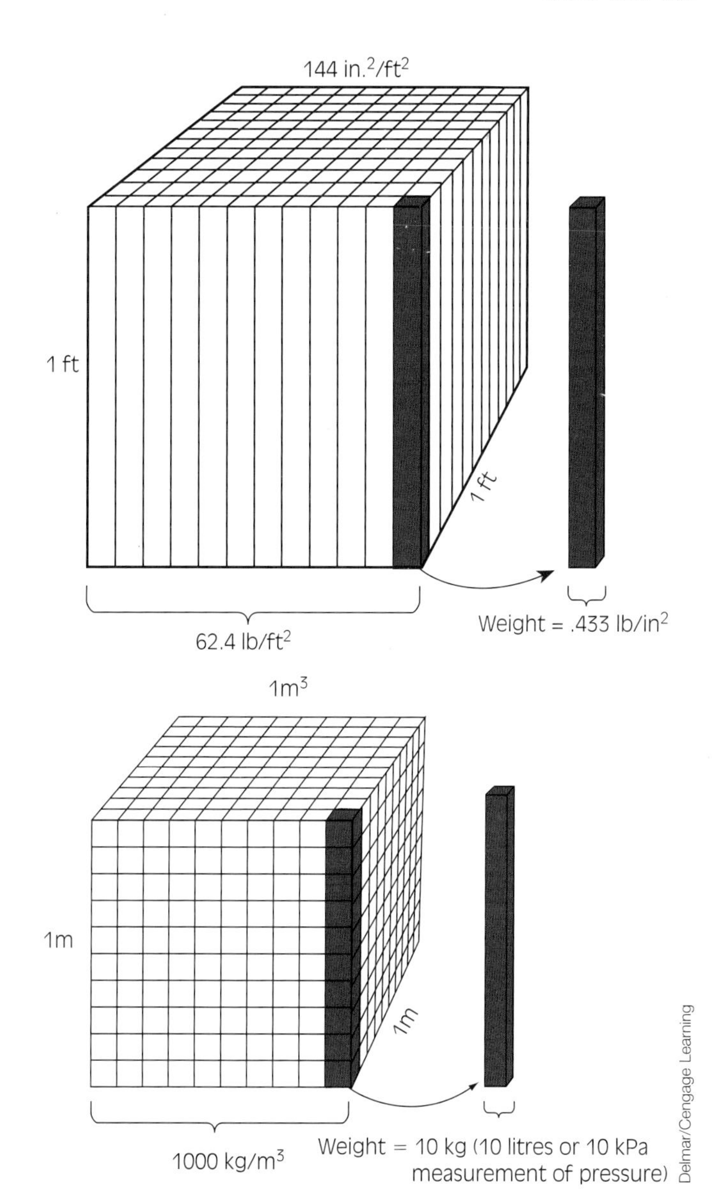

FIGURE 10-3 Pressure is expressed in pounds per unit of area (psi) or kilopascals (kPa).

Pressure Principles

The manner in which a liquid behaves while under pressure follows several basic principles. Some of these principles are straightforward and logical; others may seem confusing and perhaps illogical. In either case, these principles are important and add to the basic knowledge needed to grasp both water flow calculations and fire pump operations fully.

Principle 1

The pressure at any point in a liquid at rest is equal in every direction. One way to view this principle is that the pressure in water is exerted in every direction, downward and outward as well as upward

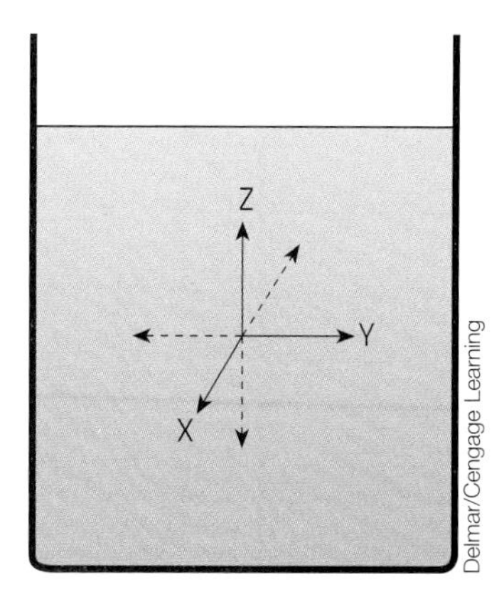

FIGURE 10-4 At any point in a liquid at rest, pressure is equal in every direction.

(**Figure 10-4**). Because liquids have weight, the most obvious direction of force in a liquid is downward. Because we know that liquids exert pressure on the walls of their containers, the lateral forces are also obvious. This outward force can be noted by filling a container, for example a milk container or plastic bag, with water. Note that the sides of the container tend to push outward. The upward pressure in a liquid is, perhaps, not as obvious. This upward pressure can be felt by pushing an empty can into a container of water. The resistance felt is the upward thrust of the water on the bottom of the empty can. This concept is the basis for buoyancy. Since each force within the body of a liquid at rest has an equal force in the opposite direction, the sum of all forces is zero.

The pressure at any point in a liquid at rest is equal in every direction.

Principle 2

The pressure of a fluid acting on a surface is perpendicular to that surface. Although the sum of all forces acting on a molecule within the body of a liquid is zero, the force created by pressure acting on a surface area does have direction. Specifically, this force is perpendicular to any surface it acts upon (**Figure 10-5**). Consider a section of 1¾-inch (45-mm) attack line. Before it is charged, the hose is flat. When the line is charged to its operating pressure, the hose becomes round because the pressure is perpendicular to the internal surface of the hose.

The pressure of a fluid acting on a surface is perpendicular to that surface.

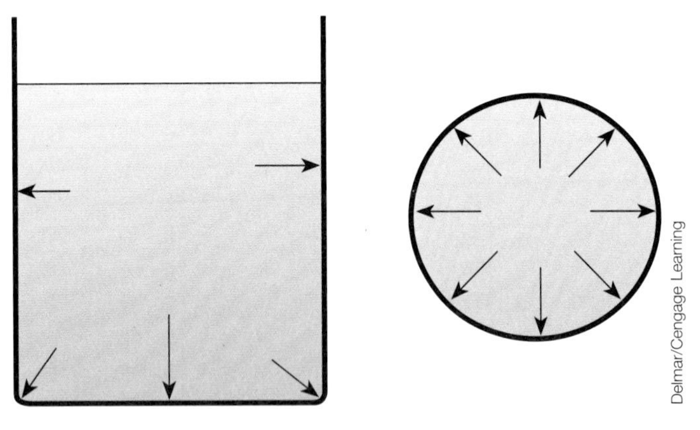

FIGURE 10-5 The pressure of water acting on a surface is perpendicular to that surface.

Principle 3

External pressure applied to a confined liquid (fluid) is transmitted equally in all directions throughout the liquid. This principle provides the basis for understanding the transmission of pressure through a network of fire hose. In essence, any change in pressure is transmitted throughout the entire hoseline. This is true even if the line extends over a great distance. Consider, for example, a pump supplying a discharge layout of 500 feet (150 m) of 2½-inch (65-mm) hose wyed to two 150-foot (45-m) sections of 1½-inch (38-mm) line, each flowing 95 gpm (380 Lpm). The combined flow rate for the two nozzles will be 190 gpm (760 Lpm). If one of the 1½-inch (38 mm) lines were closed rapidly, the pump would attempt to deliver its 190 gpm (760 Lpm) through the other 1½-inch (38-mm) line, resulting in a rapid increase in pressure. Because of principle 3, it should be fairly clear that the resultant pressure surge (**water hammer**) is transmitted throughout the entire hose lay. This means that both 1½-inch (38-mm) lines as well as the pump will experience a change in pressure. Principle 3 is one of the primary reasons for the use of pressure regulating devices on pumps.

This principle is also valid when no water is flowing. For the same example, suppose pressure gauges were evenly distributed along the entire lay, as shown in **Figure 10-6**. Suppose the pump continues to provide pressure and the nozzles of both 1½-inch (38-mm) lines are shut down. Principle 3 states that pressure will be transmitted throughout the network; therefore each of the pressure gauges would have the same reading (as long as elevation is not a factor). To restate the principle, when no water is flowing, pressure is transmitted equally and is undiminished throughout the hose lay.

This principle also provides the basis for calculating pressure in hoseline configurations that split from a supply, as in the example shown in **Figure 10-6**. Since pressure is transmitted equally in all directions, the pressure within the 2½-inch (65-mm) hose will be transmitted equally to both 1½-inch (38-mm) lines.

External pressure applied to a confined liquid (fluid) is transmitted equally throughout the liquid.

Principle 4

The pressure at any point beneath the surface of a liquid in an open container is directly proportional to its depth. An understanding that the magnitude of the downward force created by a column of water is directly proportional to its depth is important in a number of hydraulic calculations.

This principle can be illustrated by examining the downward force acting on the bottom of a vessel measuring 1 foot wide by 1 foot long by 1 foot high (1 metre wide by 1 metre long by 1 metre high). Recall from earlier in the chapter that pressure is defined as a force per unit of area. As per the previous calculation at **equation (10-5)**, the pressure that 1 cubic foot (metre) of water exerts on the bottom of its container is 62.4 lb/ft^2 (0.434 lb/in^2) (1,000 kg/m^2 or 1 kg/dm^2). Suppose we now place a second cubic foot (metre) of water on top of the first (**Figure 10-7**) and repeat the calculation for

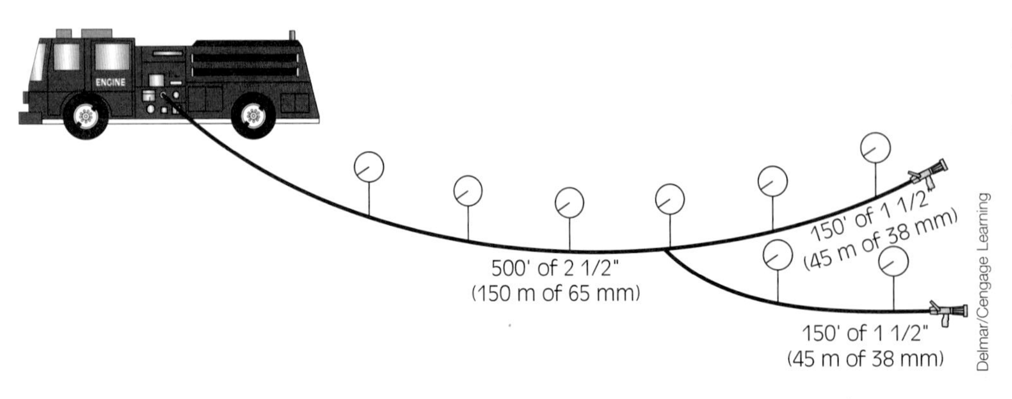

FIGURE 10-6 When no water is flowing, the pressure will be transmitted equally throughout the hose lay.

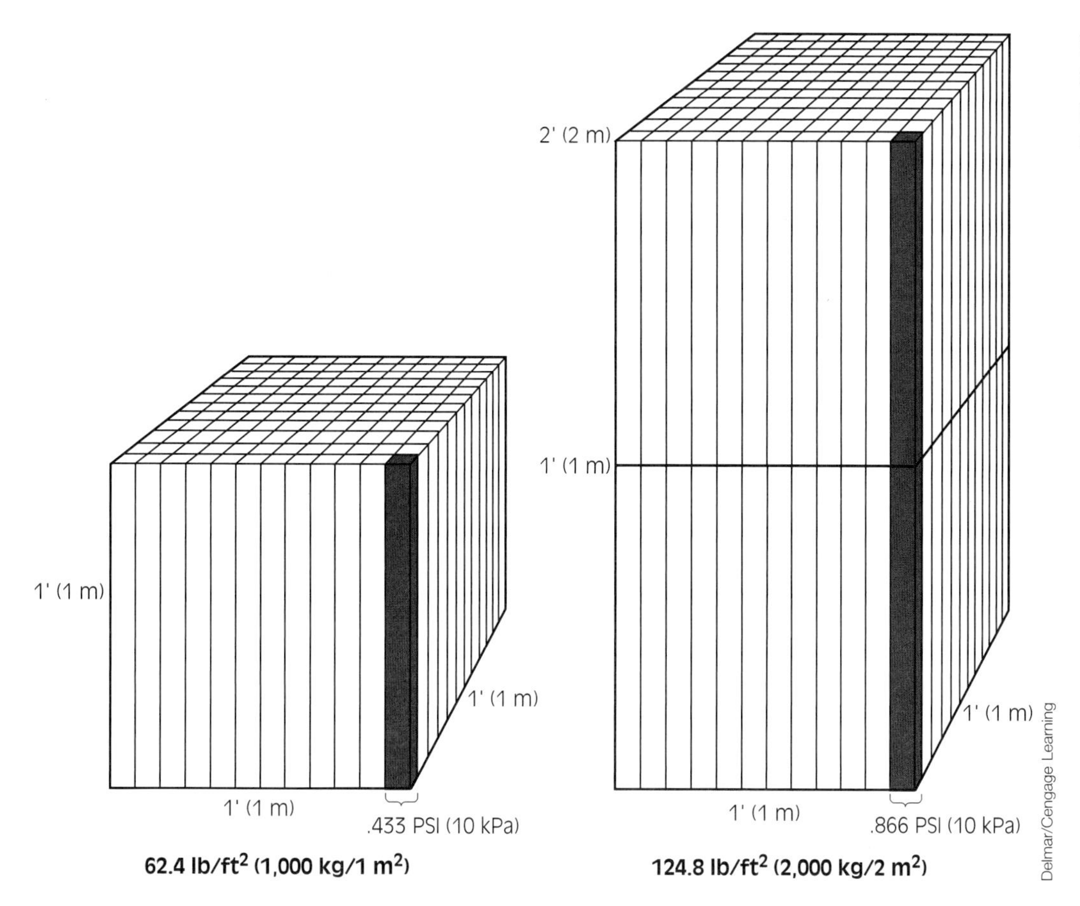

FIGURE 10-7 The pressure at a vessel's base is proportional to the height of water.

pressure. When the second cubic foot of water is added, the weight exerted at the bottom of the vessel doubles to 124.8 lb/ft^2 (0.868 lb/in.^{2}) (2,000 kg/m^2 or 2 kg/dm^2). Note that the area of the bottom of the vessel has not changed. The pressure at the bottom of the vessel, then, can be calculated as follows:

$$P = \frac{F}{A}$$
$$= \frac{124.8 \text{ lb/ft}^2}{144 \text{ in.}^2\text{/ft}}$$
$$= 0.866 \text{ lb/in.}^2$$

$$P = \frac{F}{A}$$
$$= \frac{2{,}000 \text{ kg/m}^2}{1{,}000 \text{ dm}^2\text{/m}^2}$$
$$= 2 \text{ kg/dm}^2$$

This example illustrates that when the height of the water increases, the pressure at the bottom of the vessel also increases proportionally.

The pressure at any point beneath the surface of a liquid in an open container is directly proportional to its depth.

Principle 5

The pressure exerted at the bottom of a container is independent of the shape or volume of the container. This principle is perhaps the most confusing. The key to understanding this principle is to remember that it discusses pressure rather than weight. Recall that weight is the total force of a substance over a surface area, while pressure is the weight over a specific area, typically a square inch (decimetre, or centimetre). Note that the principle states the pressure is proportional to depth as opposed to weight being proportional to depth. For example, the containers in **Figure 10-8** each have different shapes and volumes. Obviously, the total weight of water in each container will vary; however, because the level of water in each of the containers is the same, the pressure (psi or kPa) at the base will also be the same.

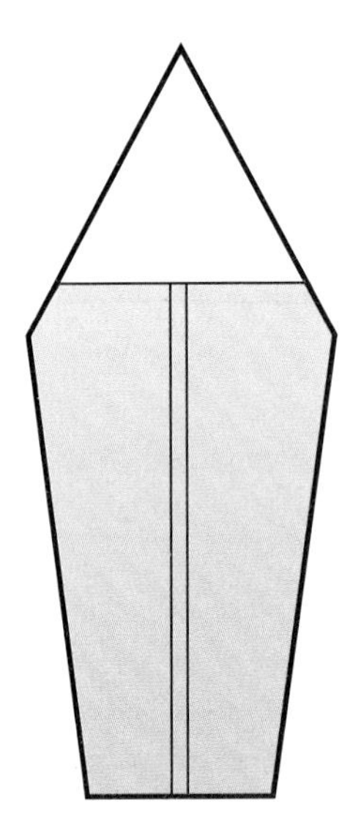

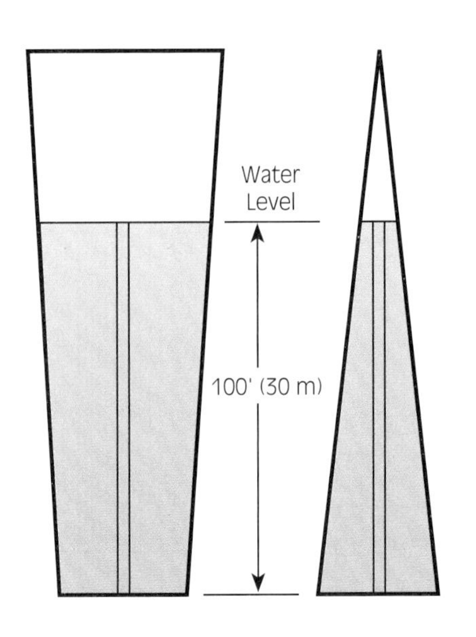

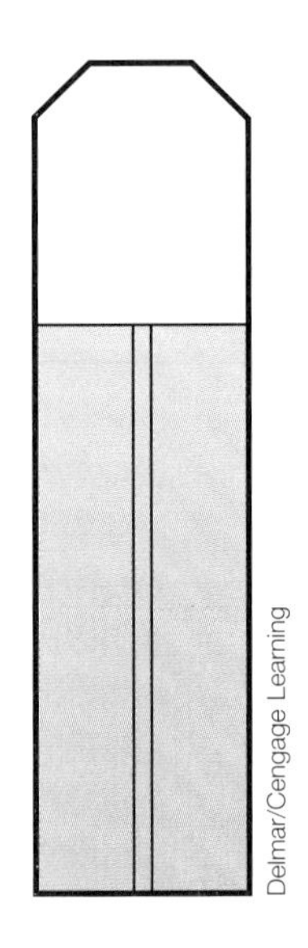

FIGURE 10-8 The pressure at the base of each vessel is the same because the height (level) of water is the same.

Imagine a 1-square-inch (decimetre) column extending from the base of the vessel to the level of the water. In the preceding principle it was determined that a 1-square-inch (dm) column 1 foot (decimetre) high has a weight or exerts a pressure of 0.434 lb/in.^{2} (1 kg/dm^2) on the base of the container. Consider a 1-square-inch (dm) column extending from the bottom to the top in each of the vessels. Because pressure on the bottom is proportional to the height of the column it supports, the pressure will increase by 0.434 lb/in.^{2} (1 kg/dm^2) for each foot (decimetre) of **head pressure**. This pressure can also be expressed as 0.434 psi per foot (0.434 lb/in.^{2}/ft) (1 kPa per decimetre).

The pressure at the bottom of the vessel can be determined by the following formula:

$$P = 0.434 \text{ lb/in.}^2/\text{ft} \times h \text{ (ft)} \qquad (10\text{-}7)$$

where

P = pressure in psi

0.434 = a constant that represents the pressure exerted by a column of water 1 inch by 1 inch by 1 foot

h = the height (depth) of the water in feet

$$P = 1 \text{ kg/dm}^2/\text{m} \times h \text{ (metres)} \qquad (10\text{-}7\text{m})$$

where

P = pressure in kPa

l = constant that represents the pressure exerted by a column of water 1 dm by 1 dm by 1 dm (1,000 dm^3 = 1 m^3 = 1,000 kg)

h = height (depth) of water in decimeters and can be converted to metres

In each of the vessels shown in **Figure 10-8**, the pressure at the bottom can be calculated as follows:

$$P = 0.434 \text{ lb/in.}^2/\text{ft} \times 100 \text{ ft}$$
$$= 43.4 \text{ lb/in.}^2$$

$$P = 1 \text{ kg/dm}^2/\text{m} \times 30 \text{ metres}$$
$$= 30 \text{ kg/dm}^2$$

The pressure exerted at the bottom of a container is independent of the shape or volume of the container.

Types of Pressures

There are many ways to measure the pressure of a system, especially for flowing systems. Consequently, there are many terms to describe specifically how pressure is measured and exactly what is being measured.

Atmospheric Pressure

Surrounding the Earth is a body of air called the atmosphere. Air has weight. Because of this weight, the atmosphere (body of air) exerts pressure on the Earth that is known as atmospheric pressure. Air is also easily compressible. Therefore, the weight of air in the upper atmosphere compresses the air in its lower layers. The result is that the higher you go in the atmosphere, the less pressure you encounter. Conversely, the lower you go in the atmosphere, the more pressure you encounter. At sea level, the atmospheric pressure is 14.7 psi (101 kPa). In higher elevations, the atmospheric pressure is reduced, while in lower elevations, the atmospheric pressure increases.

People experience these pressure differences when they feel their ears pop, such as when they use a high-speed elevator in a high-rise building or during the ascent and descent of an aircraft. For pumps, it means that drafting at high elevations, such as in the mountains, will be more difficult than at sea level or in a valley.

Gauge Pressure (psig) (kPag)

An idle pressure gauge on a pumper at sea level will read 0 psi (kPa). This reading is known as **gauge pressure** (psig or kPag). Recall that the pressure of air at sea level is 14.7 psi (101 kPa). Gauge pressure, then, is simply a pressure reading minus atmospheric pressure. For example, when the gauge reads 200 psi (1,400 kPa) (200 psig), the actual pressure is 214.7 psia (1,501 kPaa) (200 psig + 14.7 psi) (1,400 kPa + 101 kPa). Pressure gauges used on pumping apparatus typically measure psig (kPa) and are often identified as such.

Absolute Pressure (psia) (kPaa)

The measurement that includes atmospheric pressure is known as **absolute pressure** (psia or kPaa). A gauge that measures psia or kPa would have a reading of 14.7 psia at sea level. Another way of relating gauge pressure to absolute pressure is to say 14.7 psia or 101 kPa is equal to 0 psig or kPa. The use of psia or kPa is typically limited to measurements within pressurized vessels.

Vacuum (Negative Pressure)

The three previous types of pressures are considered positive pressure. **Vacuum** is considered a negative pressure. Positive pressures are measured in psi or kPa while negative pressures are measured in inches of mercury, kPa, or millimetres. Pressures less than atmospheric pressure are called vacuum and are expressed in units of inches of mercury (in. Hg) or in millimetres (mm). Typically at least one gauge on a pump panel measures vacuum, the master intake gauge, which is usually a compound gauge.

Head Pressure

Head pressure is the vertical height of a column of liquid expressed in feet (ft) or metres (m). Head pressure is also called *back pressure* or *elevation pressure.* Consider a column of water 2.31 feet (1 metre) in height (**Figure 10-9**). The pressure exerted on the bottom of vessel is calculated as follows:

$$\begin{aligned} P &= F \times h \\ &= 0.434\ \text{lb/in.}^2/\text{ft} \times 2.31\ \text{ft} \\ &= 1\ \text{lb/in.}^2\ \text{or}\ (1\ \text{psi}) \end{aligned}$$

$$\begin{aligned} &= 10\ \text{kPa/m} \times 1\ \text{m} \\ &= 10\ \text{kPa} \end{aligned}$$

Note that the units of feet (metres) cancel. The results of this calculation can also be stated as 1 psi (1 kPa) of pressure will raise water 2.31 feet (1 metre). This information provides the basis for the formula used to calculate head pressure and is stated as follows:

$$h = 2.31\ \text{ft/psi} \times p \qquad (10\text{-}8)$$

where

h = the height of water in feet
2.31 = height in feet 1 psi will raise water
p = pressure in psi

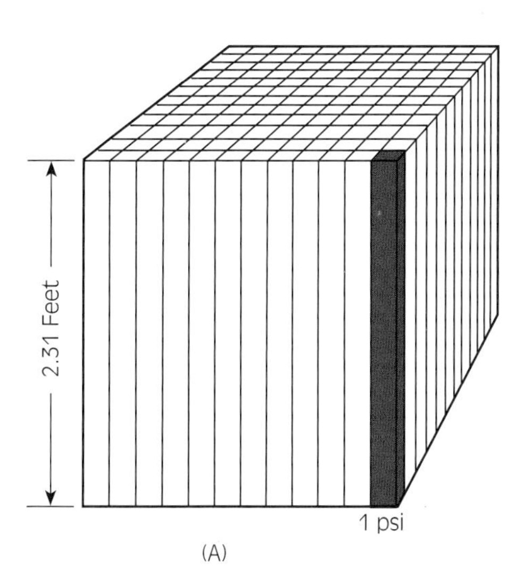

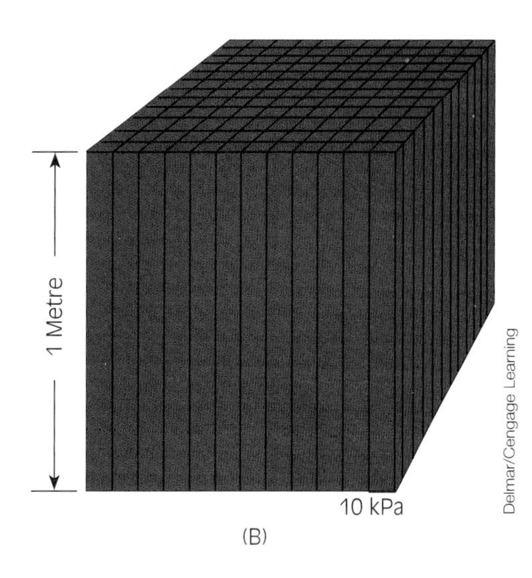

FIGURE 10-9 One psi (10 kPa) will raise a 1-square-inch (1-square-metre) column of water 2.31 feet (1 m).

$$h = 0.1 \text{ m/kPa} \times p \qquad (10\text{-}8\text{m})$$

where

h = the height of water in metres
0.1 = height in metres 1 kPa will raise water
p = pressure in kPa

Consider a vessel of water with a pressure of 43.4 psi (300 kPa) at its base. Using **equation (10-8)**, the height (head) of water can be calculated as follows:

$$\begin{aligned} h &= 2.31 \text{ ft/psi} \times p \\ &= 2.31 \text{ ft/psi} \times 43.4 \text{ psi} \\ &= 100 \text{ ft} \end{aligned}$$

Note that the units of psi cancel, leaving the answer, 100, in units of feet.

$$\begin{aligned} h &= 0.1 \text{ m/kPa} \times p \\ &= 0.1 \text{ m/kPa} \times 300 \text{ kPa} \\ &= 30 \text{ m} \end{aligned}$$

Note that the units of kpa cancel, leaving the answer, 30, in units of metres.

Positive pressures are measured in psi or kPa while negative pressures are measured in inches of mercury, kPa, or millimetres.

Static Pressure

The term *static* indicates a lack of motion or movement. **Static pressure**, then, is the pressure in a system when no water is flowing and is the stored potential energy available to push water through pipes, fittings, fire hose, and appliances. (In reality, water is always flowing in a municipal water system; however, the flow is so small compared to flows when hydrants are opened that it is assumed that normal consumption flows affect pressures.) For example, the pressure at a hydrant before it is opened to flow water is called *static pressure*. Static pressure can be measured by placing a pressure gauge (mounted on an outlet cap) on one of the outlets and turning on the hydrant. Because the gauge is mounted on an outlet cap, no water will be flowing, so the gauge will read static pressure. Static pressure can also be read on a pump panel intake gauge when water is not flowing, or prior to opening any discharges. If the pump is not engaged, the intake gauge, discharge gauge, and line gauges to any capped discharge with an open valve should all have the same reading, which will be the static pressure.

Residual Pressure

If a second outlet on the hydrant is opened, the pressure reading on the gauge at the first outlet will drop due to **friction loss** in the system. The new pressure reading is called *residual pressure*. In other words, **residual pressure** is the pressure remaining in the system after water is flowing. Residual pressure can also be read on a pump panel intake gauge once water is flowing.

Pressure Drop

The difference between the static pressure and the residual pressure when measured at the same location is called **pressure drop**. For example, if the static pressure measure is 50 psi (350 kPa) and the residual pressure measure is 35 psi (245 kPa), the pressure drop would be 15 psi (105 kPa). The reduction in pressure accounts for the loss of pressure caused by friction within the system.

During pump operations, the drop in hydrant pressure from static to residual can be used to estimate the additional flow the hydrant is capable of providing. This is accomplished by first noting the pressure on the master intake gauge on the pump panel after the hydrant is opened but before any discharge lines are opened. Next, initiate and obtain proper flow through one discharge. Again, note the pressure on the master intake gauge. Finally, determine the percentage of the drop in the two readings. Based on the percent drop in pressure, additional flows may be available from the hydrant as follows:

0–10% drop 3 times the original flow

11–15% drop 2 times the original flow

16–25% drop 1 time the original flow

The formula to determine the percentage drop is:

$$\frac{[\text{static pressure} - \text{residual pressure}] \times 100}{\text{static pressure}} \qquad (10\text{-}9)$$

For example, a static reading of 50 psi (350 kPa) was noted when the hydrant was opened and a residual reading of 40 psi (280 kPa) was noted after a 1½-inch (38-mm) line flowing 100 gpm (400 Lpm) was initiated. The drop in pressure is 10 psi (50 psi – 40 psi = 10 psi) (70 kPa [350 kPa – 280 kPa = 70 kPa]). The percent drop in pressure is 20 (10 psi ÷ 50 psi = 20%). In this case, only an additional 100 gpm

(400 Lpm) is available from the hydrant. As a note, some areas use a more conservative percent drop in pressure of 5%, 10%, and 20%.

STREETSMART

Pump operators should always note the static pressure and residual pressures so pressure drops can be calculated, allowing the operator to determine how much additional water is available.

Normal Pressure

Municipal water distribution systems are often designed to meet both consumer needs and fire protection needs. During times of high consumer demand, such as in the morning, the pressure within the system will drop. **Normal pressure** or normal operating pressure is the water pressure found in a system during normal consumption demands.

Velocity Pressure

Water pressure within a hose is converted to **velocity pressure** as it leaves a discharge opening, typically a nozzle. Velocity pressure is also called *forward pressure* or *flow pressure.* It is the forward pressure of water as it leaves an opening. Typically measured with a pitot gauge, velocity pressure can be used to calculate flow.

Pressure Gain and Loss

Previously it was stated that for every 1-foot (metre) increase in a 1-square-inch (metre) column of water, pressure will increase by 0.434 psi (10 kPa). In turn, it can be said that to increase the height of water by 1 foot (metre), an increase of 0.434 psi (10 kPa) is required. **Pressure gain and loss** is the increase and decrease in pressure as a result of an increase or decrease in elevation. This is an important concept in fire pump operations when hose lays are advanced either above or below the pump. In essence, for every foot (metre) above the pump, the pressure in a hose will decrease by 0.434 psi (10 kPa). In turn, to compensate for this loss, pressure must be increased by 0.434 psi (10 kPa) for each foot of elevation. For fireground calculations, this pressure can be rounded up to 0.5 psi.

Consider a hoseline taken to the second story of a structure. If the height of an average story is considered to be 10 feet (3.5 m), a pressure reduction of 5 psi (35 kPa) will occur ($0.5\ \text{lb/in.}^2/\text{ft} \times 10\ \text{ft} = 5\ \text{lb/in.}^2$). For every elevation above the first floor, pressure must be increased by 5 psi (35 kPa). If the hoseline is advanced to the fourth floor, the pressure must be increased by 15 psi (105 kPa). Keep in mind that the first floor is typically at ground level and no elevation loss is encountered. The same concept applies when hoselines are advanced below the pump, except that pressure will increase. Consider a line taken to the basement of a structure. In this case, pressure will increase and a reduction of 5 psi (35 kPa) is required.

SAFETY

The pump should not be operated in such a way that pressures exceed either annual service test pressures or subject components to pressures that exceed the manufacturer's specified maximum operating pressure. When elevation differences are considered, this could further reduce pump discharge pressure. For example, when appliances are located 50 feet (15 m) below the pump, the pump maximum safe operating pressure is reduced by the 25-psi (175-kPa) elevation pressure gain.

When operating in a high rise or with other significant elevation gains, a low-pressure nozzle may be required to make sure that both the correct nozzle pressure is maintained and that maximum safe operating pressures are not exceeded.

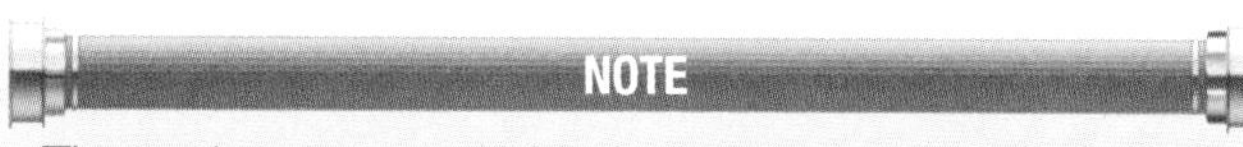

NOTE

The main purpose of fireground hydraulic calculations is to calculate the pump discharge pressure required to provide the correct nozzle pressure.

When lines are advanced up or down grades rather than floor levels of a structure, 0.5 psi (10 kPa) can be added or subtracted for every increase or decrease of 1 foot (metre). For example, a hoseline located 25 feet (7 m) above the pump would have a 12.5-psi (70-kPa) loss in pressure. If the hose is 25 feet (7 m) below the pump, a 12.5-psi (70-kPa) increase will occur.

Nozzle Pressure

Nozzles are designed and constructed to provide a specific flow, or range of flows, at a specific pressure. The designed operating pressure of a nozzle is called **nozzle pressure**. When the correct pressure exists at the nozzle, the nozzle will be provided with its designed flow. The main purpose of fireground hydraulic calculations is to calculate the pump discharge pressure required to provide the correct nozzle pressure. When flow meters are used, no calculations are necessary in that when the nozzle is provided its designed flow, the correct operating pressure exists at the nozzle. In either case, the operating pressure is called *nozzle pressure.* Most nozzles used in the fire service are designed to operate at one of the following nozzle pressures:

Pressure	Nozzle type
50 psi (350 kPa)	Smooth-bore nozzles used on handlines
75 psi (525 kPa)	Low-pressure nozzles
80 psi (560 kPa)	Smooth-bore nozzles used on master stream devices
100 psi (700 kPa)	Combination nozzles, including automatic nozzles

It should be noted that some newer combination nozzles have the ability to convert to a low-pressure nozzle while in the field.

FLOW CONCEPTS AND CONSIDERATIONS

There are several concepts basic to the discussion of fluid flow. Perhaps the most fundamental is the conservation of energy law, which states that energy is conserved in that it is neither created nor destroyed. This is not the same as saying the energy will remain the same; rather, the energy can change forms and/or can be transferred but the sum total will remain constant. Bernoulli's theorem applies this concept to fluids in motion, specifically incompressible fluids such as water. Basically, the total pressure within a system will be the same anywhere in the system. The equation can be stated as: the sum of velocity pressure, friction losses, and elevation pressure is constant within a system. The conservation of mass law states that mass cannot be created or destroyed. As it relates to fire pump operations, the same amount of water entering a hose must exit the hose.

Friction Loss

Friction is caused by rubbing and resistance to motion. It causes a reduction in energy, which results in a pressure loss. This pressure loss is called friction loss. In reality, anything water comes in contact with, including itself, causes friction loss. With greater rubbing and resistance there will be greater friction loss. When water is not in motion, as exists when a valve is closed, no rubbing or resistance exists; therefore, no friction loss exists. Put another way, zero (no) flow equals zero (no) friction loss. This is an important concept with practical applications for pump operations.

Factors affecting friction loss in fire hoses include the following:

- Rough interior lining (typical of older hose)
- Couplings, adapters, and appliances
- Bends and kinks in the hose
- Length and size of hose
- Flow, pressure, and velocity
- Incorrectly sized hose gasket

Friction loss is commonly expressed in pounds per square inch (psi) (kilopascals or kPa) and measures the reduction of pressure between two points in a system. In other words, the difference in pressure between two points in a system is the result of friction.

Laminar/Turbulent Flows

A drop in pressure will occur when water flows through a hose. When the velocity of water is low and the hose interior is smooth, turbulence in the water will be minimal. This condition is known as **laminar flow**, in which thin parallel layers of water develop and move in the same direction together (**Figure 10-10**). The outer layer moves along the interior lining of the hose, while other layers move alongside one another; therefore, varying velocities can occur among the layers. Because only the very outer layer touches the interior of the hose, friction loss is typically limited.

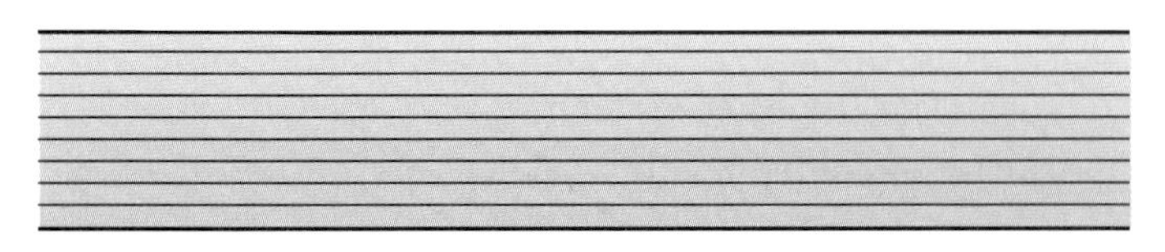

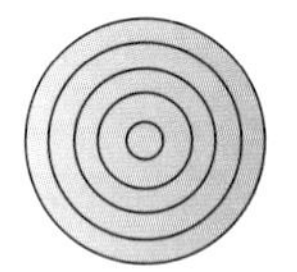
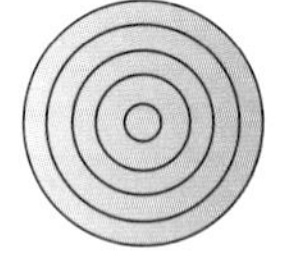

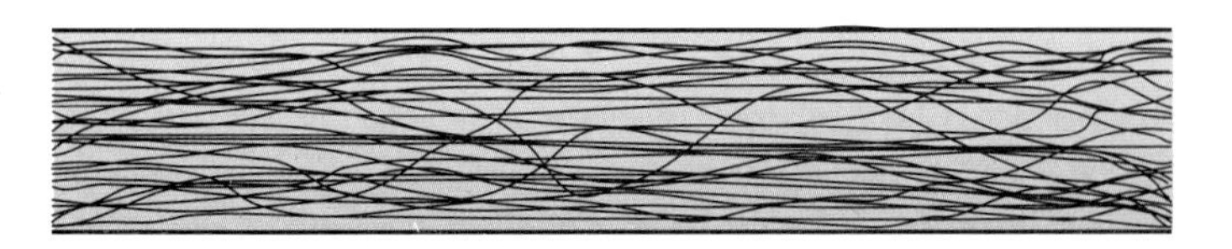

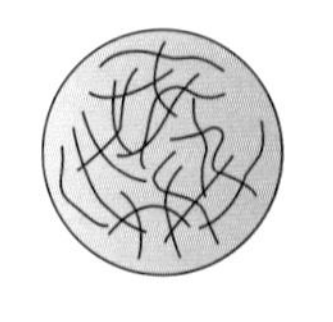

FIGURE 10-10 Laminar flow versus turbulent flow.

When the velocity is high and the hose interior is rough, turbulence in the water occurs. This critical velocity is known as **turbulent flow**, in which water moves in an erratic and unpredictable pattern. This random movement mixes the layers in the water to create a uniform velocity within the hose. Increased pressure loss occurs because more water is subjected to the interior lining of the hose. In comparison, during laminar flow only a thin layer of water touches the interior of the hose. When turbulent flow occurs, water no longer travels in a smooth straight line; rather, movement is erratic, causing increased friction. The pressures in hose are typically high enough to cause turbulent flow.

Fundamental Friction Loss Principles

Although friction loss varies with the age of hose, number of kinks or bends, and numerous other factors, several principles of friction loss remain constant.

Friction Loss Principle 1

Friction loss varies directly with hose length if all other variables are held constant. In other words, when the length of the hose doubles, friction loss doubles as well. For example, 500 feet of 3-inch hose flowing 300 gpm has a friction loss of 36 psi (150 m of 77-mm hose flowing 1,200 Lpm has a friction loss of 275 kPa) (**Figure 10-11**). If the length is doubled to 1,000 feet (300 m), the friction loss will also double to 72 psi (550 kPa). The principle explains, in part, why friction loss formulas typically calculate friction loss per 100-foot (30-m) sections of hose. (See the friction loss chart in **Appendix G**.) Because friction loss varies directly with hose length, the friction loss for 100 feet (30 m) of hose can be multiplied by the number of 100-foot (30-m) sections to determine the total friction loss in the hose.

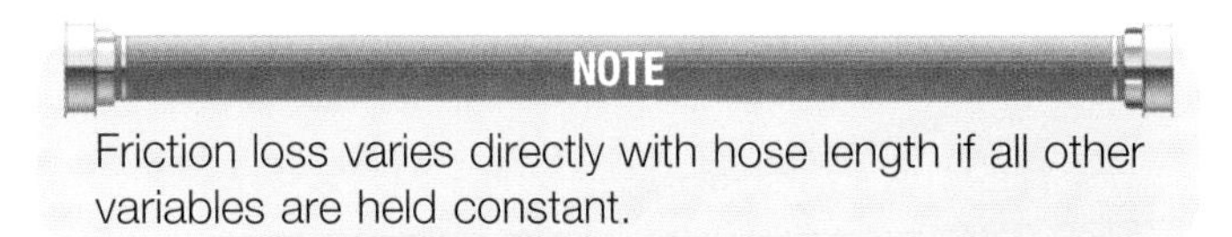

Friction loss varies directly with hose length if all other variables are held constant.

Friction Loss Principle 2

With all other variables held constant, friction loss varies approximately with the square of the flow. This rule points out that the rate of increase in friction loss is significantly greater than the increase in flow. In other words, if the flow doubles (two times as much), the friction loss will increase two times as much squared (2^2, or $2 \times 2 = 4$ times as much). If the flow triples (three times as much), the friction loss will increase nine times ($3^2 = 9$). Consider a 100-foot section of 2½-inch hose flowing 100 gpm (30 m section of 65-mm hose flowing 400 Lpm) (**Figure 10-12**). The friction loss would be 2 psi (15 kPa) while flowing 100 gpm (400 Lpm). If the flow is doubled to 200 gpm (800 Lpm), the friction loss increases four times (2^2) to 8 psi (60 kPa) ($4 \times 2 = 8$ [$4 \times 15 = 60$]). If the original flow is increased to 400 gpm (1,600 Lpm), the friction loss increases 16 times (4^2) to 32 psi (240 kPa) ($16 \times 2 = 32$ [$16 \times 15 = 240$]).

As a note, these same friction loss values can also be obtained using the formula $FL = cq^2L$, which is discussed further in **Chapters 11** and **12**.

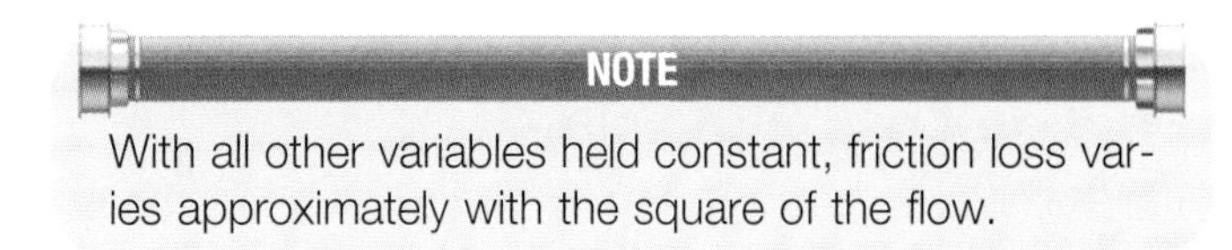

With all other variables held constant, friction loss varies approximately with the square of the flow.

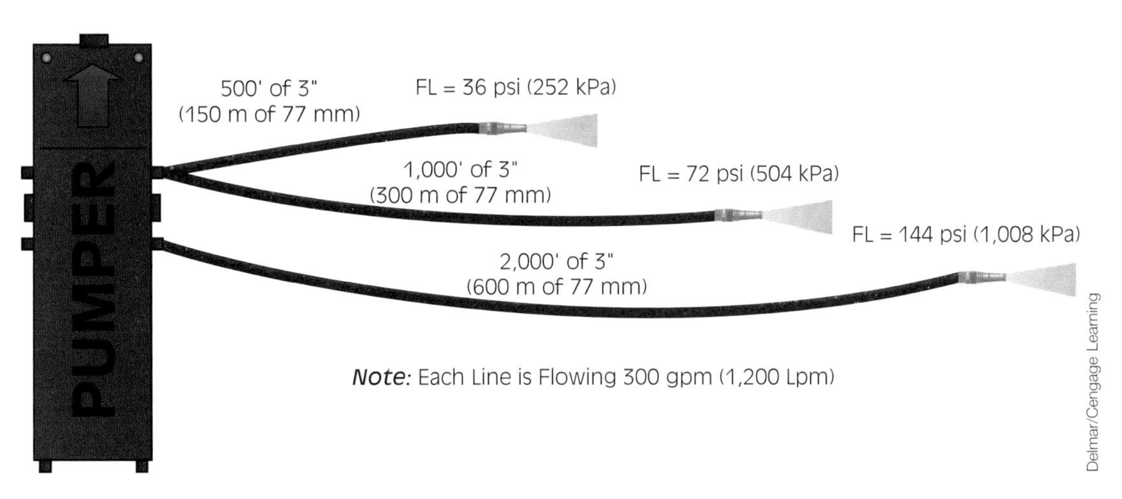

FIGURE 10-11 When the length of hose doubles, friction loss doubles as well.

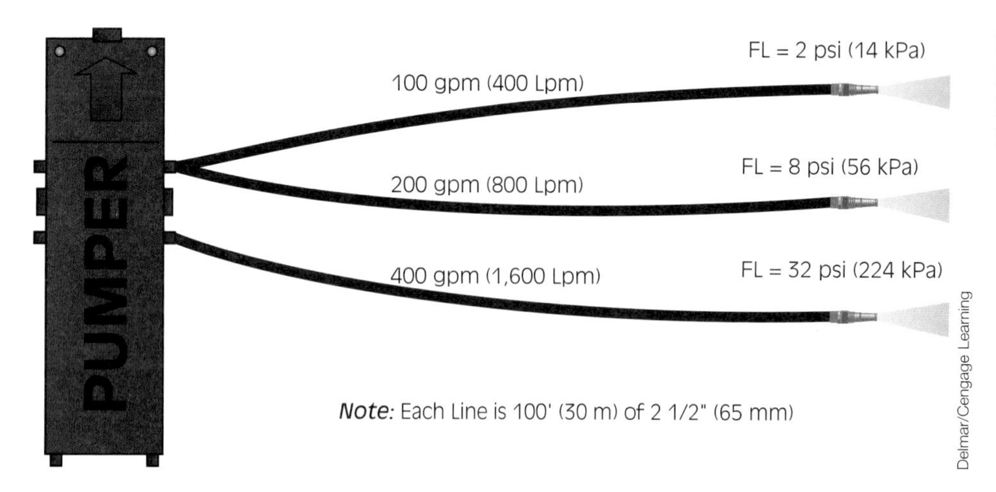

FIGURE 10-12 When flow is increased, the friction loss increases with the square of the increase in flow.

Friction Loss Principle 3

When the flow remains constant, friction loss varies inversely with the fifth power of the hose diameter. In other words, friction loss will decrease when hose diameter is increased, and friction loss will increase when hose diameter is decreased. Consider 250 gpm (1,000 Lpm) flowing through 200 feet (60 m) of different-size hose (**Figure 10-13**).

The friction loss in 2½-inch, 3-inch, and 4-inch hose is 25 psi, 10 psi, and 2.5 psi (65 mm, 77 mm, and 100 mm is 175 kPa, 70 kPa, 17.5 kPa), respectively. This principle is the reason for the introduction and use of large-diameter hose.

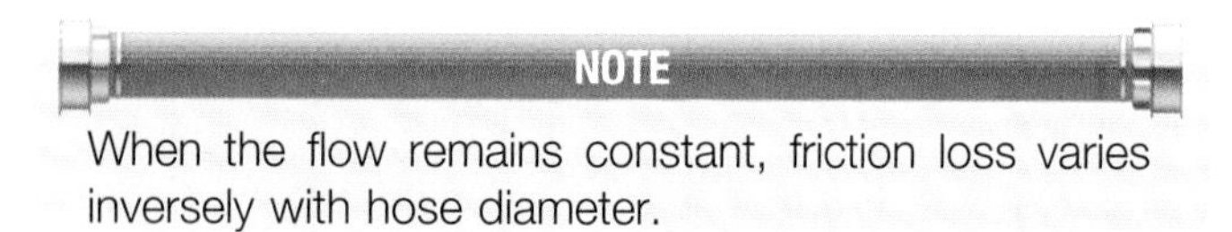

When the flow remains constant, friction loss varies inversely with hose diameter.

Friction Loss Principle 4

For any given velocity, the friction loss will be about the same regardless of water pressure. In other words, the speed (velocity) of water traveling through the hose governs friction loss rather than the pressure of water. For example, if the water within a hose is traveling 20 feet (6 m) per second, the friction loss will remain approximately the same whether the pressure is 50 psi or 150 psi (350 kPa or 1,050 kPa). Another way of stating this principle is that friction loss is independent of pressure at a given velocity.

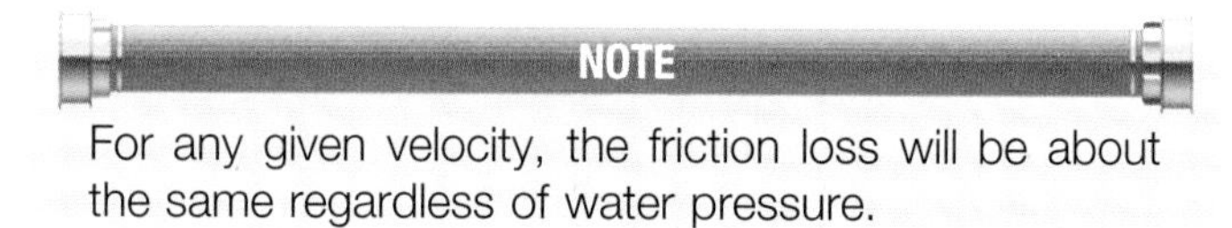

For any given velocity, the friction loss will be about the same regardless of water pressure.

NOZZLE REACTION

The basic principle behind nozzle reaction is Newton's third law of motion, which states that every action is accompanied by an equal and opposite reaction. Relating this principle to modern suppression nozzles, the forward discharge of water is accompanied by a recoil of the nozzle in the opposite direction. This tendency of

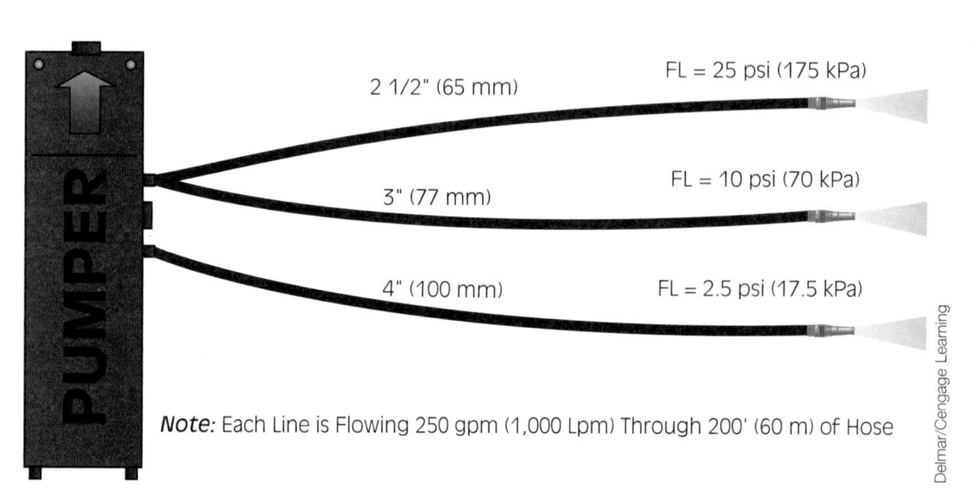

FIGURE 10-13 Friction loss decreases when hose diameter increases.

nozzles to move in the opposite direction of water flow is called nozzle reaction.

Of concern to pump operators, and perhaps even more for those operating a nozzle, is the fact that the nozzle's opposite reaction is proportional to the amount and velocity of water being discharged. Thus, greater discharge flows and pressures will increase nozzle reaction. Recall from **Chapter 6** that nozzles are designed to operate at specific nozzle pressures. In most cases, providing proper nozzle pressure will ensure that the nozzle reaction is manageable. When designed nozzle pressures are exceeded, nozzle reaction increases rapidly. Even though modern nozzles are capable of providing variable flows while maintaining proper nozzle pressure(s), nozzle reaction can still increase. An increase in flow through an automatic nozzle will maintain proper nozzle pressure while increasing nozzle reaction because more water is being discharged, creating additional force. Often, the focus of hydraulic calculations is on friction loss, flow, and nozzle pressure, with little consideration given to nozzle reaction. Nozzle operators being lifted off the ground or losing control of the nozzle are indicators of excessive and unsafe nozzle reaction.

Nozzle reaction is measured in pounds (kilograms or Newtons) (force) and can be calculated for both smooth-bore nozzles and combination nozzles. (See, for example, the nozzle reaction chart in **Appendix G**.)

Smooth-bore Nozzles

The formula for calculating nozzle reaction in smooth-bore nozzles is as follows:

$$NR = 1.57 \times d^2 \times NP \qquad (10\text{-}10)$$

where

NR = nozzle reaction in pounds
1.57 = constant
d = diameter of nozzle orifice in inches
NP = operating nozzle pressure in psi

$$NR = 0.0015 \times d^2 \times NP \qquad (10\text{-}10\text{m})$$

where

NR = nozzle reaction in Newtons
0.0015 = constant
d = diameter of hose in millimetres
NP = operating nozzle pressure in kPa

Consider a 1-inch (25-mm) smooth-bore nozzle discharging water with a nozzle pressure of 50 psi (350 kPa). The calculation for nozzle reaction is as follows:

$$\begin{aligned} NR &= 1.57 \times d^2 \times NP \\ &= 1.57 \times (1 \text{ in.})^2 \times 50 \text{ lb/in.}^2 \\ &= 1.57 \times 1 \text{ in.}^2 \times 50 \text{ lb/in.}^2 \\ &= 78.5 \text{ lb} \end{aligned}$$

$$\begin{aligned} NR &= 0.0015 \times d^2 \times NP \\ &= 0.0015 \times (25 \text{ mm})^2 \times 350 \text{ kPa} \\ &= 0.0015 \times 625 \times 350 \text{ kPa} \\ &= \frac{328}{10} \\ &= 32.8 \text{ kg} \end{aligned}$$

The nozzle reaction for this example is 78.5 pounds (32.8 kg).

Combination Nozzles

The formula for calculating nozzle reaction in combination nozzles is as follows:

$$NR = gpm \times \sqrt{NP} \times 0.0505 \qquad (10\text{-}11)$$

where

NR = nozzle reaction in pounds
gpm = gallons per minute and is sometimes written as "Q"
NP = nozzle pressure in psi
0.0505 = constant with in.^{2}/gpm as units

$$NR = 0.0156 \times Q \times \sqrt{NP} \qquad (10\text{-}11\text{m})$$

where

NR = nozzle reaction in Newtons or kg
Q = litres per minute (flow)
NP = nozzle pressure in kPa
0.0156 = metric constant

The nozzle reaction for a combination nozzle flowing 250 gpm (1,000 Lpm) with a nozzle pressure of 100 psi (700 kPa) is calculated as follows:

$$\begin{aligned} NR &= gpm \times (\sqrt{NP} \times 0.0505 \text{ in.}^2/\text{gpm}) \\ &= 250 \text{ gpm} \times (\sqrt{100} \text{ lb/in.}^2 \times 0.0505 \text{ in.}^2/\text{gpm}) \\ &= 250 \text{ gpm} \times (10 \text{ lb/in.}^2 \times 0.0505 \text{ in.}^2/\text{gpm}) \\ &= 250 \text{ gpm} \times 0.505 \text{ lb/gpm} \\ &= 126.25 \text{ lb} \end{aligned}$$

$$NR = 0.0156 \times Q \times \sqrt{NP}$$
$$= 0.0156 \times 1{,}000 \text{ Lpm} \times \sqrt{700} \text{ kPa}$$
$$= 0.0156 \times 1{,}000 \text{ Lpm} \times 26.46$$
$$= \frac{412.78 \text{ N}}{10}$$
$$= 41.28 \text{ kg}$$

Note that the square root of 100 is 10. Further note that 10 multiplied by 0.0505 is 0.505. Therefore, when combination nozzles are operated at 100-psi nozzle pressure, which they typically are, the nozzle reaction formula can be changed to:

$$NR = gpm \times 0.505 \qquad (10\text{-}12)$$

(*Note*: 0.5 for fireground use)

Using this condensed formula for the preceding example, the nozzle reaction force is calculated as follows:

$$NR = gpm \times 0.505 \text{ lb/gpm}$$
$$= 250 \times 0.505 \text{ lb/gpm}$$
$$= 126.25 \text{ lb}$$

Note that the same results are obtained.

LESSONS LEARNED

Hydraulics is a branch of science that deals with principles of water at rest (hydrostatics) and water in motion (hydrodynamics). Understanding the basic principles of hydraulics is fundamental to water flow calculations. Complicated formulas and calculations are typically not used on the fireground. Those formulas and calculations that are used on the fireground are approximations of the more complicated (scientific) formulas that describe hydraulic behavior.

A summary of water characteristics is provided followed by a listing of chapter formulas.

Basic Characteristics of Water

Density	62.4 lb/ft^3	1,000 kg/m^3
Weight	1 gallon = 8.34 lb	1 litre = 1 kg
Freezes at	32°F	0°C
Boils at	212°F	100°C
1 cubic	foot = 7.38 gallons	metre = 1,000 litres

KEY TERMS

Absolute pressure (psia or kPa) Measurement of pressure that includes atmospheric pressure, typically expressed as psia or kPa.

Atmospheric pressure The pressure exerted by the atmosphere (body of air) on the Earth.

Boiling point The temperature at which the vapor pressure of a liquid equals the surrounding pressure.

British thermal units (Btu) The amount of heat required to raise 1 pound of water 1 degree Fahrenheit (F).

Calorie The amount of heat required to raise 1 gram of water 1 degree Celsius (C); can be expressed in joules or kilojoules.

Density The weight of a substance expressed in units of mass per volume.

Evaporation The physical change of state from a liquid to a vapor.

Force Pushing or pulling action on an object.

Friction loss The reduction in energy (pressure) resulting from the rubbing of one body against another, and the resistance of relative motion between the two bodies in contact; typically expressed in pounds per square inch (psi or kPa); measures the reduction of pressure between two points in a system.

Gauge pressure (psig) Measurement of pressure that does not include atmospheric pressure, typically expressed as psig.

Head pressure The pressure exerted by the vertical height of a column of liquid expressed in feet (metres); may also be referred to as *feet (metres) of head* or just *head.*

Hydraulics The branch of science dealing with the principles and laws of fluids at rest or in motion.

Hydrodynamics The branch of hydraulics that deals with the principles and laws of fluids in motion.

Hydrostatics The branch of hydraulics that deals with the principles and laws of fluids at rest and the pressures they exert or transmit.

Laminar flow Flow of water in which thin parallel layers of water develop and move in the same direction.

Latent heat of fusion The amount of heat that is absorbed by a substance when changing from a solid to a liquid state.

Latent heat of vaporization The amount of heat absorbed when changing from a liquid to a vapor state.

Normal pressure or normal operating pressure The water flow pressure found in a system during normal consumption demands.

Nozzle pressure The designed operating pressure for a particular nozzle.

Pressure The force exerted by a substance in units of weight per area; the amount of force generated by a pump or the resistance encountered on the discharge side of a pump; typically expressed in pounds per square inch (psi) or kilopascals (kPa).

Pressure drop The difference between the static pressure and the residual pressure when measured at the same location.

Pressure gain and loss The increase or decrease in pressure as a result of an increase or decrease in elevation.

Residual pressure The pressure remaining in the system after water has been flowing through it.

Specific heat The amount of heat required to raise the temperature of a substance by 1°F. The specific heat of water is 1 BTU/lb °F (4.19 joules/gram or 4.19 kilojoules/kg).

Static pressure The pressure in a system when no water is flowing.

Turbulent flow The flow of water in an erratic and unpredictable pattern, creating a uniform velocity within the hose that increases pressure loss because more water is subjected to the interior lining of the hose.

Vacuum Measurement of pressure that is less than atmospheric pressure, typically expressed in inches of mercury (in. Hg) (kPa or millimetres).

Vapor pressure The pressure exerted on the atmosphere by molecules as they evaporate from the surface of the liquid.

Velocity pressure The forward pressure of water as it leaves an opening.

Volume Three-dimensional space occupied by an object.

Water hammer Sudden surge of pressure created by the quick opening or closing of water valves; capable of damaging water mains, valves, pumps, fire hose, and appliances.

Weight The downward force exerted on an object by the Earth's gravity, typically expressed in pounds (lb) or kilograms (kg).

CHAPTER FORMULAS

10-1 Determine the weight of water.

$$W = D\ (\text{density}) \times V\ (\text{volume})$$

10-2 Determine the volume of a rectangular vessel.

$$V = L\ (\text{length}) \times W\ (\text{width}) \times H\ (\text{height})$$

10-3 Find the volume in cubic feet (metres) for a cylinder.

$$V = 0.7854\ (\text{constant}) \times d^2\ (\text{diameter}) \times H\ (\text{height or length})$$

10-4 Find the number of gallons in a cylinder.

$$V = 6\ (\text{constant}) \times d^2\ (\text{diameter}) \times h\ (\text{height or length})$$

$$V = 0.0007854 \times d^2 \times h \qquad \text{10-4m}$$

10-5 Determine pressure.

$$P = \frac{F\ (\text{force})}{A\ (\text{surface area})}$$

10-6 Determine force.

$$F = P(pressure) \times A(surface\ area)$$

10-7 Calculate pressure at the bottom of a vessel.

$$P = 0.434\ (\text{constant in lb/in.}^2/\text{ft}) \times h\ (\text{height in feet})$$

$$P = 1\ \text{kg/dm}^2/\text{m} \times h\ (\text{metres}) \qquad \text{10-7m}$$

10-8 Calculate head.

$$h = 2.31 \text{ (height in ft 1 psi will raise water)} \times p \text{ (pressure in psi)}$$

$$h = 0.1 \text{ (height in metres 1 kPa will raise water)} \times p \text{ (pressure in kpa)} \quad \text{10-8m}$$

10-9 Determine available water (percentage method).

$$\text{Percentage drop} = \frac{[\text{static pressure} - \text{residual pressure}] \times 100}{\text{static pressure}}$$

10-10 Determine nozzle reaction for smooth-bore nozzles.

$$NR = 1.57 \text{ (constant)} \times d^2 \text{ (diameter in inches)} \times NP \text{ (nozzle pressure in psi)}$$

$$NR = 0.0015 \times d^2 \times NP \quad \text{10-10m}$$

10-11 Determine nozzle reaction for combination nozzles.

$$NR = gpm \times \sqrt{NP} \text{ (nozzle pressure in psi)} \times 0.0505 \text{ (constant with in}^2\text{/gpm as units)}$$

$$NR = 0.0156 \times Q \times \sqrt{NP} \quad \text{10-11m}$$

10-12 Determine condensed nozzle reaction for combination nozzles.

$$NR = gpm \times 0.505 \text{ (constant with lb/gpm as units)}$$

REVIEW QUESTIONS

Multiple Choice

Select the most appropriate answer.

1. The temperature at which the vapor pressure equals the surrounding pressure is known as

a. vapor pressure. **c.** flash point.
b. boiling point. **d.** vapor density.

2. The physical change of state from a liquid to a vapor is known as

a. boiling point. **c.** vapor density.
b. flash point. **d.** evaporation.

3. The density of fresh water is approximately

a. 62.4 lbs (1,000 kg).
b. 62.4 lb/ft^3 (1,000 kg/m^3).
c. 8.34 lbs (1 kg).
d. 7.48 lbs (10 kg).

4. What is the weight of water in a vessel containing 67 cubic feet (metres) of water?

a. 559.45 lbs (67 kg)
b. 501.16 lbs (6,700 kg)
c. 4,180.8 lbs (67,000 kg)
d. None of the answers is correct.

5. There are ________ gallons in a cubic foot (litres in a cubic metre).

a. 7.48 (10 L)
b. 8.35 (100 L)
c. 62.4 (1,000 L)
d. None of the answers is correct.

6. What is the weight of water in a tank containing 500 gallons (litres) of water?

a. 3,740 lbs (50 kg)
b. 4,170 lbs (500 kg)
c. 31,200 lbs (5,000 kg)
d. None of the answers is correct.

7. What is the weight of water from a fire stream flowing 125 gpm (500 Lpm) for 10 minutes?

a. 1,043.75 lbs (5,000 kg)
b. 7,800 lbs (500 kg)
c. 104,375 lbs (50,000 kg)
d. None of the answers is correct.

8. A 200-foot (60-m) section of 3-inch (77-mm) hose contains how many gallons (litres) when completely filled, with no water flowing?

a. 36 gallons (136 litres)
b. 75 gallons (280 litres)
c. 162 gallons (613 litres)
d. 3,600 gallons (13,627 litres)

9. A gauge reading at sea level of 137 psia (959 kPa) is equivalent to a gauge reading of

a. 122.3 psig (850 kPa).
b. 151.7 psig (1,060 kPa).
c. 100 psig (700 kPa).
d. None of the answers is correct.

10. Positive pressure is measured in psi (kPa), while negative pressure is measured in

a. psi.
b. psia.
c. psig.
d. inches of mercury (millimetres).

11. A pressure drop of 13% was calculated after one line was placed in service. How many additional like lines will the hydrant support?

a. 1 like line
b. 2 like lines
c. 3 like lines
d. None of the answers is correct.

12. For every 1-foot (decimetre) increase in a 1-square-inch column (decimetre) of water, pressure will increase by

a. 433 psi (3 kPa).
b. 1 psi (10 kPa).
c. 8.35 lbs (4 kg).
d. 62.4 lb/ft^3 (1,000 kg/m^3).

13. If the flow of water triples and all other variables are held constant, the friction loss in the hose will increase by

a. 3 times.
b. 6 times.
c. 9 times.
d. 12 times.

14. The tendency of nozzles to move in the opposite direction of water flow is called

a. nozzle recoil.
b. nozzle pressure.
c. nozzle velocity.
d. nozzle reaction.

Short Answer

On a separate sheet of paper, answer/explain the following questions.

1. What is the difference between hydrodynamics and hydrostatics?
2. List several physical characteristics of water.
3. What does 1 gallon (litre) of water weigh?
4. List the typical operating pressures for modern nozzles.
5. Explain the effect of pressure and temperature on vapor pressure and boiling point.
6. Explain the difference between weight and density.
7. List the formula for calculating the weight of a substance.
8. Explain how to calculate the weight of 1 gallon (litre) of water.
9. List the formula for calculating the number of gallons (litres) in a cylinder.
10. Explain the difference between weight and pressure.
11. List the formula for calculating pressure.
12. List and briefly define each of the five pressure principles.
13. What is the difference between psia and psig?
14. List the standard nozzle pressures for the majority of nozzles used in the fire service.
15. List and briefly define each of the four friction loss principles.

ACTIVITIES

1. Calculate the weight of water for a vessel containing 100 cubic feet (metres) of water.
2. A container measuring 10 feet by 10 feet by 3 feet (3 m by 3 m by 1 m) is full of water. How much does the water weigh?

3. Calculate the weight of water contained in a 750-gallon (3,000-litre) on-board water tank.
4. How many tons (tonnes) of water are being delivered into a structure when three 2½-inch (65-mm) lines are flowing 250 gpm (1,000 Lpm) for a duration of 10 minutes?
5. Calculate the weight of a section of 3-inch (77-mm) hose filled with water and measuring 200 feet (60 m) long.
6. Determine the nozzle reaction for the following smooth-bore lines:
   - **a.** ¾-inch (19-mm) tip handline NR=
   - **b.** 1-inch (25-mm) tip handline NR=
   - **c.** 1-inch (25-mm) tip monitor nozzle NR=
7. Determine the nozzle reaction for the following combination nozzle lines:
   - **a.** 100 gpm (400 Lpm) flowing NR=
   - **b.** 150 gpm (600 Lpm) flowing NR=
   - **c.** 200 gpm (800 Lpm) flowing NR=
   - **d.** 150 gpm (600 Lpm) flowing (low-pressure combination nozzle) NR=

PRACTICE PROBLEMS

1. Calculate the weight of water for the on-board water supply on at least two apparatus.
2. Which of the vessels in **Figure 10-14** has the greatest pressure at its base?
3. A gauge at the base of an open vessel indicates a pressure of 50 psi (1,050 kPa). How high is the water within the vessel?
4. Determine the additional flow available from a hydrant given the following:

   static reading = 80 psi (560 kPa)

   residual reading = 60 psi (420 kPa), flowing 250 gpm (1,000 Lpm) through a 2½-inch (65-mm) hose
5. Calculate nozzle reaction for a ¾-inch (19-mm) smooth-bore tip nozzle operating on a handline.
6. Calculate nozzle reaction for a combination nozzle operating on a handline flowing 300 gpm (1,200 Lpm).
7. Calculate the number of Btus 1 gallon of water will absorb going from 60 degrees Fahrenheit in a liquid state to 212 degrees Fahrenheit (joules 1 litre of water will absorb going from 16 degrees Celsius in a liquid state to 100 degrees Celsius) in a vapor state.

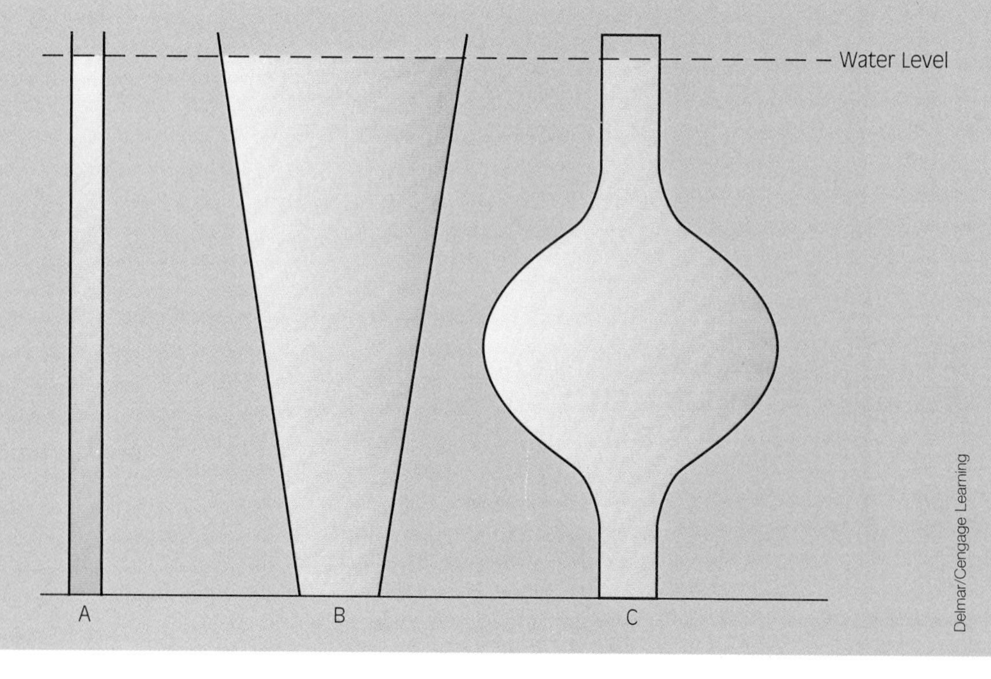

FIGURE 10-14 Which vessel has the greatest pressure at the base?

ADDITIONAL RESOURCES

Crapo, William. *Hydraulics for Firefighting*, 2nd ed. Clifton Park, NY: Cengage Learning, 2008.

Hickey, Harry E. *Hydraulics for Fire Protection*. Boston, MA: National Fire Protection Association Publications, 1980.

National Fire Protection Association. *Fire Protection Handbook*, 20th ed. Quincy, MA: National Fire Protection Association, 2008.

Richmond, Ron. *Basic Fire Hydraulics Workbook*, 1st ed. Clifton Park, NY: Cengage Learning, 2009.

Sturtevant, Thomas B. *Practical Problems in Mathematics for the Emergency Services*, 1st ed. Clifton Park, NY: Cengage Learning, 1999.

Delmar/Cengage Learning

11 Fireground Flow and Friction Loss Considerations

- Learning Objectives
- Introduction
- Needed Flow
- Available Flow
- Discharge Flow
- Friction Loss Calculations
- Appliance Friction Loss
- Elevation Gain and Loss
- Lessons Learned
- Key Terms
- Chapter Formulas
- Review Questions
- Activity
- Practice Problem
- Additional Resources

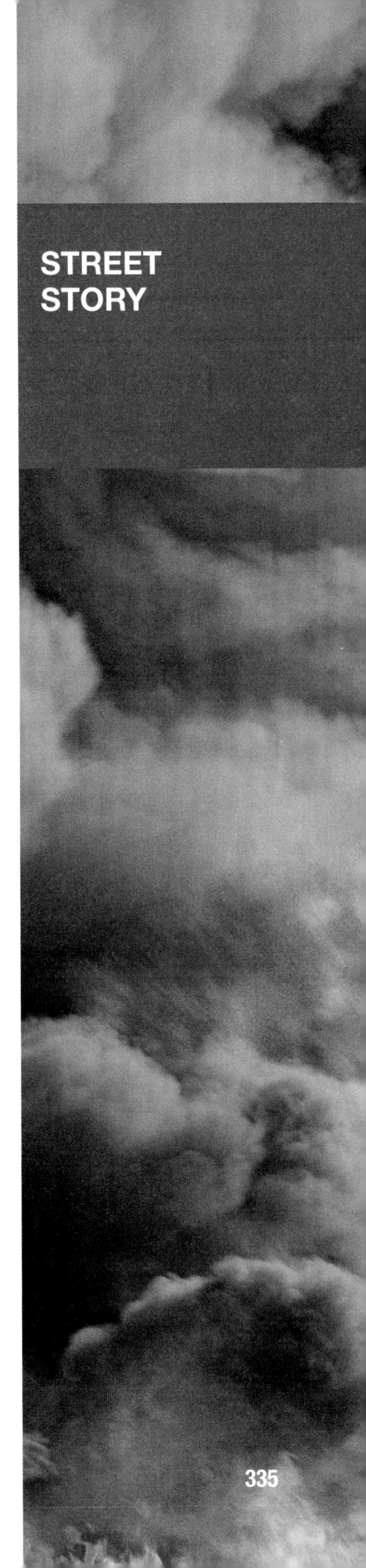

STREET STORY

This one time, we got a call for a house fire, and after consulting the map books, we quickly determined that it was in a rural water supply area, meaning no fire hydrants. That would mean doing relay operations to move water 3,000 feet (914 m) from a nearby pond to the house.

When we arrived, the house was pretty well involved. There was no one inside, so we were just going to be focusing on keeping the exposures from catching on fire. I was given the assignment to be the drafting truck, which means I had to go to the source and drop a hard sleeve into the pond to suck water from it like a straw. After I established a prime, I could send the water out to the other trucks.

The pond was five houses down, so we got the four engine companies staged a certain number of feet apart. I had to figure out how much psi (kPa) we needed to get the water pumped from one engine to the next.

You learn all this math in training—then when you actually become a driver, you end up having most of the pressures precalculated for quick operations at the fire scene. But knowing how to do the calculations is necessary. That was a case when I really did have to figure out the math, starting with how many feet of hose I had and the diameter of the hose. But I also had to account for friction loss. We had to get the water pressure to overcome friction in order to get it from one fire truck to the next, and then through the hose that's laid out too. I remember writing stuff down madly on the boards on the door of the engine!

It took us a half an hour or so to fight the fire because of all the coordination, but the pressures were right, and we were able to keep the fire in check. The math and the formulas definitely work. We can go 6 months to a year before pumping in a rural water supply area, but we try to keep up on math and do drills in the downtime, so that when the day comes it all comes back to us.

—Street Story by Tie Burtlow,
Driver, Fairfax County Fire and Rescue,
Engine 404, Fairfax, Virginia

LEARNING OBJECTIVES

After completing this chapter, the reader should be able to:

11-1. List and discuss the two basic fireground formulas used to determine needed flow.

11-2. Calculate needed fire flow using both the Iowa State and the National Fire Academy formulas.

11-3. Calculate available water when given the dimensions and velocity of a stream.

11-4. Calculate friction loss for a section of hose given the gpm (Lpm) and hose diameter.

11-5. Calculate elevation gain or loss for a given hose lay.

*The driver/operator requirements, as defined by the NFPA 1002 Standard, are identified in black; additional information is identified in blue.

INTRODUCTION

The primary objective of fire pump operations is moving water from a supply source through a pump to a discharge point. The main purpose of this activity is to extinguish a fire or protect an exposure. Simply moving water at any quantity and pressure will not guarantee success. Rather, success depends, in part, on the proper quantity of water delivered to the fire or exposure. Further, the correct pressure must occur at the nozzle to provide proper stream reach and shape. Efficient and effective fire pump operations ensure that the proper quantity and pressure are provided to the discharge point.

The first step in efficient and effective pump operations is to determine the amount of water needed to extinguish a fire. Often, little consideration is given to determining needed flow. Rather, suppression efforts are usually initiated with available resources, even though the consideration of available resources (water supply, pump, hose, and nozzle capabilities) is actually the second step to efficient and effective pump operations. Determining available flow is simply identifying the limits of each component within fire pump operations, that is, water supply, pump, hose, and nozzle. The final step is to develop discharge flows ensuring that proper pressures and volumes are provided to each discharge line.

This chapter focuses on concepts and calculations related to water flow on the fireground. First, needed and available flow are discussed, and then concepts related to developing discharge flows are presented. Finally, the discussion turns to loss of pressure within fire hoses. **Chapter 12** presents calculations for different hoseline configurations.

NOTE

Success in extinguishing a fire or protecting an exposure depends, in part, on the proper quantity of water delivered to the fire or exposure.

NEEDED FLOW

The estimated flow required to extinguish a fire is called the **needed flow**, also referred to as **required flow**. Determining the needed flow for a fire is a crucial process for several reasons, perhaps most obvious being that suppression efforts may be hindered to the point that avoidable damage may occur. For example, providing an inadequate water flow to the fire may allow the fire to continue to burn longer than necessary. In fact, suppression efforts may be so limited that they may not be able to bring the fire under control. Second, inadequate water flow and pressure may place firefighters in greater jeopardy from, for example, suppression efforts taking longer than necessary, resulting in increased exposure to fireground hazards. Finally, figuring the needed fire flow allows for the determination of the resources necessary to provide the flow. In most cases, the initial commitment of resources determines the success of suppression efforts. The time it takes to disconnect, relocate, and reconnect hoselines may be longer than the time it takes for the structure to burn out of control.

Recall from **Chapter 10** that water has a tremendous capacity to absorb heat. When water is converted to steam, it not only absorbs heat, it also helps smother the fire through dilution of air (oxygen). One means for determining the needed flow is to determine how much water is needed that, when converted from a liquid to a vapor, will fill an area with steam. The concept is that the fire will be extinguished through both heat absorption and smothering. With the information provided in **Chapter 10**, we can calculate the amount of water required to fill a room with steam. First, we determine the volume, in cubic feet (metres), of the area. Next, we divide the total volume by the expansion ratio of water (1,700) to establish the cubic feet of water required to extinguish the fire. Finally, we convert feet to gallons (metres to litres). For example, what is the needed flow for a room measuring 50 feet wide, by

60 feet long, by 9 feet tall (15 m wide, by 18 m long, by 3 m tall)? First, the volume is 27,000 ft^3 (60 ft × 50 ft × 9 ft) (810 m^3, 18 m × 15 m × 3 m). Next, the cubic feet (metres) of water is 15.8 ft^3 (.476 m^3) (27,000 ft^3/1,700) (810 m^3/1,700). Finally, the needed flow in gallons (litres) is 118.8 (476 litres) (15.8 ft^2 × 7.48 gal/ft^3) (0.476 m^3 × 1,000 L/m^3). The number 7.48 (1,000) is the number of gallons (litres) in 1 cubic foot (metre). Keep in mind that this calculation only determines the number of gallons (litres) of water that, when converted to steam, will fill an area with steam. This assumes a conversion of all water to steam and that the area is not vented. In addition, it does not address the fire load within the area. The concept of this calculation, however, is the basis for several of the current necessary flow formulas used in the fire service.

There are several formulas available for preplanning needed fire flows. One such formula was developed by the **Insurance Service Office (ISO)**. Typically used for insurance grading purposes, the formula includes factors such as type of construction, occupancy class, exposures, and area. The formula is considered to yield good estimates of needed fire flow. The basic formula reads:

$$NFF = C \times O \times (X + P) \qquad (11\text{–}1)$$

where

NFF = needed fire flow

C = construction factor that is determined using a coefficient related to class of construction and the area of the structure

O = occupancy factor

$X + P$ = exposure factor

Another formula for determining needed water flow is contained in NFPA 1142, *Water Supplies for Suburban and Rural Fire Fighting*. The focus of this standard is on areas where water must be transported to the scene. Although these formulas are not complicated, they require information that takes time to gather. In addition, several charts and tables must be referenced in order to process the information gathered.

Inevitably, situations will arise when the needed flow has not been determined. In these cases, a simple method of calculating required flow on the fireground is needed. Factors such as building construction, occupancy, and exposures cannot be included. The two widely used formulas for calculating needed flow on the fireground are the Iowa State formula and the National Fire Academy formula.

Iowa State Formula

The Iowa State formula, developed by the Fire Service Extension at Iowa State University, estimates the needed flow, in gpm (Lpm), for either an entire structure or a section of a structure. This formula is also used for preplanning. When used for preplanning, typically the needed flow is calculated for the entire structure. On the fireground, the needed flow for the specific area or areas involved is calculated. In either case, the estimated flow, if applied properly, should be able to control a typical enclosed-structure fire efficiently and effectively.

The Iowa State formula was one of the first rational approaches to determining needed fire flow. The formula is based on research of enclosed fires consisting primarily of Class A combustible material common during the 1950s and 1960s. Consequently, higher heat release materials within today's structures require higher flow rates than yielded by the formula. In addition, the formula is based on fire growth within an enclosed structure after 10 minutes from ignition. Calculated flow rates should be increased when response delays occur or when increased oxygen is available, as with ventilated structures, or when openings exist, such as hallways, connecting rooms, or windows. The calculated flow rate does not take into consideration fire spread into void spaces or exposure protection. In essence, the formula estimates the needed flow by dividing the volume of an affected area by 100. The formula can be expressed as:

$$NF = \frac{V}{100} \qquad (11\text{–}2)$$

where

NF = needed flow in gpm

V = volume of the area in cubic feet

100 = is a constant in ft^3/gpm

Or, for metric:

$$NF = V \times 1.3 \qquad (11\text{–}2m)$$

where

NF = needed fire flow in Lpm

V = volume of the area in cubic metres

1.3 = is a constant in m^3/Lpm

Consider a fire-involved floor area measuring 60 feet long by 50 feet wide with a 9-foot (18 m long by 15 m wide with a 3-m) ceiling. The volume of the room is 27,000 ft^3 (60 ft × 50 ft × 9 ft = 27,000 ft^3) (810 m^3). The needed flow is calculated as follows:

$$NF = \frac{V}{100}$$
$$= \frac{27,000}{100}$$
$$= 270 \text{ gpm}$$

$$NF = V \times 1.3$$
$$= 810 \times 1.3$$
$$= 1053 \text{ Lpm}$$

The formula can also be expressed as:

$$NF = 0.01 \text{ gpm/ft}^3 \times V \qquad (11\text{–}3)$$

When the volume is determined, simply move the decimal point two places to the left. For example, the needed flow for room volume of 35,000 ft^3 can be calculated as follows:

$$NF = 0.01 \text{ gpm/ft}^3 \times V$$
$$= 0.01 \text{ gpm/ft}^3 \times 35,000 \text{ ft}^3$$
$$= 350 \text{ gpm}$$

National Fire Academy Formula

The National Fire Academy (NFA) formula is a quick and easy calculation used to determine needed flow for a structure on the fireground. It utilizes the square footage of a structure and is expressed as:

$$NF = \frac{A}{3} \qquad (11\text{–}4)$$

where

NF = needed flow in gpm

A = area of a structure in square feet (length × width)

3 = constant in ft^2/gpm

$$NF = \frac{A}{0.07} \qquad (11\text{–}4m)$$

where

NF = needed flow in Lpm

A = area of a structure in square metres (length × width)

0.07 = constant in m^2/Lpm

Consider the structure in the previous example. The area of this structure is 3,000 ft^2 (280 m^2) (60 × 50 = 3,000). The needed flow is calculated as follows:

$$NF = \frac{A}{3}$$
$$= \frac{3,000}{3}$$
$$= 1,000 \text{ gpm}$$

$$NF = \frac{A}{0.07}$$
$$= \frac{280}{0.07}$$
$$= 4,000 \text{ Lpm}$$

When more than one floor is involved, the square footage of each floor can be added together. The needed flow can also be adjusted to compensate for the percentage of fire. For example, if 50% of the structure is involved, the needed flow would be 500 gpm (2,000 Lpm) (1,000 gpm × 0.5 = 500 gpm) (4,000 Lpm × 0.5 = 2,000 Lpm). This formula can also be expressed as:

$$NF = 0.333 \text{ gpm/ft}^2 \times A \qquad (11\text{–}5)$$

Another way of looking at this calculation is that the needed flow is equal to a third of the floor area.

Needed Flow Consideration

It must be kept in mind that formulas used to determine needed flows are only estimates. Consider the difference in needed flow derived from the previous two formulas. For the same building, the needed flow derived from the Iowa State formula (270 gpm) (1,053 Lpm) is significantly less than that derived from the NFA formula (1,000 gpm) (4,000 Lpm). Each formula is based on a set of assumptions and conditions. When the assumptions and conditions are different, the formula may be less accurate. For example, the formulas provide needed flow for assumed fire suppression tactics. Differences in assumed tactics affect the accuracy of the formula.

Flows required in excess of calculated needed flow can easily occur for several reasons. First, conditions can change from the time the needed flow was calculated during preplanning to the time of the incident. For example, the occupancy can change, additions can be constructed, water supplies may change, and combustible storage can be increased. In addition, the fireground calculation method does not include factors such as building construction and exposures.

Even so, if needed flow is calculated, the chances of efficient and effective suppression efforts are increased considerably. Furthermore, the differences between calculated flows and actual flows can be used to provide better calculations in the future.

Practice

Using the Iowa State formula and the NFA formula, calculate the needed flow for a structure that is 120 feet (36 m) long by 60 feet (18 m) wide by 10 feet (3 m) tall.

Iowa State:

$$NF = \frac{V}{100}$$
$$= \frac{(120 \times 60 \times 10)}{100}$$
$$= \frac{72,000}{100}$$
$$= 720 \text{ gpm}$$

$$NF = V \times 1.3$$
$$= (36 \times 18 \times 3) \times 1.3$$
$$= 1,944 \times 1.3$$
$$= 2,527 \text{ Lpm}$$

NFA:

$$NF = \frac{A}{3}$$
$$= \frac{(120 \times 60)}{3}$$
$$= \frac{7,200}{3}$$
$$= 2,400 \text{ gpm}$$

$$NF = \frac{A}{0.07}$$
$$= \frac{(36 \times 18)}{0.07}$$
$$= \frac{648}{0.07}$$
$$= 9,257 \text{ Lpm}$$

AVAILABLE FLOW

After the needed flow is determined, the next step is to determine the **available flow**. The amount of water that can be moved from the supply to the fire scene is called available flow. The available flow will be limited by the capabilities of each component within the pumping operation. The goal is to evaluate each of the following components to maximize the available flow.

Water Supply

Determining available flow starts with an evaluation of the water supply. Obviously, water supplies should be selected that provide the needed flow. When the needed flow has not been calculated or when the needed flow is greater than the available supply, a water supply should be selected that allows the pump to flow its rated capacity. Considerations for evaluating available water for each type of supply are discussed in **Chapter 7**.

Pump

The size of the pump is also an important consideration for evaluating available water flow. Assuming other components are of adequate capability, available water will be limited by the size of the pump. Recall that pumps are expected to provide their rated capacity at 150 psi (1,050 kPa), 70% of rated capacity at 200 psi (1,400 kPa), and 50% of rated capacity at 250 psi (1,750 kPa) (**Figure 11-1**). The pressure at which the pump is expected to operate will also limit available flow.

Hose

The size and length of hose used for both supply and discharge will affect available water. The size affects available flow in that different sizes of hose have different flow capacities. In essence, large flow requires either large-size hose or multiple lines. The length of hose also affects available water in that friction loss increases with longer hoselines. In this case, pressure is affected rather than the quantity of water. The longer the line, the greater the friction

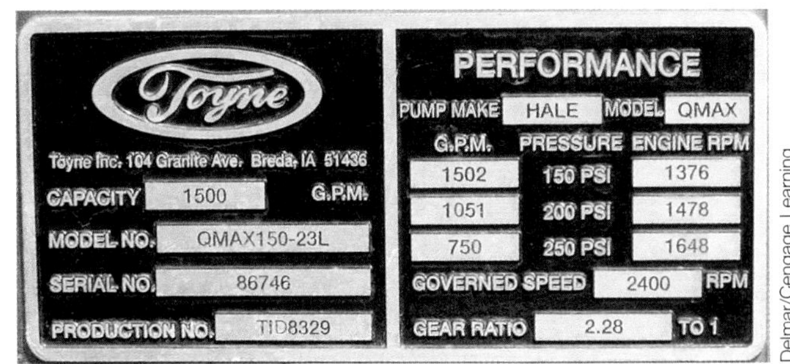

FIGURE 11-1 Higher pressures during a pumping operation can reduce the available flow from a pump.

loss, and, in turn, the greater the pressure required to provide the nozzle with its proper operating pressure. The limiting factors are the pressure rating of the hose and the maximum pressure rating of the pump. Longer lays may be impossible because the pressure needed to supply the nozzle adequately is greater than the pressure rating of the hose or the pump.

Nozzles

The type or types of nozzles used will also affect available flow. This component is perhaps the easiest to change, yet the component least evaluated. In essence, nozzles must be provided with their designed operating pressure and flow (**Figure 11-2**). The total output from all operating nozzles will equal the available flow. For example, assuming all other components are capable of providing 500 gpm (2,000 Lpm), if only four 95-gpm (380-Lpm) nozzles are available and are operating, the available flow will be 380 gpm (1,520 Lpm). When automatic nozzles are used, the gpm (Lpm) flow, within the nozzles' range of flows, must be selected. When smooth-bore nozzles are used, the gpm (Lpm) must either be looked up in a table or calculated using the following formula:

$$Q = 29.7 \times d^2 \times \sqrt{NP} \qquad (11\text{–}6)$$

where

Q = gallons per minute

29.7 = 30 is the constant most often used for fireground calculations

d = nozzle diameter in inches

NP = pressure in psi—this is also called "nozzle pressure" or "velocity pressure"

FIGURE 11-2 Nozzles are designed to operate at a specific pressure and flow (or range of flows).

Note: For fireground calculations, the $\sqrt{NP}$ for handlines, 7.07, is rounded to 7, and the $\sqrt{NP}$ for master streams, 8.94, is rounded to 9.

$$Q = 0.067 \times d^2 \times \sqrt{NP} \qquad (11-6\text{m})$$

where

Q = litres per minute

0.067 = constant

d = nozzle in diameter in millimetres

NP = pressure in kPa

For example, what is the gallon-per-minute (litre-per-minute) flow through a ¾-inch (19-mm) tip smooth-bore nozzle with 50 psi (350 kPa) at the nozzle?

$$\begin{aligned} Q &= 29.7 \times d^2 \times \sqrt{NP} \\ &= 29.7 \times 0.75^2 \times \sqrt{50} \\ &= 29.7 \times 0.56 \times 7.07 \\ &= 117.58 \text{ or rounded up to } 118 \end{aligned}$$

Note: Using the fireground calculation constants of 30 and 7 yields similar results of 117.6 or 118.

$$\begin{aligned} Q &= 0.067 \times d^2 \times \sqrt{NP} \\ &= 0.067 \times 19^2 \times \sqrt{350} \\ &= 0.067 \times 361 \times 18.7 \\ &= 0.067 \times 361 \times 18.7 \\ &= 452.3 \text{ or } 452 \text{ Lpm} \end{aligned}$$

Practice

What is the flow through a 1-inch (25-mm) smooth-bore nozzle on a master stream device?

$$\begin{aligned} Q &= 29.7 \times d^2 \times \sqrt{NP} \\ &= 29.7 \times 01^2 \times \sqrt{80} \\ &= 29.7 \times 1 \times 8.9 \\ &= 264.33 \text{ or } 264 \text{ gpm} \end{aligned}$$

$$\begin{aligned} Q &= 0.067 \times d^2 \times \sqrt{NP} \\ &= 0.067 \times 25^2 \times \sqrt{560} \\ &= 0.067 \times 625 \times 23.7 \\ &= 992.4 \text{ or } 992 \text{ Lpm} \end{aligned}$$

Note: Using the fireground calculation constants of 30 and 9 yields a slightly higher value of 270.

The available water flow is simply the amount of water that is capable of being provided to the scene. In some cases, the available water flow will be less than the needed flow. When this occurs, plans must be made to increase the available flow by securing additional water supplies or increasing the existing flow. In other cases, the available flow will be greater than the needed flow. In either case, after the supply has been secured, the number and size of discharge configurations can be manipulated to provide the available flow. In other words, the total flow a pumping operation is capable of providing is available at the discharge side of the pump. The actual discharge flow, however, may be equal to or less than the available flow.

DISCHARGE FLOW

After the needed water flow is determined and the available water flow secured, the next step is to develop discharge flows. The amount of water flowing from the discharge side of a pump through hose, appliances, and nozzles to the scene is called **discharge flow**. The most important factor in developing discharge flows is the determination of the quantity of water a specific line is expected to flow. In other words, the gallons (litres) per minute the line will be flowing. This determination is important for two reasons. First, identifying the gallons (metres) per minute a line is flowing is important for keeping track of the total flow provided to the scene as well as of the remaining flow available in the pumping operation. Second, knowing the flow of a line is essential for providing the proper operating pressure and flow required by the nozzle.

There are two methods used to ensure that nozzles are provided with the proper flow and pressure. One method uses flow meters while the other uses pressure gauges.

NOTE

The most important factor in developing discharge flows is the determination of the quantity of water a specific line is expected to flow. This is because the gpm (Lpm) flow rate is the most important factor affecting friction losses for a given hoseline.

Flow Meters

Without question, the easiest method for ensuring that nozzles are provided with the proper flow and pressure is with flow meters (**Figure 11-3**). When flow meters are used, the major consideration is knowing the rated flow of the nozzle. Flow meters measure the quantity and rate (gpm or Lpm) of water flowing through a line. Because matter cannot be created or destroyed, the same flow leaving the discharge side of the pump will also travel through the hose and discharge from the nozzle. For example, when the flow meter on the pump panel reads 100 gpm (400 Lpm), the nozzle will be discharging 100 gpm (400 Lpm). Recall that pressure is a function of restriction. If the nozzle is designed to operate at 100 gpm (400 Lpm), the restriction designed into the nozzle will create the proper operating pressure. The key, though, is knowing the required gpm (Lpm) for the nozzle.

The use of and reliance on flow meters on pumping apparatus is relatively new. Although flow meters have been around for a number of years, increased accuracy and reliability of newer flow meters has increased their popularity and use on fire apparatus. In addition, they are gaining popularity because they dramatically simplify flow development and increase flow accuracy over estimated hydraulic calculations. Some flow meters provide both gpm (Lpm) and pressure. When flow meters provide only gpm (Lpm), the operating pressure of the hose can easily be exceeded. When only one line is in operation, the master discharge gauge can be used to determine pressure. Should the flow meter fail, the master discharge gauge can again be used. When more than one line is in operation, this option is not available.

Delmar/Cengage Learning

FIGURE 11-3 Flow meters eliminate the need for friction loss calculations on the fireground.

When using flow meters, the operator must also check pressure gauges to make sure that safe operating pressures are not exceeded.

Pressure Gauges

The pressure gauge method for developing discharge flows is far more complicated (**Figure 11-4**) than the flow meter method. When pressure gauges are used, the pump operator must calculate any changes in pressure within the discharge line. Hypothetically, if the pressure did not change, the pump operator would simply increase pump speed until the discharge pressure gauge equaled the operating pressure of the nozzle. In reality this cannot occur because friction losses in the hose and appliances as well as changes in elevation always affect the pressure. The required discharge pressure to provide the nozzle with its correct operating pressure and flow must be calculated by determining friction losses in hoses and appliances and pressure gains or losses due to elevation.

(A)

Delmar/Cengage Learning

(B)

Courtesy of Ken Kroeker, Emergency Services Officer, Office of the Fire Commissioner, Manitoba Emergency Services College

FIGURE 11-4 (A) The traditional use of pressure gauges on a pump panel required pump operators to estimate/calculate friction loss in hose. (B) Metric gauge.

FRICTION LOSS CALCULATIONS

The loss of pressure within hose resulting from friction is inevitable. Scientific formulas have been developed and refined over the years that can accurately calculate friction loss. Unfortunately, most of the formulas are not practical for use on the fireground, because, perhaps obviously, they are too cumbersome and complex. In addition, the variables in the formulas cannot be adequately controlled or measured on the fireground. Fireground formulas and rules of thumb have also been developed over the years to try to estimate friction loss, each with varying levels of accuracy. When the standard hoses were 1½-inch and 2½-inch (38-mm and 45-mm) cotton-jacketed hose, the formulas and rules of thumb were accurate enough for fireground use; however, the gradual increase in different hose diameters as well as the number of manufacturers and construction methods have significantly confounded the issue of friction loss calculations. It is common for a fire department to use anywhere from three to five different sizes of various ages of hose manufactured by several different companies using different construction methods. When you add the dilemma of kinks and bends on the fireground, the complexity of calculating friction loss becomes readily obvious.

The fireground has a way of ruining even the best-laid plans.

To allow quick fireground friction loss calculations, numerous estimation formulas have been developed. Pump operators should understand that these different formulas may give different answers to the same problem. Any of the answers should, however, be close enough for use on the fireground. In addition, for consistency, a given department may adapt a given formula for use in a given situation.

In some cases, friction loss calculations are not required on the fireground. When several discharge hose layouts are consistently used, friction loss can be computed in advance. For example, must modern pumpers have preconnected attack lines. When the same diameter and length of hose and nozzle are used, the pump operator simply increases pump speed to the predesignated discharge pressure. In addition, tables and charts can be developed so that pump operators simply look up either the friction loss or the discharge pressure (**Figure 11-5**; see also **Appendix G**). In addition, a variety of slide rules (**Figure 11-6**) and even handheld friction loss calculators can be used (**Figure 11-7**). Finally, when flow meters are used, calculations are not necessary.

Even so, pump operators must still be able to calculate friction loss manually with a reasonable degree of accuracy. For one thing, such calculations are needed to develop tables and charts for specific equipment used by a fire department. Likewise, calculations are required to predetermine pump pressures on commonly used discharge lays. In addition, calculations should be made to evaluate the variety of fireground friction loss formulas and rules of thumb. Finally, and perhaps most important, the fireground has a way of ruining even the best laid plans, the situation in which manual calculations will eventually occur.

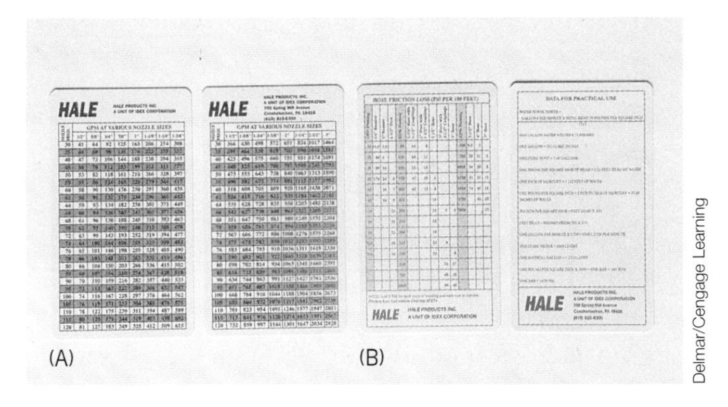

Delmar/Cengage Learning

FIGURE 11-5 A friction loss table can be used to estimate friction loss on the fireground.

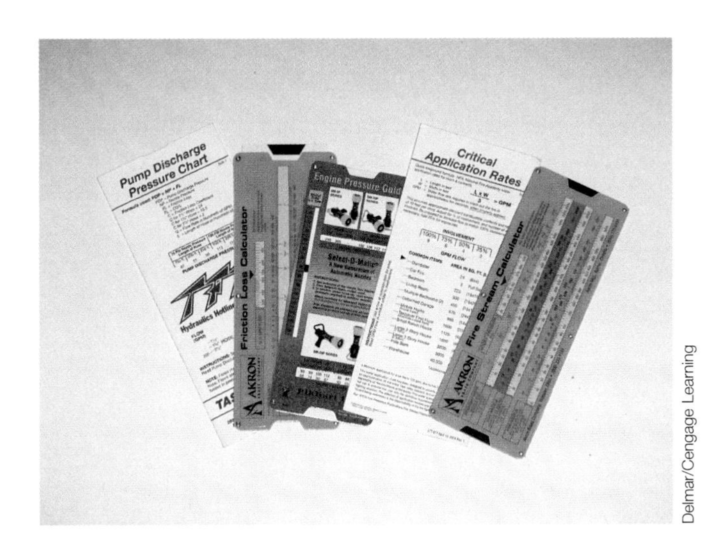

Delmar/Cengage Learning

FIGURE 11-6 Friction loss slide rules can be used to estimate friction loss on the fireground.

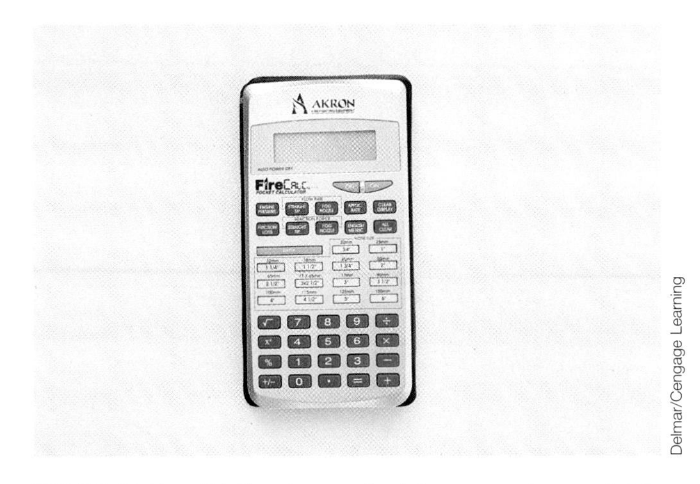

Delmar/Cengage Learning

FIGURE 11-7 Friction loss calculators are also available to simplify fireground calculations.

As previously stated, various fireground methods have been developed to calculate friction loss. The following are several of the more popular methods. It should be noted that the last method discussed, cq^2L, is considered the most accurate.

Hand Method

The hand method is a simple way to calculate friction loss in 100-foot sections of 2½-inch hose. Beginning with the thumb, each finger is assigned a number representing gpm flow (**Figure 11-8**). For example, the index finger has an assigned value of 2, representing 200 gpm, while the ring finger has an assigned value of 4, representing 400 gpm. Next, the base of each finger is assigned an even number as indicated in **Figure 11-8**. To find the friction loss in 100-foot sections of 2½-inch hose, simply select the finger representing the gpm flow and multiply the two numbers assigned to the finger. For example, the friction loss for 100 feet of 2½-inch hose flowing 300 gpm is 18 psi (3 × 6 = 18). The friction loss for 100 feet of 2½-inch hose flowing 500 gpm is 50 psi (5 × 10 = 50). The space between the fingers can be assigned 1.5, 2.5, 3.5, and so on for the gpm values and 3, 5, 7 for the gaps between figures. The friction loss for 100 feet of 2½-inch hose flowing 150 gpm is 4.5 gpm (1.5 × 3 = 4.5). Several variations to this basic theme have evolved over the years and the system has been modified for use with various hose sizes.

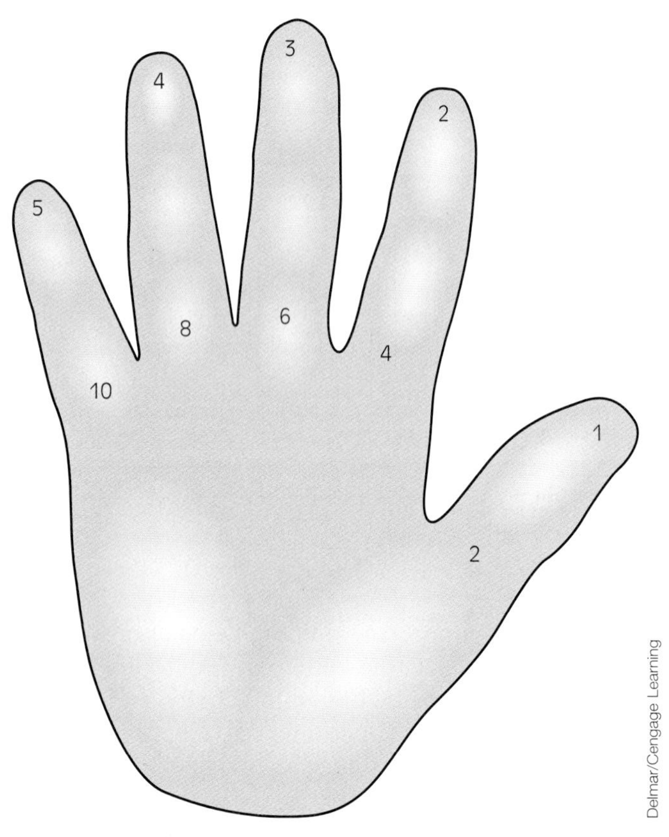

FIGURE 11-8 The "hand method" can be used to simplify fireground calculations.

Practice

Calculate the friction loss for the following hoselines:

- 100 feet of 2½-inch flowing 400 gpm:

$$FL = 32 \text{ psi } (4 \times 8)$$

- 100 feet of 2½-inch flowing 200 gpm

$$FL = 8 \text{ psi } (2 \times 4)$$

- 100 feet of 2½-inch flowing 250 gpm

$$FL = 12.5 \text{ psi } (2.5 \times 5)$$

- 200 feet of 2½-inch flowing 200 gpm

$$FL = 2 \times 4 \times 2 \text{ (number of 100-foot sections of hose)} = 16 \text{ psi}$$

Drop 10 Method

The drop 10 method is a simple rule of thumb that provides friction loss in 100-foot sections of hose, and, although less accurate than others, it can be used for more than just 2½-inch hose. This method simply subtracts 10 from the first two numbers of gpm flow. For example, the friction loss in 100 feet of hose when 500 gpm is flowing is 40 psi (50 – 10 = 40). Likewise, the friction loss in 100 feet of hose when 250 gpm is flowing is 15 psi (25 – 10 = 15).

$2q^2 + q$

This formula was developed by the National Board of Fire Underwriters based on 2½-inch cotton jacketed hose commonly used in the 1930s and 1940s. This formula was widely used for many years; the full formula for this method is written as:

$$FL = 2q^2 + q \tag{11–7}$$

where

FL = friction loss in psi per 100-foot sections of hose

2 = constant for 2½-inch cotton-jacketed hose

q = flow in hundreds of gallons per minute

The amount of water flowing, q, is given in hundreds of gpm and is calculated using the following formula:

$$q = \frac{gpm}{100} \quad \text{or as} \quad q = \frac{Q}{100} \tag{11–8}$$

where

q = hundreds of gpm

gpm = actual flow

Q = actual flow in gpm

100 = constant

Consider 250 gpm flowing through a 2½-inch line 500 feet long. The flow in hundreds of gpm (q) can be calculated as follows:

$$q = \frac{250 \text{ gpm}}{100}$$
$$= 2.5 \text{ hundreds of gpm}$$

A simple method of calculating q is to move the decimal point two places to the left.

The friction loss in 100 feet of 2½-inch hose flowing 500 gpm is calculated as follows:

$$\begin{aligned} FL &= 2q^2 + q \\ &= 2 \times \left(\frac{500}{100}\right)^2 + 5 \\ &= 2 \times (5)^2 + 5 \\ &= 2 \times 25 + 5 \\ &= 55 \end{aligned}$$

This formula provides results for 2½-inch cotton-jacketed hose for flows greater than 100 gpm. For flows less than 100 gpm, the formula is modified as follows:

$$FL = 2q + \tfrac{1}{2}q \qquad (11\text{–}9)$$

To simplify calculations further, some departments use the following:

$$FL = 2q^2 \qquad (11\text{–}10)$$

The use of hose other than 2½ inches in diameter requires a conversion factor. Typically this is accomplished by multiplying the gpm by a conversion factor to get the equivalent flow through 2½-inch hose. The equivalent flow (gpm) is then plugged into the formula to determine the friction loss per 100 feet of hose.

It should be noted that this formula provides friction loss for cotton-jacketed hose. New hose construction methods reduce friction loss as compared to older cotton-jacketed hose. The effects of newer hose construction on friction loss necessitate the use of a more accurate formula, such as $FL = cq^2L$.

Practice

Calculate the friction loss for 100 feet of 2½-inch hose flowing 200, 300, and 400 gpm.

- 200 gpm

$$\begin{aligned} FL &= 2q^2 + q \\ &= 2 \times \left(\frac{200}{100}\right)^2 + \left(\frac{200}{100}\right) \\ &= 2 \times 2^2 + 2 \\ &= 2 \times 4 + 2 \\ &= 10 \text{ psi} \end{aligned}$$

- 300 gpm

$$\begin{aligned} FL &= 2q^2 + q \\ &= 2 \times \left(\frac{300}{100}\right)^2 + \left(\frac{300}{100}\right) \\ &= 2 \times 3^2 + 3 \\ &= 2 \times 9 + 3 \\ &= 21 \text{ psi} \end{aligned}$$

- 400 gpm

$$\begin{aligned} FL &= 2q^2 + q \\ &= 2 \times \left(\frac{400}{100}\right)^2 + \left(\frac{400}{100}\right) \\ &= 2 \times 4^2 + 4 \\ &= 2 \times 16 + 4 \\ &= 36 \text{ psi} \end{aligned}$$

Condensed *q* Formula

This formula is an adaptation of the Fire Underwriters' formula ($FL = 2q^2 + q$) and is used specifically for 3-inch hose. The condensed q formula is written as:

$$FL = q^2 \qquad (11\text{–}11)$$

For example, the friction loss in 100 feet of 3-inch hose flowing 500 gpm is calculated as follows:

$$\begin{aligned} FL &= q^2 \\ &= \left(\frac{500}{100}\right)^2 \\ &= 5^2 \\ &= 25 \text{ psi} \end{aligned}$$

cq^2L

According to the NFPA Fire Protection Handbook, this formula is derived from a combination of Bernoulli's equation, the Darcy-Weibach equation, $FL = (fV^2L) \div (2gD)$, and the Continuity equation, $Q = V \times A$. The Darcy-Weibach equation is used to calculate friction loss in long, straight pipes with uniform diameter and roughness. The Continuity equation expresses a conservation of mass law stating that mass cannot be created or destroyed. In this case, it suggests that the amount of water entering and leaving a hose or pipe must be the same—the term most often used is conserved—as it exits the hose or pipe. If 500 gpm (2,000 Lpm) enters a hoseline, then 500 gpm (2,000 Lpm) must exit. Consider a garden hose flowing water: If the end of the hose is squeezed, the velocity of the water increases. You might say the water is hurrying up to get out so that the same quantity exits the hose as is entering. The resultant formula, $FL = cq^2L$, is both simple and reasonably accurate for varying hoseline diameters, so the formula can be used for both preplanning and fireground calculations. The formula is expressed as:

$$FL = c \times q^2 \times L \qquad (11\text{–}12)$$

where

FL = friction loss in psi (kPa)

c = constant for a specific hose diameter

q = gpm (Lpm) ÷ 100 (flow in hundreds of gpm [litres])

L = length of hose in hundreds of feet (hectometres)

The constant, c, takes into account all the variables associated with friction loss except flow rate (q) and length (L). The average constants for several common hose diameters are provided in **Table 11-1**. These values are average values and can be replaced with more accurate values for specific manufacturers and hose sizes.

The length of hose, L, is given in hundreds of feet (hectometres) and is calculated using the following formula:

$$L = \frac{\text{hose length}}{100} \quad (11\text{–}13)$$

where

L = hundreds of feet (hectometres)

hose length = actual length of hose

100 = constant

TABLE 11-1 Average Constants (c) for Several Common Firefighting Hose Diameters When Using the Friction Loss Equation FL = cq²L.

Hose Diameter in. (mm)	Friction Loss Constant, *c*	
	Customary	Metric
¾" (19 mm)	1,100	(1,741)
1" (25 mm)	150	(238)
1½" (38 mm)	24	(38)
1¾" (45 mm)	15.5	(24.6)
2" (50 mm)	8	(12.7)
2½" (65 mm)	2	(3.17)
3" with 2½" (77 mm with 65-mm) couplings	0.8	(1.27)
3" with 3" (77 mm with 77-mm) couplings	0.677	(1.06)
4" (100 mm)	0.2	(0.305)
5" (125 mm)	0.08	(0.138)
6" (150 mm)	0.05	(0.083)

Source: Fire Protection Handbook, 19th ed. vol. 1. Quincy, MA, National Fire Protection Association, 2003.

For example, the value of L for a 500-foot hose lay is calculated as follows:

$$L = \frac{500 \text{ ft}}{100}$$
$$= 5 \text{ hundreds of feet}$$

$$L = \frac{150 \text{ m}}{100}$$
$$= 1.5 \text{ hectometres}$$

Again, a simple method of calculating this formula is to move the decimal point two places to the left.

The following steps, then, can be used to calculate friction loss using this formula:

1. Identify the constant for the diameter of hose being used (refer to **Table 11-1**).
2. Determine the hundreds of gallons (hectolitres) of flow using the formula q = gpm (Lpm)/100.
3. Determine the hundreds of feet (metres) of hose using the formula L = hose length ÷ 100.
4. Use the answers to the previous steps in the friction loss formula, $FL = c \times q^2 \times L$.

For example, what is the friction loss in 500 feet of 2½-inch (150 m of 65-mm) hose flowing 250 gpm (1,000 Lpm)?

Step 1: $c = 2$ (from Table 11-1)

Step 2: $q = \frac{250}{100} = 2.5$

Step 3: $L = \frac{500}{100} = 5$

Step 4: $FL = 2 \times (2.5)^2 \times 5$
$= 2 \times 6.25 \times 5$
$= 62.5$

Step 1: $c = 3.17$

Step 2: $q = \frac{1,000}{100} = 10$

Step 3: $L = \frac{150}{100} = 1.5$

Step 4: $FL = 3.17 \times 10^2 \times 1.5$
$= 3.17 \times 100 \times 1.5$
$= 475.5$ or 475 kPa

To compare friction loss in different sizes of hose, use the same flow and length of hose as in the previous example and calculate the friction loss in 3- and 4-inch hose (77-mm and 100-mm).

- 3-inch hose

$$FL = c \times q^2 \times L$$
$$= 0.8 \times \left(\frac{250}{100}\right)^2 \times \left(\frac{500}{100}\right)$$
$$= 0.8 \times (2.5)^2 \times 5$$
$$= 0.8 \times 6.25 \times 5$$
$$= 25$$

- 77-mm hose with 65-mm couplings

$$FL = c \times q^2 \times L$$
$$= 1.27 \times \left(\frac{1,000}{100}\right)^2 \times \left(\frac{150}{100}\right)$$
$$= 1.27 \times 10^2 \times 1.5$$
$$= 1.27 \times 100 \times 1.5$$
$$= 190.5 \text{ or } 190 \text{ kPa}$$

- 4-inch hose

$$FL = c \times q^2 \times L$$
$$= 0.2 \times \left(\frac{250}{100}\right)^2 \times \left(\frac{500}{100}\right)$$
$$= 0.2 \times (2.5)^2 \times 5$$
$$= 0.2 \times 6.25 \times 5$$
$$= 6.25, \text{ or rounded off to } 6$$

- 100-mm hose

$$FL = c \times q^2 \times L$$
$$= 0.305 \times \left(\frac{1,000}{100}\right)^2 \times \left(\frac{150}{100}\right)$$
$$= 0.305 \times 10^2 \times 1.5$$
$$= 0.305 \times 100 \times 1.5$$
$$= 45.75 \text{ or } 46 \text{ kPa}$$

Practice

Calculate the friction loss for the following hose dimensions with each line flowing 150 gpm.

	100 feet (30 m)	250 feet (75m)	500 feet (150 m)
1½ inch (38 mm)	_______	_______	_______
1¾ inch (45 mm)	_______	_______	_______
2½ inch (65 mm)	_______	_______	_______
3 inch (77 mm)	_______	_______	_______

1½-Inch Lines

100 feet $FL = 24 \times \left(\frac{150}{100}\right)^2 \times \left(\frac{100}{100}\right)$
$$= 24 \times 1.5^2 \times 1$$
$$= 24 \times 2.25 \times 1$$
$$= 54 \text{ psi}$$

250 feet $FL = 24 \times 2.25 \times 2.5$
$$= 135 \text{ psi}$$

500 feet $FL = 24 \times 2.25 \times 5$
$$= 270 \text{ psi}$$

30-Millimetre Lines

30 m $FL = 38 \times \left(\frac{600}{100}\right)^2 \times \left(\frac{30}{100}\right)$
$$= 38 \times 6^2 \times 0.3$$
$$= 38 \times 36 \times 0.3$$
$$= 410.4 \text{ or } 410 \text{ kPa}$$

75 m $FL = 38 \times 36 \times 0.75$
$$= 1,026 \text{ kPa}$$

150 m $FL = 38 \times 36 \times 1.5$
$$= 2,052 \text{ kPa}$$

1¾-Inch Lines

100 feet $FL = 15.5 \times \left(\frac{150}{100}\right)^2 \times \left(\frac{100}{100}\right)$
$$= 15.5 \times 1.5 \times 1$$
$$= 15.5 \times 2.25 \times 1$$
$$= 34.8 \text{ or } 35 \text{ psi}$$

250 feet $FL = 15.5 \times 2.25 \times 2.5$
$$= 87.1 \text{ or } 87 \text{ psi}$$

500 feet $FL = 15.5 \times 2.25 \times 5$
$$= 174.3 \text{ or } 174 \text{ psi}$$

45-Millimetre Lines

30 m $FL = 24.6 \times \left(\frac{600}{100}\right)^2 \times \left(\frac{30}{100}\right)$
$$= 24.6 \times 6^2 \times 0.3$$
$$= 24.6 \times 36 \times 0.3$$
$$= 265.7 \text{ or } 266 \text{ kPa}$$

75 m $FL = 24.6 \times 36 \times 0.75$
$$= 664.2 \text{ or } 664 \text{ kPa}$$

150 m $FL = 24.6 \times 36 \times 1.5$
$= 1,328.4$ or $1,328$ kPa

2½-Inch Lines

100 feet $FL = 2 \times \left(\frac{150}{100}\right)^2 \times \left(\frac{100}{100}\right)$
$= 2 \times 1.5^2 \times 1$
$= 2 \times 2.25 \times 1$
$= 4.5$ or 5 psi

250 feet $FL = 2 \times 2.25 \times 2.5$
$= 11.25$ or 11 psi

500 feet $FL = 2 \times 2.25 \times 5$
$= 22.5$ or 23 psi

65-Millimetre Lines

30 m $FL = 3.17 \times \left(\frac{600}{100}\right)^2 \times \left(\frac{30}{100}\right)$
$= 3.17 \times 6^2 \times 0.3$
$= 3.17 \times 36 \times 0.3$
$= 34.2$ or 34 kPa

75 m $FL = 3.17 \times 36 \times 0.75$
$= 85.6$ or 86 kPa

150 m $FL = 3.17 \times 36 \times 1.5$
$= 171.1$ or 171 kPa

3-Inch Lines

100 feet $FL = .8 \times \left(\frac{150}{100}\right)^2 \times \left(\frac{100}{100}\right)$
$= .8 \times 1.5^2 \times 1$
$= .8 \times 2.25 \times 1$
$= 1.8$ or 2 psi

250 feet $FL = .8 \times 2.25 \times 2.5$
$= 4.5$ or 5 psi

500 feet $FL = .8 \times 2.25 \times 5$
$= 9$ psi

77-Millimetre Lines

30 m $FL = 1.27 \times \left(\frac{600}{100}\right)^2 \times \left(\frac{30}{100}\right)$
$= 1.27 \times 6^2 \times 0.3$
$= 1.27 \times 36 \times 0.3$
$= 13.7$ or 14 kPa

75 m $FL = 1.27 \times 36 \times 0.75$
$= 34.29$ or 34 kPa

150 m $FL = 1.27 \times 36 \times 1.5$
$= 68.58$ or 69 kPa

	100 feet (30 m)	250 feet (75 m)	500 feet (150 m)
1½ inch (38 mm)	54 (410)	135 (1,026)	270 (2,052)
1¾ inch (45 mm)	35 (266)	87 (664)	174 (1,328)
2½ inch (65 mm)	5 (34)	11 (86)	23 (171)
3 inch (77 mm)	2 (14)	5 (34)	9 (69)

APPLIANCE FRICTION LOSS

Recall that friction loss is a result of water rubbing against the interior of the hose. **Appliance friction loss** is the reduction in pressure resulting from increased turbulence caused by the appliance. When the interior is smooth and continuous, less friction loss will occur than when the interior is rough and with varying changes in the interior lining. Hose appliances such as wyes, Siamese, increasers, reducers, manifolds, and master stream devices change the interior lining by causing slight protrusions and indentations. These changing surfaces within the lining cause increased friction loss that must be accounted for when determining the total loss in pressure from friction. Actual friction loss varies by appliance and flow. To simplify hydraulic calculations, tables such as **Table 11-2A** and **Table 11-2B** (metric) of assumed friction are often used, and flows of less than 350 gpm in 2½-inch and larger appliances are assumed to have negligible losses. More accurate friction information can be obtained from the manufacturer or from tests conducted by a department.

ELEVATION GAIN AND LOSS

Recall that when hoselines are operated at elevation above the pump, greater pressures are required to provide the required pressure at the nozzle. In turn, when hoselines are operated at elevations below the pump, less pressure is required to provide the required nozzle pressure. Two methods are used on the fireground to calculate this gain or loss of pressure: calculations using feet (metres) above or below

TABLE 11-2A Estimated Friction Loss in Common Appliances*

Type of Appliance	Friction Loss (psi)
2½-in. to 2½-in. wye	5
2½-in. to 1½-in. wye	10
1½-in. to 1½-in. wye	15
2½-in. to 1½-in. Siamese	10
1½-in. to 1½-in. Siamese	15
Reducer	5
Increaser	5
Monitor	15
Four-way valve	15
Standpipe	25
Sprinkler	150
Combination	175
Foam eductor	200

**Various friction loss pressures are used for these appliances. Each appliance will have a specific loss in pressure based in part on age and manufacturer.*

Reproduced with permission from NFPA's Fire Protection Handbook®, Copyright 2003, 19th edition, Volume I, National Fire Protection Association.

TABLE 11-2B Estimated Friction Loss in Common Appliances (Metric)*

Any appliance flowing less than 1,400 Lpm is assumed not to have a friction loss. Appliances flowing more than 1,400 Lpm are assumed to have a 70-kPa friction loss. Add the additional friction loss listed below when these appliances are used with a flow over 1,400 Lpm.

Type of Appliance	Additional Friction Loss (kPa)
Master stream	175
Aerial device	175
Sprinkler system	350
Standpipe system	175
Foam educator	1,400

**Various friction loss pressures are used for these appliances. Each appliance will have a specific loss in pressure based in part on age and manufacturer.*

the pump level and using number of floor levels above or below the pump.

Elevation Formula in Feet (Metres)

The following formula can be used to calculate pressure gain and loss when the elevation in feet above or below the pump is known.

$$EL = 0.5 \times H \qquad (11\text{–}14)$$

where

EL = the gain or loss of elevation in psi

0.5 = pressure exerted at base of 1-in.^{3} column of water 1 foot high, expressed as psi/ft (actual pressure is .433 in.^{2}/ft)

H = height in feet

The following formula, known as the "10H rule," can be used to calculate pressure gain and loss when the elevation in metres above or below the pump is known.

$$EL = 10 \times H \qquad (11\text{–}14m)$$

where

EL = the gain or loss of elevation in kPa

H = height in metres

Elevation Gain

Consider a hoseline operating 50 feet (15 m) above the pump. The estimated pressure gain can be calculated as follows:

$$\begin{aligned} EL &= 0.5 \times H \\ &= 0.5 \text{ psi/ft} \times 50 \text{ ft} \\ &= 25 \text{ psi} \end{aligned}$$

$$\begin{aligned} EL &= 10 \text{ kPa/metre} \\ &= 10 \text{ kPa/m} \times 15 \text{ m} \\ &= 150 \text{ kPa} \end{aligned}$$

Elevation loss and elevation gain may also be referred to as forward pressure and head pressure, or combined and simply called elevation pressure, and either added or subtracted when computing total pressure losses.

Elevation Loss

Consider a hoseline operating 30 feet (9 m) below the pump. The pressure loss can be calculated as follows:

$$\begin{aligned} EL &= 0.5 \times H \\ &= 0.5 \text{ psi/ft} \times -30 \text{ ft} \\ &= -15 \text{ psi} \end{aligned}$$

$$\begin{aligned} EL &= 10 \text{ kPa/metre} \\ &= 10 \text{ kPa/m} \times -9 \text{ m} \\ &= -90 \text{ kPa} \end{aligned}$$

Elevation Formula by Floor Level

When hoselines are operated within structures, elevation gain or loss can be calculated using the number of floor levels above or below the pump, as follows:

$$EL = 5 \text{ psi} \times H \qquad (11\text{–}15)$$

$$EL = 35 \text{ kPa} \times H \qquad (11\text{–}15\text{m})$$

where

EL = the gain or loss of elevation in psi (kPa)

5 (35 kPa) = gain or loss in pressure for each floor level

H = height in number of floor levels above or below the pump. Often this is expressed as the number of floors minus one.

Elevation Gain

Consider a hoseline operating on the fifth story of a structure when the pump is on the same elevation as the first floor. The pressure gain can be calculated as follows:

$$\begin{aligned} EL &= 5 \times (5 \text{ stories} - 1) \\ &= 5 \times 4 \\ &= 20 \text{ psi} \end{aligned}$$

$$\begin{aligned} EL &= 35 \text{ kPa} \times (5 \text{ stories} - 1) \\ &= 35 \times 4 \\ &= 140 \text{ kPa} \end{aligned}$$

Elevation Loss

Consider a hoseline operating in the basement of a structure three floor levels below ground. The pressure loss can be calculated as follows:

$$\begin{aligned} EL &= 5 \times H \\ &= 5 \times -3 \\ &= -15 \text{ psi} \end{aligned}$$

OR

$$\begin{aligned} EL &= 35 \text{ kPa} \times -3 \\ &= -105 \text{ kPa} \end{aligned}$$

Practice

Calculate the elevation gain or loss for the following:

- Hoseline 75 feet (23 m) above the pump.

$$\begin{aligned} EL &= 0.5 \text{ psi/ft} \times 75 \text{ ft} \\ &= 37.5 \text{ or } 38 \text{ psi} \end{aligned}$$

$$\begin{aligned} EL &= 10 \text{ kPa/m} \times 23 \text{ m} \\ &= 230 \text{ kPa} \end{aligned}$$

- Hoseline 35 feet (10.5 m) below the pump.

$$\begin{aligned} EL &= 0.5 \text{ psi/ft} \times -35 \text{ ft} \\ &= -17.5 \text{ or } -18 \text{ psi} \end{aligned}$$

$$\begin{aligned} EL &= 10 \text{ kPa/m} \times -10.5 \\ &= -105 \text{ kPa} \end{aligned}$$

- Hoseline taken to the third floor of a structure; the first floor is at ground level.

$$\begin{aligned} EL &= 5 \text{ psi/fl} \times 2 \text{ fl} \\ &= 10 \text{ psi} \end{aligned}$$

$$\begin{aligned} EL &= 35 \text{ kPa/floor} \times (3 - 1 = 2) \\ &= 70 \text{ kPa} \end{aligned}$$

- Hoseline operating on the second level below ground; the first level is at ground level.

$$\begin{aligned} EL &= 5 \text{ psi/fl} \times -1 \text{ fl} \\ &= -5 \text{ psi} \end{aligned}$$

$$\begin{aligned} EL &= 35 \text{ kPa/floor} \times -(2 - 1 = 1) \\ &= -35 \text{ kPa} \end{aligned}$$

LESSONS LEARNED

Efficient and effective pump operations require that pump operators determine the amount of needed water flow. Ideally, this figuring is done during preplanning. When needed flow is not known, the pump operator must estimate it. The two most common formulas for estimating needed flow are the Iowa State formula and the NFA formula. Next, pump operators determine available water by considering the limits of each component within fire pump operations: the water supply, pump, hose, and nozzle. Finally, pump operators must develop appropriate discharge flows.

Proper discharge flows require that nozzles be provided with their designed operating pressure and gpm (Lpm) flow. The easiest method is to use flow meters. When pressure gauges are used, friction loss in the hose must be estimated. Friction loss can be estimated using tables, slide rules, friction loss calculators, rules of thumb, and friction loss formulas. In addition, the pump operator must account for other losses in pressure, such as losses in pressure from appliance friction loss and from changes in elevation.

KEY TERMS

Appliance friction loss The reduction in pressure resulting from increased turbulence caused by the appliance.

Available flow The amount of water that can be moved from the supply to the fire scene.

Discharge flow The amount of water flowing from the discharge side of a pump through the hose, appliances, and nozzles to the scene.

Insurance Service Office (ISO) An organization that provides underwriting information to insurance companies. Formulas the ISO develops affect the fire insurance rates property owners pay.

Needed flow The estimated flow required to extinguish a fire. Also called required flow.

Required flow See **Needed Flow**.

CHAPTER FORMULAS

Needed Fire Flow Formulas

11–1 Insurance Service Office (ISO) formula

$$NFF = C \times O \times (X + P)$$

11–2 Iowa State formula

$$NF = \frac{V}{100}$$

$$NF = V \times 1.3 \qquad \text{11–2m}$$

11–3 Iowa State formula in multiplication format

$$NF = 0.01 \text{ gpm/ft}^3 \times V$$

11–4 National Fire Academy formula

$$NF = \frac{A}{3}$$

$$NF = \frac{A}{0.07} \qquad \text{11–4m}$$

11–5 National Fire Academy Formula in multiplication format

$$NF = 0.333 \text{ gpm/ft}^2 \times A$$

11–6 Determine gpm (Lpm) through smooth-bore nozzles

$$Q = 29.7 \times d^2 \times \sqrt{NP}$$

Or, as simplified for fireground calculations

$$Q = 30 \times d^2 \times \sqrt{NP}$$

$$Q = 0.067 \times d^2 \times \sqrt{NP} \qquad \text{11–6m}$$

Friction Loss Formulas

11–7 National Board of Fire Underwriters (per 100 feet)

$$FL = 2q^2 + q$$

11–8 To determine the number of hundreds of gpm (Lpm) flowing

$$q = \frac{\text{gpm}}{100}$$

Or, as

$$q = \frac{Q}{100}$$

11–9 Flows less than 100 gpm

$$FL = 2q^2 + \tfrac{1}{2}q$$

11–10 Simplified formula

$$FL = 2q^2$$

11–11 Condensed q formula for 3-inch hose

$$FL = q^2$$

11–12 Popular friction loss formula

$$FL = c \times q^2 \times L$$

11–13 Determine number of 100 lengths of hose

$$L = \frac{\text{hose length}}{100}$$

11–14 Determine elevation gain/loss using feet (metres)

$$EL = 0.5 \times H \text{ (in units of feet)}$$

$$EL = 10 \times H \text{ (in units of metres)} \quad 11\text{–}14\text{m}$$

11–15 Determine elevation gain/loss using floors

$$EL = 5 \text{ psi} \times H \text{ (number of floor levels above or below the pump minus 1)}$$

$$EL = 35 \text{ kPa} \times H \text{ (number of floors above or below the pump minus 1)} \quad 11\text{–}15\text{m}$$

REVIEW QUESTIONS

Multiple Choice

Select the most appropriate answer.

1. Which of the following is the correct NFA formula for determining needed flow?

a. $NF = \frac{A}{3}$

b. $NF = 0.333 \times A$

c. Both a and b are correct.

d. None of the above is correct.

2. The amount of water that can be moved from the supply to the scene is called

a. needed flow.

b. available flow.

c. discharge flow.

d. None of the above is correct.

3. A flow meter on the pump panel reads 200 gpm (800 Lpm) when flowing water through a discharge line consisting of 500 feet (150 m) of 1½-inch (38-mm) hose operating on the third floor of a structure. How much water is the nozzle discharging?

a. 200 gpm (800 Lpm)

b. 210 gpm (840 Lpm)

c. 215 gpm (860 Lpm)

d. None of the above is correct.

4. In the majority of friction loss formulas, q is typically considered to represent

a. total gpm (Lpm) flow.

b. quantity of water flowing.

c. flow in hundreds of gpm (Lpm).

d. rate flow of the nozzle.

Short Answer

On a separate sheet of paper, answer/explain the following questions.

1. Explain the role of needed flow, available flow, and discharge flow in efficient and effective pump operations.
2. Using both the Iowa State formula and the NFA formula, calculate the needed flow for a two-story single-family dwelling measuring 60 feet (18 m) long by 45 feet (14 m) wide.
3. Explain the difference between discharge flow development using flow meters and pressure gauges.
4. Calculate the friction in the following hoselines:
   - **a.** 400 ft of 2½-in. hose, flowing 250 gpm (120 m of 65-mm, flowing 1,000 Lpm)
   - **b.** 400 ft of 3-in. hose, flowing 250 gpm (120 m of 77-mm, flowing 1,000 Lpm)
   - **c.** 150 ft of 1½-in. hose, flowing 125 gpm (45 m of 38-mm, flowing 500 Lpm)
   - **d.** 150 ft of 1¾-in. hose, flowing 125 gpm (45 m of 45 mm, flowing 500 Lpm)
   - **e.** 1,000 ft of 4-in. hose, flowing 500 gpm (300 m of 100 mm, flowing 2,000 Lpm)

ACTIVITY

1. Contact the manufacturers of several appliances to determine the actual friction loss of the component.

PRACTICE PROBLEM

1. You have been asked to develop a friction loss table for 100-foot (30-m) sections of 1½-inch and 1¾-inch (38-mm and 45-mm) hose for the following flows: 95, 100, 125, 150, 175, 200, and 250 gpm (380, 400, 500, 600, 700, 800, 1,000 Lpm). Provide the table.

ADDITIONAL RESOURCES

Burns, Edward, and Phelps, Burton W. "Redefining Needed Fire Flow for Structure Firefighting." *Fire Engineering*, November 1994.

Fire Protection Handbook, 20th edition, vol. 1. Quincy, MA: National Fire Protection Association, 2008.

ISO Guide for Determination of Required Fire Flow. New York: Insurance Services Office, 2008. Available online at http://www.isomitigation.com.

NFPA 1142, *Water Supplies for Suburban and Rural Fire Fighting*. National Fire Protection Association, Quincy, MA, 2007.

Richmond, Ron. *Basic Fire Hydraulics Workbook*, 1st ed. New York, Delmar, 2009.

Wiseman, John D. Jr. "How Valid Are NFF Formulas?" *Firefighter's News,* June/July 1996, pp. 78–84.

© Cengage Learning

12 Pump Discharge Pressure Calculations

- Learning Objectives
- Introduction
- Pump Discharge Pressure Calculations
- Single Lines
- Multiple Lines
- Wyed Lines
- Siamese Lines
- Standpipe Support
- Lessons Learned
- Key Terms
- Review Questions
- Activity
- Practice Problem
- Additional Resources

I work in a small department, in which you sometimes serve as the driver and engineer before you've taken the pumper course.

One of my first calls as driver/engineer was to a fire at a 10,000-square-foot (900-square-metre) home, in an area where there was no water supply. We had to bring it in, getting tankers from mutual aid departments. But I'd be the only one pumping due to limited space.

I was nervous, nervous, nervous! I had learned some things from the guys in my department, but it was just basics about how to get water from point A to point B. But here I was running a 2-inch (65-mm) hose out the back of the engine, and another crew was tying two lines off our one discharge. One was 200 feet (60 metres) and the other 150 feet (45 metres). I had learned longer hoses have more friction loss, and shorter hoses less. But having both lines coming out of a gated Y, the challenge was figuring out what pressure to pump at.

I tried to do the math but I only ended up questioning myself about whether I was doing it right or if I was even using the correct formulas. I was scared. I realized that if I didn't do my job right, the guys on the other end of the nozzle could get hurt.

I had to get water flow. So, on the fly, I started with higher psi (kPa), and radioed to the shorter line to let them know that they would have more pressure. (I figured it would be better for them to have too much than others not to have enough.) One blessing was that the house was lower than the street, so I could see what the foam coming out of our CAFS looked like. That meant I could give a little and take a little.

I was lucky that it all worked out, and no one got hurt.

After that experience, I took the first pump operations course that was available, because I knew I needed the training in the formulas. Four months later, in the class, I learned how to do pressure discharge calculations the right way. As a result, when I'm called to a fire today, I can do the math in my head, and know that I'm able to utilize every bit of pressure that faucet has. If I knew on that first fire what I know now, I would have been a lot more confident.

—Street Story by Louis Ramos,
Firefighter/Fire Inspector, Hudson Oaks Fire Department,
Parker County ESD #3, Hudson Oaks, Texas

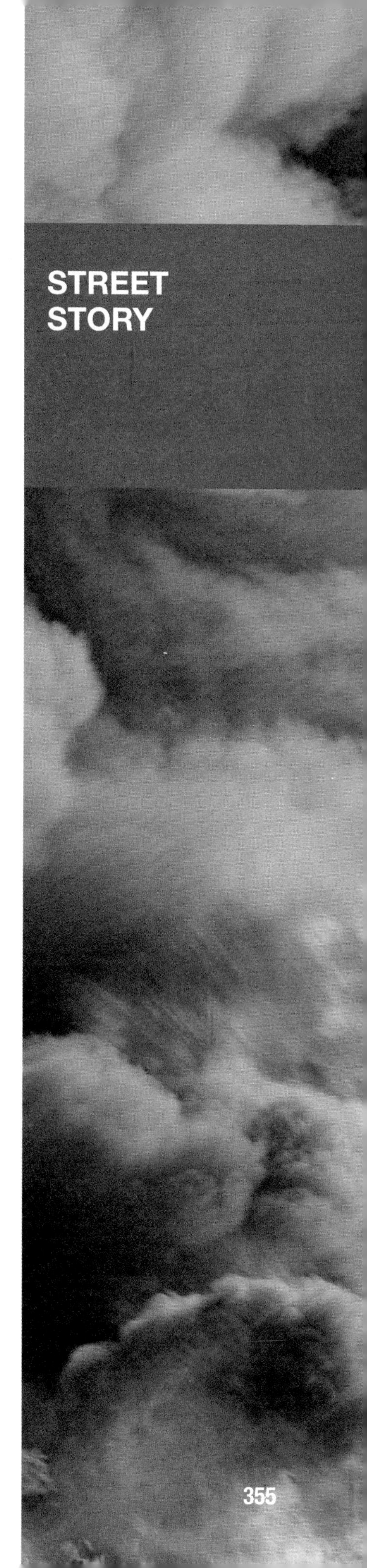

LEARNING OBJECTIVES

After completing this chapter, the reader should be able to:

12-1. Calculate pump discharge pressures for a variety of hoseline configurations using the formula $PDP = NP + FL + AFL + EL$.

12-2. Calculate the pump discharge pressure for a given sprinkler or standpipe system evolution.

*The driver/operator requirements, as defined by the NFPA 1002 Standard, are identified in black; additional information is identified in blue.

INTRODUCTION

In many respects, the sum total of this textbook boils down to this chapter: calculating pump discharge pressures. Almost every point discussed in the text can be logically tracked to the point of flowing water from a nozzle. From driving the apparatus to preventive maintenance to operating the pump, the ultimate achievement of fire pump operations is providing the proper flow and pressure to a nozzle.

As stated in **Chapter 11**, the use of flow meters simplifies the process of fire pump operations by eliminating the need to calculate friction loss. Pump operators simply increase the pump speed until the correct gallons per minute (litres per minute) for a nozzle are indicated on the flow meter. In turn, the same flow is delivered to the nozzle. When supplied with the correct gpm (Lpm), the nozzle, by design, automatically develops its proper operating pressure. When pressure gauges are used, the pump operator provides the nozzle with the correct operating pressure. In turn, the design of the nozzle automatically ensures that the correct gpm (Lpm) are flowing. The major difference between the use of pressure gauges and flow meters is the requirement to calculate pressure changes within a hose configuration.

The focus of this chapter is on calculating pump discharge pressure. The sample calculations begin with simple single lines and progress to more complicated multiple-line configurations. Friction loss constants and appliance friction loss can be obtained from either **Table 11-1** or **Table 11-2**.

NOTE

The calculated changes in pressure coupled with the nozzle pressure are the basis for determining pump discharge pressure.

PUMP DISCHARGE PRESSURE CALCULATIONS

The preceding chapters in this section of the text provide the framework for calculating pump discharge pressure (PDP). Recall that nozzles are designed to operate at a specific pressure. Further recall that friction loss in hose and appliances as well as elevation will affect pressure in hose. These changes in pressure must be compensated for to ensure that the nozzle is provided with the proper pressure. The calculated changes in pressure, or total pressure loss coupled with the nozzle pressure, are the basis for determining PDP. Pump discharge pressure is the pressure at the pump panel for a specific hose configuration and can be calculated using the following formula:

$$PDP = NP + FL + AFL + EL$$

where

PDP = pump discharge pressure

NP = nozzle pressure

FL = friction loss in hose (Any of the friction loss calculations discussed in **Chapter 11** can be used. However, for the purpose of accuracy and current use, the formula $c \times q^2 \times L$ is used.)

AFL = appliance friction loss

EL = elevation gain or loss = $0.5 \times H$ or ($EL = 5 \times$ number of floor levels above ground level) $35 \text{ kPa} \times H$ ($EL = 35 \text{ kPa} \times$ number of floor levels above or below ground level minus 1)

Calculations Considerations

Although PDP can be calculated in a variety of ways, it is best to develop a consistent method to ensure that variables are not left out. Simply stated, determine the required line pressure for each line working from the nozzle back to the pump. The pump discharge pressure should be set based on the highest required discharge pressure, with lines requiring lower discharge pressures gated back so that each line has the correct line pressure.

For the purpose of illustration, calculations in this chapter are presented in the following basic steps. Toward the end of the chapter, calculations are condensed in the interest of space and tedious minor calculations are not shown.

1. Determine the operating pressure and flow for the nozzle. One of the first steps in any PDP calculation is to determine the gpm (Lpm) to flow. When fixed- or variable-flow combination (fog) nozzles are used, the gpm (Lpm) will often be identified on the nozzle. Automatic nozzles can provide a range of flows while maintaining proper operating pressures. In this case, the nozzle flow should be chosen based on the required fire flow and required extinguishment as described in **Chapter 11**. When smooth-bore (straight-stream) nozzles are used, the flow can either be looked up on a chart (see **Appendix G**) or calculated. Recall from **Chapter 11** that the formula for determining gpm (Lpm) from a smooth-bore nozzle is as follows:

$$\text{gpm} = 30 \times d^2 \times \sqrt{NP}$$
$$\text{Lpm} = 0.067 \times d^2 \times \sqrt{NP}$$

In addition to knowing the flow for a specific nozzle, the operating pressure of the nozzle must also be known. Most nozzles have a designed operating pressure (**Table 12-1**). When calculating gpm for smooth-bore nozzles on the fireground, the square root of the nozzle pressure calculations can be replaced with 7 (18.7) for handlines and 9 (23.7) for master streams.

2. After the gpm (Lpm) is determined, the friction loss in the hose can be calculated. Recall that friction loss is affected by the diameter of a hose; therefore, friction loss must be calculated separately for each diameter hose within the lay. Several examples of this concept are presented later in this chapter.
3. Determine the appliance friction loss within the hose lay. Keep in mind that friction loss varies for specific appliances.
4. Calculate the loss or gain in pressure from changes in elevation.
5. The last step is to use the numbers obtained from the previous steps in the PDP formula. The result is the pressure required at the pump panel to provide the nozzle with its proper operating pressure and flow.

TABLE 12-1 Operating Pressures for Typical Nozzle Types

Type of Nozzle	Operating Pressure
Smooth-bore, handline	50 psi (350 kPa)
Combination (fog), low-pressure	75 psi (525 kPa)
Smooth-bore, master stream	80 psi (560 kPa)
Combination (fog) and automatic	100 psi (700 kPa)

SINGLE LINES

The easiest discharge configuration is that of a single line. Keep in mind, though, that different sizes of hose, types of nozzles, and elevation must be factored into the calculations. The following are several examples of single-line PDP calculations.

Nozzle Comparisons

The following sections compare the flow and pump discharge pressures of various nozzle types.

Combination Nozzle

Consider a 150-foot (45-m) section of 1¾-inch (45-mm) hose flowing 125 gpm (500 Lpm) with a combination (fog) nozzle (**Figure 12-1**). What is the PDP?

1. Nozzle operating pressure and flow, NP = 100 (700 kPa) and $gpm\ (Lpm)$ = 125 (500)
2. Hose friction loss:

$$\begin{aligned} FL &= c \times q^2 \times L \\ &= 15.5 \times \left(\frac{125}{100}\right)^2 \times \left(\frac{150}{100}\right) \\ &= 15.5 \times (1.25)^2 \times 1.5 \\ &= 15.5 \times 1.56 \times 1.5 \\ &= 36.27 \text{ psi} \end{aligned}$$

$$\begin{aligned} FL &= c \times q^2 \times L \\ &= 24.6 \times \left(\frac{500}{100}\right)^2 \times \left(\frac{45}{100}\right) \\ &= 24.6 \times (5)^2 \times 0.45 \\ &= 24.6 \times 25 \times 0.45 \\ &= 276.8 \text{ or } 277 \text{ kpa} \end{aligned}$$

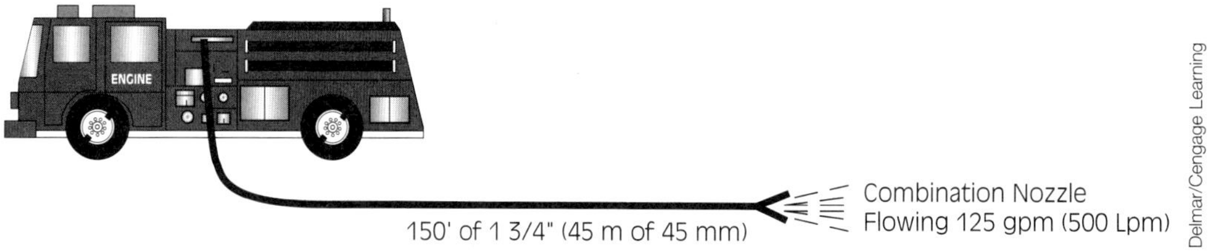

ENGINE
Combination Nozzle
Flowing 125 gpm (500 Lpm)
150' of 1 3/4" (45 m of 45 mm)
Delmar/Cengage Learning

FIGURE 12-1 Single line: combination nozzle.

3. Appliance friction loss (AFL) = no appliance
4. Elevation pressure change (EL) = no change in elevation
5. Pump discharge pressure:

$$
\begin{aligned}
PDP &= NP + FL + AFL + EL \\
&= 100 + 36 + 0 + 0 \\
&= 136 \text{ psi}
\end{aligned}
$$

$$
\begin{aligned}
PDP &= NP + FL + AFL + EL \\
&= 700 + 277 + 0 + 0 \\
&= 977 \text{ kPa}
\end{aligned}
$$

Smooth-bore Nozzle

Calculate the pump discharge pressure for a 150-foot (45-m) section of 1¾-inch (45-mm) handline equipped with a ¾-inch (19-mm) tip smooth-bore nozzle (**Figure 12-2**).

1. Nozzle operating pressure and flow, NP = 50 psi (350 kPa) and

$$
\begin{aligned}
gpm &= 30 \times d^2 \times \sqrt{NP} \\
&= 30 \times (3/4'')^2 \times \sqrt{50} \\
&= 30 \times (0.75)^2 \times \sqrt{50} \\
&= 30 \times 0.56 \times 7 \\
&= 117.6 \text{ or } 118
\end{aligned}
$$

$$
\begin{aligned}
Lpm &= 0.067 \times d^2 \times \sqrt{NP} \\
&= 0.067 \times (19)^2 \times \sqrt{350} \\
&= 0.067 \times (361) \times \sqrt{350} \\
&= 0.067 \times 361 \times 18.7 \\
&= 452.2 \text{ or } 452
\end{aligned}
$$

2. Hose friction loss:

$$
\begin{aligned}
FL &= c \times q^2 \times L \\
&= 15.5 \times \left(\frac{118}{100}\right)^2 \times \left(\frac{150}{100}\right) \\
&= 15.5 \times (1.18)^2 \times 1.5 \\
&= 15.5 \times 1.39 \times 1.5 \\
&= 32.37 \text{ or } 32 \text{ psi}
\end{aligned}
$$

$$
\begin{aligned}
FL &= c \times q^2 \times L \\
&= 24.6 \times \left(\frac{452}{100}\right)^2 \times \left(\frac{45}{100}\right) \\
&= 24.6 \times (4.52)^2 \times 0.45 \\
&= 24.6 \times 20.4 \times 0.45 \\
&= 225.8 \text{ or } 226 \text{ kPa}
\end{aligned}
$$

3. Appliance friction loss (AFL) = no appliance
4. Elevation pressure change (EL) = no change in elevation
5. Pump discharge pressure:

ENGINE
Smooth Bore Nozzle
with a 3/4 " (19 mm) Tip
150' of 1 3/4" (45 m of 45 mm)
Delmar/Cengage Learning

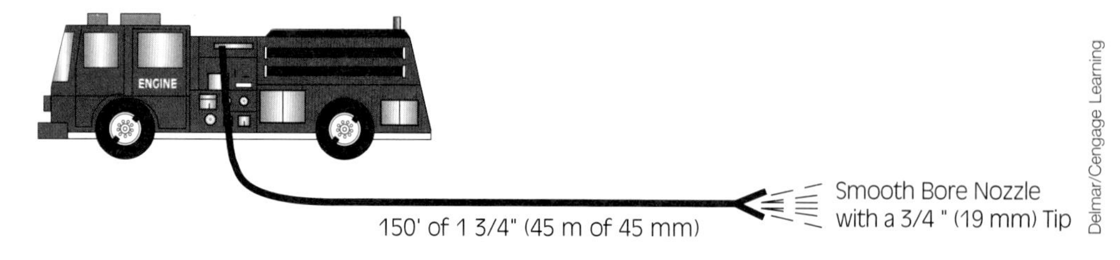

FIGURE 12-2 Single line: smooth-bore nozzle (handline).

$$PDP = NP + FL + AFL + EL$$
$$= 50 + 32 + 0 + 0$$
$$= 82 \text{ psi}$$

$$PDP = NP + FL + AFL + EL$$
$$= 350 + 226 + 0 + 0$$
$$= 576 \text{ kPa}$$

Practice

Calculate PDP for both an automatic nozzle and a 1-inch (25-mm) tip smooth-bore using the friction loss formulas (1) $2q^2 + q$, (2) hand method, and (3) cq^2L for the handline: 300 feet (90 m) of 2½-inch (65-mm) hose flowing 200 gpm (800 Lpm) for the automatic nozzle.

As mentioned earlier, a systematic approach to calculating PDP may help ensure critical steps or calculations are not overlooked. The basic steps proposed in this text are: NP (step 1) + FL (step 2) + AFL (step 3) + EL (step 4) = PDP (step 5).

		Automatic Nozzle	Smooth-bore
PDP	1. $2q^2 + q$	_______	_______
	2. hand method	_______	_______
	3. cq^2L	_______	_______

Automatic Nozzle

1. $PDP = NP + FL\ [(2q^2 + q) \times L] + AFL + EL$

Step 1: $NP = 100$

$gpm = 200$

Step 2: $FL = (2q^2 + q) \times L$

$$= \left[2 \times \left(\frac{200}{100}\right)^2 + \left(\frac{200}{100}\right)\right] \times \left(\frac{300}{100}\right)$$
$$= [2 \times (2)^2 + 2] \times 3$$
$$= (2 \times 4 + 2) \times 3$$
$$= 10 \times 3$$
$$= 30 \text{ psi}$$

Step 3: No appliance friction loss

Step 4: No elevation gain or loss

Step 5: $PDP = NP + FL\ [(2q^2 + q) \times L] + 0 + 0$

$$= 100 + 30$$
$$= 130 \text{ psi}$$

2. $PDP = NP + FL$ (hand method, see **Figure 11-8**) + AFL + EL

Step 1: $NP = 100$

$gpm = 200$

Step 2: FL = hand method $\times L$

$$= 2 \text{ (top of index finger)} \times 4 \text{ (base of index finger)} \times \left(\frac{300}{100}\right)$$
$$= 8 \times 3$$
$$= 24 \text{ psi}$$

Step 3: No appliance friction loss

Step 4: No elevation gain or loss

Step 5: $PDP = NP = FL$ (hand method, see **Figure 11-8**) + 0 + 0

$$= 100 + 24$$
$$= 124 \text{ psi}$$

3. $PDP = NP + FL\ (cq^2L) + AFL + EL$

Step 1: $NP = 100$

$gpm = 200$

$NP = 700$ kPa

$Lpm = 800$ Lpm

Step 2: $FL = c \times q^2 \times L$

$$= 2 \times \left(\frac{200}{100}\right)^2 \times \left(\frac{300}{100}\right)$$
$$= 2 \times (2)^2 \times 3$$
$$= 2 \times 4 \times 3$$
$$= 24 \text{ psi}$$

$$FL = c \times q^2 \times L$$
$$= 3.17 \times \left(\frac{800}{100}\right)^2 \times \left(\frac{90}{100}\right)$$
$$= 3.17 \times (8)^2 \times 0.9$$

$= 3.17 \times 64 \times 0.9$

$= 182.6$ or 183 kPa

Step 3: No appliance friction loss

Step 4: No elevation gain or loss

Step 5: $PDP = NP + FL\ (cq^2L) + 0 + 0$

$= 100 + 24$

$= 124$ psi

$PDP = NP + FL\ (cq^2L) + 0 + 0$

$= 700 + 183$

$= 883$ kPa

Smooth-bore Nozzle

1. $PDP = NP + FL[(2q^2 + q) \times L] + AFL + EL$

Step 1: $NP = 50$

$gpm = 29.7 \times d^2 \times \sqrt{NP}$

$= 29.7 \times 1^2 \times \sqrt{50}$

$= 29.7 \times 1 \times 7.07$

$= 209.97$ or 210 gpm

Note: Using the common fireground figures in the formula ($30 \times d^2 \times 7$) provides virtually the same answer: 210 gpm.

Step 2: $FL = (2q^2 + q) \times L$

$$= \left[2 \times \left(\frac{210}{100}\right)^2 + \left(\frac{210}{100}\right)\right] \times \left(\frac{300}{100}\right)$$

$= [2 \times (2.1)^2 + 2.1] \times 3$

$= (2 \times 4.41 + 2.1) \times 3$

$= 10.92 \times 3$

$= 32.76$ or 33 psi

Step 3: No appliance friction loss

Step 4: No elevation gain or loss

Step 5: $PDP = 50 + 33$

$= 83$ psi

2. $PDP = NP + FL$ (hand method, see **Figure 11-8**) $+ AFL + EL$

Step 1: $NP = 50$

$gpm = 210$

Step 2: $FL =$ hand method $\times\ l$

$= 2$ (top of index finger)

$\times\ 4$ (base of index finger)

$$\times \left(\frac{300}{100}\right)$$

$= 8 \times 3$

$= 24$ psi

Step 3: No appliance friction loss

Step 4: No elevation gain or loss

Step 5: $PDP = 50 + 24$

$= 74$ psi

3. $PDP = NP + FL(cq^2L) + AFL + EL$

Step 1: $NP = 50$

$gpm = 210$

$NP = 350$ kPa

$Lpm = 840$ Lpm

Step 2: $FL = c \times q^2 \times L$

$$= 2 \times \left(\frac{210}{100}\right)^2 \times \left(\frac{300}{100}\right)$$

$= 2 \times (2.1)^2 \times 3$

$= 2 \times 4.41 \times 3$

$= 26.46$ or 27 psi

$FL = c \times q^2 \times L$

$$= 3.17 \times \left(\frac{840}{100}\right)^2 \times \left(\frac{90}{100}\right)$$

$= 3.17 \times (8.4)^2 \times 0.9$

$= 3.17 \times 70.6 \times 0.9$

$= 201.4$ or 201 kPa

Step 3: No appliance friction loss

Step 4: No elevation gain or loss

Step 5: $PDP = 50 + 27$

$= 77$ psi

$PDP = 350 + 201$

$= 551$ kPa

		Automatic Nozzle	Smooth-bore
PDP	1. $2q^2 + q$	130	88
	2. hand method	124	74
	3. cq^2L	124 (883 kPa)	77 (551 kPa)

Hose Diameter Comparisons

The following calculations for 2½-inch (65-mm) and 3-inch lines (77-mm), with all other variables held constant, provide a good illustration of the difference hose size makes on the PDP.

2½-inch (65-mm) Line

Pumper A is flowing 350 gpm (1,400 Lpm) through 500 feet (150 m) of 2½-inch (65-mm) hose equipped with an automatic nozzle (**Figure 12-3**). Calculate the discharge pressure.

1. Nozzle operating pressure and flow, NP = 100 psi (700 kPa) and gpm (Lpm) = 350 (1,400)
2. Hose friction loss:

$$\begin{aligned} \mathrm{FL} &= c \times q^2 \times L \\ &= 2 \times \left(\frac{350}{100}\right)^2 \times \left(\frac{500}{100}\right) \\ &= 2 \times (3.5)^2 \times 5 \\ &= 2 \times 12.25 \times 5 \\ &= 122.5 \text{ or } 123 \end{aligned}$$

$$\begin{aligned} FL &= c \times q^2 \times L \\ &= 3.17 \times \left(\frac{1,400}{100}\right)^2 \times \left(\frac{150}{100}\right) \\ &= 3.17 \times (14)^2 \times 1.5 \\ &= 3.17 \times 196 \times 1.5 \\ &= 931.9 \text{ or } 932 \text{ kPa} \end{aligned}$$

3. Appliance friction loss (AFL) = no appliance
4. Elevation pressure change (EL) = no change in elevation
5. Pump discharge pressure:

$$\begin{aligned} PDP &= NP + FL + AFL + EL \\ &= 100 + 123 + 0 + 0 \\ &= 223 \end{aligned}$$

$$\begin{aligned} PDP &= NP + FL + AFL + EL \\ &= 700 + 932 + 0 + 0 \\ &= 1,632 \text{ kPa} \end{aligned}$$

3-inch (77-mm) Line

Pumper B has the same hose lay configuration except that 3-inch (77-mm) hose is being used instead of 2½-inch (65-mm) hose (refer again to **Figure 12-3**). What is the PDP?

1. Nozzle operating pressure and flow, NP = 100 psi (700 kPa) and gpm (Lpm) = 350 (1,400)
2. Hose friction loss:

$$\begin{aligned} FL &= c \times q^2 \times L \\ &= 0.8 \times \left(\frac{350}{100}\right)^2 \times \left(\frac{150}{100}\right) \\ &= 0.8 \times (3.5)^2 \times 5 \end{aligned}$$

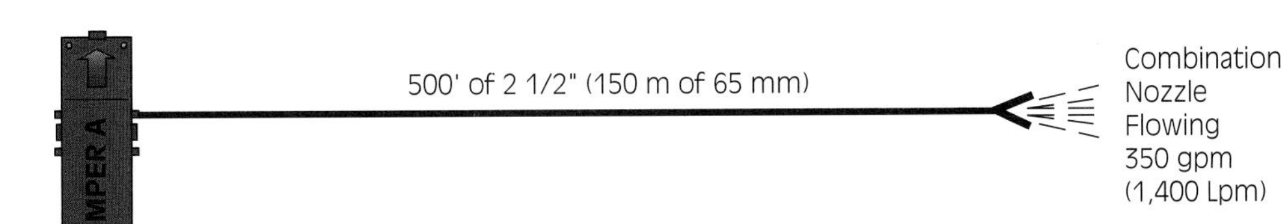

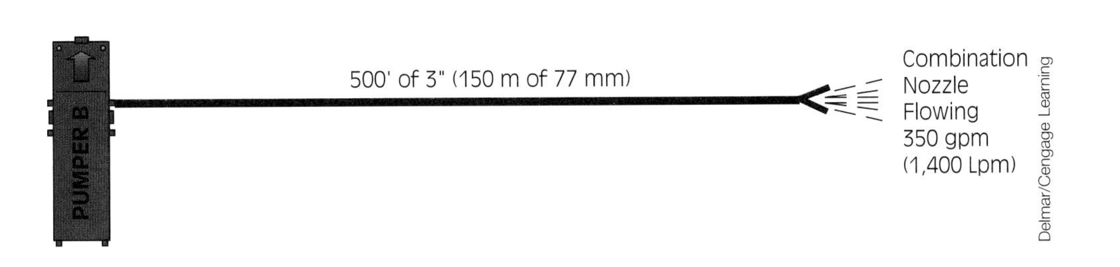

FIGURE 12-3 Single line: comparison of PDP for 2½-inch (65-mm) and 3-inch (77-mm) hose when length, nozzle, and flow are held constant.

$$= 0.8 \times 12.25 \times 5$$
$$= 49$$

$$FL = c \times q^2 \times L$$
$$= 1.27 \times \left(\frac{1,400}{100}\right)^2 \times \left(\frac{150}{100}\right)$$
$$= 1.27 \times (14)^2 \times 1.5$$
$$= 1.27 \times 196 \times 1.5$$
$$= 373.3 \text{ or } 373 \text{ kPa}$$

3. Appliance friction loss (AFL) = no appliance

4. Elevation pressure change (EL) = no change in elevation

5. Pump discharge pressure:

$$PDP = NP + FL + AFL + EL$$
$$= 100 + 49 + 0 + 0$$
$$= 149$$

$$PDP = NP + FL + AFL + EL$$
$$= 700 + 373 + 0 + 0$$
$$= 1,073 \text{ kPa}$$

Note the significant difference in pump discharge pressure between pumper A and B. Because of the high discharge pressure, pumper A can deliver less than 70% of its rated capacity, but pumper B is capable of delivering 100% of its rated capacity.

Practice

Calculate PDP for a 2½-inch (65-mm) hose using the FL formulas in the following table for the master stream line: 200 feet (60 m) of hose with a 1½-inch (38-mm) tip smooth-bore master stream nozzle. Also calculate the same 3-inch (77-mm) line except with hose using the condensed Q and cq^2L formula (no monitor nozzle).

		2½-inch (65-mm)	3-inch (65-mm)
PDP	1. $2q^2 + q$	_______	_______
	2. Hand method	_______	_______
	3. Drop 10	_______	_______
	4. cq^2L	_______	_______
	5. Condensed Q	_______	_______

2½-Inch (65-mm) Hoseline

1. $PDP = NP + FL[(2q^2 + q) \times L] + AFL + EL$

Step 1: $NP = 80$

$$\text{gpm} = 29.7 \times d^2 \times \sqrt{NP}$$
$$= 29.7 \times (1.5)^2 \times \sqrt{80}$$
$$= 29.7 \times 2.25 \times 8.94$$
$$= 597.4 \text{ or } 597 \text{ psi}$$

Note: Using the rounded figures in the formula ($30 \times d^2 \times 9$) provides a slightly higher value of 607.5 gpm.

Step 2: $FL = (2q^2 + q) \times L$

$$= \left[2 \times \left(\frac{597}{100}\right)^2 + \left(\frac{597}{100}\right)\right] \times \left(\frac{200}{100}\right)$$
$$= [2 \times (5.97)^2 + 5.97] \times 2$$
$$= (2 \times 35.6 + 5.97) \times 2$$
$$= 77.17 \times 2$$
$$= 154.34 \text{ or } 154 \text{ psi}$$

Step 3: No appliance friction loss

Step 4: No elevation gain or loss

Step 5: $PDP = NP + FL\ [(2q^2 + q) \times L] + 0 + 0$

2. $PDP = NP + FL$ (hand method, see **Figure 11-8**) $+ AFL + EL$

Step 1: $NP = 80$

$$gpm = 597$$

Step 2: $FL = \text{hand method} \times L$

$= 6$ (top of first finger on second hand)

$\times 12$ (base of first finger on second hand)

$$\times \left(\frac{200}{100}\right)$$
$$= 6 \times 12 \times 2$$
$$= 144 \text{ psi}$$

Step 3: No appliance friction loss

Step 4: No elevation gain or loss

Step 5: $PDP = 80 + 144$

$$= 224 \text{ psi}$$

3. $PDP = NP + FL(\text{Drop } 10) + AFL + EL$

Step 1: $NP = 80$

$gpm = 597$

Step 2: $FL = 59 - 10 \times L \left(\frac{200}{100}\right)$

$= 49 \times 2$

$= 98$ psi

Step 3: No appliance friction loss

Step 4: No elevation gain or loss

Step 5: $PDP = 80 + 98$

$= 178$ psi

4. $PDP = NP + FL(cq^2L) + AFL + EL$

Step 1: $NP = 800$

$gpm = 597$

$NP = 560$

$Lpm = 2,388$

Step 2: $FL = c \times q^2 \times L$

$$= 2 \times \left(\frac{597}{100}\right)^2 \times \left(\frac{200}{100}\right)$$

$= 2 \times (5.97)^2 \times 2$

$= 2 \times 35.6 \times 2$

$= 142.4$ or 142 psi

$FL = c \times q^2 \times L$

$$= 3.17 \times \left(\frac{2,388}{100}\right)^2 \times \left(\frac{60}{100}\right)$$

$= 3.17 \times (23.88)^2 \times 0.6$

$= 3.17 \times 570.2 \times 0.6$

$= 1,084.5$ or $1,085$ kPa

Step 3: No appliance friction loss

Step 4: No elevation gain or loss

Step 5: $PDP = 80 + 142$

$= 222$ psi

$PDP = 560 + 1085$

$= 1,645$ kPa

3-Inch (77-mm) Line

1. $PDP = NP + FL(cq^2L) + AFL + EL$

Step 1: $NP = 80$

$gpm = 597$

$NP = 560$ kPa

$Lpm = 2,388$

Step 2: $FL = c \times q^2 \times L$

$$= 0.8 \times \left(\frac{597}{100}\right)^2 \times \left(\frac{200}{100}\right)$$

$= 0.8 \times (5.97)^2 \times 2$

$= 0.8 \times 35.6 \times 2$

$= 56.9$ or 57 psi

$FL = c \times q^2 \times L$

$$= 1.27 \times \left(\frac{2,388}{100}\right)^2 \times \left(\frac{60}{100}\right)$$

$= 1.27 \times (23.88)^2 \times 0.6$

$= 1.27 \times (570.2) \times 0.6$

$= 434$ kPa

Step 3: No appliance friction loss

Step 4: No elevation gain or loss

Step 5: $PDP = 80 + 57$

$= 137$ psi

$PDP = 560 + 434$

$= 994$ kPa

2. $PDP = NP + FL(q^2L) + AFL + EL$

Step 1: $NP = 80$

$gpm = 597$

Step 2: $FL = q^2 \times L$

$$= \left(\frac{597}{100}\right)^2 \times \left(\frac{200}{100}\right)$$

$= (5.97)^2 \times 2$

$= 35.6 \times 2$

$= 71.2$ or 71 psi

Step 3: No appliance friction loss

Step 4: No elevation gain or loss

Step 5: $PDP = 80 + 71$

$= 151$ psi

		2½-inch (65-mm)	3-inch (77-mm)
PDP	1. $2q^2 + q$	234	
	2. Hand method	224	
	3. Drop 10	178	
	4. cq^2L	222 (1,645)	137 (994)
	5. Condensed Q		151

Elevation Comparisons

Recall from **Chapter 11** that changes in elevation affect the pressure within hose. The following calculations illustrate elevation calculations using both the change in grade method and the floor level method.

Elevation Gain, Change in Grade

Consider 300 feet (90 m) of 1¾-inch (45-mm) line with a combination nozzle flowing 100 gpm (400 Lpm) when the line is taken up a hill to an elevation of 50 feet (15 m) above the pump (**Figure 12-4**). What is the PDP?

1. Nozzle operating pressure and flow, NP = 100 psi (700 kPa) and *gpm (Lpm)* = 100 (400)

2. Hose friction loss:

$$FL = c \times q^2 \times L$$

$$= 15.5 \times \left(\frac{100}{100}\right)^2 \times \left(\frac{300}{100}\right)$$

$$= 15.5 \times 1^2 \times 3$$

$$= 15.5 \times 1 \times 3$$

$$= 46.5 \text{ or } 47$$

$$FL = c \times q^2 \times L$$

$$= 24.6 \times \left(\frac{400}{100}\right)^2 \times \left(\frac{90}{100}\right)$$

$$= 24.6 \times (4)^2 \times 0.9$$

$$= 24.6 \times 16 \times 0.9$$

$$= 354.2 \text{ or } 354 \text{ kPa}$$

3. Appliance friction loss (AFL) = no appliance

4. Elevation pressure change:

$$EL = 0.5 \times \text{h}$$

$$= 0.5 \times 50 \text{ ft}$$

$$= 25$$

$$EL = 10 \text{ kPa} \times \text{h}$$

$$= 10 \text{ kPa} \times 15 \text{ m}$$

$$= 150 \text{ kPa}$$

5. Pump discharge pressure:

$$PDP = NP + FL + AFL + EL$$

$$= 100 + 47 + 0 + 25$$

$$= 172$$

$$PDP = NP + FL + AFL + EL$$

$$= 700 + 354 + 0 + 150$$

$$= 1,204 \text{ kPa}$$

Elevation Loss, Change in Grade

Consider 300 feet (90 m) of 1¾-inch (45-mm) line with a combination nozzle flowing 100 gpm (400 Lpm) when the line is taken down a hill to an

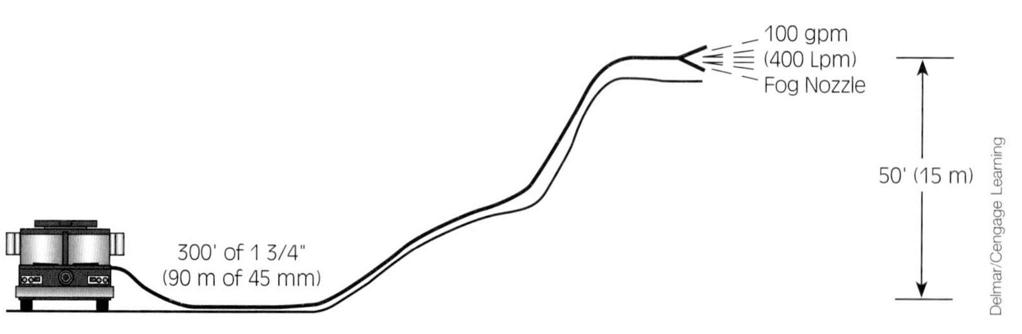

FIGURE 12-4 Single line: elevation gain.

elevation of 50 feet (15 m) below the pump (**Figure 12-5**). What is the pump discharge pressure?

1. Nozzle operating pressure and flow, NP = 100 psi (700 kPa) and gpm (Lpm) = 100 (400)

2. Hose friction loss:

$$FL = c \times q^2 \times L$$
$$= 15.5 \times \left(\frac{100}{100}\right)^2 \times \left(\frac{300}{100}\right)$$
$$= 15.5 \times 1^2 \times 3$$
$$= 15.5 \times 1 \times 3$$
$$= 46.5 \text{ or } 47$$

$$FL = c \times q^2 \times L$$
$$= 24.6 \times \left(\frac{400}{100}\right)^2 \times \left(\frac{90}{100}\right)$$
$$= 24.6 \times (4)^2 \times 0.9$$
$$= 24.6 \times 16 \times 0.9$$
$$= 354.2 \text{ or } 354 \text{ kPa}$$

3. Appliance friction loss (AFL) = no appliance

4. Elevation pressure change:

$$EL = 0.5 \times h$$
$$= 0.5 \times -50 \text{ ft}$$
$$= -25$$

$$EL = 10 \text{ kPa} \times h$$
$$= 10 \text{ kPa} \times -15 \text{ m}$$
$$= -150 \text{ kPa}$$

5. Pump discharge pressure:

$$PDP = NP + FL + AFL + EL$$
$$= 100 + 47 + 0 + -25$$
$$= 122$$

$$PDP = NP + FL + AFL + EL$$
$$= 700 + 354 + 0 + -150$$
$$= 904 \text{ kPa}$$

Elevation Gain, Floor Level

Consider 200 feet (60 m) of 1½-inch (38-mm) line with a ¾-inch (19-mm) smooth-bore nozzle taken to the third floor of a structure (**Figure 12-6**). What is the PDP?

1. Nozzle operating pressure and flow, NP = 50 psi (560 kPa) and

$$gpm = 30 \times d^2 \times \sqrt{NP} \text{ (using fireground formula)}$$
$$= 30 \times (0.75)^2 \times \sqrt{50}$$
$$= 30 \times 0.56 \times 7$$
$$= 117.6, \text{ or } 118$$

$$Lpm = 0.067 \times d^2 \times \sqrt{NP}$$
$$= 0.067 \times (19)^2 \times \sqrt{350}$$
$$= 0.067 \times 361 \times 18.7$$
$$= 452.3 \text{ or } 452 \text{ Lpm}$$

2. Hose friction loss:

$$FL = c \times q^2 \times L$$
$$= 24 \times \left(\frac{118}{100}\right)^2 \times \left(\frac{200}{100}\right)$$

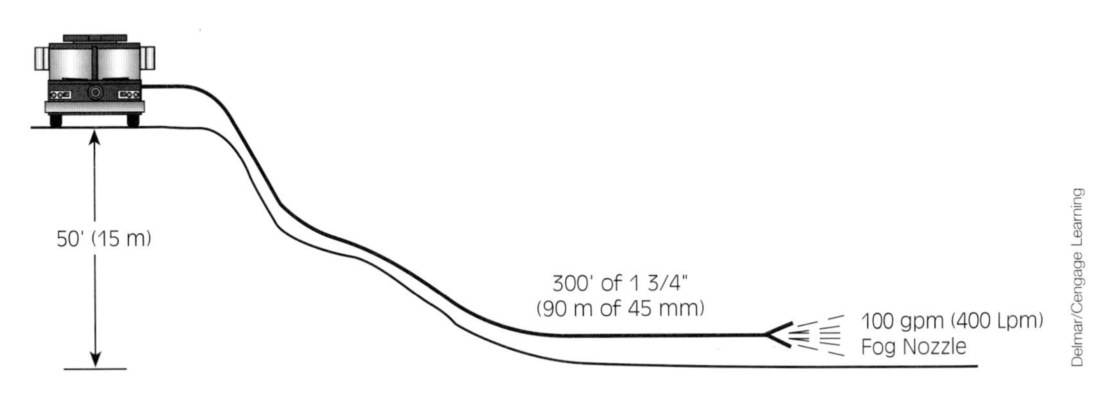

FIGURE 12-5 Single line: elevation loss.

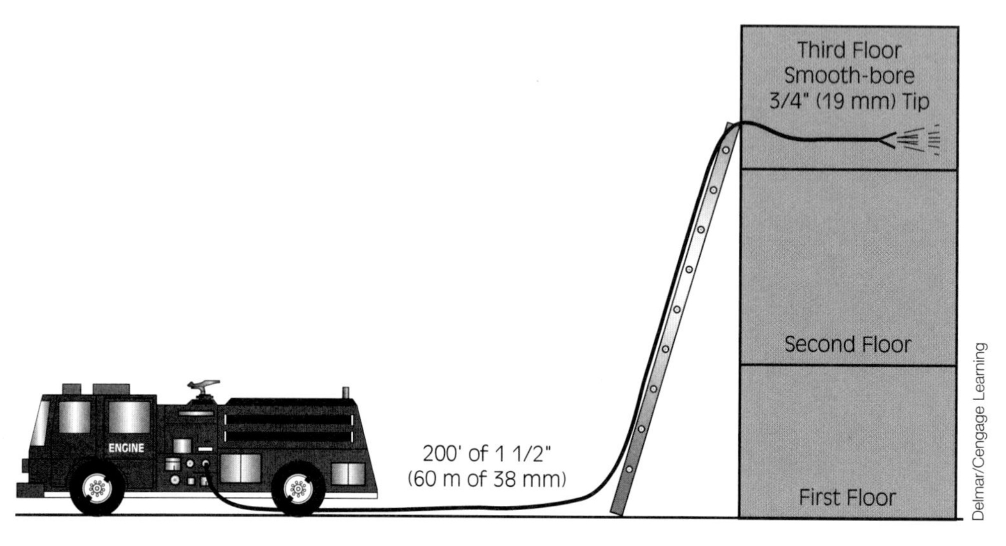

FIGURE 12-6 Single line: elevation gain in a structure.

$$= 24 \times (1.18)^2 \times 2$$
$$= 24 \times 1.39 \times 2$$
$$= 66.72 \text{ or } 67$$

$$FL = c \times q^2 \times L$$
$$= 38 \times \left(\frac{452}{100}\right)^2 \times \left(\frac{60}{100}\right)$$
$$= 38 \times (4.52)^2 \times 0.6$$
$$= 38 \times 20.43 \times 0.6$$
$$= 465.8 \text{ or } 466 \text{ kPa}$$

3. Appliance friction loss (AFL) = no appliance

4. Elevation pressure change:

$$EL = 5 \times \text{number of levels minus 1}$$
$$= 5 \times 2$$
$$= 10$$

$$EL = 35 \text{ kPa} \times \text{number of levels minus 1}$$
$$= 35 \text{ kPa} \times 2$$
$$= 70 \text{ kPa}$$

5. Pump discharge pressure:

$$PDP = NP + FL + AFL + EL$$
$$= 50 + 67 + 0 + 10$$
$$= 127$$

$$PDP = NP + FL + AFL + EL$$
$$= 350 + 466 + 0 + 70$$
$$= 886 \text{ kPa}$$

Practice

Calculate PDP using the FL formulas and elevations in the following table for 300 feet (90 m) of 3-inch (77-mm) hose flowing 250 gpm (1,000 Lpm) through an automatic nozzle.

	+15 ft (+4.5 m)	5th Floor (first level = 1)	−30 ft (−9 m)
PDP 1. cq^2L	____	____	____
2. Condensed Q	____	____	____

+15 ft (+ 4.5 m) (elevation gain)

1. $PDP = NP + FL(cq^2L) + AFL + EL$

Step 1: $NP = 100$

$$gpm = 250$$
$$NP = 700 \text{ kPa}$$
$$Lpm = 1{,}000$$

Step 2: $FL = c \times q^2 \times L$

$$= 0.8 \times \left(\frac{250}{100}\right)^2 \times \left(\frac{300}{100}\right)$$
$$= 0.8 \times (2.5)^2 \times 3$$
$$= 0.8 \times 6.25 \times 3$$
$$= 15 \text{ psi}$$

$$FL = c \times q^2 \times L$$
$$= 1.27 \times \left(\frac{1{,}000}{100}\right)^2 \times \left(\frac{90}{100}\right)$$

$$= 1.27 \times (10)^2 \times 0.9$$
$$= 1.27 \times 100 \times 0.9$$
$$= 114.3 \text{ or } 114 \text{ kPa}$$

Step 3: No appliance friction loss

Step 4: $EL = 0.5 \times h$
$$= 0.5 \times 15$$
$$= 7.5 \text{ or } 8 \text{ psi}$$

$$EL = 10 \text{ kPa} \times h$$
$$= 10 \text{ kPa} \times 4.5 \text{ m}$$
$$= 45 \text{ kPa}$$

Step 5: $PDP = NP + FL\ (cq^2L) + AFL + EL$
$$= 100 + 15 + 8$$
$$= 123 \text{ psi}$$

2. $PDP = NP + FL(q^2L) + AFL + EL$

Step 1: $NP = 100$

$gpm = 250$

Step 2: $FL = q^2 \times L$
$$= \left(\frac{250}{100}\right)^2 \times \left(\frac{300}{100}\right)$$
$$= (2.5)^2 \times 3$$
$$= 6.25 \times 3$$
$$= 18.75 \text{ or } 19 \text{ psi}$$

Step 3: No appliance friction loss

Step 4: $EL = 0.5 \times h$
$$= 0.5 \times 15$$
$$= 7.5 \text{ or } 8 \text{ psi}$$

Step 5: $PDP = NP + FL\ (q^2L) + AFL + EL$
$$= 100 + 19 + 8$$
$$= 127 \text{ psi}$$

5th Floor (first floor = 1)

1. $PDP = NP + FL(cq^2L) + AFL + EL$

Step 1: $NP = 100$

$gpm = 250$

$NP = 700 \text{ kPa}$

$Lpm = 1,000$

Step 2: $FL = c \times q^2 \times L$
$$= 0.8 \times \left(\frac{250}{100}\right)^2 \times \left(\frac{300}{100}\right)$$
$$= 0.8 \times (2.5)^2 \times 3$$
$$= 0.8 \times 6.25 \times 3$$
$$= 15 \text{ psi}$$

$$FL = c \times q^2 \times L$$
$$= 1.27 \times \left(\frac{1000}{100}\right)^2 \times \left(\frac{90}{100}\right)$$
$$= 1.27 \times (10)^2 \times 0.9$$
$$= 1.27 \times 100 \times 0.9$$
$$= 114.3 \text{ or } 144 \text{ kPa}$$

Step 3: No appliance friction loss

Step 4: $EL = 5 \times \text{number of levels minus}$
$$= 5 \times 4$$
$$= 20 \text{ psi}$$

$$EL = 35 \text{ kPa} \times \text{number of floors minus } 1$$
$$= 35 \text{ kPa} \times 4$$
$$= 140 \text{ kPa}$$

Step 5: $PDP = NP + FL\ (cq^2L) + AFL + EL$
$$= 100 + 15 + 20$$
$$= 135$$

$$PDP = NP + FL\ (cq^2L) + AFL + EL$$
$$= 700 + 114 + 140$$
$$= 954 \text{ kPa}$$

2. $PDP = NP + FL(q^2L) + AFL + EL$

Step 1: $NP = 100$

$gpm = 250$

Step 2: $FL = q^2 \times L$
$$= \left(\frac{250}{100}\right)^2 \times \left(\frac{300}{100}\right)$$

$$= (2.5)^2 \times 3$$
$$= 6.25 \times 3$$
$$= 18.75 \text{ or } 19 \text{ psi}$$

Step 3: No appliance friction loss

Step 4: $EL = 5 \times \text{number of levels}$
$$= 5 \times 4$$
$$= 20 \text{ psi}$$

Step 5: $PDP = NP + FL\ (q^2L) + AFL + EL$
$$= 100 + 19 + 20$$
$$= 139 \text{ psi}$$

–30 Feet (–9 m) (elevation loss)

1. $PDP = NP + FL(cq^2L) + AFL + EL$

Step 1: $NP = 100$
$$gpm = 250$$
$$NP = 700 \text{ kPa}$$
$$Lpm = 1{,}000$$

Step 2: $FL = c \times q^2 \times L$
$$= 0.8 \times \left(\frac{250}{100}\right)^2 \times \left(\frac{300}{100}\right)$$
$$= 0.8 \times (2.5)^2 \times 3$$
$$= 0.8 \times 6.25 \times 3$$
$$= 15 \text{ psi}$$

$$FL = c \times q^2 \times L$$
$$= 1.27 \times \left(\frac{1{,}000}{100}\right)^2 \times \left(\frac{90}{100}\right)$$
$$= 1.27 \times (10)^2 \times 0.9$$
$$= 1.27 \times 100 \times 0.9$$
$$= 114.3 \text{ or } 144 \text{ kPa}$$

Step 3: No appliance friction loss

Step 4: $EL = 0.5 \times h$
$$= 0.5 \times -30$$
$$= -15 \text{ psi}$$

$$EL = 10 \text{ kPa} \times h$$
$$= 10 \text{ kPa} \times -9$$
$$= -90 \text{ kPa}$$

Step 5: $PDP = NP + FL\ (cq^2L) + AFL + EL$
$$= 100 + 15 + -15$$
$$= 100 \text{ psi}$$

$$PDP = NP + FL\ (cq^2L) + AFL + EL$$
$$= 700 + 114 - 90$$
$$= 724 \text{ kPa}$$

2. $PDP = NP + FL(q^2L) + AFL + EL$

Step 1: $NP = 100$
$$gpm = 250$$

Step 2: $FL = q^2 \times L$
$$= \left(\frac{250}{100}\right)^2 \times \left(\frac{300}{100}\right)$$
$$= (2.5)^2 \times 3$$
$$= 6.25 \times 3$$
$$= 18.75 \text{ or } 19 \text{ psi}$$

Step 3: No appliance friction loss

Step 4: $EL = 0.5 \times h$
$$= 0.5 \times -30$$
$$= -15 \text{ psi}$$

Step 5: $PDP = NP + FL\ (q^2L) + AFL + EL$
$$= 100 + 19 + -15$$
$$= 104 \text{ psi}$$

	+15 ft (+4.5 m)	5th Floor (first level = 1)	–30 ft (–9 m)
PDP 1. cq^2L	123 psi (859)	135 psi (954)	100 psi (724)
2. Condensed Q	127 psi	139 psi	104 psi

Different Hose Sizes

When different sizes of hose diameters are used within the same lay, friction loss must be calculated separately for each diameter of hose. This type of lay typically uses medium-diameter hose to overcome distance and then is reduced to a smaller attack line. Following is an example of how to calculate single line lays utilizing two different sizes of hose: 3-inch to 1½-inch (77-mm to 38-mm) hose. Consider a 700-foot (210-m) lay consisting of 550 feet (165 m) of 3-inch (77-mm) line and 150 feet (45 m)

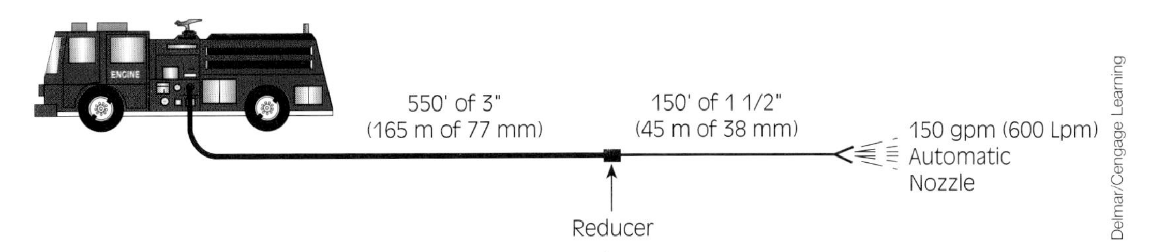

FIGURE 12-7 Single line: change of hose diameter.

of 1½-inch (38-mm) line. The line is equipped with an automatic nozzle flowing 150 gpm (600 Lpm) (**Figure 12-7**). What is the PDP?

1. Nozzle operating pressure and flow, NP = 100 (700) and $gpm\ (Lpm)$ = 150 (600).
2. Hose friction loss, in this case, must be calculated separately for the 3-inch (77-mm) line and the 1½-inch (38-mm) line. FL_s represents the 3-inch (77-mm) supply line, while FL_a represents the 1½-inch (38-mm) attack line.

$$\begin{aligned} FL_s &= c \times q^2 \times L \\ &= 0.8 \times (1.5)^2 \times 5.5 \\ &= 0.8 \times 2.25 \times 5.5 \\ &= 9.9, \text{ or } 10 \text{ psi} \end{aligned}$$

$$\begin{aligned} FL_s &= c \times q^2 \times L \\ &= 1.27 \times (6)^2 \times 1.65 \\ &= 1.27 \times 36 \times 1.65 \\ &= 75.4 \text{ or } 75 \text{ kPa} \end{aligned}$$

$$\begin{aligned} FL_a &= c \times q^2 \times L \\ &= 24 \times (1.5)^2 \times 1.5 \\ &= 24 \times 2.25 \times 1.5 \\ &= 81 \text{ psi} \end{aligned}$$

$$\begin{aligned} FL_a &= c \times q^2 \times L \\ &= 38 \times (6)^2 \times 0.45 \\ &= 38 \times 36 \times 0.45 \\ &= 615.6 \text{ or } 616 \text{ kPa} \end{aligned}$$

3. Appliance friction loss (AFL) = 5 psi (reducer) (0 kPa)
4. Elevation pressure change (EL) = no change in elevation
5. Pump discharge pressure:

$$\begin{aligned} PDP &= NP + FL_s + FL_a + AFL + EL \\ &= 100 + 10 + 81 + 5 + 0 \\ &= 196 \end{aligned}$$

$$\begin{aligned} PDP &= NP + FL_s + FL_a + AFL + EL \\ &= 700 + 75 + 616 + 0 + 0 \\ &= 1391 \text{ kPa} \end{aligned}$$

Practice

Calculate PDP using the FL formula cq^2L for 500 feet (45 m) of 2½-inch (65-mm) hose reduced to 100 feet (30 m) of 1½-inch (38-mm) hose with a ¾-inch (19-mm) tip smooth-bore nozzle.

$$PDP = NP + FL + AFL + EL$$

Step 1: $NP = 50$

$$\begin{aligned} gpm &= 29.7 \times d^2 \times \sqrt{NP} \\ &= 29.7 \times 75^2 \times \sqrt{50} \\ &= 29.7 \times .56 \times 7.07 \\ &= 117.58 \text{ or } 118 \text{ gpm} \end{aligned}$$

Note: Using the rounded figures in the formula ($30 \times d^2 \times 7$) provides virtually the same answer: 117.6 or 118.

$$\begin{aligned} NP &= 350 \text{ kPa} \\ Lpm &= 0.067 \times d^2 \times \sqrt{NP} \\ &= 0.067 \times 19^2 \times \sqrt{350} \\ &= 0.067 \times 361 \times 18.7 \\ &= 452.3 \text{ or } 453 \text{ Lpm} \end{aligned}$$

Step 2: FL

FL_a : 500 feet of 2½-inch hose

$$\begin{aligned} &= 2 \times \left(\frac{118}{100}\right)^2 \times \left(\frac{500}{100}\right) \\ &= 2 \times (1.18)^2 \times 5 \\ &= 2 \times 1.39 \times 5 \\ &= 13.9 \text{ or } 14 \text{ psi} \end{aligned}$$

$$
\begin{aligned}
FL_a &: 150 \text{ m } 65\text{-mm hose}\\
&= 3.17 \times \left(\frac{453}{100}\right)^2 \times \left(\frac{150}{100}\right)\\
&= 3.17 \times (4.53)^2 \times 1.5\\
&= 3.17 \times 20.5 \times 1.5\\
&= 97.5 \text{ or } 98 \text{ kPa}
\end{aligned}
$$

$$
\begin{aligned}
FL_b &: 100 \text{ feet of } 1\tfrac{1}{2}\text{-inch hose}\\
&= 24 \times \left(\frac{118}{100}\right)^2 \times \left(\frac{100}{100}\right)\\
&= 24 \times (1.18)^2 \times 1\\
&= 24 \times 1.39 \times 1\\
&= 33.36 \text{ or } 33 \text{ psi}
\end{aligned}
$$

$$
\begin{aligned}
FL_b &: 30 \text{ m of } 38\text{-mm hose}\\
&= 38 \times \left(\frac{453}{100}\right)^2 \times \left(\frac{30}{100}\right)\\
&= 38 \times (4.53)^2 \times 0.3\\
&= 38 \times 20.5 \times 0.3\\
&= 233.7 \text{ or } 234 \text{ kPa}
\end{aligned}
$$

$$
\begin{aligned}
FL_{a+b} &= 14 + 33\\
&= 47 \text{ psi}
\end{aligned}
$$

$$
\begin{aligned}
FL_{a+b} &= 98 + 234\\
&= 332 \text{ kPa}
\end{aligned}
$$

Step 3: Reducer = 5 psi

$= 0$ kPa (less than 1,400 Lpm)

Step 4: No elevation gain or loss

Step 5:

$$
\begin{aligned}
PDP &= NP + FL + AFL + EL\\
&= 50 + 47 + 5\\
&= 102 \text{ psi}
\end{aligned}
$$

$$
\begin{aligned}
PDP &= NP + FL + AFL + EL\\
&= 350 + 332 + 0\\
&= 682 \text{ kPa}
\end{aligned}
$$

MULTIPLE LINES

Multiple-line calculations range from simple to complex. When the pump is supplying more than one line, lower-pressure lines must be feathered and the total pump discharge in gpm (Lpm) is the sum of the discharges for each individual line. The following are examples of multiple-line PDP calculations.

Multiple Like Lines

When the pump is supplying two or more lines that are the same in size and flow, only one line is calculated. Because the other line is the same, the results of the calculations would be the same. When initiating flow for multiple like lines, the pump operator need only increase the PDP to the calculated pressure of the one line.

For the remaining calculations in this chapter, formulas are gradually omitted and calculations gradually condensed by omitting mention of obvious steps.

Two Like Lines

Consider two sections of 3-inch (77-mm) hose 1,000 feet (300 m) long each with a 1¼-inch (32-mm) tip handheld smooth-bore nozzle (**Figure 12-8**). What is the PDP for each line?

1. Nozzle operating pressure and flow, $NP = 50$ (350 kPa) and:

$$
\begin{aligned}
gpm &= 30 \times d^2 \times \sqrt{NP}\\
&= 30 \times (1.25)^2 \times \sqrt{50}\\
&= 30 \times 1.56 \times 7\\
&= 327.6 \text{ or } 328
\end{aligned}
$$

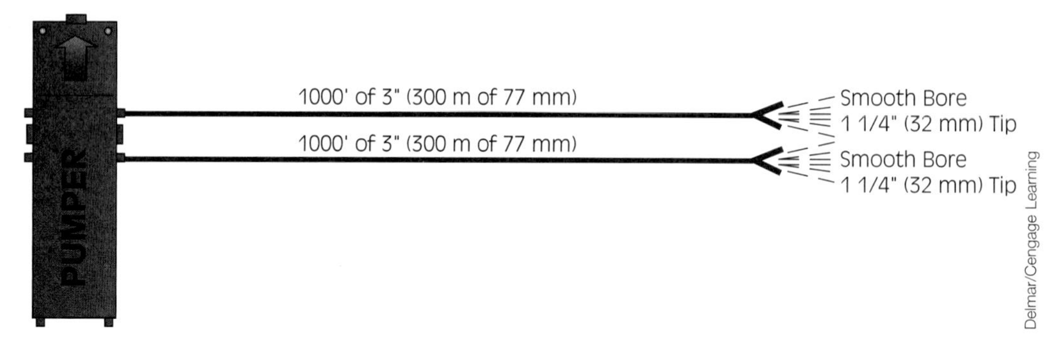

FIGURE 12-8 Multiple lines: two like lines.

$$\begin{aligned} Lpm &= 0.067 \times d^2 \times \sqrt{NP} \\ &= 0.067 \times (32)^2 \times \sqrt{350} \\ &= 0.067 \times 1,024 \times 18.7 \\ &= 1,282.9 \text{ or } 1,283 \text{ Lpm} \end{aligned}$$

2. Hose friction loss:

$$\begin{aligned} FL &= c \times q^2 \times L \\ &= 0.8 \times (3.28)^2 \times 10 \\ &= 0.8 \times 10.76 \times 10 \\ &= 86.08 \text{ or } 86 \end{aligned}$$

$$\begin{aligned} FL &= c \times q^2 \times L \\ &= 1.27 \times (12.83)^2 \times 3 \\ &= 1.27 \times 164.6 \times 3 \\ &= 627.1 \text{ or } 627 \text{ kPa} \end{aligned}$$

3. Appliance friction loss (AFL) = no appliances

4. Elevation pressure change (EL) = no change in elevation

5. Pump discharge pressure:

$$\begin{aligned} PDP &= NP + FL + AFL + EL \\ &= 50 + 86 = 136 \end{aligned}$$

$$\begin{aligned} PDP &= NP + FL + AFL + EL \\ &= 350 + 627 = 977 \text{ kPa} \end{aligned}$$

Both lines should be pumped at 136 psi (977 kPa) to obtain a nozzle pressure of 50 psi (350 kPa). Regardless of how many lines are flowing, as long as they are all the same size and flow, the discharge pressure will be the same. Keep in mind that the elevation must also be the same.

Multiple Lines of Different Sizes and Flows

When the pump is supplying multiple lines of different sizes or flows, each line must be calculated separately.

Two Lines of Different Size

Consider a pump supplying two lines. Line A is 500 feet (150 m) of 3-inch (77-mm) hose flowing 250 gpm (1,000-Lpm) through a fog nozzle. Line B is 200 feet (60 m) of 2½-inch (65-mm) hose flowing 300 gpm (1,200 Lpm) through an automatic nozzle (**Figure 12-9**). What is the PDP for both lines?

1. Nozzle operating pressure and flow, *NP* = 100 psi (700 kPa) for both line A and line B, and *gpm (Lpm)* = 250 (1,000) for line A and 300 (1,200) for line B.

2. Hose friction loss:

Line A

$$\begin{aligned} FL &= 0.8 \times (2.5)^2 \times 5 \\ &= 0.8 \times 6.25 \times 5 \\ &= 25 \end{aligned}$$

$$\begin{aligned} FL &= 1.27 \times (10)^2 \times 1.5 \\ &= 1.27 \times 100 \times 1.5 \\ &= 190.5 \text{ or } 190 \text{ kPa} \end{aligned}$$

Line B

$$\begin{aligned} FL &= 2 \times 32 \times 2 \\ &= 2 \times 9 \times 2 \\ &= 36 \end{aligned}$$

$$\begin{aligned} FL &= 3.17 \times (12)^2 \times 0.6 \\ &= 3.17 \times 144 \times 0.6 \\ &= 273.9 \text{ or } 274 \text{ kPa} \end{aligned}$$

3. Appliance friction loss (AFL) = no appliances in either line

4. Elevation pressure change (EL) = no change in elevation in either line

5. Pump discharge pressure:

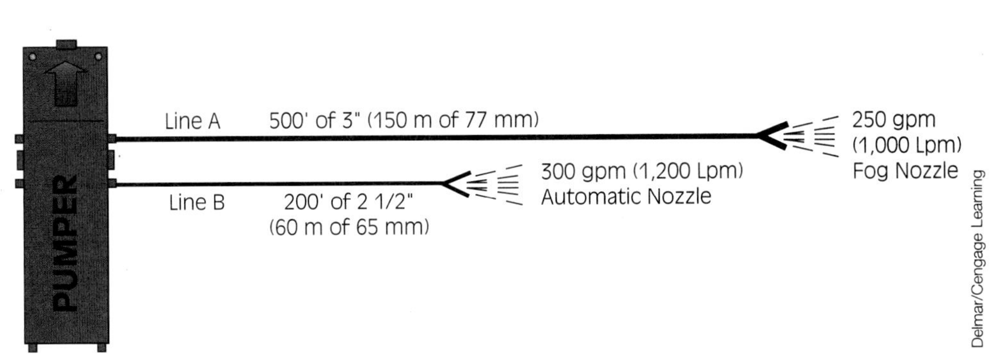

FIGURE 12-9 Multiple lines: different hose diameter and length.

Line A

$$PDP = 100 + 25$$
$$= 125$$

$$PDP = 700 + 190$$
$$= 890 \text{ kPa}$$

Line B

$$PDP = 100 + 36$$
$$= 136$$

$$PDP = 700 + 274$$
$$= 890 \text{ kPa}$$

Two Lines of Different Flow

Consider a pump supplying two lines of equal length (250 ft) (75 m) and diameter (1½ in.) (38 mm). However, line A is flowing 80 gpm (320 Lpm) and line B is flowing 125 gpm (500 Lpm) both through automatic nozzles (**Figure 12-10**). What is the PDP for each line?

1. $NP = 100$

$NP = 700$

2. Hose friction loss:

Line A

$$FL = 24 \times 0.8^2 \times 2.5$$
$$= 24 \times 0.64 \times 2.5$$
$$= 38.4 \text{ or } 38$$

$$FL = 38 \times 3.2^2 \times 0.75$$
$$= 38 \times 10.24 \times 0.75$$
$$= 291.84 \text{ or } 292 \text{ kPa}$$

Line B

$$FL = 24 \times (1.25)^2 \times 2.5$$
$$= 24 \times 1.56 \times 2.5$$
$$= 93.6 \text{ or } 94$$

$$FL = 38 \times 5^2 \times 0.75$$
$$= 38 \times 25 \times 0.75$$
$$= 712.5 \text{ or } 713 \text{ kPa}$$

3. No appliance friction loss.

4. No elevation gain or loss.

5. Pump discharge pressure:

Line A

$$PDP = 100 + 38$$
$$= 138$$

$$PDP = 700 + 292$$
$$= 992 \text{ kPa}$$

Line B

$$PDP = 100 + 94$$
$$= 194$$

$$PDP = 700 + 713$$
$$= 1,413 \text{ kPa}$$

Three Lines of Various Configurations

Determine the PDP for each of the lines shown in **Figure 12-11**.

1. $NP = 100$.

2. Step 2: Hose friction loss:

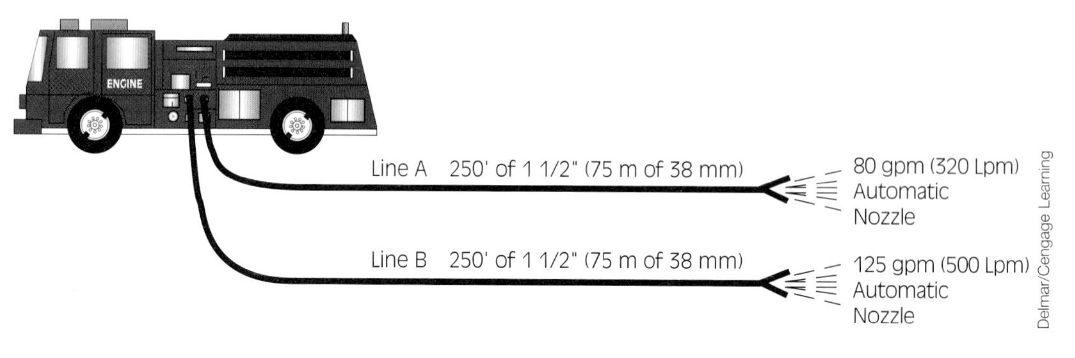

FIGURE 12-10 Multiple lines: different flow.

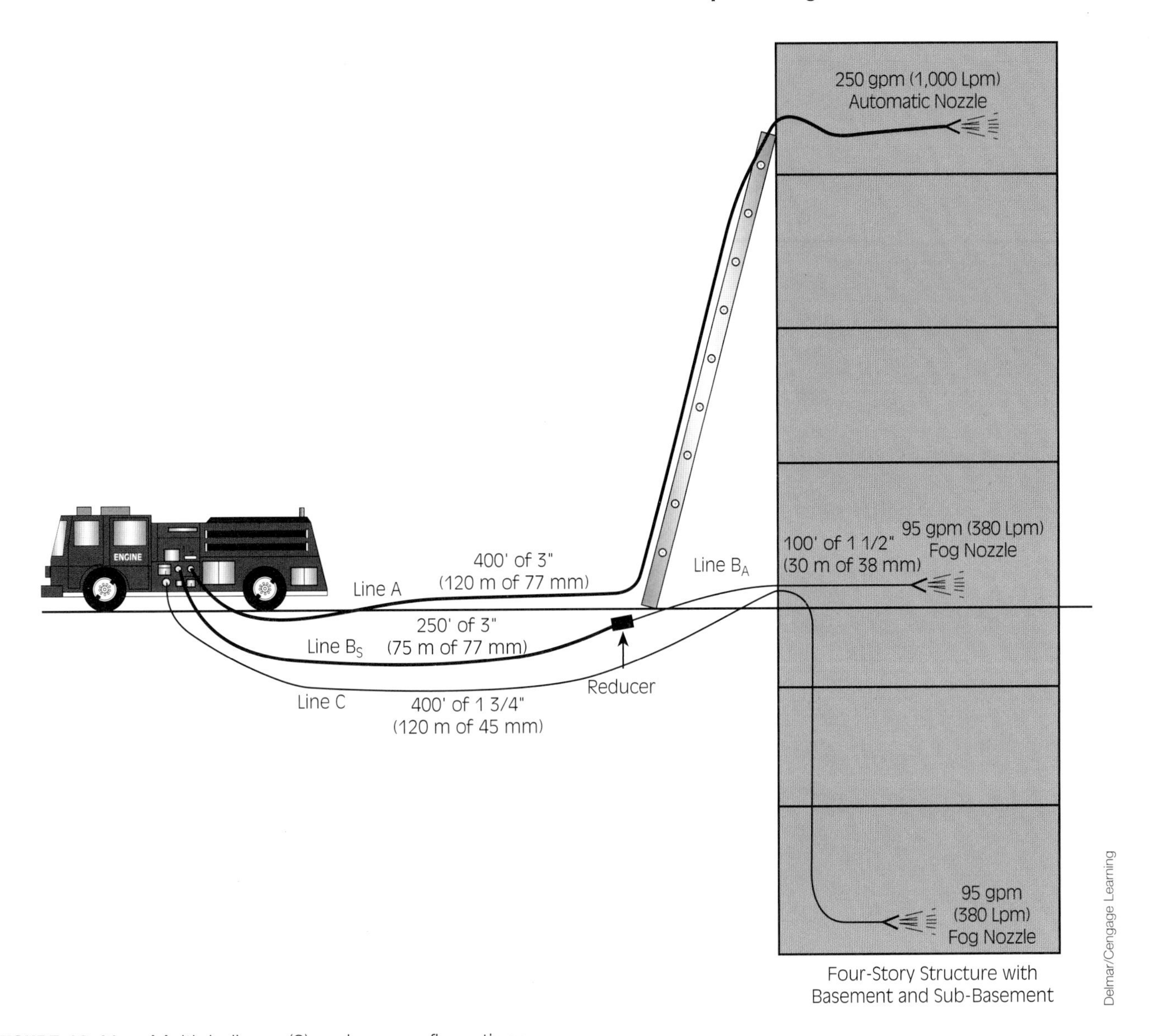

FIGURE 12-11 Multiple lines: (3) various configurations.

Line A

$$FL = 0.8 \times 2.5^2 \times 4$$
$$= 8 \times 6.25 \times 4$$
$$= 20$$

$$FL = 1.27 \times 10^2 \times 1.2$$
$$= 1.27 \times 100 \times 1.2$$
$$= 152.4 \text{ or } 152 \text{ kPa}$$

Line B

$$FL_s = 0.8 \times (0.95)^2 \times 2.5$$
$$= 0.8 \times 0.9 \times 2.5$$
$$= 1.8 \text{ or } 2$$

$$FL_s = 1.27 \times 3.8^2 \times 0.75$$
$$= 1.27 \times 14.4 \times 0.75$$
$$= 13.75 \text{ or } 14 \text{ kPa}$$

$$FL_a = 24 \times (0.95)^2 \times 1$$
$$= 24 \times 0.9 \times 1$$
$$= 21.6 \text{ or } 22$$

$$FL_a = 38 \times 3.8^2 \times 0.3$$
$$= 38 \times 14.4 \times 0.3$$
$$= 164.6 \text{ or } 165 \text{ kPa}$$

Line C

$$FL = 15 \times (0.95)^2 \times 4$$
$$= 15 \times 0.9 \times 4$$
$$= 54$$

$$FL = 24.6 \times 3.8^2 \times 1.2$$
$$= 24.6 \times 14.4 \times 1.2$$
$$= 426.2 \text{ or } 426 \text{ kPa}$$

3. Appliance friction loss: line A $AFL = 0$; line B $AFL = 5$; and line C $AFL = 0$.
4. Elevation pressure change:

Line A

$$EL = 5 \times 3$$
$$= 15$$

$$EL = 35 \text{ kPa} \times 3$$
$$= 105 \text{ kPa}$$

Line B

$$EL = 0$$
$$= -15$$

$$EL = 0$$
$$= -105 \text{ kPa}$$

Line C

$$EL = 5 \times (-3)$$

5. Pump discharge pressure:

Line A

$$PDP = NP + FL + EL$$
$$= 100 + 20 + 15$$
$$= 135$$

$$PDP = NP + FL + EL$$
$$= 700 + 152 + 105$$
$$= 957 \text{ kPa}$$

Line B

$$PDP = NP + FL_s + FL_a + AFL$$
$$= 100 + 2 + 22 + 5$$
$$= 129$$

$$PDP = NP + FL_s + FL_a + AFL$$
$$= 700 + 14 + 165 + 0$$
$$= 879 \text{ kPa}$$

Line C

$$PDP = NP + FL + EL$$
$$= 100 + 54 + -15$$
$$= 139$$

$$PDP = NP + FL + EL$$
$$= 700 + 426 + -105$$
$$= 1{,}021 \text{ kPa}$$

Practice

Calculate PDP for the following pumping operation using cq^2L. Two pumpers are operating on the fire scene. Pumper 1 is attacking the fire with two 200-foot 1¾-inch (60-m 45-mm) lines with automatic nozzles flowing 125 gpm (500 Lpm) and operating on the third level of a structure. Pumper 2 is covering exposures with three lines. The first master stream $line_a$ is 400 feet (120 m) in length consisting of 200 feet of 3-inch (60 m of 77-mm) reduced to 200 feet of 2½ inch (60 m of 65-mm) with a combination nozzle flowing 300 gpm (1,200 Lpm) and operating on the second level. The second master stream $line_b$ is 600 feet of 3-inch (180 m of 77-mm) with a 1½-inch (38-mm) tip on the smooth-bore master stream, which is 10 feet (3 m) below the pumper. The third line is the same as $line_b$ with the exception that it is 25 feet (7.5 m) below the pumper. All master streams are monitors.

Pumper 1: *PDP = NP + FL + AFL + EL*

Step 1: Equal lines, calculate for one line

$$NP = 100$$
$$gpm = 125$$

$$NP = 700 \text{ kPa}$$
$$Lpm = 500$$

$$\text{Step 2: } FL = 15.5 \times \left(\frac{125}{100}\right)^2 \times \left(\frac{200}{100}\right)$$
$$= 15.5 \times (1.25)^2 \times 2$$
$$= 15.5 \times 1.56 \times 2$$
$$= 48.36 \text{ or } 48 \text{ psi}$$

$$FL = 24.6 \times \left(\frac{500}{100}\right)^2 \times \left(\frac{60}{100}\right)$$
$$= 24.6 \times (5)^2 \times 0.6$$

$= 24.6 \times 25 \times 0.6$

$= 369\ \text{kPa}$

Step 3: *AFL*, monitor nozzle = 15 psi (175 kPa for monitor)

Step 4: $EL = 5 \times 2$

$= 10\ \text{psi}$

$EL = 35\ \text{kPa} \times 2$

$= 70\ \text{kPa}$

Step 5: $PDP = NP + FL + AFL + EL$

$= 100 + 48 + 15 + 10 = 173\ \text{psi}$

$PDP = NP + FL + AFL + EL$

$= 700 + 369 + 175 + 70 = 1,314\ \text{kPa}$

Pumper 2: *PDP* = *NP* + *FL* + *AFL* + *EL*

Step 1: *NP* and *gpm*

$\text{Line}_a\ NP = 100$

$gpm = 300$

$\text{Line}_a\ NP = 700\ \text{kPa}$

$Lpm = 1,200$

$\text{Line}_b\ NP = 80$

$gpm = 30 \times 1.5^2 \times 9$

$gpm = 607.5\ \text{or}\ 608$

$\text{Line}_b\ NP = 560\ \text{kPa}$

$Lpm = 0.067 \times 38^2 \times 23.66$

$Lpm = 2,289\ \text{or}\ 2,290$

$\text{Line}_c\ NP = 80$

$gpm = 608$

$\text{Line}_c\ NP = 560\ \text{kPa}$

$Lpm = 2,290$

Step 2: *FL*

Line_a 3-inch line (77-mm)

$$FL = 0.8 \times \left(\frac{300}{100}\right)^2 \times \left(\frac{200}{100}\right)$$

$= 0.8 \times (3)^2 \times 2$

$= 0.8 \times 9 \times 2$

$= 14.4\ \text{or}\ 14\ \text{psi}$

$$FL = 1.27 \times \left(\frac{1,200}{100}\right)^2 \times \left(\frac{60}{100}\right)$$

$= 1.27 \times (12)^2 \times 0.6$

$= 1.27 \times 144 \times 0.6$

$= 109.7\ \text{or}\ 110\ \text{kPa}$

Line_a 2½-inch (65-mm) line

$$FL = 2 \times \left(\frac{300}{100}\right)^2 \times \left(\frac{200}{100}\right)$$

$= 2 \times (3)^2 \times 2$

$= 2 \times 9 \times 2$

$= 36\ psi$

$$FL = 3.17 \times \left(\frac{1,200}{100}\right)^2 \times \left(\frac{60}{100}\right)$$

$= 3.17 \times (12)^2 \times 0.6$

$= 3.17 \times 144 \times 0.6$

$= 273.8\ \text{or}\ 274\ \text{kPa}$

Line_a

$\text{Line}_a\ FL = 14 + 36$

$= 50\ \text{psi}$

$FL = 110 + 274$

$= 384\ \text{kPa}$

Line_b

$$FL = 0.8 \times \left(\frac{608}{100}\right)^2 \times \left(\frac{600}{100}\right)$$

$= 0.8 \times (6.08)^2 \times 6$

$= 0.8 \times 36.96 \times 6$

$= 177.4\ \text{or}\ 177\ \text{psi}$

$$FL = 1.27 \times \left(\frac{2,290}{100}\right)^2 \times \left(\frac{180}{100}\right)$$

$= 1.27 \times (22.9)^2 \times 1.8$

$= 1.27 \times 524.4 \times 1.8$

$= 1,198.7\ \text{or}\ 1,200\ \text{kPa}$

$Line_c$

$$FL = 177 \text{ psi}$$

$$FL = 1,200 \text{ kPa}$$

Step 3: Appliance FL

$Line_a$ = 5 psi (70 kPa) (reducer) + 15 psi (175 kPa) (monitor nozzle) = 20 psi (210 kPa)
$Lines_{b \text{ and } c}$ = 15 psi (175 kPa) per line (monitor nozzles)

Step 4: Elevation

$Line_a$

$$= 5 \times 1$$
$$= 5 \text{ psi}$$
$$= 35 \text{ kPa} \times 1$$
$$= 35 \text{ kPa}$$

$Line_b$

$$= 0.5 \times -10$$
$$= -5 \text{ psi}$$
$$= 10 \text{ kPa} \times -3$$
$$= -30 \text{ kPa}$$

$Line_c$

$$= 0.5 \times -25$$
$$= -12.5 \text{ or } -13 \text{ psi}$$
$$= 10 \text{ kPa} \times -7.5 \text{ m}$$
$$= -75 \text{ kPa}$$

Step 5: $PDP = NP + FL + AFL + EL$

$Line_a$ $= 100 + 50 + 20 + 5$
$= 175$ psi
$= 700 + 384 + 210 + 35$
$= 1,329$ kPa

$Line_b$ $= 80 + 177 + 15 + -5$
$= 267$ psi
$= 560 + 1,200 + 175 + -30$
$= 1,905$ kPa

$Line_c$ $= 80 + 177 + 15 + -13$
$= 259$ psi
$= 560 + 1,200 + 175 + -75$
$= 1,860$ kPa

Pumper 1 PDP would be 173 psi (1,314 kPa) (like lines pumped at same pressure because they are like lines).

Pumper 2 PDP would be 267 psi (1,905 kPa) ($line_b$) and the other two lines would be gated down to 175 psi ($line_a$) (1,329 kPa) and 259 psi (1,860 kPa) ($line_c$).

WYED LINES

Wyed lines are hose configurations in which one hoseline supplies two or more separate lines. Typically, one larger line supplies two or more smaller lines, for example, a 2½- or 3-inch (65- or 77-mm) supply line wyed to two or more 1½- or 1¾-inch (38-mm or 45-mm) attack lines. When calculating wyed lines, the supply line is calculated separately from the attack lines. Because the supply line feeds each of the wyed attack lines, the flow through the supply line will be the sum total of the flow through each of the wyed attack lines. If the attack lines are of equal size and flow, simply calculate friction loss for one of the lines.

Simple Wyed Configuration

Calculate the PDP for the wyed line shown in **Figure 12-12**.

1. Nozzle operating pressure and flow, NP = 100 psi (700 kPa). The flow through each of the 1¾-inch (45-mm) attack lines is 125 gpm (500 Lpm), while the flow through the 3-inch (77-mm) supply line is 250 gpm (1,000 Lpm). The attack line flow is designated as gpm (Lpm), and the flow through the supply line will be gpm (Lpm). This distinction is necessary for separately calculating the friction loss in the supply and attack lines. Because both attack lines are the same (like lines), only one line is calculated. gpm_s = 250 (1,000 Lpm) and gpm_a = 125 (500 Lpm)
2. Hose friction loss, in this case, must be calculated separately for the 3-inch (77-mm) line, FL_s, and the 1¾-inch (45-mm) line, FL_a. L_s represents the length of the 3-inch (77-mm) supply line, while L_a represents the length of the 1¾-inch (45-mm) attack line.

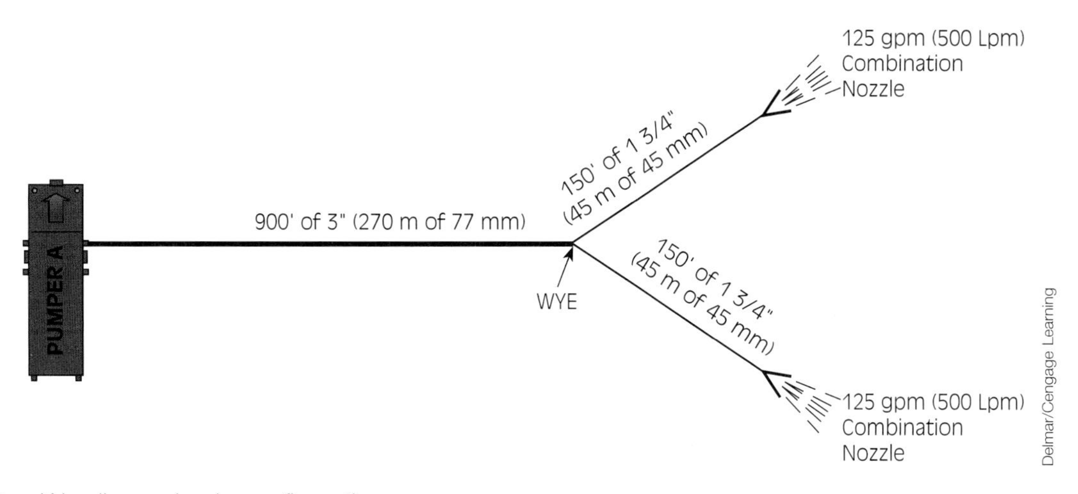

FIGURE 12-12 Wye lines: simple configurations.

$$FL_s = c \times \left(\frac{gpm_s}{100}\right)^2 \times L_s$$
$$= 0.8 \times (2.5)^2 \times 9$$
$$= 0.8 \times 6.25 \times 9$$
$$= 45$$

$$FL_s = c \times \left(\frac{Lpm_s}{100}\right)^2 \times L_s$$
$$= 1.27 \times (10)^2 \times 2.7$$
$$= 1.27 \times 100 \times 2.7$$
$$= 342.9 \text{ or } 343 \text{ kPa}$$

$$FL_a = C \times \left(\frac{gpm_a}{100}\right)^2 \times L_a$$
$$= 15.5 \times (1.25)^2 \times 1.5$$
$$= 15.5 \times 1.56 \times 1.5$$
$$= 36.27 \text{ or } 36$$

$$FL_a = C \times \left(\frac{Lpm_a}{100}\right)^2 \times L_a$$
$$= 24.6 \times (5)^2 \times 0.45$$
$$= 24.6 \times 25 \times 0.45$$
$$= 276.6 \text{ or } 277 \text{ kPa}$$

3. Appliance friction loss (AFL) = 10 psi (2½ inches to 1½ inches) (70 kPa if flowing more than 1,400 Lpm). Note that 3-inch (77-mm) hose typically utilizes 2½-inch (65-mm) couplings, and 1¾-inch (45-mm) hose typically uses 1½-inch (38-mm) couplings.
4. No elevation loss or gain.
5. Pump discharge pressure:

$$PDP = NP + FL_s + FL_a + AFL$$
$$= 100 + 45 + 36 + 10$$
$$= 191$$

$$PDP = NP + FL_s + FL_a + AFL$$
$$= 700 + 343 + 277 + 0$$
$$= 1,320 \text{ kPa}$$

Practice

Calculate PDP for the following pumping operation using cq^2L. A pumper is supporting two wyed lines as follows:

- Line 1 = 400 feet of 3-inch (120 m of 77-mm) line wyed to two 150 feet of 1¾-inch (45 m of 45-mm) line with an automatic nozzle flowing 125 gpm (500 Lpm).
- Line 2 = 400 feet of 2½-inch (120 mm of 65-mm) line wyed to two 150 feet of 1½-inch (45 m of 38-mm) line with an automatic nozzle flowing 125 gpm (500 Lpm).

$$PDP = NP + FL + AFL + EL$$

Line 1

Step 1: $NP = 100$

$gpm_a = 125$

$gpm_s = 250$ (combined flow for the two attack lines)

$NP = 700$ kPa

$Lpm_a = 500$

$Lpm_s = 1,000$ (both lines)

Step 2: FL_a 150 feet of 1¾-inch line

$$= 15.5 \times \left(\frac{125}{100}\right)^2 \times \left(\frac{150}{100}\right)$$
$$= 15.5 \times (1.25)^2 \times 1.5$$
$$= 15.5 \times 1.56 \times 1.5$$
$$= 36.27 \text{ or } 36 \text{ psi}$$

FL_a 45 m of 45-mm hose

$$= 24.6 \times \left(\frac{500}{100}\right)^2 \times \left(\frac{45}{100}\right)$$
$$= 24.6 \times (5)^2 \times 0.45$$
$$= 24.6 \times 25 \times 0.45$$
$$= 276.8 \text{ or } 277 \text{ kPa}$$

FL_s 400 feet of 3-inch line

$$= 0.8 \times \left(\frac{250}{100}\right)^2 \times \left(\frac{400}{100}\right)$$
$$= 0.8 \times (2.5)^2 \times 4$$
$$= 0.8 \times 6.25 \times 4$$
$$= 20 \text{ psi}$$

FL_s 120 m of 77-mm hose

$$= 1.27 \times \left(\frac{1,000}{100}\right)^2 \times \left(\frac{120}{100}\right)$$
$$= 1.27 \times (10)^2 \times 1.2$$
$$= 1.27 \times 100 \times 1.2$$
$$= 152.4 \text{ or } 152 \text{ kPa}$$

$$FL_{a+s} = 36 + 20$$
$$= 56 \text{ psi}$$

$$FL_{a+s} = 277 + 152$$
$$= 429 \text{ kPa}$$

Step 3: 10 psi (2½ to 1½ wye) = 0 (less than 1,400 Lpm)

Step 4: No elevation gain or loss

Step 5: $PDP = NP + FL + AFL + EL$

$$= 100 + 56 + 10 = 166 \text{ psi}$$
$$= 700 + 429 + 0 = 1,129 \text{ kPa}$$

Line 2

Step 1: $NP = 100$

$gpm_a = 125$

$gpm_s = 250$

$NP = 700$ kPa

$Lpm = 500$

$Lpm = 1,000$

Step 2: FL_a 150 feet of 1½-inch line

$$= 24 \times \left(\frac{125}{100}\right)^2 \times \left(\frac{150}{100}\right)$$
$$= 24 \times (1.25)^2 \times 1.5$$
$$= 24 \times 1.56 \times 1.5$$
$$= 56.16 \text{ or } 56 \text{ psi}$$

FL_a 45 m of 38-mm line

$$= 38 \times \left(\frac{500}{100}\right)^2 \times \left(\frac{45}{100}\right)$$
$$= 38 \times (5)^2 \times 0.45$$
$$= 38 \times 25 \times 0.45$$
$$= 427.5 \text{ or } 428 \text{ kPa}$$

FL_b 400 feet of 2½-inch line

$$= 2 \times \left(\frac{250}{100}\right)^2 \times \left(\frac{400}{100}\right)$$
$$= 2 \times (2.25)^2 \times 4$$
$$= 2 \times 5.06 \times 4$$
$$= 4.48 \text{ or } 40 \text{ psi}$$

$$FL = FL_a + FL_s$$
$$= 56 + 40$$
$$= 96 \text{ psi}$$

FL_b 120 m of 65-mm line

$$= 3.17 \times \left(\frac{1,000}{100}\right)^2 \times \left(\frac{120}{100}\right)$$
$$= 3.17 \times (10)^2 \times 1.2$$
$$= 3.17 \times 100 \times 1.2$$

$$FL = FL_a + FL_s$$
$$= 428 + 380$$
$$= 808 \text{ kPa}$$

Step 3: 2½ to 1½ wye = 10 psi = 0 kPa (flowing less than 1,400 Lpm)
Step 4: No elevation gain or loss

Step 5: $PDP = NP + FL + AFL + EL$
$$= 100 + 96 + 10$$
$$= 206 \text{ psi}$$

$$PDP = NP + FL + AFL + EL$$
$$= 700 + 808 + 0$$
$$= 1,508 \text{ kPa}$$

Pump Discharge Pressure:

- Line 1: 166 psi (1,129 kPa)
- Line 2: 206 psi (1,508 kPa)

Complicated Wyed Configurations

Complicated wye configurations include variables such as elevation and sometimes unequal lines on the downstream side of the wye. When calculating PDP for complicated configurations, be sure to indicate each line clearly and recheck figures to help reduce the chance of error.

Practice

Calculate the pump discharge pressure for the wyed line shown in **Figure 12-13**.

1. Nozzle operating pressure and flow, NP = 100 psi (700 kPa) (both of the nozzles are combination [fog] nozzles) and

- Line A gpm_s = 200 (800 Lpm) (supply line), gpm_a = 100 (400 Lpm) (wyed attack like line)
- Line B gpm_s = 400 (1,600 Lpm) (supply line), gpm_a = 250 (1,000 Lpm) (wyed attack line B1), gpm_a = 150 (600 Lpm) (wyed attack line B2)

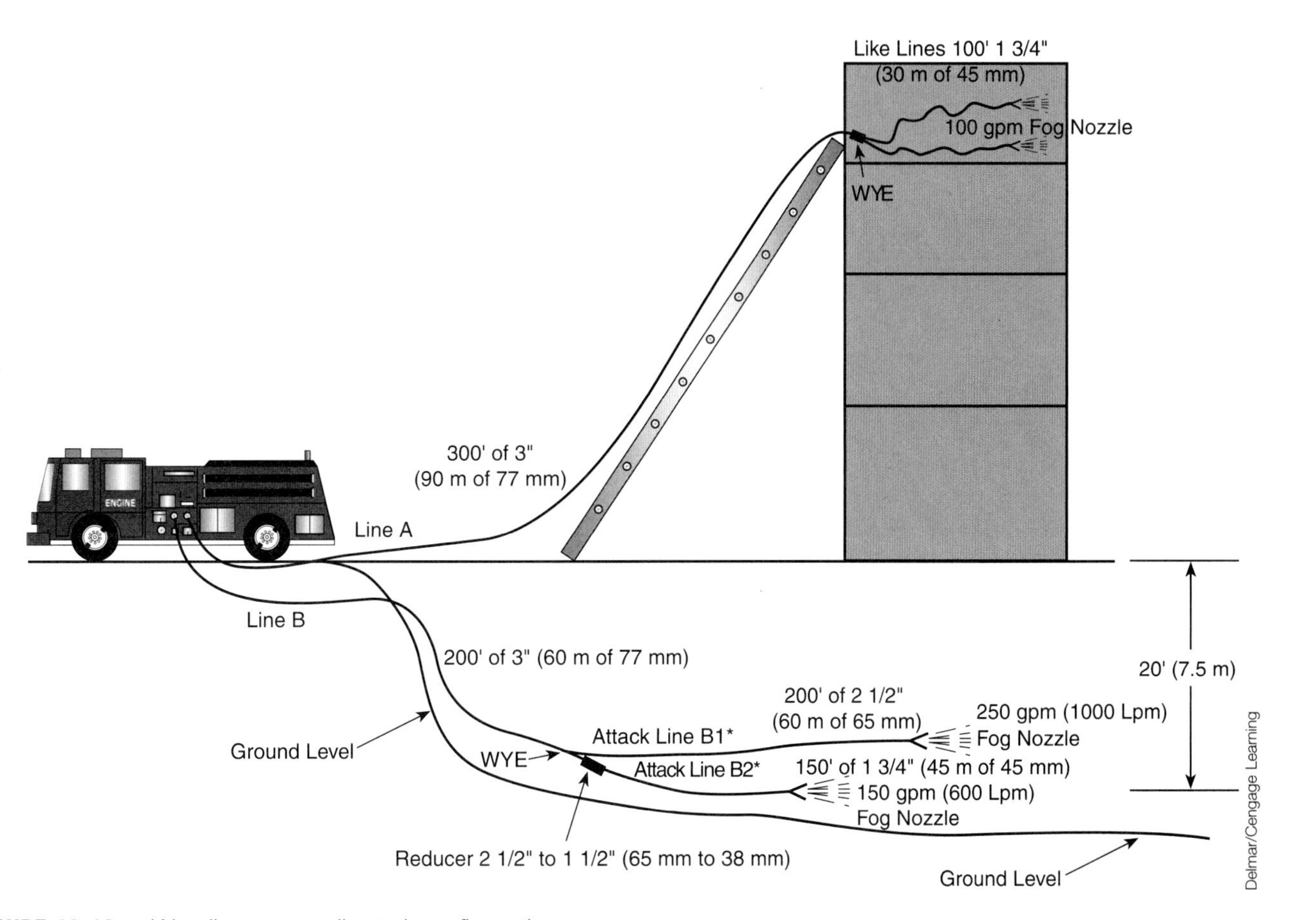

FIGURE 12-13 Wye lines: complicated configurations.

Note that the two lines downstream from the wye are unequal (different size and flow); therefore, friction loss calculations must be made for both lines.

2. Hose friction loss:

Line A

Supply Line

$$FL_s = c \times \left(\frac{gpm_s}{100}\right)^2 \times L_s$$
$$= 0.8 \times 2^2 \times 3$$
$$= 0.8 \times 4 \times 3$$
$$= 9.6, \text{ or } 10$$

$$FL_s = c \times \left(\frac{Lpm_s}{100}\right)^2 \times L_s$$
$$= 1.27 \times (8)^2 \times 0.9$$
$$= 1.27 \times 64 \times 0.6$$
$$= 73.15 \text{ or } 73 \text{ kPa}$$

Attack Line

$$FL_a = c \times \left(\frac{gpm_{sa}}{100}\right)^2 \times L_a$$
$$= 15.5 \times 1^2 \times 1$$
$$= 15.5 \times 1 \times 1$$
$$= 15.5 \text{ or } 16$$

$$FL_a = c \times \left(\frac{Lpm_{sa}}{100}\right)^2 \times L_a$$
$$= 24.6 \times (4)^2 \times 0.3$$
$$= 24.6 \times 16 \times 0.3$$
$$= 118 \text{ kPa}$$

Line B

Supply Line

$$FL_s = 0.8 \times 4^2 \times 2$$
$$= 0.8 \times 16 \times 2$$
$$= 25.6 \text{ or } 26$$

$$FL_s = 1.27 \times (16)^2 \times 0.6$$
$$= 1.27 \times 256 \times 0.6$$
$$= 195 \text{ kPa}$$

*Attack Line B1**

$$FL_a = 2 \times (2.5)^2 \times 2$$
$$= 2 \times 6.25 \times 2$$
$$= 25$$

$$FL_a = 3.17 \times (10)^2 \times 0.6$$
$$= 3.17 \times 100 \times 0.6$$
$$= 190.2 \text{ or } 190 \text{ kPa}$$

*Attack Line B2**

$$FL_a = 15.5 \times (1.5)^2 \times 1.5$$
$$= 15.5 \times 2.25 \times 1.5$$
$$= 52.3 \text{ or } 53$$

$$FL_a = 24.6 \times (6)^2 \times 0.45$$
$$= 24.6 \times 36 \times 0.45$$
$$= 398.5 \text{ or } 399 \text{ kPa}$$

3. Appliance friction loss:

- Line A: $AFL = 10$ psi (2½ inches to 1½-inch wye) (0 kPa because flow is less than 1,400 Lpm)
- Line B: $AFL = 10$ psi (2½ inches to 2½-inch wye and a 2½ inches to 1½-inch reducer for the second attack line) (70 kPa)(greater than 1,400 Lpm)

4. Elevation pressure change:

Line A

$$EL = 5 \times \text{number of stories minus } 1$$
$$= 5 \times 3$$
$$= 15$$

$$EL = 35 \text{ kPa} \times \text{number of stories minus } 1$$
$$= 35 \text{ kPa} \times 3$$
$$= 105 \text{ kPa}$$

Line B

$$EL = h \times 5$$
$$= -20 \times .5$$
$$= -10$$

$$EL = 10 \text{ kPa} \times h$$
$$= 10 \text{ kPa} \times -6$$
$$= -60 \text{ kPa}$$

5. Pump discharge pressure:

Line A

$$PDP = NP + FL + FL_a + AFL + EL$$
$$= 100 + 10 + 16 + 10 + 15$$
$$= 151$$

$$PDP = NP + FL + FL_a + AFL + EL$$
$$= 700 + 73 + 118 + 0 + 105$$
$$= 996 \text{ kPa}$$

Line B

$$PDP = NP + FL + {}^*FL_a + AFL = EL$$
$$= 100 + 26 + 53 + 10 - \ 10$$
$$= 179$$

$$PDP = NP + FL + {}^*FL_a + AFL = EL$$
$$= 700 + 195 + 399 + 70 - \ 60$$
$$= 1,304 \text{ kPa}$$

Note that the two attack lines downstream of the wye in **Figure 12-13** (designated by the asterisks) require two different pressures. The highest-pressure line is included in the PDP calculation. The second attack line is gated down (feathered) to the lower required pressure. This is similar to having two lines of different pressure at the pump panel. Obviously, this would require a pressure gauge on, or just after the wye. Since gauges on wyes are typically not common, uneven lines downstream of the wye should be avoided.

SIAMESE LINES

Siamese lines are hose configurations in which two or more separate lines supply one line, monitor nozzle, fixed system, or a pump in a relay or similar situation. Typically, two smaller or equal lines supply one larger or equal line, for example, several 2½-inch (65-mm) supply lines are Siamesed to a 3-inch (77-mm) line. When calculating Siamese lines, the supply line is calculated separately from the attack line. The flow through the line downstream of the Siamese will be divided among the supply lines. If the supply lines are of equal size and flow, simply calculate friction loss for one of the lines.

Calculate the PDP for the Siamese line shown in **Figure 12-14**.

1. Nozzle operating pressure and flow, NP = 80 psi (560 kPa) and

$$gpm_a = 30 \times (1.375)^2 \times \sqrt{80}$$
$$= 30 \times 1.89 \times 9$$
$$= 510.3, \text{ or } 510 \text{ (total flow)}$$
$$gpm_s = 255 \text{ (flow through each of the supply lines)}$$

$$Lpm_a = 0.067 \times (35)^2 \times \sqrt{560}$$
$$= 0.067 \times 1225 \times 23.66$$
$$= 1,942 \text{ Lpm (total flow)}$$
$$Lpm_s = 971 \text{ (flow through each of the supply lines)}$$

Note that flow through the attack line will be divided between the supply lines.

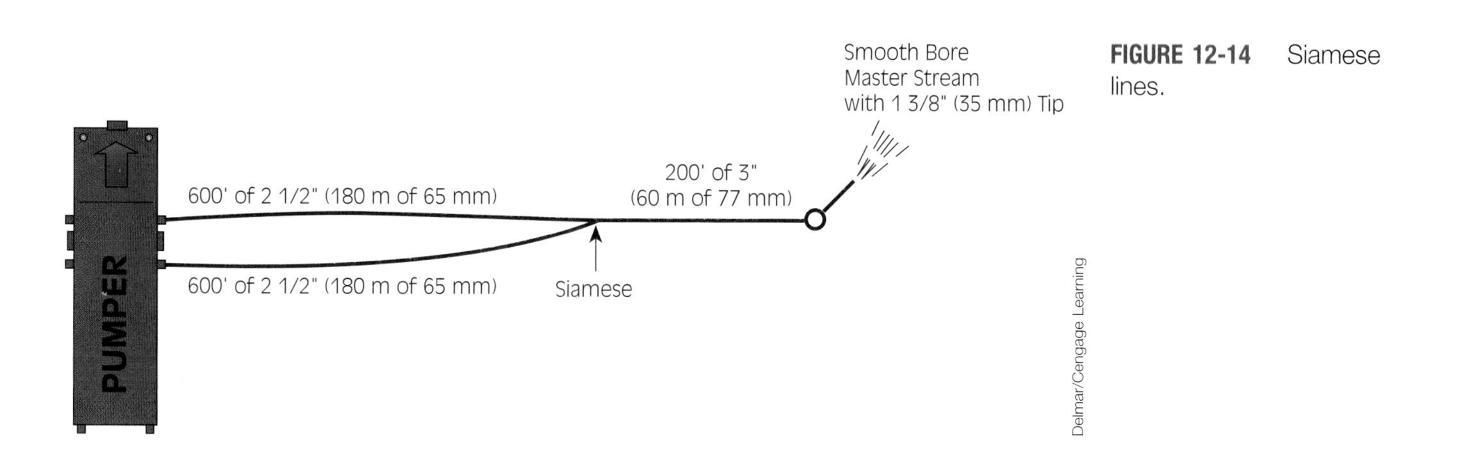

FIGURE 12-14 Siamese lines.

2. Hose friction loss:

$$FL_s = c \times \left(\frac{gpm_s}{100}\right)^2 \times L_s$$
$$= 2 \times 2.55^2 \times 6$$
$$= 2 \times 6.5 \times 6$$
$$= 78$$

$$FL_s = c \times \left(\frac{Lpm_s}{100}\right)^2 \times L_s$$
$$= 0.789 \times (9.71)^2 \times 1.8$$
$$= 0.789 \times 94.3 \times 1.8$$
$$= 133.9 \text{ or } 134 \text{ kPa}$$

$$FL_a = c \times \left(\frac{gpm_a}{100}\right)^2 \times L_a$$
$$= 0.8 \times 5.1^2 \times 2$$
$$= 0.8 \times 26 \times 2$$
$$= 41.6 \text{ or } 42$$

$$FL_a = c \times \left(\frac{Lpm_a}{100}\right)^2 \times L_a$$
$$= 1.27 \times (19.42)^2 \times 0.6$$
$$= 1.27 \times 377 \times 0.6$$
$$= 287.3 \text{ or } 287 \text{ kPa}$$

3. Appliance friction loss: AFL_a = 15 psi (70 kPa) (Siamese) and AFL_b = 15 (175 kPa) (monitor nozzle)

4. No elevation gain or loss.

5. Pump discharge pressure:

$$PDP = NP + FL_s + FL_a + AFL_a + AFL_b$$
$$= 80 + 78 + 42 + 15 + 15 = 230$$
$$= 560 + 134 + 287 + 70 + 175$$
$$= 1,226 \text{ kPa}$$

Practice

Calculate PDP for a Siamese hose configuration with three 500-foot (150-m) sections of 2½-inch (65-mm) line leading to a 200-foot (60-m) section of 3-inch (77-mm) line with a 1½-inch (38-mm) tip on a smooth-bore monitor master stream nozzle.

$$PDP = NP + FL + AFL + EL$$

Step 1: $NP = 80$ psi

$$gpm = 30 \times d^2 \times \sqrt{80}$$
$$= 30 \times 1.5^2 \times 9$$
$$= 30 \times 2.25 \times 9$$
$$= 607.5 \text{ or } 608 \text{ gpm}$$

$gpm = 203$ (combined flow divided by the number of lines supplying the Siamese)

$NP = 560$ kPa

$$Lpm = 0.067 \times d^2 \times \sqrt{560}$$
$$= 0.067 \times 38^2 \times 23.66$$
$$= 0.067 \times 1,444 \times 23.66$$
$$= 2,289 \text{ Lpm}$$

$Lpm = 763$ (combined flow divided by the number of lines supplying the Siamese)

Step 2: FL

FL_a: 200 feet of 3-inch hose

$$= 0.8 \times \left(\frac{608}{100}\right)^2 \times \left(\frac{200}{100}\right)$$
$$= 0.8 \times (6.08)^2 \times 2$$
$$= 0.8 \times 36.96 \times 2$$
$$= 59 \text{ psi}$$

FL_a: 60 m of 77-mm line

$$= 1.27 \times \left(\frac{2,289}{100}\right)^2 \times \left(\frac{60}{100}\right)$$
$$= 1.27 \times (22.89)^2 \times 0.6$$
$$= 1.27 \times 524 \times 0.6$$
$$= 399.3 \text{ or } 400 \text{ kPa}$$

FL_s: 500 feet of 2½-inch line

$$= 2 \times \left(\frac{203}{100}\right)^2 \times \left(\frac{500}{100}\right)$$
$$= 2 \times (2.03)^2 \times 5$$
$$= 2 \times 4.12 \times 5$$
$$= 41.2 \text{ or } 41 \text{ psi}$$

FL_s : 150 m of 65-mm line

$$= 0.347 \times \left(\frac{763}{100}\right)^2 \times \left(\frac{150}{100}\right)$$

$$= 0.347 \times (7.63)^2 \times 1.5$$

$$= 0.347 \times 58.2 \times 1.5$$

$$= 30.3 \text{ or } 30 \text{ kPa}$$

$$FL_{a+s} = 59 + 41$$

$$= 100 \text{ psi}$$

$$FL_{a+s} = 400 + 30$$

$$= 430 \text{ kPa}$$

Step 3: AFL = 10 (70 kPa) (Siamese) + 15 (175 kPa) (monitor) = 25 psi (210 kPa)

Step 4: No elevation gain or loss

Step 5: $PDP = NP + FL + AFL + EL$

$$= 80 + 100 + 25$$

$$= 205 \text{ psi}$$

$$PDP = NP + FL + AFL + EL$$

$$= 560 + 430 + 210 + 0$$

$$= 1,200 \text{ kPa}$$

A handy fireground rule of thumb used for Siamese lines is:

- Two-line Siamese: use 25% of the friction loss for one line.
- Three-line Siamese: use 10% of the friction loss for one line.

STANDPIPE SUPPORT

Calculate the PDP required to support the standpipe configuration illustrated in **Figure 12-15**.

1. Nozzle operating pressure and flow, NP = 100 psi (700 kPa) and gpm = 100 (Lpm = 400)

2. Hose friction loss:

$$FL_a = c \times q^2 \times L$$

$$= 24 \times \left(\frac{100}{100}\right)^2 \times \left(\frac{150}{100}\right)$$

$$= 24 \times 1 \times 1.5$$

$$= 36 \text{ psi}$$

$$FL_a = c \times q^2 \times L$$

$$= 38 \times \left(\frac{400}{100}\right)^2 \times \left(\frac{45}{100}\right)$$

$$= 38 \times (4)^2 \times 0.45$$

$$= 273.6 \text{ or } 274 \text{ kPa}$$

$$FL_s = 0.8 \times \left(\frac{100}{100}\right)^2 \times \left(\frac{300}{100}\right)$$

$$= 0.8 \times 1 \times 3$$

$$= 2.4 \text{ or } 2 \text{ psi}$$

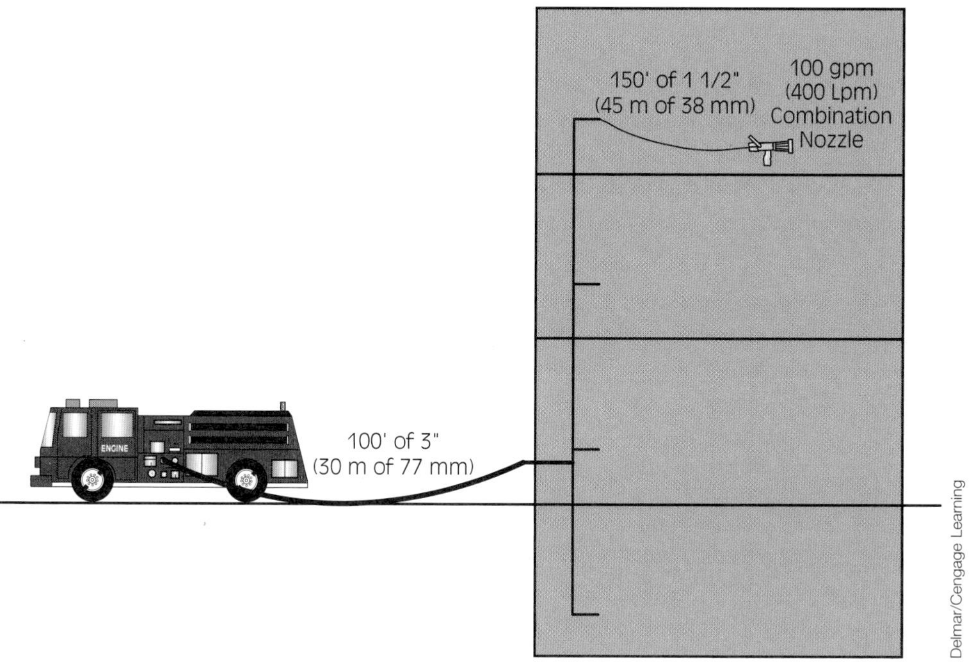

FIGURE 12-15 Standpipe support.

$$FL_s = 1.27 \times \left(\frac{400}{100}\right)^2 \times \left(\frac{30}{100}\right)$$
$$= 1.27 \times 16 \times 0.3$$
$$= 6 \text{ kPa}$$

3. Appliance friction loss (AFL) = 25 = 175 kPa

4. Elevation pressure changes:

$$EL = 5 \times \text{number of levels minus 1}$$
$$= 5 \times 2$$
$$= 10 \text{ psi}$$

$$EL = 35 \text{ kPa} \times \text{number of levels minus 1}$$
$$= 35 \text{ kPa} \times 2$$
$$= 70 \text{ kPa}$$

5. Pump discharge pressure:

$$PDP = NP + (FLA + FL) + AFL + EL$$
$$= 100 + 36 + 2 + 25 + 10$$
$$= 173 \text{ psi}$$

$$PDP = NP + (FLA + FL) + AFL + EL$$
$$= 700 + 274 + 6 + 175 + 70$$
$$= 1{,}225 \text{ kPa}$$

Recall from **Chapter 9** that, unless otherwise indicated, the supply line for sprinkler systems should be pumped at 150 psi (1,050 kPa).

LESSONS LEARNED

Hose lay configurations can be as simple as a single line or as complicated as three lines with wyes, Siamese, and elevation gain or loss. Regardless of the method used, pump operators must strive to provide nozzles with the proper flow and pressure by calculating the appropriate PDP.

KEY TERMS

Siamese lines Hose configurations in which two or more separate lines supply one line, monitor nozzle, fixed system, or a pump in a relay or similar situation.

Wyed lines Hose configurations in which one hoseline supplies two or more separate lines. Typically, one larger line supplies two or more smaller lines, for example, a 2½- or 3-inch (65- or 77-mm) supply line wyed to two or more 1½- or 1¾-inch (38-mm or 45-mm) attack lines. When calculating wyed lines, the supply line is calculated separately from the attack lines.

REVIEW QUESTIONS

Short Answer

For each of the following, review the figure referenced and provide the requested information on a separate sheet of paper.

1. Single line with smooth-bore nozzle, no elevation; see **Figure 12-16**.
   a. Quantity of water flowing:
   b. Nozzle pressure:
   c. Friction loss per 100 feet (30 m):
   d. Pump discharge pressure:
2. Two lines of same flow with different lengths, no elevation; see **Figure 12-17**.
   a. Total friction loss for Line A:
   b. Total friction loss for Line B:
   c. PDP for Line A:
   d. PDP for Line B:

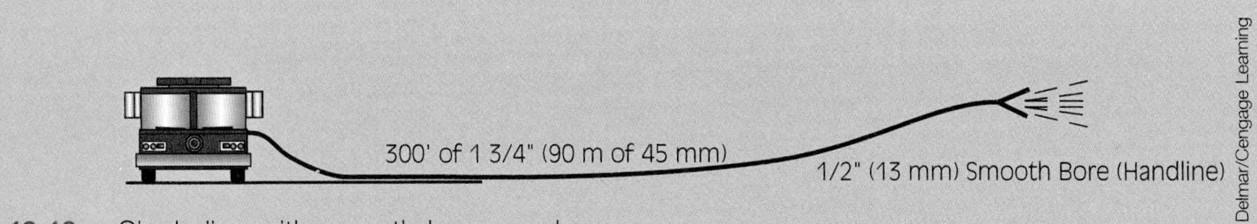

FIGURE 12-16 Single line with smooth-bore nozzle.

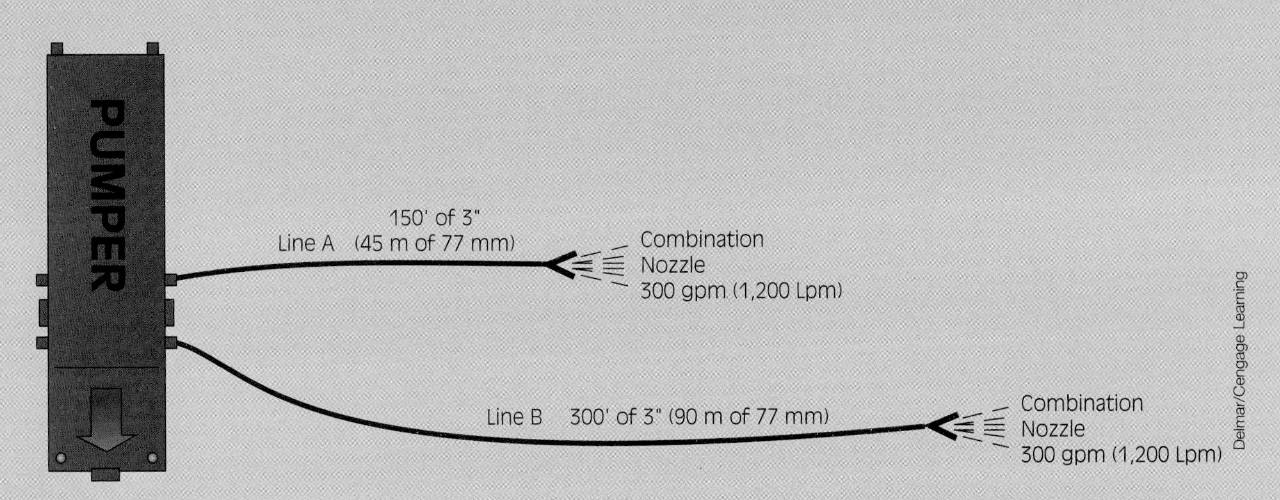

FIGURE 12-17 Two lines of same flow with different lengths.

3. Two lines of same length with different flows, no elevation; see **Figure 12-18**.

a. FL Line A:

b. FL Line B:

c. PDP Line A:

d. PDP Line B:

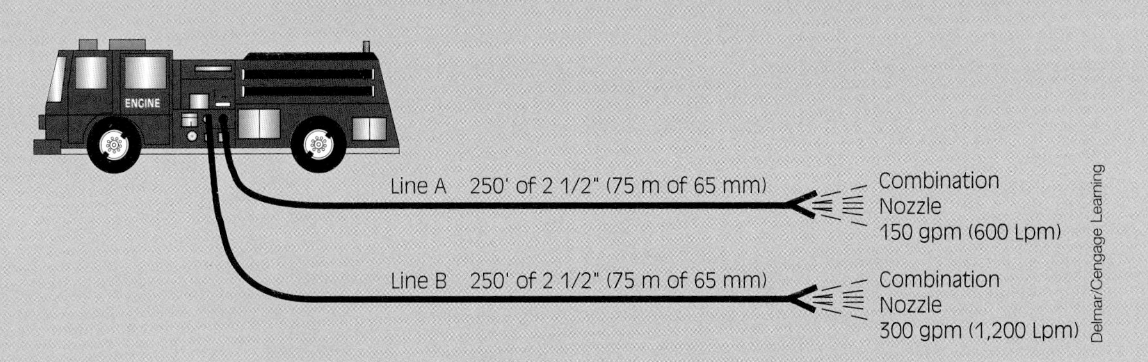

FIGURE 12-18 Two lines of same length with different flows.

4. Multiple lines of same length and flow with different sizes of hose, no elevation; see **Figure 12-19**.

a. Friction loss (*FL*):

Line A =

Line B =

Line C =

Line D =

b. Pump discharge pressure (PDP):

Line A =

Line B =

Line C =

Line D =

5. Single line wyed to two like lines, no elevation; see **Figure 12-20**.

a. *FL* in the 3-inch (77-mm) hose:

b. *FL* in the 1¾-inch (45-mm) hose:

c. PDP:

6. Two like wyed lines, no elevation; see **Figure 12-21**.

a. Line A

FL for 2½-inch line (65-mm):

FL for 1½-inch line (38-mm):

PDP:

b. Line B

FL for 2½-inch line (65 mm):

FL for 1½-inch line (38 mm):

PDP:

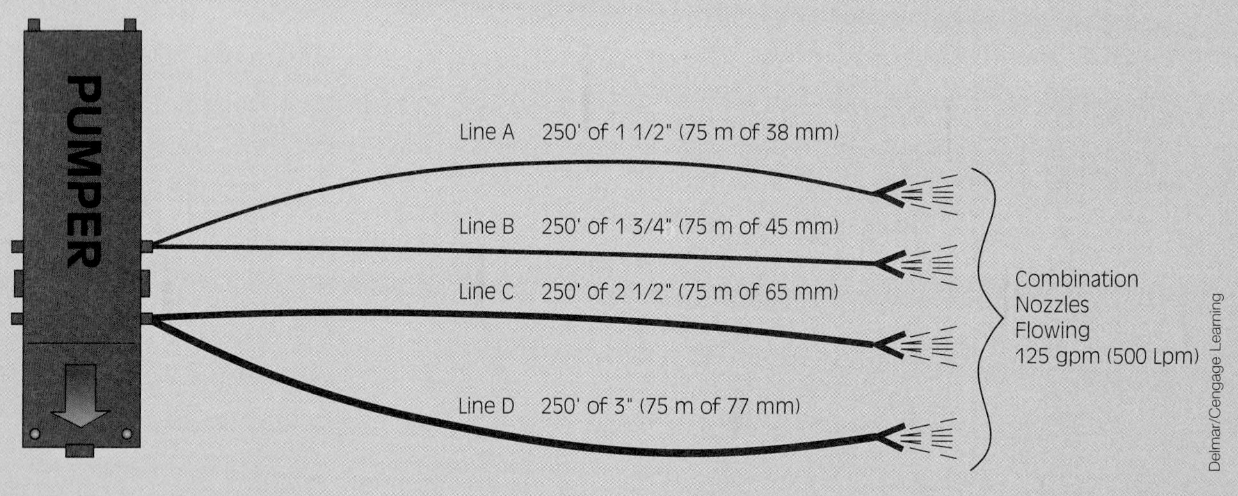

FIGURE 12-19 Multiple lines of same length and flow with different sizes of hose.

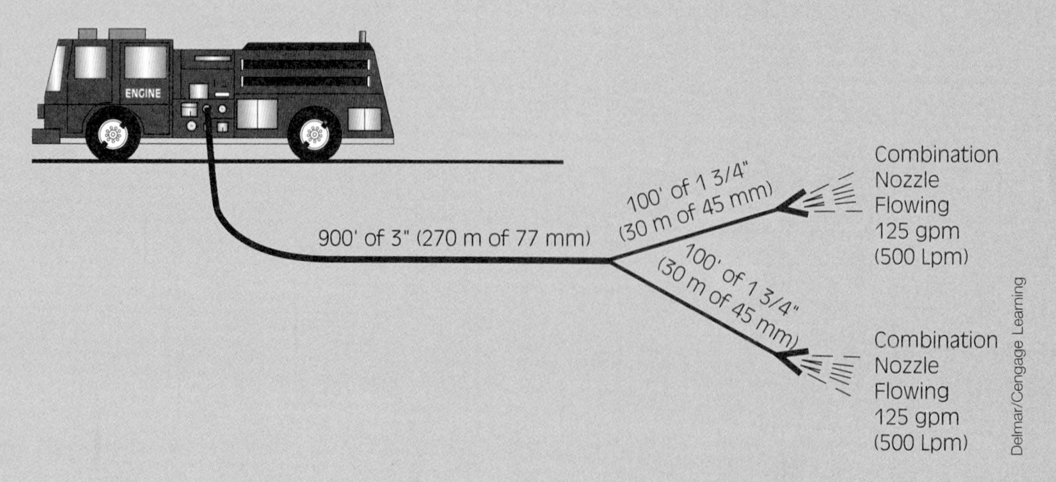

FIGURE 12-20 Single line wyed to like lines.

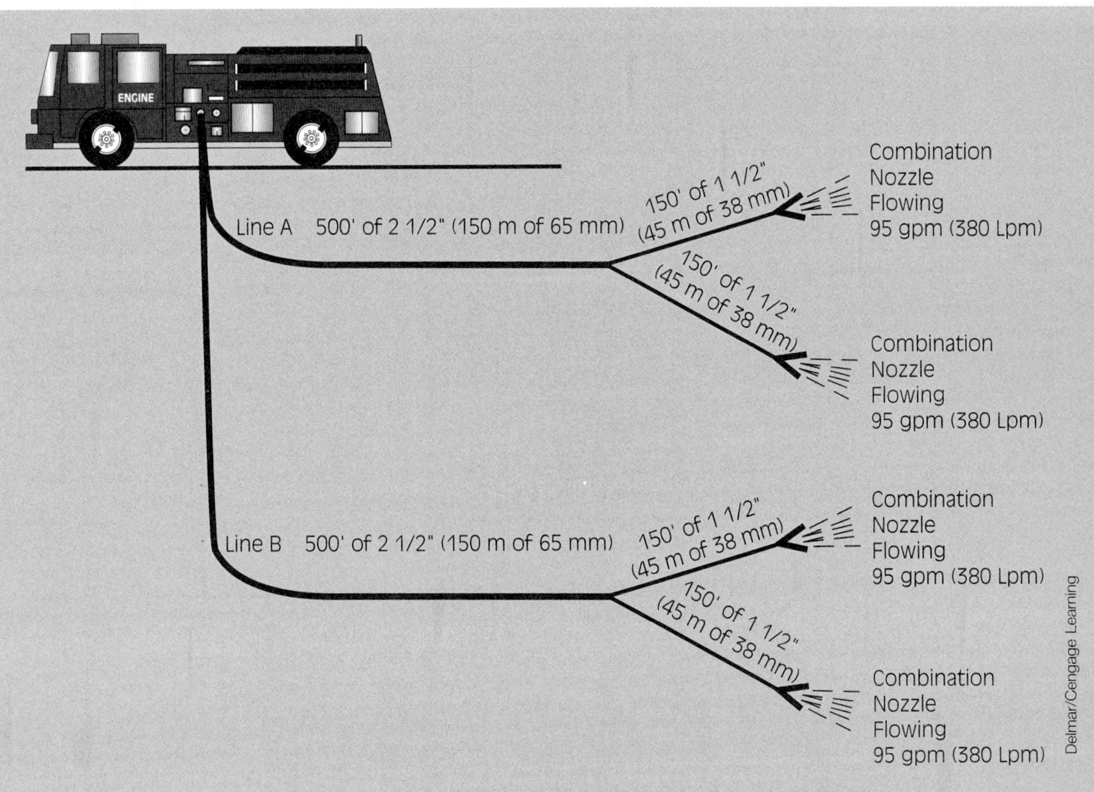

FIGURE 12-21 Two like wyed lines.

7. Siamese line, no elevation; see **Figure 12-22**.

a. Friction loss (FL_s) for the lines supplying the Siamese:

b. Friction loss (FL_a) for the line supplied by the Siamese:

c. Pump discharge pressure (PDP) for the hose configuration:

8. Two lines, one a Siamese and the other a single line, each of the same total length and flow, no elevation; see **Figure 12-23**.

a. PDP Line A:

b. PDP Line B:

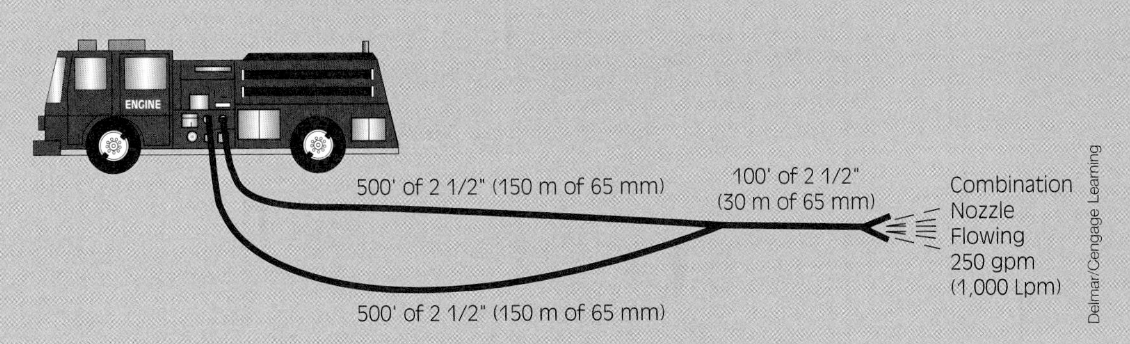

FIGURE 12-22 Siamese line.

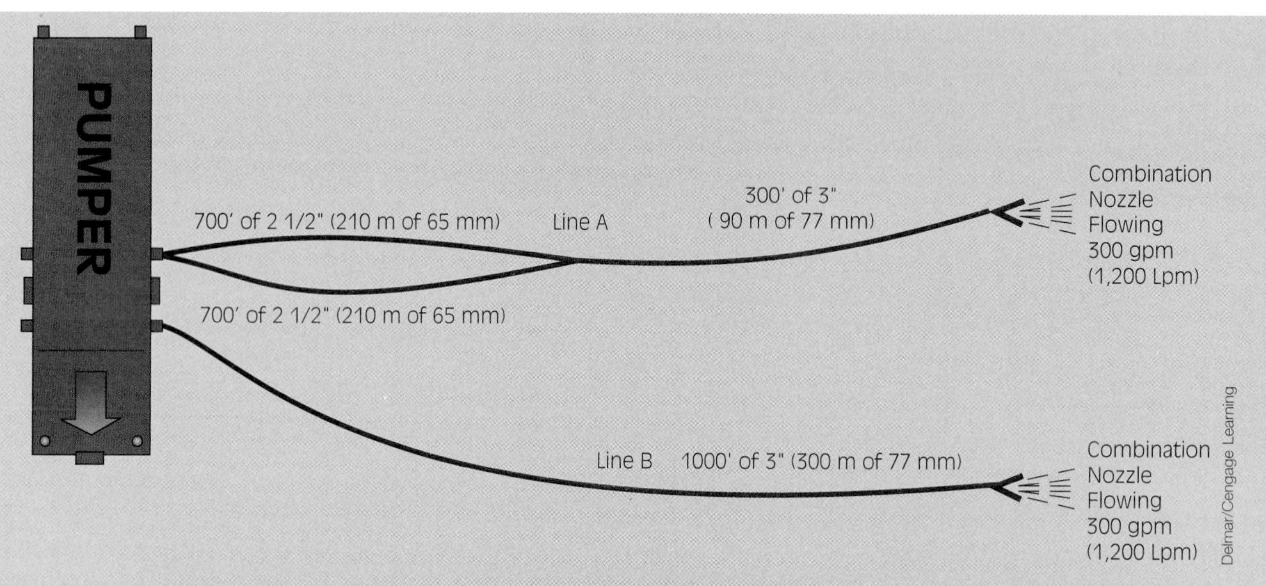

FIGURE 12-23 Two lines, one a Siamese and the other a single line, each with the same total length and flow.

9. Two Siamese lines, no elevation; see **Figure 12-24**.

a. Which Siamese line has the lower PDP?

10. Monitor nozzle supplied by three lines, no elevation; see **Figure 12-25**.

a. PDP:

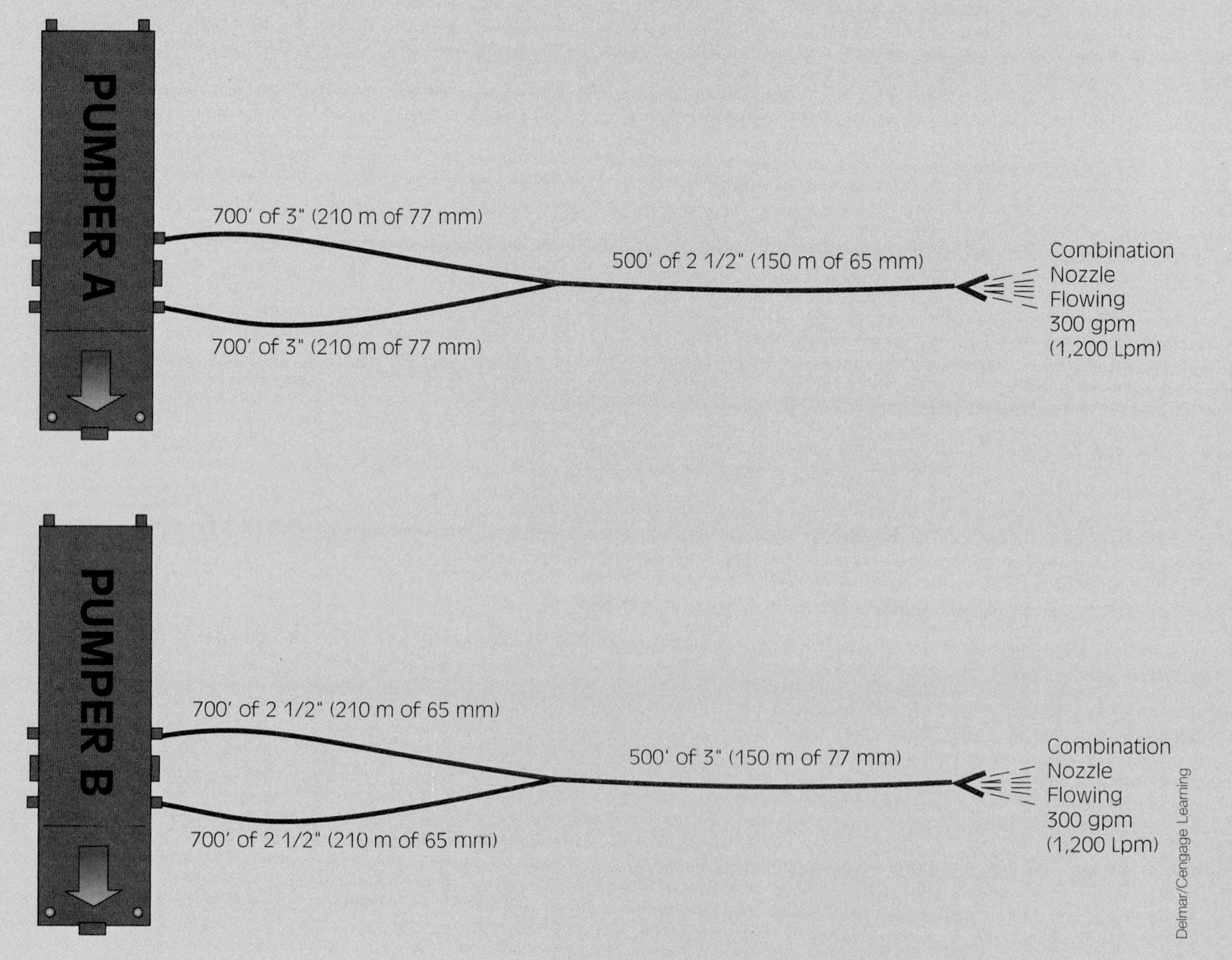

FIGURE 12-24 Two Siamese lines.

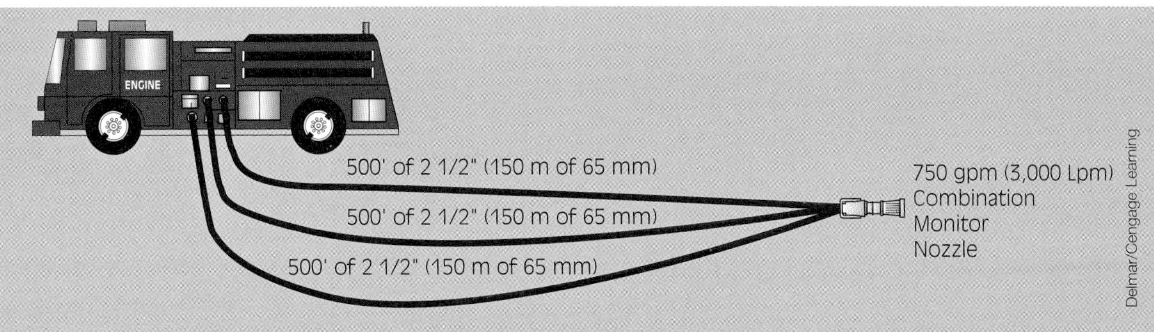

FIGURE 12-25 Monitor nozzle supplied by three lines.

11. One 3-inch (77-mm) handline operating on the third floor; see **Figure 12-26**.

a. Pressure change caused by elevation:

b. Flow through nozzle (gpm) (Lpm):

c. PDP:

12. Multiple lines with elevation; see **Figure 12-27**.

a. PDP Line A:

b. PDP Line B:

13. Complicated hose lay; see **Figure 12-28**.

a. PDP Line A:

b. PDP Line B:

On a separate sheet of paper, draw the described hose lay and determine the PDP for each of the following, being sure to identify each component properly.

14. 500 feet of 2½-inch (150 m of 65-mm) line flowing 325 gpm (1,300 Lpm) through an automatic nozzle.

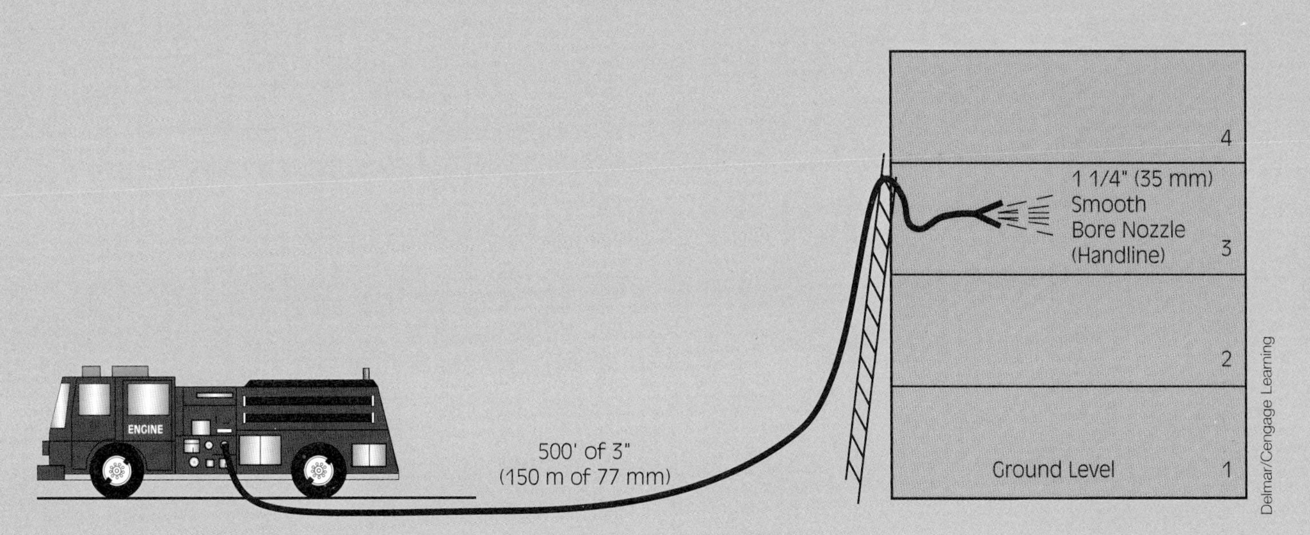

FIGURE 12-26 One 3-inch (77-mm) handline operating on the third floor.

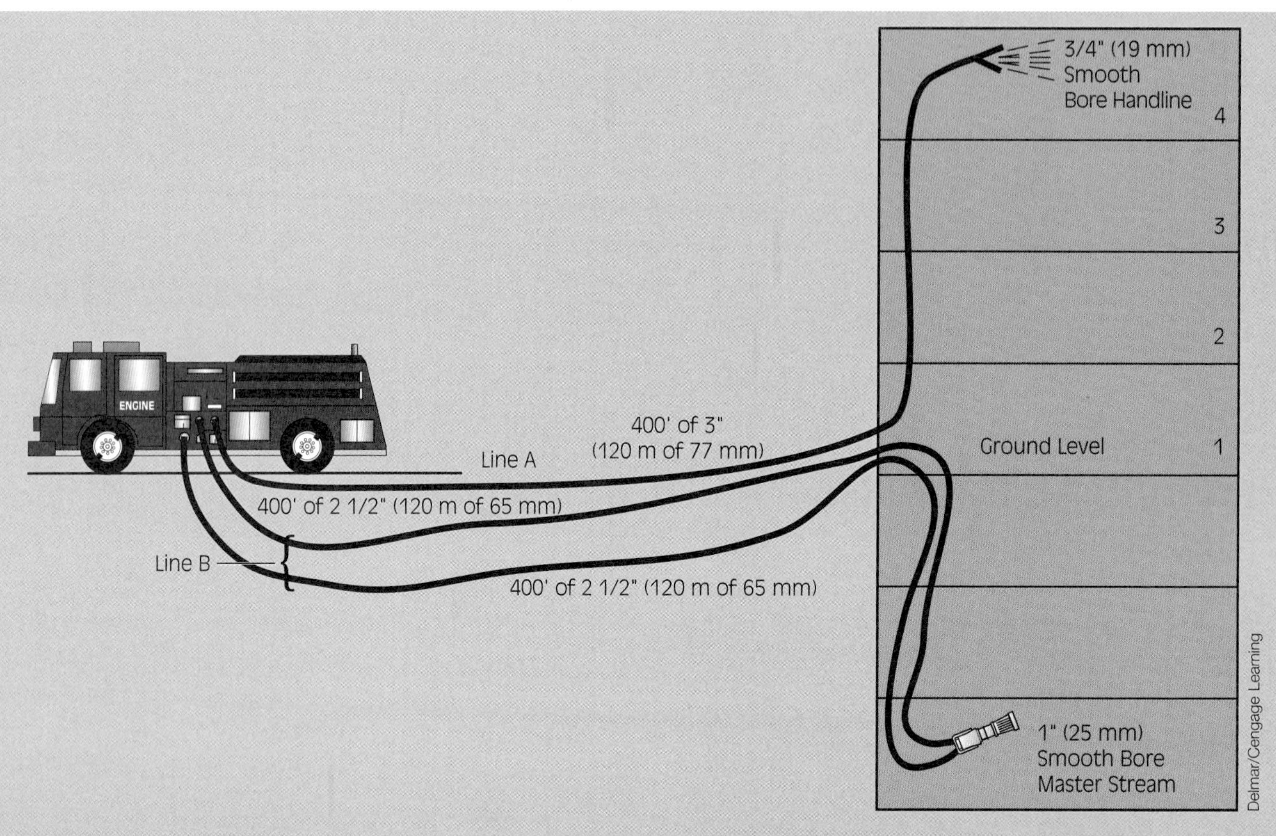

FIGURE 12-27 Multiple lines with elevation.

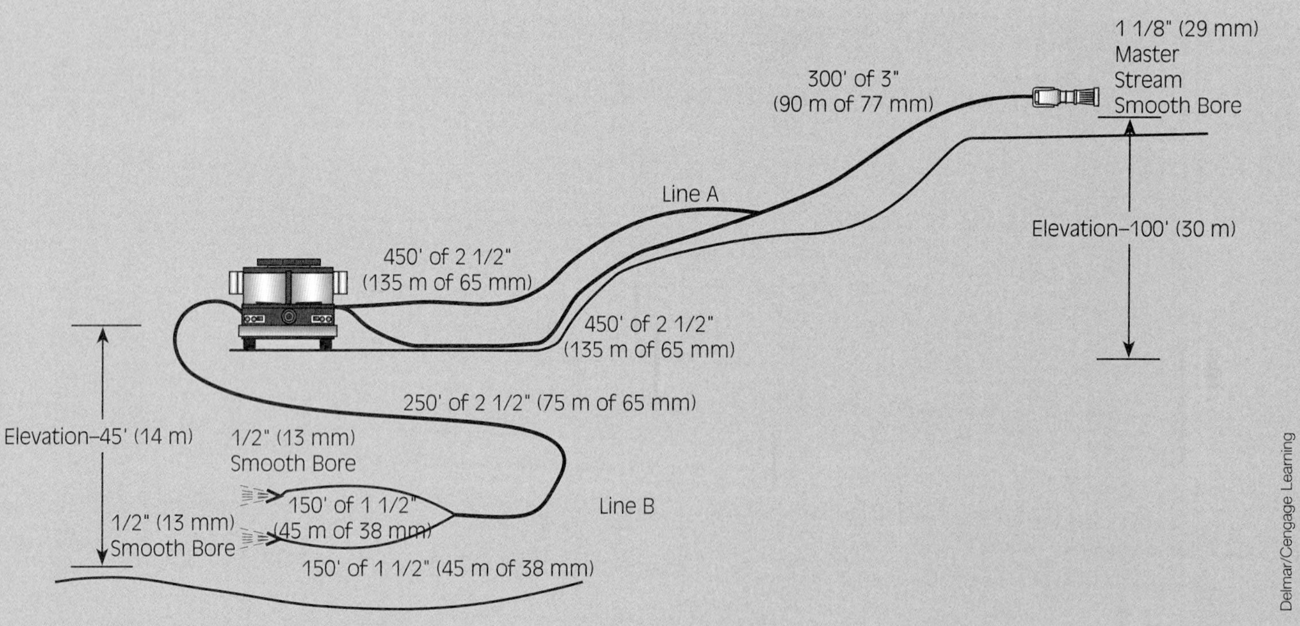

FIGURE 12-28 Complicated hose lay.

15. Two lines, one consisting of 300 feet of 3-inch (90 m of 77-mm) hose flowing 250 gpm (1,000 Lpm) through a fog nozzle, and the second consisting of 300 feet of 1½-inch (90 m of 38-mm) hose flowing 125 gpm (500 Lpm) through an automatic nozzle.
16. One 900-foot (270-m) line of 3-inch (77-mm) hose wyed to two 150-foot (45-m) sections of 1¾-inch (45-mm) hose, each flowing 200 gpm (800 Lpm) through an automatic nozzle.
17. A Siamese lay consisting of two 750-foot (225-m) sections of 3-inch (77-mm) hose to 150 feet of 3-inch (45 m of 77-mm) hose flowing 500 gpm (2,000 Lpm) through a master stream combination nozzle.
18. Three 400-foot (120-m) lines of 3-inch (77-mm) hose attached to a master stream flowing 1,000 gpm (4,000 Lpm) through a combination nozzle.
19. Two lines consisting of a wyed line (line A) and a Siamese line (line B). Line A is 400 feet of 3-inch (120 m of 77-mm) hose wyed to two 200-foot lines of 1¾-inch (60 m of 45-mm) hose with ¾-inch (19-mm) tips taken to the sixth floor of a structure. Line B is two 500-foot lines of 3-inch (150-m of 77-mm) hose Siamesed to 350 feet of 3-inch (105 m of 77-mm) hose with a 1-inch (25-mm) tip taken to the second basement level in a structure.

ACTIVITY

1. Review a response district and identify four structures. For each of the structures, determine four likely hose lay configurations, two of which are simple and two of which are complex. Finally, determine PDP for each hose lay configuration.

PRACTICE PROBLEM

1. For a given structure, develop a pumping operation in which the hoseline configurations include all the variables presented in this chapter.

ADDITIONAL RESOURCES

Richmond, Ron. *Basic Fire Hydraulics Workbook*, 1st ed. New York, Delmar, 2009.

Appendix A

NFPA STANDARDS RELATED TO PUMP OPERATIONS

NFPA 11	*Standard for Low-, Medium-, and High-Expansion Foam Systems*
NFPA 11A	*Standard for Medium- and High-Expansion Foam Systems*
NFPA 13	*Standard for the Installation of Sprinkler Systems*
NFPA 13D	*Standard for the Installation of Sprinkler Systems in One- and Two-Family Dwellings and Manufactured Homes*
NFPA 13E	*Recommended Practice for Fire Department Operations in Properties Protected by Sprinkler and Standpipe Systems*
NFPA 13R	*Standard for the Installation of Sprinkler Systems in Residential Occupancies up to and Including Four Stories in Height*
NFPA 14	*Standard for the Installation of Standpipe and Hose Systems*
NFPA 18	*Standard on Wetting Agents*
NFPA 291	*Recommended Practice for Fire Flow Testing and Marking of Hydrants*
NFPA 414	*Standard for Aircraft Rescue and Fire-Fighting Vehicles*
NFPA 1001	*Standard for Fire Fighter Professional Qualifications*
NFPA 1002	*Standard for Fire Apparatus Driver/Operator Professional Qualifications*
NFPA 1071	*Emergency Vehicle Technician Professional Qualifications*
NFPA 1081	*Standard for Industrial Fire Brigade Member Professional Qualifications*
NFPA 1142	*Standard on Water Supplies for Suburban and Rural Fire Fighting*
NFPA 1145	*Guide for the Use of Class A Foams in Manual Structural Fire Fighting*
NFPA 1150	*Standard on Foam Chemicals for Fires in Class A Fuels*
NFPA 1201	*Standard for Providing Emergency Services to the Public*
NFPA 1403	*Standard on Live Fire Training Evolutions*
NFPA 1410	*Standard on Training for Initial Emergency Scene Operations*
NFPA 1451	*Standard for a Fire Service Vehicle Operations Training Program*
NFPA 1500	*Standard on Fire Department Occupational Safety and Health Program*
NFPA 1561	*Standard on Emergency Services Incident Management System*
NFPA 1582	*Standard on Comprehensive Occupational Medical Program for Fire Departments*
NFPA 1901	*Standard for Automotive Fire Apparatus*
NFPA 1906	*Standard for Wildland Fire Apparatus*
NFPA 1911	*Standard for the Inspection, Maintenance, Testing, and Retirement of In-Service Automotive Fire Apparatus*

NFPA 1914 *Standard for Testing Fire Department Aerial Devices*

NFPA 1915 *Standard for Fire Apparatus Preventive Maintenance Program*

NFPA 1961 *Standard on Fire Hose*

NFPA 1962 *Standard for the Inspection, Care, and Use of Fire Hose, Couplings, and Nozzles and the Service Testing of Fire Hose*

NFPA 1963 *Standard for Fire Hose Connections*

NFPA 1964 *Standard for Spray Nozzles*

NFPA 1965 *Standard for Fire Hose Appliances*

AN EXCERPT: NFPA® 1002, STANDARD FOR FIRE APPARATUS DRIVER/OPERATOR PROFESSIONAL QUALIFICATIONS, 2009 EDITION, CHAPTERS 4 AND 5

CHAPTER 4 GENERAL REQUIREMENTS

4.1 General

Prior to operating fire department vehicles, the fire apparatus driver/operator shall meet the job performance requirements defined in Sections 4.2 and 4.3.

4.2 Preventive Maintenance

4.2.1*

Perform routine tests, inspections, and servicing functions on the systems and components specified in the following list, given a fire department vehicle, its manufacturer's specifications, and policies and procedures of the jurisdiction, so that the operational status of the vehicle is verified:

(1) Battery(ies)
(2) Braking system
(3) Coolant system
(4) Electrical system
(5) Fuel
(6) Hydraulic fluids
(7) Oil
(8) Tires
(9) Steering system
(10) Belts
(11) Tools, appliances, and equipment

(A) Requisite Knowledge.

Manufacturer specifications and requirements, policies, and procedures of the jurisdiction.

(B) Requisite Skills.

The ability to use hand tools, recognize system problems, and correct any deficiency noted according to policies and procedures.

4.2.2

Document the routine tests, inspections, and servicing functions, given maintenance and inspection forms, so that all items are checked for operation and deficiencies are reported.

(A) Requisite Knowledge.

Departmental requirements for documenting maintenance performed and the importance of keeping accurate records.

(B) Requisite Skills.

The ability to use tools and equipment and complete all related departmental forms.

4.3 Driving/Operating

4.3.1*

Operate a fire department vehicle, given a vehicle and a predetermined route on a public way that incorporates the maneuvers and features, specified in the following list, that the driver/operator is expected to encounter during normal operations, so that the vehicle is operated in compliance with all applicable state and local laws, departmental rules and regulations, and the requirements of NFPA 1500, Section 4.2:

(1) Four left turns and four right turns
(2) A straight section of urban business street or a two-lane rural road at least 1.6 km (1 mile) in length
(3) One through-intersection and two intersections where a stop has to be made
(4) One railroad crossing
(5) One curve, either left or right
(6) A section of limited-access highway that includes a conventional ramp entrance and exit and a section of road long enough to allow two lane changes
(7) A downgrade steep enough and long enough to require down-shifting and braking
(8) An upgrade steep enough and long enough to require gear changing to maintain speed
(9) One underpass or a low clearance or bridge

(A) Requisite Knowledge.

The effects on vehicle control of liquid surge, braking reaction time, and load factors; effects of high center of gravity on roll-over potential, general steering reactions, speed, and centrifugal force; applicable laws and regulations; principles of skid avoidance, night driving, shifting, and gear patterns; negotiating intersections, railroad crossings, and bridges; weight and height limitations for both roads and bridges; identification and operation of automotive gauges; and operational limits.

(B) Requisite Skills.

The ability to operate passenger restraint devices; maintain safe following distances; maintain control of the vehicle while accelerating, decelerating, and turning, given road, weather, and traffic conditions; operate under adverse environmental or driving surface conditions; and use automotive gauges and controls.

4.3.2*

Back a vehicle from a roadway into restricted spaces on both the right and left sides of the vehicle, given a fire department vehicle, a spotter, and restricted spaces 3.7 m (12 ft) in width, requiring 90-degree right-hand and left-hand turns from the roadway, so that the vehicle is parked within the restricted areas without having to stop and pull forward and without striking obstructions.

(A) Requisite Knowledge.

Vehicle dimensions, turning characteristics, spotter signaling, and principles of safe vehicle operation.

(B) Requisite Skills.

The ability to use mirrors and judge vehicle clearance.

4.3.3*

Maneuver a vehicle around obstructions on a roadway while moving forward and in reverse, given a fire department vehicle, a spotter for backing, and a roadway with obstructions, so that the vehicle is maneuvered through the obstructions without stopping to change the direction of travel and without striking the obstructions.

(A) Requisite Knowledge.

Vehicle dimensions, turning characteristics, the effects of liquid surge, spotter signaling, and principles of safe vehicle operation.

(B) Requisite Skills.

The ability to use mirrors and judge vehicle clearance.

4.3.4*

Turn a fire department vehicle 180 degrees within a confined space, given a fire department vehicle, a spotter for backing up, and an area in which the vehicle cannot perform a U-turn without stopping and backing up, so that the vehicle is turned 180 degrees without striking obstructions within the given space.

(A) Requisite Knowledge.

Vehicle dimensions, turning characteristics, the effects of liquid surge, spotter signaling, and principles of safe vehicle operation.

(B) Requisite Skills.

The ability to use mirrors and judge vehicle clearance.

4.3.5*

Maneuver a fire department vehicle in areas with restricted horizontal and vertical clearances, given a fire department vehicle and a course that requires the operator to move through areas of restricted horizontal and vertical clearances, so that the operator accurately judges the ability of the vehicle to pass through the openings and so that no obstructions are struck.

(A) Requisite Knowledge.

Vehicle dimensions, turning characteristics, the effects of liquid surge, spotter signaling, and principles of safe vehicle operation.

(B) Requisite Skills.

The ability to use mirrors and judge vehicle clearance.

4.3.6*

Operate a vehicle using defensive driving techniques under emergency conditions, given a fire department vehicle and emergency conditions, so that control of the vehicle is maintained.

(A) Requisite Knowledge.

The effects on vehicle control of liquid surge, braking reaction time, and load factors; the effects of high center of gravity on roll-over potential, general steering reactions, speed, and centrifugal force; applicable laws and regulations; principles of skid avoidance, night driving, shifting, gear patterns; and automatic braking systems in wet and dry conditions; negotiation of intersections, railroad crossings, and bridges; weight and height limitations for both roads and bridges; identification and operation of automotive gauges; and operational limits.

(B) Requisite Skills.

The ability to operate passenger restraint devices; maintain safe following distances; maintain control of the vehicle while accelerating, decelerating, and turning, given road, weather, and traffic conditions; operate under adverse environmental or driving surface conditions; and use automotive gauges and controls.

4.3.7*

Operate all fixed systems and equipment on the vehicle not specifically addressed elsewhere in this standard, given systems and equipment, manufacturer's specifications and instructions, and departmental policies and procedures for the systems and equipment, so that each system or piece of equipment is operated in accordance with the applicable instructions and policies.

(A) Requisite Knowledge.

Manufacturer's specifications and operating procedures, and policies and procedures of the jurisdiction.

(B) Requisite Skills.

The ability to deploy, energize, and monitor the system or equipment and to recognize and correct system problems.

CHAPTER 5 APPARATUS EQUIPPED WITH FIRE PUMP

5.1* General

The requirements of FireFighter 1 as specified in NFPA 1001 (or the requirements of Advanced Exterior Industrial Fire Brigade Member or Interior Structural Fire Brigade Member as specified in NFPA 1081) and the job performance requirements defined in Sections 5.1 and 5.2 shall be met prior to qualifying as a fire department driver/operator—pumper.

5.1.1

Perform the routine tests, inspections, and servicing functions specified in the following list in addition to those in 4.2.1, given a fire department pumper, its manufacturer's specifications, and policies and procedures of the jurisdiction, so that the operational status of the pumper is verified:

(1) Water tank and other extinguishing agent levels (if applicable)
(2) Pumping systems
(3) Foam systems

(A) Requisite Knowledge.

Manufacturer's specifications and requirements, and policies and procedures of the jurisdiction.

(B) Requisite Skills.

The ability to use hand tools, recognize system problems, and correct any deficiency noted according to policies and procedures.

5.2 Operations

5.2.1

Produce effective hand or master streams, given the sources specified in the following list, so that the pump is engaged, all pressure control and vehicle safety devices are set, the rated flow of the nozzle is achieved and maintained, and the apparatus is continuously monitored for potential problems:

(1) Internal tank

(2) *Pressurized source

(3) Static source

(4) Transfer from internal tank to external source

(A) Requisite Knowledge.

Hydraulic calculations for friction loss and flow using both written formulas and estimation methods, safe operation of the pump, problems related to small-diameter or dead-end mains, low-pressure and private water supply systems, hydrant coding systems, and reliability of static sources.

(B) Requisite Skills.

The ability to position a fire department pumper to operate at a fire hydrant and at a static water source, power transfer from vehicle engine to pump, draft, operate pumper pressure control systems, operate the volume/pressure transfer valve (multistage pumps only), operate auxiliary cooling systems, make the transition between internal and external water sources, and assemble hose lines, nozzles, valves, and appliances.

5.2.2

Pump a supply line of 65 mm (2½ in.) or larger, given a relay pumping evolution the length and size of the line and the desired flow and intake pressure, so that the correct pressure and flow are provided to the next pumper in the relay.

(A) Requisite Knowledge.

Hydraulic calculations for friction loss and flow using both written formulas and estimation methods, safe operation of the pump, problems related to small-diameter or dead-end mains, low-pressure and private water supply systems, hydrant coding systems, and reliability of static sources.

(B) Requisite Skills.

The ability to position a fire department pumper to operate at a fire hydrant and at a static water source, power transfer from vehicle engine to pump, draft, operate pumper pressure control systems, operate the volume/pressure transfer valve (multistage pumps only), operate auxiliary cooling systems, make the transition between internal and external water sources, and assemble hose lines, nozzles, valves, and appliances.

5.2.3

Produce a foam fire stream, given foam-producing equipment, so that properly proportioned foam is provided.

(A) Requisite Knowledge.

Proportioning rates and concentrations, equipment assembly procedures, foam system limitations, and manufacturer's specifications.

(B) Requisite Skills.

The ability to operate foam proportioning equipment and connect foam stream equipment.

5.2.4

Supply water to fire sprinkler and standpipe systems, given specific system information and a fire department pumper, so that water is supplied to the system at the correct volume and pressure.

(A) Requisite Knowledge.

Calculation of pump discharge pressure; hose layouts; location of fire department connection; alternative supply procedures if fire department connection is not usable; operating principles of sprinkler systems as defined in NFPA 13, NFPA 13D, and NFPA 13R; fire department operations in sprinklered properties as defined in NFPA 13E; and operating principles of standpipe systems as defined in NFPA 14.

(B) Requisite Skills.

The ability to position a fire department pumper to operate at a fire hydrant and at a static water source, power transfer from vehicle engine to pump, draft, operate pumper pressure control systems, operate the volume/pressure transfer valve (multistage pumps only), operate auxiliary cooling systems, make the transition between internal and external water sources, and assemble hose line, nozzles, valves, and appliances.

Appendix C

SAMPLE GENERIC STANDARD OPERATING PROCEDURES

FIRE DEPARTMENT NAME

Generic SOP

Safe Operation and Movement of Emergency Vehicles

PURPOSE: The purpose of this standard operating procedure is to establish responsibility and procedures for the safe operation and movement of vehicles assigned to the department. This standard is intended to meet the requirements of the National Fire Protection Association (NFPA) 1500, *Standard on Fire Department Occupational Safety and Health Program*. This policy does not imply a comprehensive list of items that could occur.

RESPONSIBILITIES: The chief, officers, and all drivers assigned to the department are responsible for ensuring the safe operation and movement of emergency vehicles.

PROCEDURES: All drivers must maintain a healthy vigilance to safety and awareness of the hazards associated with driving emergency vehicles. **The following procedures will be followed to ensure the safe operation and movement of the department vehicles:**

1. During emergency response, installed audible and visual warning devices shall be operated until arrival on-scene. Speed will be governed by weather and traffic conditions. Drivers are responsible for maintaining control of emergency vehicles at all times by operating emergency vehicles no faster than conditions permit.
2. During emergency response, drivers will bring the vehicle to a complete stop: when directed by a law enforcement officer, at red traffic light or stop sign (may proceed when safe), when encountering stopped school bus with flashing warning light (may proceed when school bus driver indicates it is safe), at unguarded railroad grade crossing (may proceed after verifying tracks are clear). Never, ever assume you have the right of way because you are using the emergency warning lights and sirens.
3. Operation of any department vehicle under other than emergency situations will be in accordance with existing state, county, and city traffic regulations. All personnel shall be seated with safety devices properly engaged prior to vehicle movement.
4. When backing any department vehicle, at least one spotter is required, two when personnel are available or conditions require. Spotters shall wear hearing protection devices when vehicles utilize loud backup warning devices. Vehicles will be backed at slow speeds. If eye contact is lost between the driver and spotter, the vehicle will be stopped immediately.
5. When parking a vehicle with the engine running, the operator will ensure the emergency brake is set and chocks are properly placed at the wheel. When parking the vehicle with the engine off, place the vehicle in gear for standard transmissions or in park for automatic transmissions and set the emergency brake.

FIRE DEPARTMENT NAME

Generic SOP

Securing Tools and Equipment

PURPOSE: The purpose of this standard operating procedure is to establish responsibility and procedures for securing tools and equipment on department vehicles. This standard is intended to meet the requirements of the National Fire Protection Association (NFPA) 1500, *Standard on Fire Department Occupational Safety and Health Program*. This policy does not imply a comprehensive list of items that could occur.

RESPONSIBILITIES: All firefighters are responsible for compliance with the procedures as outlined in this operating instruction.

PROCEDURES: The following procedures shall be followed to ensure that tools and equipment are properly secured:

1. All hand tools and equipment will be secured in a compartment or appropriate location on each vehicle. All compartment doors will be completely closed and/or latched prior to any vehicle being moved.
2. A walk-around of the vehicle will be completed after each vehicle checkout and before returning from a response to ensure that all compartments are properly latched.
3. All exterior mounted equipment will be secured to the vehicles by safety straps or latches. Breathing apparatus and spare air bottles will be secured by safety straps.
4. All personnel will ensure that ground ladders are securely mounted in the brackets provided on the vehicles.
5. All vehicle operators will make a check of all safety straps during vehicle checkout and when returning from a response. If straps are not serviceable, immediately notify the [officer].

FIRE DEPARTMENT NAME

Generic SOP

Emergency Vehicles Inspections

PURPOSE: The purpose of this standard operating procedure is to establish responsibility and procedures for the inspection of vehicles assigned to the department for the safety and health of members and other persons involved in those activities. This standard is intended to meet the requirements of the National Fire Protection Association (NFPA) 1500, *Standard on Fire Department Occupational Safety and Health Program*. This policy does not imply a comprehensive list of items that could occur.

RESPONSIBILITIES: The [officer] and all firefighters assigned to the department are responsible for ensuring that the inspection of all fire equipment is performed on at least a weekly basis, within 24 hours after use, and when any major repair and/or modification is made.

PROCEDURES: The [officer] and the fire chief are required to ensure that vehicles and equipment of the department are in full operational readiness. This requirement necessitates constant vigilance to detail and prompt attention to the effectiveness of the inspection program.

1. Preventive maintenance inspections will be performed in accordance with the vehicle inspection form (checklist) for each department vehicle. Discrepancies found during the inspection shall be documented and reported immediately to the [officer]. These forms shall be maintained as documentation for open discrepancies (items not corrected) as well as to document closed discrepancies (items corrected).
2. The [officer] will ensure that all vehicle inspection forms are properly maintained and that on the first of each month open entries from previous months are transcribed to the new monthly form. The [officer] will routinely review vehicle inspection forms to determine trends and preventive maintenance intervention, if necessary.
3. Only qualified individuals shall perform vehicle inspections and document inspection results. Safety will be maintained as a priority during the inspection process.

SAMPLES OF STATE LAWS FOR EMERGENCY VEHICLE DRIVERS

CALIFORNIA CODES

Vehicle Code

Division 1 **Words and Phrases Defined** (Sections 100–680)

100. Unless the provision or context otherwise requires, these definitions shall govern the construction of this code.

165. An authorized emergency vehicle is:

(a) Any publicly owned and operated ambulance, lifeguard, or lifesaving equipment or any privately owned or operated ambulance licensed by the Commissioner of the California Highway Patrol to operate in response to emergency calls.

(b) Any publicly owned vehicle operated by the following persons, agencies, or organizations:

(1) Any federal, state, or local agency, department, or district employing peace officers as that term is defined in Chapter 4.5 (commencing with Section 830) of Part 2 of Title 3 of the Penal Code, for use by those officers in the performance of their duties.

(2) Any forestry or fire department of any public agency or fire department organized as provided in the Health and Safety Code.

(c) Any vehicle owned by the state, or any bridge and highway district, and equipped and used either for fighting fires, or towing or servicing other vehicles, caring for injured persons, or repairing damaged lighting or electrical equipment.

(d) Any state-owned vehicle used in responding to emergency fire, rescue or communications calls and operated either by the Office of Emergency Services or by any public agency or industrial fire department to which the Office of Emergency Services has assigned the vehicle.

(e) Any vehicle owned or operated by any department or agency of the United States government when the vehicle is used in responding to emergency fire, ambulance, or lifesaving calls or is actively engaged in law enforcement work.

(f) Any vehicle for which an authorized emergency vehicle permit has been issued by the Commissioner of the California Highway Patrol.

165.5. No act or omission of any rescue team operating in conjunction with an authorized emergency vehicle as defined in Section 165, while attempting to resuscitate any person who is in immediate danger of loss of life, shall impose any liability upon the rescue team or the owners or operators of any authorized emergency vehicle, if good faith is exercised. For the purposes of this section, "rescue team" means a special group of physicians and surgeons, nurses, volunteers, or employees of the owners or operators of the authorized emergency vehicle who have been trained in cardiopulmonary resuscitation and have been designated by the owners or operators of the emergency vehicle to attempt to resuscitate persons who are in immediate danger of loss of life in cases of emergency.

This section shall not relieve the owners or operators of any other duty imposed upon them by law for the designation and training of members of a rescue team or for any provisions regarding maintenance of equipment to be used by the rescue team.

Members of a rescue team shall receive such training in a program approved by, or conforming to, standards prescribed by an emergency medical care committee established pursuant to Article 3 (commencing with Section 1797.270) of Chapter 4 of Division 2.5 of the Health and Safety Code, or a voluntary area health planning agency established pursuant to Section 437.7 of the Health and Safety Code.

676.5. A "water tender vehicle" is a vehicle designed to carry not less than 1,500 gallons of water and used primarily for transporting and delivering water to be applied by other vehicles or pumping equipment at fire emergency scenes.

Division 11 **Rules of the Road** (Sections 21050–21062)
Chapter 1 Obedience to and Effect of Traffic Laws
Article 2 Effects of Traffic Laws

21055. The driver of an authorized emergency vehicle is exempt from:

Chapter 2 Traffic Signs, Signals, and Markings (commencing with Section 21350),

Chapter 3 Driving, Overtaking, and Passing (commencing with Section 21650),

Chapter 4 Right-of-Way (commencing with Section 21800),

Chapter 5 Pedestrians' Rights and Duties (commencing with Section 21950),

Chapter 6 Turning and Stopping and Turning Signals (commencing with 22100),

Chapter 7 Speed Laws (commencing with Section 22348),

Chapter 8 Special Stops Required (commencing with Section 22450),

Chapter 9 Stopping, Standing, and Parking (commencing with Section 22500), and

Chapter 10 Removal of Parked and Abandoned Vehicles (commencing with Section 22650) of this division, and

Division 16.5 Off-Highway Vehicles, Chapter 5 Off-Highway Vehicle Operating Rules, Article 3 Speed Laws (commencing with Section 38305) and Article 4 Turning and Starting (commencing with Section 38312), under all of the following conditions:

(a) If the vehicle is being driven in response to an emergency call or while engaged in rescue operations or is being used in the immediate pursuit of an actual or suspected violator of the law or is responding to, but not returning from, a fire alarm, except that fire department vehicles are exempt whether directly responding to an emergency call or operated from one place to another as rendered desirable or necessary by reason of an emergency call and operated to the scene of the emergency or operated from one fire station to another or to some other location by reason of the emergency call.

(b) If the driver of the vehicle sounds a siren as may be reasonably necessary and the vehicle displays a lighted red lamp visible from the front as a warning to other drivers and pedestrians.

A siren shall not be sounded by an authorized emergency vehicle except when required under this section.

21056. Section 21055 does not relieve the driver of a vehicle from the duty to drive with due regard for the safety of all persons using the highway, nor protect him from the consequences of an arbitrary exercise of the privileges granted in that section.

MAINE REVISED STATUTES ANNOTATED

Title 29-A. Motor Vehicles

Chapter 19. Operation, Subchapter I. Rules of the Road

§2054. Emergency and auxiliary lights; sirens: privileges

1. Definitions. As used in this section, unless the context otherwise indicates, the following terms have the following meanings.

A. "Ambulance" means any vehicle designed, constructed and routinely used or intended to be used for the transportation of ill or injured persons and licensed by Maine Emergency Medical Services pursuant to Title 32, chapter 2-B.

B. "Authorized emergency vehicle" means any one of the following vehicles:

(1) An ambulance;

(5) A Department of Conservation vehicle used for forest fire control;

(9) An emergency medical service vehicle;

(10) A fire department vehicle;

(11) A hazardous material response vehicle;

(16) A vehicle operated by a municipal fire inspector, a municipal fire chief, an assistant or deputy chief or a town forest fire warden;

D. "Emergency light" means a light, other than standard equipment lighting such as headlights, taillights, directional signals, brake lights, clearance lights, parking lights and license plate lights, that is displayed on a vehicle and used to increase the operator's visibility of the road or the visibility of the vehicle to other operators and pedestrians.

F. "Fire vehicle" means any vehicle listed under paragraph B, subparagraph (5) or (16).

G. "Hazardous material response vehicle" means a vehicle equipped for and used in response to reports of emergencies resulting from actual or potential releases, spills or leaks of, or other exposure to, hazardous substances that is authorized by a mutual aid agreement pursuant to Title 37-B, section 795, subsection 3 and approved by the local emergency planning committee or committees whose jurisdiction includes the area in which the vehicle operates.

2. Authorized lights. Authorized lights are governed as follows.

A. Only an ambulance, an emergency medical service vehicle, a fire department vehicle ... may be equipped with a device that provides for alternate flashing of the vehicle's headlights.

F. Only vehicles listed in this paragraph ... may be equipped with, display or use a red auxiliary or emergency light.

(1) Emergency lights used on an ambulance, and emergency medical service vehicle, a fire department vehicle, a fire vehicle or a hazardous material response vehicle must emit a red light or a combination of red and white lights.

(2) The municipal officers or a municipal official designated by the municipal officers, with the approval of the fire chief, may authorize an active member of a municipal or volunteer fire department to use a flashing red signal light not more than 5 inches in diameter on a vehicle. The light may be displayed but may be used only while the member is en route to or at the scene of a fire or other emergency. The light must be mounted as near as practicable above the registration plate on the front of the vehicle or on the dashboard. A light mounted on the dashboard must be shielded so that the emitted light does not interfere with the operator's vision.

(3) Members of an emergency medical service licensed by Maine Emergency Medical Services may display and use on a vehicle a flashing red signal light of the same proportion, in the same location and under the same conditions as those permitted municipal and volunteer firefighters, when authorized by the chief official of the emergency medical service.

3. Sirens. A bell or siren may not be installed or used on any vehicle, except an authorized emergency vehicle.

4. Right-of-way. An authorized emergency vehicle operated in response to, but not returning from, a call or fire alarm... has the right-of-way when emitting a visual signal using an emergency light and an audible signal using a bell or siren.

5. Exercise of privileges. The operator of an authorized emergency vehicle when responding to, but not upon returning from, an emergency call or fire alarm...may exercise the privileges set forth in this subsection. The operator of an authorized emergency vehicle may:

A. Park or stand, notwithstanding the provisions of this chapter;

B. Proceed past a red signal, stop signal or stop sign, but only after slowing down as necessary for safe operation;

C. Exceed the maximum speed limits as long as life or property is not endangered, except that employees of the Department of Corrections may not exercise this privilege;

D. Disregard regulations governing direction of movement or turning in specified directions; and

E. Proceed with caution past a stopped school bus that has red lights flashing only:
   (1) After coming to a complete stop; and
   (2) When signaled by the school bus operator to proceed.

6. **Emergency lights and audible signals.** The operator of an authorized emergency vehicle who is exercising the privileges granted under subsection 5 shall use an emergency light authorized by subsection 2. The operator of an authorized emergency vehicle who is exercising the privileges granted under subsection 5, paragraphs B, C, D, and E shall sound a bell or siren when reasonably necessary to warn pedestrians and other operators of the emergency vehicle's approach.

7. **Duty to drive with due regard for safety.** Subsections 4, 5, and 6 do not relieve the operator of an authorized emergency vehicle from the duty to drive with due regard for the safety of all persons, nor do those subsections protect the operator from the consequences of the operator's reckless disregard for the safety of others.

MINNESOTA STATUTES

Transportation 160–174A

169 Traffic Regulations

169.01 **Definitions**

Subdivision 1. Terms.

For the purposes of this chapter, the terms defined in this section shall have the meanings ascribed to them.

Subdivision 5. Authorized emergency vehicle.

"Authorized emergency vehicle" means any of the following vehicles when equipped and identified according to law:

(1) a vehicle of a fire department;

(2) a publicly owned police vehicle or a privately owned vehicle used by a police officer for police work under agreement, express or implied, with the local authority to which the officer is responsible;

(3) a vehicle of a licensed land emergency ambulance service, whether publicly or privately owned;

(4) an emergency vehicle of a municipal department or a public service corporation, approved by the commissioner of public safety or the chief of police of a municipality;

(5) any volunteer rescue squad operating pursuant to Law 1959, chapter 53;

(6) a vehicle designated as an authorized emergency vehicle upon a finding by the commissioner of public safety that designation of that vehicle is necessary to the preservation of life or property or to the execution of emergency governmental functions.

169.03 **Emergency vehicles; exemptions; application.**

Subdivision 1. Scope.

The provisions of this chapter applicable to the drivers of vehicles upon the highways shall apply to the drivers of all vehicles owned or operated by the United States, this state, or any county, city, town, district, or any other political subdivision of the state, subject to such specific exemptions as are set forth in this chapter with reference to authorized emergency vehicles.

Subdivision 2. Stops.

The driver of any authorized emergency vehicle, when responding to an emergency call, upon approaching a red or stop signal or any stop sign shall slow down as necessary for safety, but may proceed cautiously past such red or stop sign or signal after sounding siren and displaying red lights.

Subdivision 3. One-way roadways.

The driver of any authorized emergency vehicle, when responding to any emergency call, may enter against the run of traffic on any one-way street, or highway where there is authorized division of traffic, to facilitate traveling to the area in which an emergency has been reported; and the provisions of this section shall not affect any cause of action arising prior to its passage.

Subdivision 4. Parking at emergency scene.

An authorized emergency vehicle, when at the scene of a reported emergency, may park or stand, notwithstanding any law or ordinance to the contrary.

Subdivision 5. Course of duty.

No driver of any authorized emergency vehicle shall assume any special privilege under this chapter except when such vehicle is operated in response to any emergency call or in the immediate pursuit of an actual or suspected violator of the law.

169.17 **Emergency vehicles.**

The speed limitations set forth in sections 169.14 to 169.17 do not apply to authorized emergency vehicles when responding to emergency calls, but the drivers thereof shall sound audible signal by siren and display at least one lighted red light to the front. This provision does not relieve the driver of an authorized emergency vehicle from the duty to drive with due regard for the safety of persons using the street, nor does it protect the driver of an authorized emergency vehicle from the consequence of a reckless disregard of the safety of others.

Appendix E

SAMPLE PUMP MANUFACTURERS' TROUBLESHOOTING GUIDES

Condition	Possible Cause	Suggested Remedy
Pump fails to prime or loses prime	Air leaks	Clean and tighten all intake connections. Make sure intake hoses and gaskets are in good condition. Use the following procedure to locate air leaks: **1.** Connect intake hose to pump and attach intake cap to end of hose. **2.** Close all pump openings. **3.** Open priming valve and operate primer until vacuum gauge indicates 22 in. Hg. (If primer fails to draw specified vacuum, it may be defective, or leaks are too large for primer to handle.) **4.** Close priming valve and shut off primer. If vacuum drops more than 10 in. Hg in 5 minutes, serious air leaks are indicated. With engine stopped, air leaks are frequently audible. If leaks cannot be heard, apply engine oil to suspected points and watch for break in film or oil being drawn into pump. Completely fill water tank (if so equipped). Connect intake hose to hydrant or auxiliary pump. Open one discharge valve and run in water until pump is completely filled and all air is expelled. Close discharge valve, apply pressure to system, and watch for leaks or overflowing water tank. A pressure of 100 psi is sufficient. DO NOT EXCEED RECOMMENDED PRESSURE If pump has not been operated for several weeks, packing may be dried out. Close discharge and drain valves and cap intake openings. Operate primer to build up a strong vacuum in pump. Run pump slowly and apply oil to impeller shaft near packing gland. Make sure packing is adjusted properly.

(Continues)

Condition	Possible Cause	Suggested Remedy
	Dirt on intake strainer	Remove all leaves, dirt, and other foreign material from intake strainer. When drafting from shallow water source with mud, sand, or gravel bottom, protect intake strainer in one of the following ways: **1.** Suspend intake strainer from a log or other floating object to keep it off the bottom. Anchor float to prevent it from drifting into shallow water. **2.** Remove top from a clean barrel. Sink barrel so open end is below water surface. Place intake strainer inside barrel. **3.** Make an intake box, using fine mesh screen. Suspend intake strainer inside box.
	No oil in priming tank	With rotary primer, oil is required to maintain a tight rotor seal. Check priming tank oil supply and replenish, if necessary.
	Defective priming system	A worn or damaged priming valve may leak and cause pump to lose prime. Consult primer instructions for priming valve repair.
	Improper clearance in rotary gear or vane primer	After prolonged service, wear may increase primer clearance and reduce efficiency. Refer to primer instructions for adjusting primer clearance.
	Engine speed too low	Refer to instructions supplied with primer for correct priming speeds. Speeds much higher than those recommended do not accelerate priming, and may actually damage priming pump.
	Bypass line open	If a bypass line is installed between the pump discharge and water tank to prevent pump from overheating with all discharge valves closed, look for a check valve in the line. If valve is stuck open, clean it, replace it, or temporarily block off line until a new valve can be obtained.
	Lift too high	Do not attempt lifts exceeding 22 feet except at low altitudes and with equipment in new condition.
	End of intake hose not submerged deep enough	Although intake hose might be immersed enough for priming, pumping large volumes of water may produce whirlpools, which will allow air to be drawn into intake hose. Whenever possible, place end of intake hose at least 2 feet below water source.
	High point in intake line	If possible, avoid placing any part of intake hose higher than pump inlet. If high point cannot be prevented, close discharge valve as soon as pressure drops, and prime again. This procedure will usually eliminate air pockets in intake line, but it may have to be repeated several times.
	Primer not operated long enough	Refer to instructions supplied with primer for required priming time. The maximum time for priming should not exceed 45 seconds for lifts up to 10 feet.
Insufficient capacity A. Engine and pump speed too low at full throttle	Insufficient engine power	Engine requires maintenance. Check engine in accordance with manufacturer's instructions supplied with truck. Engine operated at high altitudes and/or high air temperatures. Engine power decreases with an increase in altitude or air temperature, except for turbo-charged engines. Adjusting carburetor or changing carburetor jets (or injector nozzles) may improve engine performance. Consult with engine manufacturer.

Condition	Possible Cause	Suggested Remedy
	Discharge relief valve set improperly	If relief valve is set to relieve below desired operating pressure, water will bypass and reduce capacity. Adjust relief valve in accordance with instructions supplied with valve.
Insufficient capacity A. Engine and pump speed too low at full throttle (continued)	Transfer valve set improperly (Does not apply to single-stage pumps.)	Place transfer valve in VOLUME (parallel) position when pumping more than two-thirds rated capacity. When shifting transfer valve, make sure it travels all the way into new position. Failure of transfer valve to move completely into new position will seriously impair pump efficiency.
	Truck transmission in too high a gear	Consult vehicle instructions for correct pump gear. Pump usually works best with transmission in direct drive. If truck is equipped with an automatic transmission, be sure transmission is in pumping gear.
Insufficient capacity B. Engine and pump speed higher than specified for desired pressure and volume	Transfer valve set improperly (Does not apply to single-stage pumps.)	Place transfer valve in VOLUME (parallel) position when pumping more than two-thirds rated capacity. When shifting transfer valve, make sure it travels all the way into new position. Failure of transfer valve to move completely into new position will seriously impair pump efficiency.
	Pump impeller(s) or wear rings badly worn	Install undersize wear rings if impeller-to-wear-ring clearance is within limits indicated in MAINTENANCE INSTRUCTIONS. If not, install new impeller(s) and wear rings.
	Intake strainer, intake screens, or impeller vanes fouled with debris	Remove intake strainer and hose, and clear away all debris. Pressure backwash (preferably in parallel or "volume" position) will usually clear impeller vanes when pump is stopped.
	Intake hose defective	On old intake hoses, the inner liner sometimes becomes so rough it causes enough friction loss to prevent pump from drawing capacity. Sometimes, the liner will separate from the outer wall and collapse when drafting. It is usually impossible to detect liner collapse, even with a light. Try drafting with a new intake hose; if pump then delivers capacity, it may be assumed that previous hose was defective.
	Intake hose too small	When pumping at higher-than-normal lifts, or at high altitudes, use a larger or additional intake hoses.
Insufficient capacity C. Engine speed higher than specified for desired pressure and volume	Truck transmission in too low a gear	Consult vehicle instructions for correct pumping gear. Pump usually works best with transmission in direct drive. (Check both engine and pump speed, if possible, to be sure transmission is in "direct.")
Insufficient pressure	Pump speed too low	In general, the above causes and remedies for low pump capacity will also apply to low pump pressure. Check pump speed with a tachometer. If pump speed is too low, refer to engine manufacturer's instructions for method of adjusting engine speed governor.
	Pump capacity limits pump pressure	Do not attempt to pump greater volume of water at the desired pressure than the pump is designed to handle. Exceeding pump capacity may cause a reduction in pressure. Exceeding maximum recommended pump speed will produce cavitation, and will seriously impair pump efficiency.

(Continues)

Condition	Possible Cause	Suggested Remedy
	Flap valve stuck open	When pump is in PRESSURE (series), discharge will bypass to first stage intake. Operate pump at 75 psi pressure, and rapidly switch transfer valve back and forth between positions. If this fails, try to reach valve with a stick or wire and work it free.
Relief valve malfunction A. Pressure not relieved when discharge valves are closed	Sticky pilot valve	Disassemble and clean. Replace noticeably worn parts.
	Plugged tube lines	Disconnect lines and inspect.
Relief valve malfunction B. Pressure will not return to original setting after dis-charge valves are reopened	Sticky pilot valve	Disassemble and clean. Replace noticeably worn parts.
	Sticky main valve	Disassemble and clean. Replace noticeably worn parts.
	Incorrect installation	Check all lines to be sure installation instructions have been followed.
Relief valve malfunction C. Fluctuating pressure	Sticky pilot valve	Disassemble and clean. Replace noticeably worn parts.
	Water surges (relief valve)	Pressure fluctuation can result from a combination of intake and dis-charge conditions involving the pump, relief valve, and engine. When the elasticity of the intake and discharge system and the response rate (reaction time) of the engine, pilot valve, and relief valve are such that the system never stabilizes, fluctuation results. With the proper combi-nation of circumstances, fluctuation can occur regardless of the make or type of equipment involved. Changing one or more of these factors enough to disrupt this timing should eliminate fluctuation.
Relief valve malfunction D. Slow response	Plugged filter or line	Clean lines and filter.

4. CORRECTIVE MAINTENANCE

Trouble Analysis

Table 4-1 lists the symptoms of some common problems and possible corrective measures. Before calling Hale for assistance, eliminate problem causes using **Table 4-1**. If you cannot correct a problem, before calling Hale Customer Service (215/825-6300) for assistance, please have the following information ready:

- Pump model and serial number
- Pump configuration information
- Observed symptoms and under what circumstances the symptoms occur

TABLE 4-1 Hale Midship Pump Trouble Analysis

Condition	Possible Cause	Suggested Correction
PUMP WILL NOT ENGAGE		
Standard Transmission with Manual Pump Shift	Clutch not fully disengaged or malfunction in shift linkage	Check clutch disengagement. Drive shaft must come to a complete stop before attempting pump shift.
Automatic Transmission with Manual Pump Shift	Automatic transmission not in neutral position	Repeat recommended shift procedures with transmission in neutral position.
Standard Transmission with Power Shift System	Insufficient air or vacuum supply to shift system	Repeat recommended shift procedures.
		Check system for loss of vacuum or air supply. If an inadequate supply exists, perform the following emergency pump shift procedures after bringing the apparatus to a complete stop, applying the parking brake, setting the transmission to neutral and chocking the wheels. • Turn the engine off. • Place the in-cab shift control in neutral. (Neutral is exactly in the middle of the "road" and "pump" position.) • A hole is provided in the shifting shaft. Insert a screwdriver or punch in the hole to accomplish emergency shifting. • Complete an emergency shift by manually pulling the shaft for "pump" gear or pushing the shaft for "road" gear. • If the shift cannot be completed manually, turn the driveshaft slightly by hand to align the internal gears and repeat the preceding step. Note: Some apparatus offer a manual shift handle or separate cable for activation.
Automatic Transmission with Power Shift System	Automatic transmission not in neutral position	Repeat recommended shift procedures with transmission in neutral position.
	Pump shift attempted before vehicle was completely stopped	Release braking system momentarily. Then reset and repeat recommended shifting procedures.
	Premature application of parking brake system (before truck comes to a complete stop)	Release braking system momentarily. Then reset and repeat recommended shifting procedures.

(Continues)

TABLE 4-1 Hale Midship Pump Trouble Analysis *(continued)*

Condition	Possible Cause	Suggested Correction
	Insufficient air or vacuum in shift system	Repeat recommended shift procedures. Check system for loss of air or vacuum. Check for leak in system. Employ manual override procedures if necessary. See Standard Transmission with Power Shift System.
	Air or vacuum leaks in shift system	Attempt to locate and repair leak(s). Leakage, if external, may be detected audibly. Leakage could be internal and not as easily detected.
PUMP LOSES PRIME OR IT WILL NOT PRIME	No lubricant in priming lubricant tank	Fill priming lubricant tank with Hale-approved lubricant.
	Electric priming system	No recommended engine speed is required to operate the electric primer; however, 1000 engine RPM will maintain truck electrical system while providing enough speed for initial pumping operation.
NOTE: Weekly priming pump operation is recommended to promote good operation.	Defective priming system	Check priming system by performing dry vacuum test per NFPA standards. If pump is tight, but primer pulls less than 22 inches of vacuum, it could indicate excessive wear in the primer.
	Defective priming valve (electric)	Replace the sealing rings if defective. Lubricate the rings. Priming valve shirk open will allow loss of prime. Also, it will permit unnecessary running of electric priming motor. Ensure complete priming valve closure; dismantle and lubricate if necessary.
	Suction lift too high	Do not attempt lifts exceeding 22 feet except at low elevations.
	Blocked suction strainer	Remove obstruction from suction hose strainer. Do not allow suction hose and strainer to rest on bottom of water supply.
	Suction connections	Clean and tighten all suction connections. Check suction hose and hose gaskets for possible defects.
	Primer not operated long enough	Proper priming procedures should be followed. Do not release the primer control before assuring a complete prime. Open the discharge valve slowly during completion of prime to ensure same. NOTICE: Do not run the primer over 45 seconds. If prime is not achieved in 45 seconds, stop and look for causes (for example, air leaks or blocked suction).
	Air trap in suction line	Avoid placing any part of the suction hose higher than the suction intake. Suction hose should be laid with continuous decline to water supply. If trap in hose is unavoidable, repeated priming may be necessary to eliminate air pocket in suction hose.
	Pump pressure too low when nozzle is opened	Prime the pump again, and maintain higher pump pressure while opening discharge valve slowly.

Condition	Possible Cause	Suggested Correction
	Air leaks	Attempt to locate and correct air leaks using the following procedure. **1.** Perform dry vacuum test on pump per NFPA standards with 22 inches minimum vacuum required with loss not to exceed 10 inches of vacuum in 5 minutes. **2.** If a minimum of 22 inches of vacuum cannot be achieved, the priming device or system may be defective, or the leak is too big for the primer to overcome (such as an open valve). The loss of vacuum indicates leakage and could prevent priming or cause loss of prime. **3.** Attempt above dry prime and shut engine off. Audible detection of a leak is often possible. **4.** Connect the suction hose from the hydrant or the discharge of another pumper to pressurize the pump with water, look for visible leakage, and correct. A pressure of 100 PSI should be sufficient. Do not exceed pressure limitations of pump, pump accessories, or piping connections. **5.** Check pump packing during attempt to locate leakage. If leakage is in excess of recommendations, adjust accordingly, following instructions in Section 3. **6.** The suction side relief valves can leak. Plug the valve outlet connection, and re-test.
INSUFFICIENT PUMP CAPACITY	Insufficient engine power	Engine power check or tune up may be required for peak engine and pump performance.
	Transfer valve not in proper "Volume" position	Two-stage pumps only. Place transfer valve in "Volume" position (parallel) when pumping more than two-thirds rated capacity. For pressure above 200 PSI, pump should be placed in "Pressure" (series) position.
	Relief valve improperly set	If relief valve control is set for too low a pressure, it will allow relief valve to open and bypass water. Reset relief valve control per the procedures in Section 3. Other bypass lines (such as foam system or inline valves) may reduce pump capacity or pressure.
	Engine governor set incorrectly	Engine governor, if set for too low a pressure when on automatic, will decelerate engine speed before desired pressure is achieved. Reset the governor per the manufacturer's procedures.
	Truck transmission in wrong gear or clutch is slipping	Recheck the pumping procedure for the recommended transmission or gear range; see Section 3 for assistance. Use mechanical speed counter on the pump panel to check speed against possible clutch or transmission slipping or inaccurate tachometer. (Check the truck manual for the proper speed counter ratio.)
	Air leaks	See Air leaks under "PUMP LOSES PRIME OR IT WILL NOT PRIME"
INSUFFICIENT PRESSURE	Check similar causes for insufficient capacity	Recheck pumping procedures for recommended transmission gear or range. Use mechanical speed

(Continues)

TABLE 4-1 Hale Midship Pump Trouble Analysis *(continued)*

Condition	Possible Cause	Suggested Correction
		counter on pump panel to check actual speed against possible clutch or transmission slippage or inaccurate tachometer. (Check the truck manual for proper speed counter ratio.)
	Transfer valve not in "Pressure" position	Two-stage pumps only. For desired pump pressure above 200 PSI, transfer valve should be in "Pressure" position.
ENGINE SPEEDS TOO HIGH FOR REQUIRED CAPACITY OR PRESSURE	Impeller blockage	Blockage in the impeller can prevent loss of both capacity and pressure. Back-flushing of pump from discharge to suction may free blockage. Removal of one-half of the pump body may be required; this is considered a major repair.
	Worn pump impeller(s) and clearance rings	Installation of new parts required.
	Blockage of suction hose entry	Clean suction hose strainer of obstruction, and follow recommended practices for laying suction hose–keep off the bottom of the water supply but at least 2 feet below the surface of the water.
	Defective suction hose	Inner line of suction hose may collapse when drafting and is usually undetectable. Try a different suction hose on same pump test mode for comparison against original hose and results.
	Lift too high, suction hose too small	Higher-than-normal lift (10 feet) will cause higher engine speeds, high vacuum, and rough operation. Larger suction will assist above condition.
	Truck transmission in wrong range or gear	Check recommended procedures for correct transmission selection; see Section 3 and truck manual.
RELIEF VALVE DOES NOT RELIEVE PRESSURE WHEN VALVES ARE CLOSED	Incorrect setting of control (pilot) valve	Check and repeat proper procedures for setting relief valve system; see Section 3.
	Relief valve inoperative	Possibly in need of lubrication. Remove relief valve from pump; dismantle; clean and lubricate. Weekly use of the relief valve is recommended.
RELIEF VALVE DOES NOT RECOVER AND RETURN TO ORIGINAL PRESSURE SETTING AFTER OPENING VALVES	Dirt in system causing sticky or slow reaction	Relief valve dirty or sticky. Follow instructions for disassembling, cleaning, and lubricating the relief valve. Blocked relief valve. Clean the valve with a small wire or straightened paper clip.
RELIEF VALVE OPENS WHEN CONTROL VALVE IS LOCKED OUT	Drain hole in housing, piston, or sensing valve blocked	Clean the hole with a small wire or straightened paper clip. Dismantle and clean the sensing valve.
UNABLE TO ATTAIN PROPER SETTING ON RELIEF VALVE	Wrong procedures	Check instructions for setting the relief valve; reset the valve.
	Blocked strainer	Check and clean the strainer in the supply line from the pump discharge to the control valve. Check the truck manual for the exact location. Check and clean tubing lines related to the relief valve and control valve.
	Foreign matter in the control valve	Remove the control valve and clean it.

Condition	Possible Cause	Suggested Correction
	Hunting condition	Insufficient water supply coming from the pump to the control valve. Check the strainer in the relief valve system. Foreign matter in the control valve. Remove the control valve, and clean it.
LEAK AT PUMP PACKING	Packing out of adjustment or worn	Adjust the pump packing per the procedure in Section 3 of this manual (8 to 10 drops per minute leakage preferred). Replace pump packing per Section 3 of this manual. Packing replacement is recommended every 2 or 3 years, depending on usage.
WATER IN PUMP GEARBOX	Leak coming from above pump	Check all piping connections and tank overflow for possible spillage falling directly on the pump gearbox. Follow the procedures in Section 3 of this manual for adjustment or replacement of packing. Excess packing leakage permits the flushing of water over the gearbox casing to the input shaft area. Induction of this excessive water may occur through the oil seal or speedometer connection. Inspect the oil seal; replace it if necessary.
DISCHARGE VALVES DIFFICULT TO OPERATE	Lack of lubrication	Recommended weekly lubrication of discharge and suction valve. Use a good grade of petroleum-based or silicon grease.
	Valve in need of more clearance	Add gasket to the valve cover (per the truck manual). Multigasket design allows additional gaskets for more clearance and free operation. NOTE: Addition of too many gaskets to the valve will permit leakage.
REMOTE CONTROL DIFFICULT TO OPERATE	Lack of lubrication	Lubricate the remote control linkages and collar with oil.

Appendix F

SAMPLE VEHICLE INSPECTION FORMS

NASHVILLE FIRE DEPARTMENT
DAILY APPARATUS REPORT

Caution: Any Engine noise, overheating, malfunction of brakes or steering, check out of service and notify shop immediately. **ON 24 HOUR CALL.**

09 ____________ Company ______________ Make ______________ Month ____________

Beginning of Month Mileage: ______________ Beginning Engine Hours: ______________

Date Last Oil Change ______________ Mileage Last Oil Change ______________

DAILY CHECKLIST

TO BE CHECKED DAILY ______	Power Tools ______
Oil ______	Air Pressure ______
Coolant Level ______	Air - Horn - Siren ______
Fan and Alternator Belts or All Belts ______	Gauges - Dash Lights ______
Fuel ______	Pump Panel Gauges and Light ______
Transmission Fluid ______	Valves - Handles ______
Power Steering Fluid ______	Leaks - Oil - Fuel ______
Batteries - Clamps ______	Unusual Noises ______
Tires - Lug Nuts ______	Generator ______
Wipers - Blades ______	Hydraulic ______
Windshield - Windows ______	Ladder Rungs ______
Doors - Handles ______	Cables ______
Headlights - Turn Signals ______	Intercom ______
Warning Lights - Spots ______	Fly Nozzle ______
Seat Belts ______	Outriggers ______

Engineers/Relief Drivers responsible for checking the above items. Sign date Below.

CHECKED BY

Day	Driver	Day	Driver	Day	Driver	Day	Driver
1		9		17		25	
2		10		18		26	
3		11		19		27	
4		12		20		28	
5		13		21		29	
6		14		22		30	
7		15		23		31	
8		16		24			

MONTHLY REPORT

		HOSE LAID OUT	
Actual Fire Runs	______		
1st Responder Emt Runs	______	5" Hose Used	______ Feet
Special Duty Runs	______	3" Hose Used	______ Feet
Miscellaneous Runs	______	2 ½" Hose Used	______ Feet
TOTAL ALL RUNS	______	1 ¾" Hose Used	______ Feet

TOTAL PUMP TIME ______ **TOTAL HOURS** ______
TOTAL MILES ______

COMPANY OFFICER SIGNATURE

______ A ______ B ______ C

FUEL REPORT

Date	Operator	Fuel	Oil
	TOTALS		

Date	Operator	Fuel	Oil
	TOTALS		

TOTAL FUEL [ ] [ ] TOTAL OIL

APPARATUS SERVICE REPORT

Report Date:	Person Notified	Comments	Repairs Completed By:	Date:

Equipment Lost or In Need of Repair

Date	Item	Status Replace/Fixed	Reported By	Received By

- Fill out information at first of month at top front.
- End of month, complete and send to shop
- Report stays with apparatus at all times

Engineers/Relief Drivers - It is your responsibility to see that this report is filled out properly and completely.

SOUTH PORTLAND FIRE DEPARTMENT
VEHICLE INSPECTION REPORT

VEHICLE ID #:__________ WEEK ENDING:___/___/___

CHECK MARK = NO PROBLEM X = PROBLEM 0 = REPAIRED

DAILY VEHICLE INSPECTIONS

	SUN.	MON.	TUES.	WEDS.	THURS.	FRI.	SAT.
OPERATOR NAME:	1	1	1	1	1	1	1
FUEL LOG -- # OF GALS:	2	2	2	2	2	2	2
OIL LEVEL:	3	3	3	3	3	3	3
TRANSMISSION FLUID:	4	4	4	4	4	4	4
FLUID LEAKS:	5	5	5	5	5	5	5
COOLANT / HOSES:	6	6	6	6	6	6	6
BATTERIES:	7	7	7	7	7	7	7
ELECTRICAL SYSTEM:	8	8	8	8	8	8	8
LIGHTS:	9	9	9	9	9	9	9
SIREN/HORNS:	10	10	10	10	10	10	10
RADIOS REC/TRANS:	11	11	11	11	11	11	11
AIR SYSTEM:	12	12	12	12	12	12	12
BRAKES:	13	13	13	13	13	13	13
EMERG/MAXI BRAKE:	14	14	14	14	14	14	14
WHEELS/TIRES:	15	15	15	15	15	15	15
GLASS/WIPERS:	16	16	16	16	16	16	16
WATER TANK:	17	17	17	17	17	17	17
PORTABLE EQUIPMENT:	18	18	18	18	18	18	18
SCBA:	19	19	19	19	19	19	19
MIRRORS:	20	20	20	20	20	20	20
BODY DAMAGE:	21	21	21	21	21	21	21

SATURDAY WEEKLY INSPECTION

ENDING MILEAGE:__________ ENDING HOURS - ENGINE:__________ ENDING HOURS - AERIAL:__________

UNDER CARRIAGE INSPECTION

SPRINGS ______ AIRLINES ______ FUEL LINES ______

SHOCKS ______ EXHAUST SYSTEM______ OIL / FLUID LEAKS ______

PUMP CHECK ___ OPEN & CLOSE ALL GATES AND INTAKE VALVES ___

PUMP SHIFT _____ PRIMING PUMP _____ CHANGE OVER VALVE _____ PRESSURE RELIEF _____ GUAGES _____

THROTTLE CONTROL _____ PRIMING PUMP TANK _____ TANK TO PUMP _____ TANK FILL _____

MAIN PUMP DRAIN _____ LEAKING VALVES _____ BOOSTER REEL _____ PUMP PACKING LEAKING Y N

TANK LEVEL INDICATOR - WATER _____ FOAM -_____ FOAM PUMP ______ FOAM SYSTEM______

AERIAL DEVICE INSPECTION PERFORM ALL OPERATIONS

HYDRAULIC LEVEL _____ CONTROLS _____ HYDRAULIC PRESSURE _____ PTO _____ LADDER GLIDES _____

HYDRAULIC LINES _____ CABLES _____ OUTRIGGERS / JACKS _____ HYDRAULIC LEAKS _____

PLEASE DATE AND INITIAL ALL COMMENTS TO ALLOW FOR FOLLOW-UP ---- PLEASE USE REVERSE SIDE FOR COMMENTS

KNOXVILLE FIRE DEPARTMENT

DAILY AND WEEKLY VEHICLE CHECKLIST

FIRE COMPANY __________ APPARATUS __________ DATE __________

ANY ENGINE NOISE, OVERHEATING, MALFUNCTION OF BRAKES OR STEERING, CHECK OUT OF SERVICE IMMEDIATELY. NOTIFY THE SHOP.

Apparatus weekly check to be done each Monday.
Any minor repairs, call the shop and get on work list.
Any major repairs, call the shop immediately.

Marking code: - OK Repairs needed - 0 Adjustment made - X

ENGINE.	Sun.	Mon.	Tue.	Wed.	Thu.	Fri.	Sat.
1. Fuel level							
2. Oil level							
3. Radiator water level							
4. Unusual noises							
5. Engine (clean)							
6. Leaks (water, fuel, oil)							

ELECTRICAL SYSTEM:	Sun.	Mon.	Tue.	Wed.	Thu.	Fri.	Sat.
1. Lights (all)							
2. Gauges (all)							
3. Windshield wipers							
4. Switches (all)							
5. Battery and cables (clean)							
6. Battery water level							

BATTERY CHECK: NO. 1

	Specific Gravity						Charge Time	Amps.
Sun.								
Mon.								
Tue.								
Wed.								
Thu.								
Fri.								
Sat.								

BATTERY CHECK: NO. 2

Sun.								
Mon.								
Tue.								
Wed.								
Thu.								
Fri.								
Sat.								

TIRES:	Sun.	Mon.	Tue.	Wed.	Thu.	Fri.	Sat.
Visual conditions (cuts, tread irregularity, pressure, and lugs)							
BRAKES:							
1. Air Brakes (pressure drop 15 lbs. per application, report to the shop)							
2. Air Brakes (effectiveness)							
3. Air Brakes (drain water from cylinder)							
4. Hydraulic Brakes (pedal travel)							
5. Hydraulic Brakes (fluid level)							
6. Parking Brake							

AERIAL EQUIPMENT:	Sun.	Mon.	Tue.	Wed.	Thu.	Fri.	Sat.
1. Auxiliary generator (oil, fuel, operation)							
2. Cables and ladders and towers							
3. Hydraulic hoses (leaks and condition)							
4. Bolts and nuts (loose or missing)							
5. Physical defects							
DRIVE TRAIN:							
1. Clutch action and free travel 1" to 2½"							
2. Gear shift or noise in transmission							
3. Differential noise							
MISCELLANEOUS:							
1. Exterior mounted equipment and brackets							
2. Interior equipment and brackets							
3. Check water level in booster tank							
4. Clean all tools and equipment							
5. Check body, door latches, compartments							
6. Clean apparatus							
7. Operate nozzles							
8. Check breathing equipment operation							
9. Check radio operation							

NOTE: When any item is marked 0 explain under remarks.

REMARKS: Condition - Date - Who notified - By Whom

Current speedometer reading ______________ Tachometer ______________

Speedometer reading when lubricated ______________ Tachometer ______________

OFFICER IN CHARGE:	DRIVER IN CHARGE:
A Shift ______________	______________
B Shift ______________	______________
C Shift ______________	______________

FD85-447(1000)

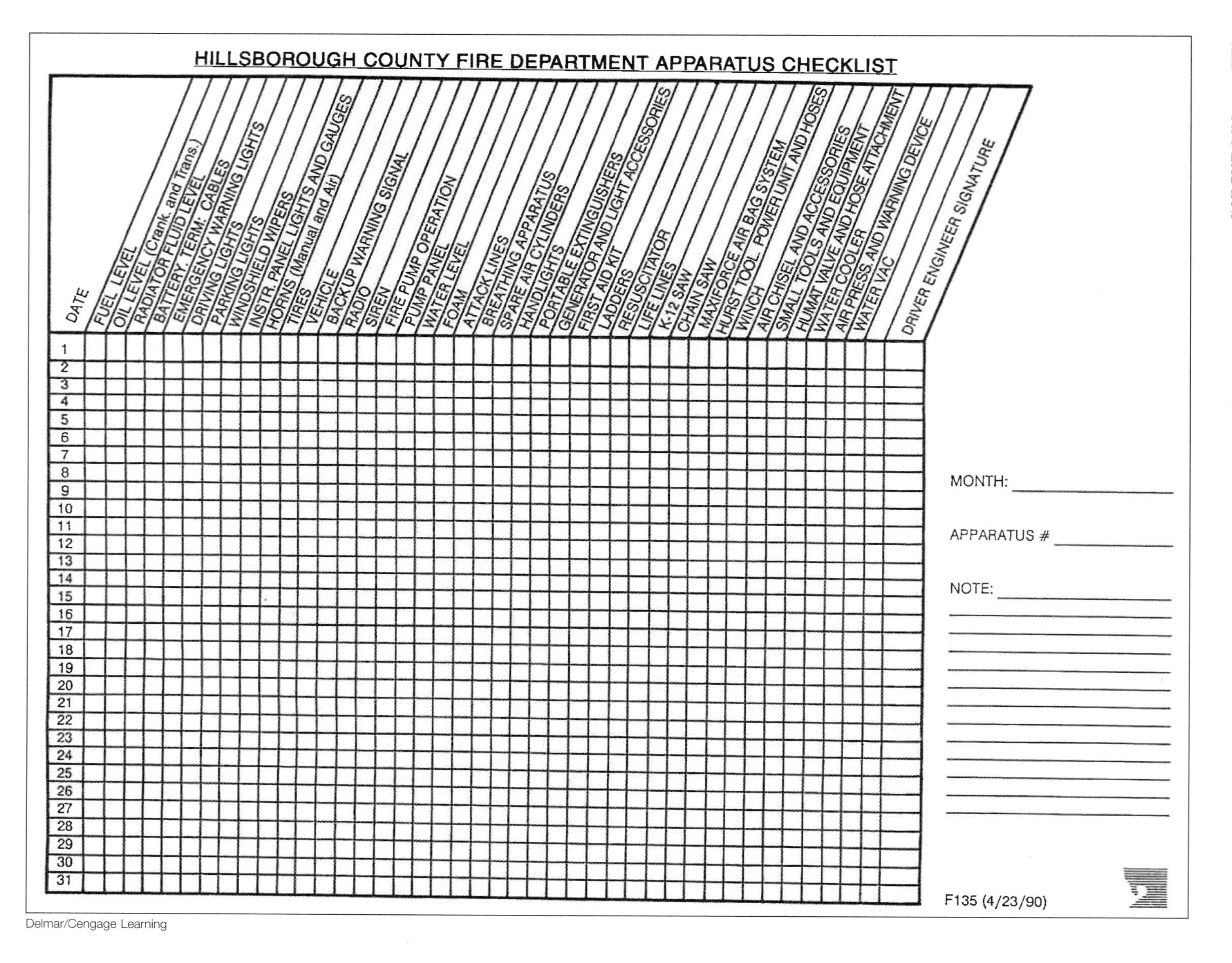

HILLSBOROUGH COUNTY FIRE DEPARTMENT APPARATUS CHECKLIST

DATE	FUEL LEVEL	OIL LEVEL (Crank. and Trans.)	RADIATOR FLUID LEVEL	BATTERY. TERM: CABLES	EMERGENCY WARNING LIGHTS	DRIVING LIGHTS	PARKING LIGHTS	WINDSHIELD WIPERS	INSTR. PANEL LIGHTS AND GAUGES	HORNS (Manual and Air)	TIRES	VEHICLE	BACKUP WARNING SIGNAL	RADIO	SIREN	FIRE PUMP OPERATION	PUMP PANEL	WATER LEVEL	FOAM	ATTACK LINES	BREATHING APPARATUS	SPARE AIR CYLINDERS	HANDLIGHTS	PORTABLE EXTINGUISHERS	GENERATOR AND LIGHT ACCESSORIES	FIRST AID KIT	LADDERS	RESUSCITATOR	LIFE LINES	K-12 SAW	CHAIN SAW	MAXIFORCE AIR BAG SYSTEM	HURST TOOL. POWER UNIT AND HOSES	WINCH	AIR CHISEL AND ACCESSORIES	SMALL TOOLS AND EQUIPMENT	HUMAT VALVE AND HOSE ATTACHMENT	WATER COOLER	AIR PRESS. AND WARNING DEVICE	WATER VAC	DRIVER ENGINEER SIGNATURE
1																																									
2																																									
3																																									
4																																									
5																																									
6																																									
7																																									
8																																									
9																																									
10																																									
11																																									
12																																									
13																																									
14																																									
15																																									
16																																									
17																																									
18																																									
19																																									
20																																									
21																																									
22																																									
23																																									
24																																									
25																																									
26																																									
27																																									
28																																									
29																																									
30																																									
31																																									

MONTH: ____________

APPARATUS # ____________

NOTE: ____________

F135 (4/23/90)

Delmar/Cengage Learning

Appendix G

CHARTS FOR FRICTION LOSS, NOZZLE REACTION, AND NOZZLE DISCHARGE

Friction Loss in Rubber-lined Fire Hose

Loss of Pressure in psi per 100 Feet of Fire Hose

Flow in gpm	Size of Hose 1 ½	1 ¾	2 ½	Flow in gpm	Size of Hose 2 ½	3	4	5	6
20	1	1		260	14	5	1		
30	2	1		270	15	6	1	1	
40	4	2		280	16	6	2	1	
50	6	4	1	290	17	7	2	1	
60	9	6	1	300	18	7	2	1	
70	12	8	1	310	19	8	2	1	
80	15	10	1	320	20	8	2	1	1
90	19	13	2	330	22	9	2	1	1
100	24	16	2	340	23	9	2	1	1
110	29	19	2	350	25	10	2	1	1
120	35	22	3	360	26	10	3	1	1
130	41	26	3	370	27	11	3	1	1
140	47	30	4	380	29	12	3	1	1
150	54	35	5	390	30	12	3	1	1
160	61	40	5	400	32	13	3	1	1
170	69	45	6	425	36	14	4	1	1
180	78	50	6	450	41	16	4	2	1
190	87	56	7	475	45	18	5	2	1
200	96	62	8	500	50	20	5	2	1
210		68	9	525	55	22	6	2	1
220		75	10	550	61	24	6	2	2
230		82	11	575	66	26	7	3	2
240		89	12	600	72	29	7	3	2
250		97	13	650	85	34	8	3	2

3 inch with 2½-inch couplings

$FL = c \times g^2$

where

FL = friction loss per 100' of hose

c = coefficient

g = gallons per minute divided by 100

Note: Actual friction loss will vary depending on type of manufacturer and age/condition of hose.

Friction Loss in Rubber-lined Fire Hose

Loss of Pressure in Kilopascals per 30 Metres of Fire Hose

	Size of Hose				Size of Hose				
Flow in Lpm	38 mm	45 mm	65 mm	Flow in Lpm	65 mm	77 mm	100 mm	125 mm	150 mm
80	7	5		1,040	103	41	10	4	3
120	16	11	1	1,080	111	44	11	5	3
160	29	19	2	1,120	119	48	12	5	3
200	46	30	4	1,160	128	51	12	6	3
240	66	43	6	1,200	137	55	13	6	4
280	89	58	8	1,240	146	59	14	6	4
320	117	76	10	1,280	156	62	15	7	4
360	148	96	12	1,320	166	66	16	7	4
400	182	118	15	1,360	176	70	17	8	5
440	220	143	18	1,400	186	75	18	8	5
480	263	170	22	1,440	197	79	19	9	5
520	308	200	26	1,480	208	83	20	9	5
560	358	231	30	1,520	220	88	21	10	6
600	410	266	34	1,560	231	93	22	10	6
640	467	302	39	1,600	244	98	23	11	6
680	527	341	44	1,700	275	110	26	12	7
720	591	383	50	1,800	308	123	30	13	8
760	659	426	55	1,900	343	138	33	15	9
800	730	472	61	2,000	380	152	37	17	10
840	804	521	67	2,100	419	168	40	18	11
880	883	572	74	2,200	460	184	44	20	12
920	965	625	81	2,300	503	202	48	22	13
960	1,050	680	88	2,400	548	220	53	24	14
1,000	1,140	738	95	2,600	643	258	62	28	17

77 mm with 65-mm couplings

$FL = c \times q^2$

where

FL = friction loss per 30 m of hose

c = coefficient

q = litres per minute divided by 100

Note: Actual friction loss will vary depending on type of manufacturer and age/condition of hose.

Nozzle Reaction

Combination Nozzles			Smooth-bore Nozzles		
gpm	75 psi	100 psi	Nozzle Diameter	50 psi	80 psi
20	9	10	1/8"	1	2
40	17	20	¼"	5	8
60	26	30	3/8"	11	18
80	35	40	7/16"	15	24
95	42	48	½	20	31
100	44	51	5/8"	31	49
120	52	61	¾"	44	71
125	55	63	7/8"	60	96
140	61	71	15/16"	69	110
160	70	81	1"	79	126
180	79	91	1 1/8"	99	159
200	87	101	1¼"	123	196
220	96	111	1 3/8"	148	237
240	105	121	1½"	177	283
250	109	126	1 5/8"	207	332
260	114	131	1¾"	240	385
280	122	141	1 7/8"	276	442
300	131	152	2"	314	502

$NR = 0505 \times g \times \sqrt{NP}$

where

NR = nozzle reaction

0.0505 = constant

g = gallons per minute (gpm)

NP = nozzle pressure in psi

$NR\ 1.57 \times d^2 \times NP$

where

NR = nozzle reaction

1.57 = constant

d = nozzle diameter in inches

NP = nozzle pressure in psi

Nozzle Reaction Measured in Newtons

Combination Nozzles			Smooth-bore Nozzles		
Lpm	**525 kPa**	**700 kPa**	**Nozzle Diameter**	**350 kPa**	**560 kPa**
80	29	33	3 mm	5	8
160	57	66	6 mm	19	30
240	86	99	9 mm	43	68
320	114	132	11 mm	64	102
380	136	157	13 mm	89	142
400	143	165	16 mm	134	215
480	172	198	19 mm	190	303
500	179	206	22 mm	254	407
560	200	231	24 mm	302	484
640	229	264	25 mm	328	525
720	257	297	29 mm	442	706
800	286	330	32 mm	538	860
880	315	363	34 mm	607	971
960	343	396	38 mm	758	1,213
1,000	357	413	41 mm	883	1,412
1,040	372	429	45 mm	1,063	1,701
1,120	400	462	48 mm	1,210	1,935
1,200	429	495	50 mm	1,312	2,100

$NR = 0.0156 \times Q \times \sqrt{NP}$

where

- NR = nozzle reaction
- 0.0156 = constant
- Q = litres per minute
- NP = nozzle pressure in kilopascals

$NR = 0.0015 \times d^2 \times NP$

where

- NR = nozzle reaction
- 0.0015 = constant
- d = nozzle diameter in millimetres
- NP = nozzle pressure in kilopascals

Note: The Newton is the unit of force in the SI system (International System of Units); it is equal to the amount of net force required to accelerate a mass of 1 kilogram at a rate of 1 metre per second.

Smooth-bore Discharge Table

Discharge Given in Gallons per Minute (gpm)

Nozzle Pressure in psi	Nozzle Diameter																	
	1/8"	1/4"	3/8"	7/16"	1/2"	5/8"	3/4"	7/8"	15/16"	1"	1 1/8"	1 1/4"	1 3/8"	1 1/2"	1 5/8"	1 3/4"	1 7/8"	2"
20	2	8	19	25	33	52	75	102	117	133	168	208	251	299	351	407	467	531
22	2	9	20	27	35	54	78	107	122	139	176	218	263	313	368	427	490	557
24	2	9	20	28	36	57	82	111	128	145	184	227	275	327	384	446	512	582
26	2	9	21	29	38	59	85	116	133	151	192	237	286	341	400	464	532	606
28	2	10	22	30	39	61	88	120	138	157	199	246	297	354	415	481	553	629
30	3	10	23	31	41	64	92	125	143	163	206	254	308	366	430	498	572	651
32	3	11	24	32	42	66	95	129	148	168	213	263	318	378	444	515	591	672
34	3	11	24	33	43	68	97	133	152	173	219	271	327	390	457	530	609	693
36	3	11	25	34	45	70	100	136	157	178	226	278	337	401	471	546	626	713
38	3	11	26	35	46	72	103	140	161	183	232	286	346	412	483	561	644	732
40	3	12	26	36	47	73	106	144	165	188	238	293	355	423	496	575	660	751
42	3	12	27	37	48	75	108	147	169	192	244	301	364	433	508	589	677	770
44	3	12	28	38	49	77	111	151	173	197	249	308	372	443	520	603	693	788
46	3	13	28	39	50	79	113	154	177	201	255	315	381	453	532	617	708	806
48	3	13	29	39	51	80	116	158	181	206	260	322	389	463	543	630	723	823
50	3	13	30	40	53	82	118	161	185	210	266	328	397	473	555	643	738	840
52	3	13	30	41	54	84	120	164	188	214	271	335	405	482	566	656	753	857
54	3	14	31	42	55	85	123	167	192	218	276	341	413	491	576	668	767	873
56	3	14	31	43	56	87	125	170	195	222	281	347	420	500	587	681	781	889
58	4	14	32	43	57	88	127	173	199	226	286	353	428	509	597	693	795	905
60	4	14	32	44	58	90	129	176	202	230	291	359	435	518	607	705	809	920
62	4	15	33	45	58	91	132	179	206	234	296	365	442	526	618	716	822	935
64	4	15	33	45	59	93	134	182	209	238	301	371	449	535	627	728	835	950

68	4	15	34	47	61	96	138	188	215	245	310	383	463	551	647	750	861	980
70	4	16	35	48	62	97	140	190	218	248	314	388	470	559	656	761	874	994
72	4	16	35	48	63	98	142	193	221	252	319	394	476	567	665	772	886	1,008
74	4	16	36	49	64	100	144	196	225	255	323	399	483	575	675	782	898	1,022
76	4	16	36	50	65	101	146	198	228	259	328	405	490	583	684	793	910	1,036
78	4	16	37	50	66	102	148	201	231	262	332	410	496	590	693	803	922	1,049
80	4	17	37	51	66	104	149	203	233	266	336	415	502	598	701	814	934	1,063
82	4	17	38	51	67	105	151	206	236	269	340	420	508	605	710	824	946	1,076
84	4	17	38	52	68	106	153	208	239	272	345	425	515	612	719	834	957	1,089
86	4	17	39	53	69	108	155	211	242	275	349	430	521	620	727	843	968	1,102
88	4	17	39	53	70	109	157	213	245	279	353	435	527	627	736	853	979	1,114
90	4	18	40	54	70	110	158	216	248	282	357	440	533	634	744	863	991	1,127
92	4	18	40	55	71	111	160	218	250	285	361	445	539	641	752	872	1,002	1,139
94	4	18	40	55	72	112	162	220	253	288	364	450	544	648	760	882	1,012	1,152
96	5	18	41	56	73	114	164	223	256	291	368	455	550	655	768	891	1,023	1,164
98	5	18	41	56	74	115	165	225	258	294	372	459	556	662	776	900	1,034	1,176
100	5	19	42	57	74	116	167	227	261	297	376	464	562	668	784	910	1,044	1,188

$gpm = 29.7 \times d^2 \times \sqrt{NP}$

where

gpm = gallons per minute

29.7 = constant

d = nozzle diameter in inches

NP = nozzle pressure in psi

Smooth-bore Discharge Table

Discharge Given in Litres per Minute (Lpm)

Nozzle Pressure in kPa	Nozzle Diameter (in millimetres) mm																	
	3	6	9	11	13	16	19	22	24	25	29	32	34	38	41	45	48	50
140	7	29	64	96	134	203	286	384	457	495	667	812	916	1,145	1,334	1,605	1,827	1,982
154	7	30	67	101	141	213	300	402	479	520	699	851	961	1,200	1,398	1,684	1,916	2,079
168	8	31	70	105	147	222	313	420	500	543	730	889	1,004	1,254	1,460	1,759	2,000	2,171
182	8	33	73	109	153	231	326	437	521	565	760	926	1,045	1,305	1,519	1,830	2,083	2,260
196	8	34	76	113	159	240	339	454	540	586	789	960	1,084	1,354	1,577	1,899	2,161	2,345
210	9	35	79	117	164	249	350	470	559	607	817	994	1,122	1,402	1,632	1,966	2,237	2,427
224	9	36	81	121	169	257	362	485	578	627	843	1,027	1,159	1,448	1,686	2,030	2,310	2,507
238	9	37	84	125	175	265	373	500	595	646	869	1,058	1,195	1,493	1,738	2,093	2,381	2,584
252	10	38	86	129	180	272	384	515	613	665	894	1,089	1,230	1,536	1,788	2,154	2,451	2,659
266	10	39	89	132	185	280	394	529	629	683	919	1,119	1,263	1,578	1,837	2,213	2,518	2,732
280	10	40	91	136	189	287	405	543	646	701	943	1,148	1,296	1,619	1,885	2,270	2,583	2,803
294	10	41	93	139	194	294	415	556	662	718	966	1,176	1,328	1,659	1,931	2,326	2,647	2,872
308	11	42	95	142	199	301	424	569	677	735	989	1,204	1,359	1,698	1,977	2,381	2,709	2,940
322	11	43	97	145	203	308	434	582	693	751	1,011	1,231	1,390	1,736	2,021	2,435	2,770	3,006
336	11	44	99	149	208	314	443	594	707	768	1,033	1,258	1,420	1,773	2,064	2,487	2,830	3,070
350	11	45	101	152	212	321	452	607	722	783	1,054	1,284	1,449	1,810	2,107	2,538	2,888	3,134
364	12	46	103	155	216	327	461	619	736	799	1,075	1,309	1,478	1,846	2,149	2,589	2,945	3,196
378	12	47	106	158	220	333	470	630	750	814	1,096	1,334	1,506	1,881	2,190	2,638	3,001	3,257
392	12	48	107	161	224	340	479	642	764	829	1,116	1,358	1,533	1,916	2,230	2,686	3,056	3,316
406	12	49	109	163	228	346	487	653	778	844	1,135	1,382	1,561	1,949	2,269	2,734	3,110	3,375
420	12	49	111	166	232	352	496	665	791	858	1,155	1,406	1,587	1,983	2,308	2,781	3,164	3,433
434	13	50	113	169	236	357	504	676	804	872	1,174	1,429	1,614	2,016	2,346	2,826	3,216	3,489
448	13	51	115	172	240	363	512	686	817	886	1,193	1,452	1,639	2,048	2,384	2,872	3,267	3,545

476	13	53	118	177	247	374	528	707	842	914	1,229	1,497	1,690	2,111	2,457	2,960	3,368	3,654
490	14	53	120	179	251	380	535	535	854	927	1,247	1,519	1,714	2,141	2,493	3,003	3,417	3,708
504	14	54	122	182	254	385	543	728	866	940	1,265	1,540	1,739	2,172	2,528	3,046	3,466	3,760
518	14	55	124	185	258	390	550	738	878	953	1,282	1,561	1,763	2,202	2,563	3,088	3,513	3,812
532	14	56	125	187	261	396	558	748	890	966	1,300	1,582	1,786	2,232	2,598	3,129	3,561	3,863
546	14	56	127	189	265	401	565	758	902	978	1,317	1,603	1,810	2,261	2,632	3,170	3,607	3,914
560	14	57	128	192	268	406	572	767	913	991	1,333	1,624	1,834	2,289	2,665	3,211	3,653	3,964
574	14	58	130	194	271	411	579	777	925	1,003	1,350	1,644	1,856	2,318	2,698	3,251	3,698	4,013
588	15	59	132	197	275	416	587	786	936	1,015	1,366	1,664	1,878	2,346	2,731	3,290	3,743	4,062
602	15	59	133	199	278	421	593	796	947	1,027	1,383	1,683	1,900	2,374	2,763	3,329	3,788	4,110
616	15	60	135	201	281	426	600	805	958	1,039	1,398	1,703	1,922	2,401	2,795	3,367	3,831	4,157
630	15	61	136	203	284	431	607	814	969	1,051	1,414	1,722	1,944	2,428	2,827	3,405	3,875	4,204
644	15	61	138	206	287	435	614	823	979	1,063	1,430	1,741	1,966	2,455	2,858	3,443	3,917	4,251
658	15	62	139	208	290	440	620	832	990	1,074	1,445	1,760	1,987	2,482	2,889	3,480	3,960	4,297
672	16	63	141	210	294	445	627	841	1,000	1,086	1,461	1,779	2,008	2,508	2,920	3,517	4,002	4,342
686	16	63	142	212	297	449	633	849	1,011	1,097	1,476	1,797	2,029	2,534	2,950	3,554	4,043	4,387
700	16	64	144	215	300	454	640	858	1,021	1,108	1,491	1,815	2,049	2,560	2,980	3,590	4,084	4,432

$Lpm = 0.067 \times d^2 \times \sqrt{NP}$

where

Lpm = litres per minute

0.067 = constant

d = nozzle diameter in millimetres (mm)

NP = nozzle pressure in kilopascals (kPa)

HYDRAULICS CALCULATION INFORMATION SHEET

Engine Pressure Formula

$PDP = NP + FL + AFL +/- EL$
where
PDP = pump discharge pressure (psi/kPa)
NP = nozzle pressure (psi/kPa)
FL = friction loss for hose (psi/kPa)
AFL = appliance friction loss (psi/kPa)
EL = elevation gain/loss (psi/kPa)

(NP) Nozzle Pressure

Type of Nozzle	Operating Pressure
Smooth-bore, handline	50 psi/350 kPa
Combination (fog), low pressure	75 psi/525 kPa
Smooth-bore, master stream	80 psi/560 kPa
Combination (fog), automatic	100 psi/700 kPa

(FL) Friction Loss Formula

$FL = c \times g^2 \times l$
$FL = c \times Q^2 \times l$
where
FL = friction loss
c = coefficient for specific hose
g/Q = flow rate in hundreds of gpm (g/100); liters (L/100)
l = length of hose in hundreds of feet (l/100); metres (l/100)

Nozzle Reaction Formula

Smooth-bore: $NR = 1.57 \times d^2 \times NP$
$NR = 0.0015 \times d^2 \times NP$
Combination: $NR = 0.0505 \times g \times \sqrt{NP}$
or *$NR = gpm \times 0.5$
$NR = 0.0156 \times Q \times \sqrt{NP}$
NR = nozzle reaction (pounds/Newtons)
d = nozzle diameter (inches/millimetres)
g = flow (gpm), Q = litres per minute
NP = nozzle pressure (psi/kPa)
1.57 and 0.0505 = constants
0.0015 and 0.0156 = constants
*condensed formula for combination nozzles operated at 100 psi NP

Note: The Newton is the unit of force in the SI system (International System of Units); it is equal to the amount of net force required to accelerate a mass of 1 kilogram at a rate of 1 metre per second
10 Newtons = 1 kilogram (kg)

Rated Capacity Pumps

100% @ 150 psi/1,050 kPa
70% @ 200 psi/1,400 kPa
50% @ 250psi/1,750 kPa

(c) Coefficient Information

1½"	24	38 mm (1 1/2")	= 38
1¾"	15.5	45 mm (1 3/4")	= 24.6
2"	8	50 mm (2")	= 12.7
2½"	2	65 mm (2 1/2")	= 3.17
*3"	0.8	*77 mm (3")	= 1.27
4"	0.2	100 mm (4")	= 0.305
5"	0.08	125 mm (5")	= 0.138
6"	0.05	150 mm (6")	= 0.083
*with 2½-in. couplings		*with 65-mm couplings	

Estimating Remaining Hydrant Flow

0–10% drop — 3 more like lines
11–15% drop — 2 more like lines
16–25% drop — 1 more like line
% drop = $(SP - RP)/SP$
where
SP = static pressure (psi/kPa)
RP = residual pressure (psi/kPa)

(AFL) Appliance Friction Loss

2-½" to 2-½" wye	5 psi
2-½" to 1-½" wye	10 psi
1-½" to 1-½" wye	15 psi
2-½" to 2-½" siamese	10 psi
1-½" to 1-½" siamese	15 psi
reducer	5 psi
increaser	5 psi
monitor	15 psi
four-way valve	15 psi
standpipe	25 psi

There is no friction loss calculated in these appliances if the flow is less than 1,400 Lpm. If the flow is greater than 1,400 Lpm the friction loss is 70 kPa.

Master stream device	= 175 kPa
Aerial devices	= 175 kPa
Sprinkler systems	= 350 kPa
Standpipe systems	= 175 kPa

Flow, Smooth-Bore Nozzles

$gpm = 29.7 \times d^2 \times \sqrt{NP}$
where
gpm = gallons per minute
29.7 = constant
d = nozzle diameter (inches)
NP = nozzle pressure (psi)

$Lpm = 0.067 \times d^2 \times \sqrt{NP}$
where
gpm = gallons per minute
Lpm = litres per minute
29.7 = constant, 0.067 = constant
d = nozzle diameter (millimetres)
NP = nozzle pressure (kPa)

(EL) Elevation

Floor Method
add/subtract 5 psi/35 kPa for
each floor level above/below the first
Elevation Method
0.5 psi per foot
10 kPa per metre

Miscellaneous Information

1 gallon of water = 8.35 lbs (8.5)
1 cubic foot of water = 7.48 gallons
density of water = 62.4 lb/ft3
1 psi will raise water 2.31 feet

one litre of water = 1 kilogram (kg)
one cubic metre of water = 1,000 litres
density of water = 1,000 kg/cubic metre
10 kPa will raise water 1 metre or divide the height in metres by 0.1

Distance Between Pumpers

$(PDP - 20) \times 100/FL$
where
PDP = pump discharge pressure
20 = residual pressure (psi) @ next pump
100 = 100'-section of hose
FL = friction loss (psi) per 100' of hose

$(PDP - 140) \times 30/FL$
where
PDP = pump discharge pressure
140 = residual pressure (kPa) @ next pump
30 = 30-m section of hose
FL = friction loss (kPa) per 30 m of hose

Appendix I

METRIC CONVERSION CHARTS

Volume

To Convert	Into	Multiply by
Ounces (oz)	Millilitres (ml)	29.57000
Pints (pt)	Millilitres (ml)	473.20000
Pints (pt)	Litres (L)	0.47320
Quarts (qt)	Millilitres (ml)	946.40000
Quarts (qt)	Litres (L)	0.94640
Gallons (gal)	Litres (L)	3.78500
Millilitres (ml)	Ounces (oz)	0.03380
Litres (L)	Pints (pt)	2.11300
Litres (L)	Quarts (qt)	1.05700
Litres (L)	Gallons (gal)	0.26420

Mass

To Convert	Into	Multiply by
Ounces (oz)	Grams (g)	28.35000
Pounds (lb)	Grams (g)	453.60000
Pounds (lb)	Kilograms (kg)	0.45360
Tons (t)	Kilograms (kg)	1016.00000
Grams (g)	Ounces (oz)	0.03527
Grams (g)	Pounds (lb)	0.00221
Kilograms (kg)	Ounces (oz)	35.27000
Kilograms (kg)	Pounds (lb)	2.20500

Length

To Convert	Into	Multiply by
Inches (in.)	Millimetres (mm)	25.40000
Inches (in.)	Centimetres (cm)	2.54000
Feet (ft)	Centimetres (cm)	30.48000
Feet (ft)	Millimetres (mm)	3.04800
Yards (yd)	Metres (m)	0.91400
Miles (mi)	Kilometres (km)	1.60900
Millimetres (mm)	Inches (in.)	0.03900
Centimetres (cm)	Inches (in.)	0.39400
Metres (m)	Feet (ft)	3.28100
Metres (m)	Yards (yd)	1.09400
Kilometres (km)	Miles (mi)	0.62140

Area

To Convert	Into	Multiply by
Square inches (in.^{2})	Square centimetres (cm^2)	6.45200
Square inches (in.^{2})	Square millimetres (mm^2)	645.20000
Square feet (ft^2)	Square millimetres (mm^2)	92900.00000
Square feet (ft^2)	Square metres (m^2)	0.09300
Square yards (yd^2)	Square metres (m^2)	0.83600
Square miles (mi^2)	Square kilometres (km^2)	2.59000
Square centimetres (cm^2)	Square inches (in.^{2})	0.15500
Square millimetres (mm^2)	Square inches (in.^{2})	0.00155
Square metres (m^2)	Square feet (ft^2)	10.76000
Square metres (m^2)	Square yards (yd^2)	1.19600
Square kilometres (km^2)	Square miles (mi^2)	0.38600

Pressure

To Convert	Into	Multiply by
Pounds per square inch (psi)	Kilopascals (kPa)	6.89500
Pounds per square inch (psi)	Atmosphere	0.06895
Kilopascals (kPa)	Pounds per square inch (psi)	0.14500
Kilopascals (kPa)	Atmosphere	0.01000
Atmosphere	Pounds per square inch (psi)	14.70000
Atmosphere	Kilopascals (kPa)	101.30000

Flow

To Convert	Into	Multiply by
Gallons per minute (gpm)	Litres per second (Lps)	0.06308
Gallons per minute (gpm)	Litres per minute (Lpm)	3.78500
Litres per second (Lps)	Gallons per minute (gpm)	15.85000

Hose Sizes (metric equivalent)

Inches	Millimetres
1	25
1½	38
1¾	45
2	50
2½	65
3	77
3½	90
4	100
5	125
6	150

Tip Sizes (metric equivalent)

Inches	Millimetres
¾	19
⅞	22
1	25
1⅛	29
1¼	32
1⅜	35
1½	38
1¾	45
2	50
2¼	57
2½	65
3	77

Pressures (metric equivalent)

psi	Kilopascals (kPa)
20	140
40	280
60	420
80	560
100	700
120	840
140	980
160	1,120
180	1,260
200	1,400
220	1,540
240	1,680
260	1,820
280	1,960
300	2,100

Flow Rate (metric equivalent)

Gallons per Minute (gpm)	Litres per Minute (Lpm)
20	80
40	160
60	240
80	320
100	400
120	480
140	560
160	640
180	720
200	800
220	880
240	960
260	1,040
280	1,120
300	1,200

Glossary

Absolute pressure (psia or kPa) Measurement of pressure that includes atmospheric pressure, typically expressed as psia or kPa.

Adapter Appliance used to connect mismatched couplings.

Aerial A fire apparatus using mounted ladders and other devices for reaching areas beyond the length of ground ladders.

Aircraft rescue and firefighting (ARFF) Apparatus designed for fighting aircraft fires at or near an airport.

Anti-lock braking system (ABS) A safety system that monitors wheel lock-up (wheels that stop turning or skid while the vehicle is still in motion) and, when wheel lock-up occurs, automatically pumps (quickly releases and then reapplies) the brakes. The pumping action under this condition increases vehicle control and reduces stopping distance.

Appliance friction loss The reduction in pressure resulting from increased turbulence caused by the appliance.

Appliances Accessories and components used to support varying hose configurations.

Atmospheric pressure The pressure exerted by the atmosphere (body of air) on the Earth.

Attack hose 1½-in. to 3-in. (38-mm to 77-mm) hose used to combat fires beyond the incipient stage.

Authorized emergency vehicles Legal terminology for vehicles used for emergency response, such as fire department apparatus, ambulances, rescue vehicles, and police vehicles equipped with appropriate identification and warning devices.

Auxiliary cooling system A system used to maintain the engine temperature within operating limits during pumping operations.

Auxiliary pumps Pumps other than the main pump or priming pump that are either permanently mounted or carried on an apparatus.

Available flow The amount of water that can be moved from the supply to the fire scene.

Back-flushing A process in which a pressurized water line is connected to a discharge while pump intakes are opened. The procedure allows removal of debris that has become trapped within the pump. The procedure is performed when the pump is not running.

Boiling point The temperature at which the vapor pressure of a liquid equals the surrounding pressure.

Bourdon tube gauge The most common pressure gauge found on an apparatus, consisting of a small curved tube linked to an indicating needle.

Braking distance The distance of travel from the time the brake is depressed until the vehicle comes to a complete stop.

British thermal units (Btu) The amount of heat required to raise 1 pound of water 1 degree Fahrenheit (F).

Calorie The amount of heat required to raise 1 gram of water 1 degree Celsius (C); can be expressed in joules or kilojoules.

Cavitation The formation and collapse of vapor pockets when certain conditions exist during pumping operations. Cavitation causes damage to the pump and should be avoided.

Centrifugal force The tendency of a body to move away from the center when rotating in a circular motion.

Closed relay Relay operation in which water is contained within the hose and pump from the time it enters the relay until it leaves the relay at the discharge point; excessive pressure and flow is controlled at each pump within the system.

Combination nozzle A nozzle designed to provide both a straight stream and a wide fog pattern; the type most widely used in the fire service.

Compound gauge A pressure gauge that reads both positive pressure (psi) (kPa) above atmospheric pressure (psi) (kPa) and negative pressure below atmospheric pressure (in. or mm of Hg).

Control valves Devices used by a pump operator to open, close, and direct water flow pressure and below atmospheric pressure (in. or mm Hg).

Couplings A set or pair of connection devices attached to a fire hose that allow the hose to be connected to additional lengths of hoses, adapters, and other firefighting appliances.

Density The weight of a substance expressed in units of mass per volume.

Discharge The point at which water leaves the pump; also called the pressure side.

Discharge flow The amount of water flowing from the discharge side of a pump through the hose, appliances, and nozzles to the scene.

Discharge maintenance The process of ensuring that pressures and flows on the discharge side of the pump are properly initiated and maintained.

Drafting The process of pumping water from a static source such as a lake or dump tank where the water's surface is lower than the pump's intake. It requires the use of atmospheric pressure to push water into the pump.

Dry barrel hydrant A hydrant operated by a single control valve in which the barrel does not normally contain water; typically used in areas where freezing is a concern.

Dry hydrant A piping system for drafting from a static water source with a fire department connection at one end and a strainer at the water end.

Dual pumping A hydrant that directly supplies two pumps through intakes.

Dump site Location where tankers operating in a shuttle unload their water.

Dump tank See **Portable dump tank**.

Eductor A specialized device used in foam operations that utilizes the Venturi principle to draw the foam into the water stream.

Enhanced roll stability systems Systems that recognize when a vehicle is likely to roll over and can independently control power and braking to individual wheels in such a way as to maximize the chances of the apparatus responding as the operator desires while minimizing rollovers.

Evaporation The physical change of state from a liquid to a vapor.

Feathering or gating The process of partially opening or closing control valves to regulate pressure and flow for individual lines.

Fill site Location where tankers operating in a shuttle receive their water.

Finished foam The final foam that is applied after the foam solution has been aerated.

Fire pump operations The systematic movement of water from a supply source through a pump to a discharge point with a resultant increase in pressure.

Floating pumps Pumps placed in a static water source to pump water to the apparatus intake, avoiding restrictions the apparatus would have with maximum lift.

Flow The rate and quantity of water delivered by a pump, typically expressed in gallons per minute (gpm) or litres per minute (Lpm).

Flow meter A device used to measure the quantity and rate of water flow in gallons (Litres) per minute (gpm) (Lpm).

Foam concentrate The material purchased from the manufacturer prior to dilution.

Foam solution Foam concentrate that has been mixed with water but not yet mixed with air; what is in a hoseline after the eductor and before an aerating nozzle.

Force Pushing or pulling action on an object.

Forward lay Supply hoseline configuration in which the apparatus stops at the hydrant and a supply line is laid to the fire.

Four-way hydrant valve Appliance used to increase hydrant pressure without interrupting the flow.

Friction loss The reduction in energy (pressure) resulting from the rubbing of one body against another, and the resistance of relative motion between the two bodies in contact; typically expressed in pounds per square inch (psi or kPa); measures the reduction of pressure between two points in a system.

Front crankshaft method (pump engagement) Method of driving a pump in which power is transferred directly from the crankshaft located at the front of an engine to the pump; this method of power transfer is used when the pump is mounted on the front of the apparatus and allows for either stationary or mobile operation.

Gating or feathering The process of partially opening or closing control valves to regulate pressure and flow for individual lines.

Gauge pressure (psig) Measurement of pressure that does not include atmospheric pressure, typically expressed as psig.

Head pressure The pressure exerted by the vertical height of a column of liquid expressed in feet (metres); may also be referred to as *feet (metres) of head* or just *head*.

Hose Flexible conduit used to convey water; also see **Attack hose**, **Supply hose**, and **Suction hose**.

Hose bridge Device used to allow vehicles to move across a hose without damaging the hose.

Hose clamp Device used to control the flow of water in a hose.

Hose jacket Device used to temporarily minimize flow loss from a leaking hose or coupling.

Hydraulics The branch of science dealing with the principles and laws of fluids at rest or in motion.

Hydrodynamics The branch of hydraulics that deals with the principles and laws of fluids in motion.

Hydrostatics The branch of hydraulics that deals with the principles and laws of fluids at rest and the pressures they exert or transmit.

Impeller A disk mounted on a shaft that spins within the pump casing.

Indicators Devices other than pressure gauges and flow meters (such as tachometer, oil pressure, pressure regulator, and on-board water level) used to monitor and evaluate a pump and related components.

Inspecting Determining the condition and operational status of equipment by sight, sound, or touch.

Instrumentation Devices such as pressure gauges, flow meters, and indicators used to monitor and evaluate the pump and related components.

Insurance Services Office (ISO) An organization that calculates a Public Protection Classification (PPC) based upon what a fire department does; fire insurance rates within a fire district are based upon this classification.

Intake The point at which water enters the pump; also called the supply side or suction.

Intake pressure relief valve Pressure regulating system that protects against excessive pressure buildup on the intake side of the pump.

Jet siphon Device that helps move water quickly without generating a lot of pressure and that is used to move water from one portable tank to another or to assist with the quick off-loading of tanker water.

Job performance requirements (JPRs) Objectives that describe what competencies an individual is expected to have; the prerequisite knowledge and skills required to complete the JPRs are also included when appropriate.

Laminar flow Flow of water in which thin parallel layers of water develop and move in the same direction.

Landmark An aid to positioning an apparatus for maneuvers; it could be marking tape on a wall, floor pavement marking, or a natural item such as a tree.

Large-diameter hose Hose with a diameter of 3½ inches (77 mm) or larger.

Latent heat of fusion The amount of heat that is absorbed by a substance when changing from a solid to a liquid state.

Latent heat of vaporization The amount of heat absorbed when changing from a liquid to a vapor state.

Laws Rules that are legally binding and enforceable.

Main pump Primary working pump permanently mounted on an apparatus.

Manifolds Devices that provide the ability to connect numerous smaller lines from a large supply line.

Manufacturer's inspection recommendations Those items recommended by the manufacturer to be included in apparatus inspections.

Mobile data terminals (MDTs) An apparatus-mounted computer providing access to 9-1-1 dispatch information and perhaps other functions such as fire preplans, Internet, maps, and reporting software; may also be called a mobile communications terminal (MCT).

Municipal supply A water supply distribution system provided by a local government and consisting of mains and hydrants.

Needed flow The estimated flow required to extinguish a fire. Also called required flow.

Net pump discharge pressure (NPDP) The difference between the intake pressure and the discharge pressure. It is the amount of pressure the pump adds.

NH A common thread used in the fire service to attach hose couplings and appliances.

Normal pressure or normal operating pressure The water flow pressure found in a system during normal consumption demands.

Nozzle flow The amount of water flowing from a nozzle; also used to indicate the rated flow or flows of a nozzle.

Nozzle pressure The designed operating pressure for a particular nozzle.

Nozzle reach The distance water travels after leaving a nozzle.

Nozzle reaction The tendency of the nozzle to move in the opposite direction of water flow.

Nurse feeding A method of water supply where one tanker pumps the water in its booster tank into another tanker or engine.

On-board supply The water carried in a tank on the apparatus.

Open relay Relay operation in which water is not contained within the entire relay system, excessive pressure is controlled by intake relief valves and pressure regulators, and dedicated discharge lines are used to allow water to exit the relay at various points in the system.

Perception distance The distance the apparatus travels from the time a hazard is seen until the brain recognizes it as a hazard.

Portable dump tank A temporary reservoir used in tanker shuttle operations that provides the means to unload water from a tanker for use by a pump.

Positive displacement pump Moves a specified quantity of water through the pump chamber with each stroke or cycle; it is capable of pumping air, and therefore is self-priming, but must have a pressure relief provision if pumping or hoses have shut-off nozzles or valves; examples of positive displacement pumps are the gear pump, the piston pump, the rotary lobe pump; and the rotary vane pump.

Pressure The force exerted by a substance in units of weight per area; the amount of force generated by a pump or the resistance encountered on the discharge side of a pump; typically expressed in pounds per square inch (psi) or kilopascals (kPa).

Pressure drop The difference between the static pressure and the residual pressure when measured at the same location.

Pressure gain and loss The increase or decrease in pressure as a result of an increase or decrease in elevation.

Pressure gauge Device used to measure positive pressure in pounds per square inch (psi) (kilopascals or kPa).

Pressure governor A pressure regulating system that protects against excessive pressure buildup by controlling the speed of the pump engine to maintain a steady pump pressure.

Pressure regulating systems Devices used to control sudden and excessive pressure buildup during pumping operations.

Pressure relief device A pressure regulating system component or valve that protects against excessive pressure buildup by diverting excess water flow from an area of excessive pressure such as the discharge side of the pump or intake valve back to the intake side of the pump or to the atmosphere.

Preventive maintenance Proactive steps taken to ensure the operating status and readiness of the apparatus, pump, and related components.

Primary water supply A water supply used for a sprinkler system or standpipe system that exists on the establishment's property. If no water supply exists on the establishment's property, the public or only water supply is considered the primary water supply.

Priming The process of replacing air in a pump with water.

Priming pump Positive displacement pump permanently mounted on an apparatus and used to prime the main pump.

PTO method (pump engagement) Method of driving a pump in which power is transferred from just before the transmission to the pump through a PTO; a method of power transfer that allows either stationary or mobile operation of the pump.

Pump A mechanical device that raises and transfers liquids from one point to another.

Pump discharge pressure (PDP) The amount of pressure on the discharge side of the pump. It is the pressure read on the pump discharge gauge.

Pump engagement The process or method of providing power to the pump.

Pump gear Indicates that power from the transmission will be transferred to the pump on a split-shaft pump.

Pump operator The individual responsible for operating the fire pump, driving the apparatus, and conducting preventive maintenance.

Pump operators People who have had training, knowledge, experience, and demonstrated proof of competence in operating a fire pump as defined by NFPA 1002. For pump operators, the proof of competence is generally shown in terms of certification agencies, including departmental, state, and independent groups such as the National Board on Fire Service Professional Qualifications (PRO BOARD) and the International Fire Service Accreditation Congress (IFSAC).

Pump panel The central location for controlling and monitoring the pump and related components.

Pump peripherals Those components directly or indirectly attached to the pump that are used to control and monitor the pump and related components.

Pump speed The rate at which a pump is operating, typically expressed in revolutions per minute (RPM).

Pump-and-roll An operation where water is discharged while the apparatus is in motion.

Pumper The basic unit of the fire apparatus, an automotive fire apparatus that has a permanently mounted fire pump with a rated discharge rate of at least 750 gpm (3,000 Lpm), a water tank, and a hose body; also called a "triple-combination pumper." The unit is designed for sustained pumper operations during structural firefighting and is capable of supporting associated fire department operations.

Qualified mechanics People who have had training, knowledge, experience, and demonstrated proof of competence in performing mechanical repairs. For fire apparatus mechanics, the proof of competence is generally shown in terms of manufacturer certifications, Automotive Service Excellence (ASE) certifications, and Emergency Vehicle Technician (EVT) certifications.

Rated capacity The flow of water at specific pressures that a pump is expected to provide.

Reaction distance The distance of travel from the time the brain sends the message to depress the brakes until the brakes are actually depressed.

Recirculate line A line that carries a relatively small amount of water but enough water to make sure water is flowing through the pump, preventing it from overheating.

Relay operation Water supply operations where two or more pumpers are connected in-line to move water from a source to a discharge point.

Relay valves Work like a four-way hydrant valve in that water can initially flow through them and subsequently an additional pumper can connect to boost pressure without interrupting flow; the only difference from a four-way hydrant valve is that the side that would connect to a hydrant on a four-way hydrant valve instead connects to a supply line.

Repair To restore or replace components that have become unserviceable, or not meeting their manufacturers' specifications for whatever reason, including damage and wear.

Required flow The estimated flow of water needed for a specific incident.

Residual pressure The pressure remaining in the system after water has been flowing through it.

Reverse lay Supply hoseline configuration where the apparatus stops at the scene; drops attack lines, equipment, and personnel; and then advances to the hydrant laying a supply line.

Road gear Indicates that power from the transmission will be transferred to the drive axle(s) on a split-shaft pump.

Roll stability control (RSC) systems Systems that recognize when a vehicle is likely to roll over and can independently control power and braking to individual wheels in such a way as to maximize the chances of the apparatus responding as the operator desires while minimizing rollovers.

Safety-related components Those items that affect the safe operation of the apparatus and pump, and that should be included in apparatus inspections.

Secondary water supply A public or municipal water supply that backs up the primary water supply. Normal operating procedures should use the secondary water supply whenever possible to avoid depleting the primary water supply.

Servicing The act of performing maintenance to keep equipment working as intended.

Shuttle cycle time The total time it takes for a tanker in a shuttle operation to dump water and return with another load; includes the time it takes to fill the tanker, to dump the water, and the travel time between the fill and dump stations.

Shuttle flow capacity The volume of water a tanker shuttle operation can provide without running out of water.

Siamese Appliance used to combine two or more hoselines into a single hoseline.

Siamese lines Hose configurations in which two or more separate lines supply one line, monitor nozzle, fixed system, or a pump in a relay or similar situation.

Slippage The leaking of water between the surfaces of the internal moving parts of a pump.

Smooth-bore nozzle Nozzle designed to produce a compact, solid stream of water with extended reach.

Soft sleeve Shorter section of hose used when the pump is close to a pressurized water source such as a hydrant.

Spanner wrench Tool used to connect and disconnect hose and appliance couplings.

Specific heat The amount of heat required to raise the temperature of a substance by 1°F. The specific heat of water is 1 BTU/lb °F (4.19 joules/gram or 4.19 kilojoules/kg).

Speed The rate at which a pump is operating, typically expressed in revolutions per minute (RPM).

Split lay A hose lay that includes both a forward lay and a reverse lay component.

Split-shaft method (pump engagement) Method of driving a pump in which a sliding clutch gear transfers power to either the road transmission or to the pump transmission; this method of power transfer is used for stationary pumping only.

Spotter An individual used to assist in backing up an apparatus.

Standard operating procedures (SOPs) Specific information and instructions on how a task or assignment is to be accomplished.

Standards Guidelines that are not legally binding or enforceable by law unless they are adopted as such by a governing body.

Static pressure The pressure in a system when no water is flowing.

Static source Water supply that generally requires drafting operations, such as ponds, lakes, swimming pools, and rivers.

Static source hydrants or dry hydrants Pre-piped lines that extend into a static source.

Storz couplings The most popular of the nonthreaded hose couplings; they are also quarter-turn and sexless couplings.

Stream shape The configuration of water droplets (shape of the stream) after leaving a nozzle.

Suction hose Special noncollapsible hose used for drafting operations; also called hard suction hose.

Supply hose Used with pressurized water sources and operated at a maximum pressure of 185 psi (1,295 kPa).

Supply layout The required supply hose configuration necessary to secure the water supply efficiently and effectively.

Supply reliability The extent to which the supply will consistently provide water.

Systematic Orderly process or following of a prescribed procedure—for example, a systematic inspection is used to make sure nothing is accidentally missed.

Tandem pumping A hydrant that directly supplies one pumper and then discharges to the second pumper's intake.

Tanker shuttle Water supply operations in which the apparatus are equipped with large tanks to transport water from a source to the scene.

Testing Verifying the condition and operational status of equipment by measurement of its characteristics and comparing those measurements to the required specifications.

Throttle control Device used to control the engine speed, which in turn controls the speed of the pump, when engaged, from the pump panel.

Tiller apparatus Aerial apparatus in which the rear wheels are steered from the back of the apparatus by a tiller operator.

Total stopping distance Measured from the time a hazard is detected until the vehicle comes to a complete stop.

Traction Friction between the tires and road surface.

Transfer valve Control valve used to switch between the pressure and volume modes on two-stage centrifugal pumps.

Turbulent flow The flow of water in an erratic and unpredictable pattern, creating a uniform velocity within the hose that increases pressure loss because more water is subjected to the interior lining of the hose.

Vacuum Measurement of pressure that is less than atmospheric pressure, typically expressed in inches of mercury (in. Hg) (kPa or millimetres).

Vapor pressure The pressure exerted on the atmosphere by molecules as they evaporate from the surface of the liquid.

Velocity pressure The forward pressure of water as it leaves an opening.

Venturi principle Process that creates a low-pressure area in the induction chamber of an eductor to allow foam to be drawn into and mixed with the water stream.

Volume Three-dimensional space occupied by an object.

Volute An increasing void space in a pump that converts velocity into pressure and directs water from the impeller to the discharge.

Water availability The quantity, flow, pressure, and accessibility of a water supply.

Water hammer A surge in pressure created by the sudden increase or decrease of water during pumping operations. It can damage hoses, appliances, pumps, water mains, and anything else in direct contact with the water.

Water thieves Similar to gated wyes, water thieves are used to connect additional smaller hoselines from an existing larger hoseline.

Weight The downward force exerted on an object by the Earth's gravity, typically expressed in pounds (lb) or kilograms (kg).

Wet barrel hydrant A hydrant operated by individual control valves; it contain water within the barrel at all times; only used where freezing is not a concern.

Wheel chock Device placed in front of or behind wheels to guard against inadvertent movement of the apparatus.

Wildland Apparatus designed for fighting wildland fires; contain small pumps with pump-and-roll capabilities and limited on-board water supplies.

Wye Appliance used to divide one hoseline into two or more hoselines. The wye lines may be the same size or a smaller size, and the wye may or may not have gate control valves to control the water flow.

Wyed lines Hose configurations in which one hoseline supplies two or more separate lines. Typically, one larger line supplies two or more smaller lines, for example, a 2½- or 3-inch (65- or 77-mm) supply line wyed to two or more 1½- or 1¾-inch (38-mm or 45-mm) attack lines. When calculating wyed lines, the supply line is calculated separately from the attack lines.

Index

NOTES

NOTES

NOTES

NOTES

NOTES

NOTES

NOTES

NOTES

NOTES

NOTES